# 100% Clean, Renewable Energy and Storage for Everything

Transitioning the world from fossil fuels to 100 percent clean, renewable energy and storage for everything is seen internationally as necessary to address global warming, air pollution, and energy insecurity.

This textbook lays out the science, technology, economics, policy, and social aspects of transitioning from fossil fuels to 100 percent clean, renewable energy sources for all energy purposes. It discusses the electricity- and heat-generating technologies needed; the electricity, heat, cold, and hydrogen storage technologies needed; how to keep the electric power grid stable; and how to address non-energy sources of emissions. It also describes how to develop science-based clean, renewable energy plans for cities, states, and countries, as well as the history of the 100 Percent Movement, which evolved from a collaboration among scientists, cultural leaders, businesspeople, and community leaders.

The text includes numerous worked-out example problems and Transition Highlight boxes that illustrate real-world successes. Online resources consist of lecture slides, answers to the end-of-chapter questions, and a list of extra resources.

**Mark Z. Jacobson** is Director of the Atmosphere/Energy Program and Professor of Civil and Environmental Engineering at Stanford University. He is also a Senior Fellow of the Woods Institute for the Environment and of the Precourt Institute for Energy. He received a B.S. in Civil Engineering, an A.B. in Economics, and an M.S. in Environmental Engineering from Stanford in 1988. He received an M.S. and PhD in Atmospheric Sciences in 1991 and 1994, respectively, from UCLA and joined the faculty at Stanford in 1994. He has published three textbooks and over 165 peer-reviewed journal articles. He received the 2005 American Meteorological Society Henry G. Houghton Award and the 2013 American Geophysical Union Ascent Award for his work on black carbon climate impacts and the 2013 Global Green Policy Design Award for developing state and country energy plans. In 2015, he received a Cozzarelli Prize from the *Proceedings of the National Academy of Sciences* for his work on the grid integration of 100 percent wind, water and solar energy systems. In 2018, he received the Judi Friedman Lifetime Achievement Award "For a distinguished career dedicated to finding solutions to large-scale air pollution and climate problems." In 2019, he was selected as "one of the world's 100 most influential people in climate policy" by Apolitical. He has served on an advisory committee to the U.S. Secretary of Energy, appeared in a TED talk, appeared on the *Late Show with David Letterman* to discuss converting the world to clean, renewable energy, and cofounded The Solutions Project. His work is the scientific basis of the energy portion of the U.S. Green New Deal and laws to go to 100 percent renewable energy in cities, states, and countries worldwide.

'Engineering professors of the world: Are you teaching a course on climate change, or planning one? If you are, this is the textbook you should be adopting. Civil, mechanical, electrical, materials, and chemical engineering aspects of the energy transition are exhaustively addressed. And this book has soul: today's engineering student feels the need to do something about climate change, and this book empowers them.'

Anthony R. Ingraffea, Department of Civil and Environmental Engineering, Cornell University

'Mark Jacobson's new book - *100% Clean, Renewable Energy and Storage for Everything* - provides the most authoritative look yet at the future of energy beyond fossil fuels. The text is clearly written, authoritative, and thoroughly referenced. This will make a great textbook for courses on energy and climate change, but is also a must read for all of us interested in the transition to a renewable future.'

Robert W. Howarth, Department of Ecology and Evolutionary Biology, Cornell University

'Professor Jacobson's work on the possibilities for renewable energy have opened eyes around the globe. Where people once saw barriers, increasingly they see possibilities and openings, and this book consolidates that new understanding.'

Bill McKibben, Middlebury College

'A great book! It shows why problems of air pollution and global warming can be solved by using renewable energies. It explains very clearly all aspects of a secure and climate-friendly full supply of renewable energies, using comprehensive scientific facts and clear practical examples. It should be used as a standard textbook in all energy economics lectures, worldwide! It is highly relevant not only for students but for all those interested in energy economics in these times of unsolved challenges caused by climate change and pollution. A book that everyone should read!'

Claudia Kemfert, German Institute for Economic Research

'The world's major crises need radical and comprehensive solutions, with 100% clean renewable energy systems at the core of any health, climate, peace, or prosperity plan. Mark Jacobson shows in a brilliant and scientifically profound way why such a worldwide transformation is necessary and how it can be realized. A powerful work that leaves no more excuses for political inaction.'

Hans-Josef, Former German Parliamentarian and founder of German solar tariffs

'Mark Jacobson shines a bright light illuminating the path forward, painstakingly detailing - with numbers and facts - how we can decarbonize our energy infrastructure, take action on climate, create a cleaner environment and sustain a healthy, green economy. At a time when there is far too much doom-and-gloom over our prospects for averting climate catastrophe, read this book, take action, and be part of the battle to preserve a healthy, livable planet.'

Michael E. Mann, Penn State University

# 100% Clean, Renewable Energy and Storage for Everything

**Mark Z. Jacobson**
Stanford University, California

**CAMBRIDGE**
UNIVERSITY PRESS

University Printing House, Cambridge CB2 8BS, United Kingdom

One Liberty Plaza, 20th Floor, New York, NY 10006, USA

477 Williamstown Road, Port Melbourne, VIC 3207, Australia

314–321, 3rd Floor, Plot 3, Splendor Forum, Jasola District Centre, New Delhi – 110025, India

79 Anson Road, #06–04/06, Singapore 079906

Cambridge University Press is part of the University of Cambridge.

It furthers the University's mission by disseminating knowledge in the pursuit of
education, learning, and research at the highest international levels of excellence.

www.cambridge.org
Information on this title: www.cambridge.org/jacobson
DOI: 10.1017/9781108786713

© Cambridge University Press 2021

First published 2021
Reprinted 2021

Printed in Singapore by Markono Print Media Pte Ltd

*A catalogue record for this publication is available from the British Library.*

ISBN 978-1-108-47980-6 Hardback
ISBN 978-1-108-79083-3 Paperback
Additional resources for this publication at www.cambridge.org/jacobson.

To Dionna, Daniel, and Jessica
and all others who will inherit the
Earth and everything that is in it

# CONTENTS

# 9

# Evolution of the 100 Percent Movement and Policies Needed for a WWS Solution 347

# PREFACE

The world is on a path to transition away from fossil fuels to clean, renewable energy in order to address environmental pollution, global warming, and energy insecurity. However, unless such a transition occurs quickly, efficiently, and most everywhere, the world risks substantially increased mortality, warming, and economic instability compared with today. This textbook lays out the scientific, technological, economic, political, and social aspects of how to transition the world rapidly to entirely clean, renewable energy for all purposes.

Evolving out of a course I teach at Stanford University, this book includes a description of how to transition the world's current combustion-based energy to 100 percent clean, renewable **wind, water, and solar** (**WWS**) electricity and heat for all energy purposes; how to store electricity, heat, cold, and hydrogen; how to keep the electric power grid stable; and how to address non-energy sources of emissions.

Whereas many textbooks teach about clean, renewable technologies, this one also teaches about what is needed to transition towns, cities, states, provinces, and countries entirely to clean renewables and storage. It also describes how to develop science-based clean, renewable energy plans for cities, states, and countries. These plans have been used to justify 100 percent renewable and zero emissions laws and policies, including the Green New Deal, in many countries, states, and cities. They have also been used to justify 100 percent renewable commitments by many international companies. The textbook further discusses the history of the 100 Percent Movement, which evolved from a collaboration among scientists, cultural leaders, businesspeople, and community leaders. Finally, it discusses progress to date in transitioning to 100 percent WWS and policies needed to complete the transition.

## Motivation for Transitioning

This book is motivated by the fact that air pollution, global warming, and energy security are three of the most significant problems facing the world today. Most scientists recognize that solutions to these problems must be implemented rapidly. Every year that indoor and outdoor air pollution continues, 4 to 9 million more children and adults die from it. If at least 80 percent of emissions that cause global warming are not eliminated by 2030, and if 100 percent are not eliminated by 2050 or sooner, globally averaged temperatures will likely rise at least 1.5 °C above those in the early 1900s. This will likely trigger more glacier and sea-ice melting, sea level rise, coastal flooding, severe storminess, wildfires, air pollution mortality, heat-related mortality, drought, famine, agricultural shifts, climate migration, species extinction, coral reef damage, and more. In addition, if limited-resource fossil fuels are not replaced with sustainable clean, renewable energy, energy prices will probably rise dramatically, causing economic, social, and political instability worldwide.

The solution to these problems is to transition world energy in all energy sectors to 100 percent WWS electricity and heat, combined with storage, and to address non-energy sources of emissions.

The main idea behind the solution comes from the fact that air pollution health and climate problems result from the same source: combustion. That is, combustion of fossil fuels, biofuels, bioenergy, and open biomass. To solve the problems, it is necessary to move away from combustion by electrifying and providing direct heat without combustion. For the electricity and heat to remain clean and available for millennia to come while not creating other risks, they need to originate from clean, renewable, and sustainable sources – namely, WWS.

WWS includes energy from **wind** (onshore wind, offshore wind, and airborne wind electricity), **water** (hydro, tidal and ocean current, wave, and geothermal electricity and geothermal heat), and **sunlight** (solar photovoltaic [PV] electricity, concentrated solar power [CSP] electricity and heat, and captured solar heat [solar thermal]). WWS needs to power **all energy sectors**, which means electricity, transportation, building heating/cooling, industry, agriculture/forestry/fishing, and the military. Whereas human-designed energy systems cause about 95 percent of **anthropogenic** (human-produced) air pollution and 75 percent of anthropogenic greenhouse gas emissions, this book also discusses methods to address non-energy anthropogenic emissions that affect air pollution and climate. The book additionally describes technologies available for electricity, heat, cold, and hydrogen storage.

Many solutions to date that have focused on the climate problem have included some technologies that are less helpful than WWS technologies. This book describes such technologies. The reason they are less helpful is that they raise costs to consumers and society, slow solutions to pollution and warming due to their long planning-to-operation times, increase emissions relative to WWS sources, and/or create risks that WWS sources don't have. Given the limited time and funding to solve pollution, climate, and security problems, it is essential to focus on known, effective solutions. Money spent on less-useful options will permit more damage to occur.

Some technologies that are clean and renewable are not discussed here because it seems they will not be commercially available in the next decade. One example is a technology that takes advantage of salinity and temperature gradients in the ocean to generate electricity. If such a technology does come to fruition, it could be included as a WWS resource.

Why 100 percent clean, renewable energy and storage for everything? Why not 50 percent, 80 percent, or 99 percent. The first reason is that the health plus climate cost of every tonne of air pollution, down to the last tonne, is so enormous that it outweighs other uses of the money required to remove the pollution. More important, one more person should not die or become ill from air pollution. Species extinction, global-warming-driven wildfires, supercharged hurricanes, and smog should no longer occur. Gas wells, coal mines, oil pipelines, gas stations, coal-fired power plants, gas storage reservoirs, diesel cars, jet fuel airplanes, and bunker fuel ships should no longer be needed. Nuclear power plant meltdowns and nuclear waste pileups should no longer occur. We don't want to drink chemicals in our water due to oil, gas, coal, or uranium mining leaks. We don't want to see more wars over fossil fuels. We don't want any more oil spills devastating the oceans, lakes, or rivers. Blackouts due to reliance on centralized power plants should be a thing of the past. Plus, we want to eliminate high energy prices that arise from fuel shortages and the need to transport fossil fuels long distances.

Aside from the fact that it is technically and economically possible to transition everything to 100 percent clean,

renewable energy and storage, 99 percent is not an ambitious goal to shoot for. Did Magellan aspire to circumnavigate 99 percent of his way around the Earth? Did the Apollo 11 crew aspire to reach 99 percent of its way to the moon? No. One hundred percent is the goal because that may be the best society can do and may result in the cleanest air and most stable climate for future generations. Societies often strive for the best and safest.

Can society reach the goal of 100 percent? This book examines the science, engineering, economic, social, and political aspects of transitioning towns, cities, states, countries, businesses, and the world to 100 percent clean, renewable WWS energy and storage for everything. Such a transition will address air pollution, global warming, and energy security simultaneously. The book also examines ways to reduce major types of non-energy emissions. It concludes that a transition among all energy and non-energy sectors worldwide is technically and economically possible. The main obstacles are social and political.

## Intended Audience, Level, and Scope

This book is written to be accessible to everyone concerned with renewable energy and storage, including those studying Renewable Energy, Sustainability, Environmental Sciences and Engineering, Earth Sciences, Climate Sciences, Atmospheric Sciences, Electrical Engineering, Mechanical Engineering, Geography, Health Sciences, Economics, Business, and Policy departments as well as researchers, professionals, policymakers, advocates, and interested readers in many areas.

The book assumes no prior knowledge of, yet provides, needed information about, energy systems, electromagnetism, thermodynamics, dynamical meteorology, radiation transfer, mechanical engineering, aerodynamics, economics, weather, climate, and air pollution. Readers will therefore be able to understand the operations of the following: wind turbines, solar photovoltaics, concentrated solar power systems, hydropower systems, pumped hydropower storage systems, batteries, flywheels, gravitational mass electricity storage, underground thermal energy storage systems, thermal mass storage in buildings, generators, hydrogen fuel cells, heat pumps, electric vehicles, hydrogen fuel cell vehicles, arc furnaces, induction furnaces, resistance furnaces, dielectric heaters, and transmission/distribution systems.

The book also gives information about how to determine wind and solar resources, the maximum wind and solar potentials of the world, the impacts of wind turbines on global temperatures and hurricanes, the effects of tilting and tracking solar panels on electricity output, the efficiencies of wind turbines and solar photovoltaics, and the rates at which different electricity storage technologies can ramp up.

By the end of the book, readers will understand why all the technologies covered throughout the book will help to solve the air pollution, climate, and energy security problems we face and why other technologies are not so useful.

Readers will be able to calculate the private (business) and economic (social) costs of energy technologies and of energy systems. They will understand the methods of matching electricity, heat, cold, and hydrogen demand with clean, renewable supply and storage over time at low cost. Finally, they will understand the origin of the 100 percent clean, renewable energy movement, the progress made to date in transitioning the world to 100 percent, and the policies needed to complete the transition.

## Structure

The book is structured in the order that I teach the material in a course. It starts by defining the air pollution, global warming, and energy insecurity problems we seek to solve (Chapter 1). Chapter 2 discusses WWS electricity- and heat-generating technologies; transportation technologies; building heating and cooling technologies; high-temperature industrial heat technologies; and appliances and machines needed for a transition. It further discusses energy efficiency measures, electricity storage, heat and cold storage, and hydrogen storage. Finally, it discusses methods of addressing non-energy sources of greenhouse gas and aerosol particle pollution. Chapter 3 goes into depth about why we don't need natural gas as a bridge fuel, fossil fuels with carbon capture, nuclear power, biomass (with or without carbon capture), biofuels, synthetic direct air capture, or geoengineering.

Because a 100 percent WWS world is mostly electrified, Chapter 4 focuses on electricity basics. Because solar photovoltaics (PV) and wind power will likely comprise the largest share of energy generation in a WWS world, Chapter 5 discusses solar PV and solar radiation, and Chapter 6 discusses onshore and offshore wind. Chapter 7 moves on to discuss steps in developing a 100 percent WWS roadmap for a country, state, or city. Chapter 8 explains how to match power demand with 100 percent WWS supply plus storage.

Finally, Chapter 9 outlines my personal journey toward 100 percent; the 100 Percent Movement that has arisen around the WWS roadmaps; laws and commitments that have been implemented to date due to them; and the policies needed in the future to finally solve the problems of air pollution, global warming, and energy in security.

## Pedagogical Features

The book is supported by a comprehensive set of pedagogical features:

- A short introduction to each chapter and an end-of-chapter Summary to clarify each chapter's objectives and to ensure understanding of the material discussed
- Highlighted key terms and clear definitions throughout the book
- Numerical examples in each chapter that explain how to apply important equations
- Abundant tables, diagrams, and photographs that illustrate important and interesting aspects of the field
- Problems and Exercises in the Further Resources section that consolidate and extend student understanding
- A list of Recommended Readings in the Further Resources section at the end of each chapter
- A Glossary of Acronyms and a list of conversion constants and units in the Appendices

In order to assist with teaching information in the textbook, teaching materials have been developed to accompany the course and are available online. Such materials include a teaching guide, a model syllabus, lecture slides for each chapter, a solution manual, and a test bank.

## Acknowledgments

Several colleagues and students helped to edit, provide data, or provide figures for this text. I would like to thank (in alphabetical order) Mary A. Cameron, Stephen J. Coughlin, Mark A. Delucchi, Joshua Eichman, Catherine Hay, Ken Hnottavange-Telleen, Robert W. Howarth, Anthony R. Ingraffea, Scott M. Katalenich, Willett Kempton, Indu P. Manogaran, Gilbert M. Masters, Matthew McCarville, Julian Rey, John Ribeiro-Broomhead, Daniel Sambor, Yanbo Shu, Anna-Katharina von Krauland, Jon Wank, and Tim W. Yeskoo. Special thanks also to Anna-Katharina von Krauland for creating several graphics for the text.

# CHAPTER

# 1

# What Problems Are We Trying to Solve?

Three of the most significant problems affecting the world today are air pollution, global warming, and energy insecurity. This chapter discusses each of these problems, in turn. It starts by discussing the magnitude of the global air pollution health problem today, the sources of the pollution, and how transitioning to clean, renewable energy can solve this problem. It then discusses the difference between the greenhouse effect and global warming and quantifies the major contributors to each. It describes the strength of each warming component, including gases, particles, and direct heat emissions. It also describes how cooling chemicals in the atmosphere mask part of global warming. It then discusses the impacts of global warming. Finally, the chapter describes four types of energy insecurity problems the world faces with the current energy system and how transitioning to WWS can help to solve those problems.

## 1.1 The Air Pollution Tragedy

Air pollution has affected human health since fire was first used for cooking and heating. Today, pollution arises due to the burning of wood, vegetation, dung, waste, liquid biofuels, coal, natural gas, oil, gasoline, kerosene, diesel, jet fuel, and chemicals. Although some air pollution today is from natural sources, most is from human burning of fuels to produce energy. Because combustion to produce energy is the major source of air pollution, eliminating such combustion is key to eliminating air pollution.

### 1.1.1 Health Risks from Air Pollution

Worldwide, air pollution kills 4 to 9 million people each year and injures hundreds of million more through illness. In fact, air pollution is the second leading cause of death worldwide. As discussed in Section 7.6.2, these air pollution deaths and illnesses are from heart disease, stroke, respiratory infection, lung cancer, complications from asthma, and pneumonia (GBD, 2015; WHO, 2017a, 2017c; Burnett et al., 2018). Around 20 percent of the deaths are of children under the age of five years. Many such children live in homes in which wood, dung, coal, or waste is burned for home heating and cooking. Their little lungs absorb a high concentration of particles from such burning. Many of these children die of pneumonia because their immune systems become weakened due to the assault of air pollutants on their respiratory system. Most of the casualties are in developing countries, where indoor burning still occurs on a large scale. The worldwide cost of all air pollution mortality and morbidity is estimated to be around $20 to $25 trillion per year today (Jacobson et al., 2017).

### 1.1.2 Sources of Air Pollution

The sources of air pollution mortality and morbidity are gases and small particles in the atmosphere emitted primarily during combustion of fossil fuels, biomass, and biofuels for energy; the burning of forests, woodlands, and savannah for land clearing or ritual; the burning of agricultural crops to improve harvesting; the burning of agricultural residue to clear land; wildfires resulting from arson, human carelessness, or lightning; lightning itself; volcanic eruptions; bacterial metabolism; and wind-blown emissions of road dust, soil dust, sea spray, plant debris, pollen, spores, bacteria, and viruses. Whereas soil dust emissions are often considered natural, human erosion of land at the edges of deserts has increased soil dust emissions; thus, a portion of it, too, is caused by humans. Similarly, more than 90 percent of wildfires today are due to human carelessness or arson.

Around half the mass of air pollution particles emitted worldwide is in natural particles, including volcanic particles and wind-driven particles (sea spray, soil dust, plant debris, pollen, spores, and bacteria). However, natural particles are mostly large; thus, they don't penetrate deep into the lungs of people or animals. On the other hand, combustion particles, which are almost all from human (**anthropogenic**) sources today, are mostly small and penetrate deep into the lungs. In addition, most of these combustion particles emitted by humans are discharged near where people live; thus, these are the particles that most people breathe.

### 1.1.3 How Transitioning the Energy Infrastructure Can Address the Air Pollution Tragedy

In sum, around 80 percent of air pollution mortality is due to particles emitted by human activity, 10 percent is due to gases from human activity, and 10 percent is from naturally emitted particles and gases. Given that 90 percent of human mortality from air pollution is due to anthropogenic emissions, and by far most of such emissions relate to energy creation, changing the energy infrastructure of the world to eliminate combustion will largely eliminate most air pollution mortality and morbidity worldwide. This is one of the goals of a transition to clean, renewable energy and storage for everything.

## 1.2 Global Warming

**Greenhouse gases** are naturally and anthropogenically emitted gases in the atmosphere that are generally transparent to **solar radiation** (sunlight) but selectively absorb certain wavelengths of **thermal-infrared (TIR) radiation** (**heat**) emitted by the surface of the Earth. Once a

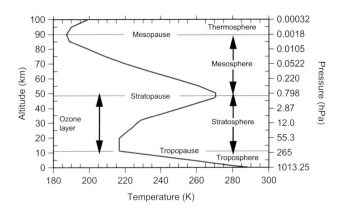

**Figure 1.1 Globally averaged temperature structure in the lowest 100 km of the Earth's atmosphere.**

greenhouse gas molecule absorbs a specific wavelength of TIR radiation, the molecule vibrates. It then converts the vibrational energy back to TIR radiation at a different wavelength, which depends on temperature. Finally, the molecule, which is no longer vibrating, re-emits the new wave in a random direction. The probability that the new wave is emitted downward is the same as the probability that it is emitted upward.

Due to this mechanism, greenhouse gases in the lowest part of the atmosphere, the troposphere (Figure 1.1), trap a portion of the TIR (heat) radiation emitted by the surface of the Earth. The trapping increases near-surface air and ground temperatures. Because some heat is trapped near the surface, that heat does not reach higher altitudes, so the addition of greenhouse gases to the atmosphere causes temperature to decrease with increasing height in the troposphere.

The more greenhouse gas that is present, the less TIR radiation that reaches the upper atmosphere (stratosphere, mesosphere, or thermosphere in Figure 1.1). Therefore, **anthropogenic greenhouse gases emitted near the surface of the Earth warm the surface, lower troposphere, and mid-troposphere at the expense of cooling the upper troposphere and stratosphere**, relative to no anthropogenic emissions. This is because the additional heat trapped by the anthropogenic greenhouse gases would have otherwise radiated to the upper atmosphere where natural background greenhouse gases there would have absorbed it, heating the upper atmosphere.

## 1.2.1 The Natural Greenhouse Effect

Global warming must be distinguished from the **natural greenhouse effect**, which is the historic buildup of naturally emitted greenhouse gases in the Earth's atmosphere before humans arrived on the scene.

In the absence of any greenhouse gases in its atmosphere, the Earth's temperature can be approximated by considering a balance among the solar radiation reaching the Earth, the reflectivity of the Earth and atmosphere, and the TIR leaving the Earth, which depends on temperature. Such a balance results in an **equilibrium**

temperature of the Earth of about 255 K (–18 °C). At this temperature life could not have evolved on Earth. Fortunately, the actual average temperature of the Earth's surface prior to the Industrial Revolution was about 288 K (+15 °C) (Figure 1.1). Temperatures near this have allowed life to evolve. The difference between the pre–Industrial Revolution temperature and the equilibrium temperature is the natural greenhouse effect. It is due to the historic buildup of natural greenhouse gases.

Table 1.1 indicates that, of the 33 K (33 °C) temperature rise above the Earth's equilibrium temperature due to the natural greenhouse effect, about 66 percent is due to **water vapor** (**H$_2$O**) and about 25 percent is due to background **carbon dioxide** (**CO$_2$**). Several other chemicals contribute to the remainder.

Table 1.1 **Primary contributors to the natural greenhouse effect and global warming on Earth.**

| Contributor | Contribution to the 33-K natural greenhouse effect (%) | Contribution to the 2.4-K gross global warming through 2018 before cooling subtracted (%) |
|---|---|---|
| Water vapor (H$_2$O) | 66 | 0.23 |
| Carbon dioxide (CO$_2$) | 25 | 45.7 |
| Black carbon (BC)+brown carbon (BrC) | 0.2 | 16.3 |
| Methane (CH$_4$) | 0.6 | 12.0 |
| Halogens | 0.0029* | 9.0 |
| Ozone (O$_3$) | 6.2 | 8.8 |
| Nitrous oxide (N$_2$O) | 1.4 | 4.3 |
| Carbon monoxide (CO) | 0.032 | <0.01 |
| Molecular oxygen (O$_2$) | 0.6 | 0 |
| Urban heat island effect (UHI) | 0 | 3.0 |
| Anthropogenic heat flux (AHF) | 0 | 0.7 |

Gross global warming (2.4 K) is warming before cooling (1.2 K) due to cooling aerosol particles is subtracted out. Net global warming (1.2 K) is gross warming minus cooling.
* The halogen contributing to the natural greenhouse effect is methyl chloride from the oceans. Halogens contributing to global warming are all synthetic compounds (Section 1.2.3.3). Percentage contributions to the natural greenhouse effect are from Jacobson (2014). Percentage contributions to global warming are from Figure 1.2.

## 1.2.2 Global Warming

**Global warming** (also referred to as **anthropogenic global warming**, or **AGW**) is the net rise, caused by human activities, in globally averaged near-surface air and ground temperatures above and beyond those due to the natural greenhouse effect. The net rise in temperatures is due to four major warming processes that result in a gross global warming partially offset by one major cooling process (Figure 1.2).

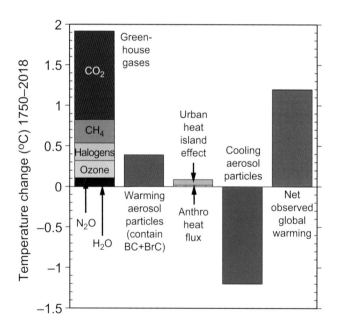

Figure 1.2 **Estimated primary contributors to net observed global warming from 1750 to 2018.**

Results are based on global model calculations (Jacobson and Ten Hoeve, 2012; Jacobson, 2014; Russell et al., 2018) and radiative forcing data for greenhouse gases from Myhre et al. (2013). Greenhouse gases are discussed in Section 1.2.1. Warming aerosol particles include black carbon (BC) and brown carbon (BrC) from fossil-fuel, biofuel, and open biomass burning sources (Section 1.2.2). Cooling aerosol particle components include primarily sulfate, nitrate, chloride, ammonium, sodium, potassium, calcium, magnesium, weakly absorbing organic carbon, and water. Section 1.2.4 describes the urban heat island effect. Its contribution to warming is from Jacobson and Ten Hoeve (2012). Section 1.2.3 discusses anthropogenic heat fluxes. Their contribution to warming is from Jacobson (2014). Net observed warming is from NASA (2018) using November 2016 to October 2017 average surface temperatures versus 1900 to 1930 average temperatures. Of all the warming in the figure, 45.7 percent is due to $CO_2$, 16.3 percent is due to BC +BrC, 12 percent is due to $CH_4$, 9.0 percent is due to halogens, 8.8 percent is due to ozone, 4.3 percent is due to $N_2O$, 3.0 percent is due to the urban heat island effect, 0.7 percent is due to anthropogenic heat flux, and 0.23 percent is due to anthropogenic $H_2O$.

The four major processes that contribute to global warming are anthropogenic greenhouse gas emissions, anthropogenic emissions of aerosol particles containing sunlight-absorbing chemicals, anthropogenic heat emissions, and the urban heat island effect. The process causing cooling is the anthropogenic emission of non-absorbing or weakly absorbing aerosol particles that cool climate by reflecting sunlight to space and by thickening clouds, which are largely reflective. The strongly absorbing particles are referred to as **warming particles**, and the weakly or non-absorbing particles are referred to as **cooling particles**.

Warming particles include primarily those containing black carbon (BC) and brown carbon (BrC). Cooling particles contain primarily sulfate, nitrate, chloride, ammonium, sodium, potassium, calcium, magnesium, weakly absorbing organic carbon, and water. The sources of warming particles generally differ from those of cooling particles. Also, because cooling particles tend to be more soluble in water than warming particles, cooling particles enhance cloudiness, thus cool climate, more than do warming particles. Warming particles, on the other hand, tend to heat clouds, helping to burn them off.

Because all aerosol particles together are the leading cause of air pollution mortality, reducing both cooling and warming particles is desirable from a public health perspective. However, Figure 1.2 indicates that cooling particles cause more cooling than warming particles cause warming globally. As such, if emissions of all warming and cooling particles are eliminated together without eliminating other sources of heat, global warming will worsen.

Similarly, since cooling particles mask half of global warming (Figure 1.2), eliminating only cooling particles will roughly double net global warming.

One strategy to address global warming and human health simultaneously is to eliminate only warming particles. The downside is that this strategy permits continued global warming and ocean acidification due to greenhouse gases and air pollution mortality due to cooling particles.

**Thus, Figure 1.2 suggests that the best strategy for addressing human health and climate simultaneously is to eliminate greenhouse gases, cooling particles, and warming particles simultaneously.** This will also reduce some anthropogenic heat flux, the main source of which is fossil-fuel combustion.

This book is about understanding and implementing that strategy – eliminating all anthropogenic emissions of greenhouse gases, warming particles, and cooling particles at the same time. This strategy will be accomplished by transitioning the world's energy to 100 percent wind, water, and solar plus storage for everything and by eliminating remaining non-energy emissions.

### 1.2.3 Anthropogenic Greenhouse Gases

The primary anthropogenic greenhouse gases contributing to global warming are $CO_2$, **methane** ($CH_4$), **nitrous oxide** ($N_2O$), ozone ($O_3$), halogens, and $H_2O$. The first three of these are introduced briefly next, followed by the remaining three.

#### 1.2.3.1 Carbon Dioxide, Methane, and Nitrous Oxide

The **volume mixing ratio** of a gas is the number of molecules of the gas per million molecules of air (parts-per-million-volume, ppmv) or per billion molecules of air (parts-per-billion-volume, ppbv). Since the start of the **Industrial Revolution** in the mid-1700s, the volume mixing ratio of the well-mixed (on the large scale) greenhouse gas $CO_2$ has increased by about 50 percent, from a natural background level of 275 ppmv to near 416 ppmv in 2019. The mixing ratio of CH$_4$ has increased by about 150 percent, from 0.715 ppmv to 1.8 ppmv, and that of N$_2$O has increased by almost 600 percent, from 0.27 ppmv to 1.86 ppmv in 2018 (Winther et al., 2018). Whereas the total mass of tropospheric O$_3$ has increased by about 43 percent, the mass of stratospheric ozone has decreased by about 5 percent (Myhre et al., 2013, Section 8.2.3.1). All of these greenhouse gases, except for ozone, are emitted both anthropogenically and naturally. However, their atmospheric growth since the Industrial Revolution is due almost exclusively to anthropogenic sources.

Next, ozone and halogens are briefly discussed, followed by a summary of the relative contributions of the different major greenhouse gases to global warming and a discussion of anthropogenic water vapor.

### 1.2.3.2 Ozone

**Ozone** is not emitted. Instead, it forms by chemical reaction from other chemicals that are emitted. Ozone in the atmosphere can be separated into tropospheric ozone and stratospheric ozone. The **troposphere** is a layer of the atmosphere that extends from the surface of the Earth to the **tropopause** (see Figure 1.1), which is a boundary that separates the troposphere from the stratosphere. The altitude of the tropopause above the surface is around 6 to 8 km near the poles, increasing to 18 km at the equator. In Figure 1.1, the tropopause is drawn with a mean value of about 12 km. The troposphere is characterized by a decrease in temperature with increasing altitude, except in the bottom 1 km (0.2 to 1.5 km), referred to as the **boundary layer**, where the temperature may increase or decrease with increasing altitude between night and day and under different meteorological conditions.

The **stratosphere** is the atmospheric layer between the tropopause and 48 km above the surface (**stratopause**). Ninety percent of all ozone in the current atmosphere resides in the stratosphere, and almost 10 percent resides in the troposphere. The stratosphere is characterized by a large **temperature inversion**, which is defined as an atmospheric layer where temperature increases with increasing altitude. The inversion is caused by ozone, as described next.

Stratospheric ozone densities peak between 25 and 32 km, in the middle of the stratosphere. Peak stratospheric temperatures, though, occur at the top of the stratosphere because this is the altitude at which the most intense UV wavelengths first enter the stratosphere. Ozone at the top

of the stratosphere absorbs these short wavelengths, preventing them from penetrating further down. Although the ozone concentration at the top of the stratosphere is low, each ozone molecule that absorbs a short UV wavelength converts that radiation into kinetic energy and thus moves fast. The resulting increase in the average kinetic energy of air molecules in the upper stratosphere increases the air temperature there (since temperature is proportional to the average kinetic energy among all molecules). Because the shortest UV wavelengths do not penetrate further down, temperatures increase less and less between the top and bottom of the stratosphere.

Ozone is a greenhouse gas. However, since preindustrial times, the mass of tropospheric ozone has increased by about 43 percent, whereas since the late 1970s, that of stratospheric ozone has declined by about 5 percent.

The increase in tropospheric ozone is due to the anthropogenic emissions of **nitric oxide (NO)**, **nitrogen dioxide** (NO$_2$), CH$_4$, other organic gases, and **carbon monoxide (CO)** due to fossil-fuel combustion, biofuel combustion, and open biomass burning. In the presence of UV radiation, these gases chemically produce ozone in the troposphere. Such gases usually break down or are removed from the atmosphere before they get to the stratosphere, so have less of an effect in the stratosphere.

Ozone in the stratosphere is produced by UV radiation breakdown of **molecular oxygen (O$_2$)** into two **atomic oxygen (O)** atoms followed by reaction of an O atom with an O$_2$ molecule, which gives O$_3$. Ozone is lost by natural chemical reaction as well. Ozone's natural production and loss lead to a natural equilibrium amount of it in the stratosphere. However, since 1978, stratospheric ozone mass has declined about 5 percent. The decrease is due to the emissions of certain halogens into the atmosphere starting in 1928, the transport of those halogens to the middle stratosphere, which took up to 50 years, the breakdown of those halogens into chlorine (Cl) and bromine (Br) atoms by UV radiation in the middle and upper stratosphere, and the chemical destruction of stratospheric ozone by its reaction with Cl and Br.

### 1.2.3.3 Halogens

**Halogens** are a group of synthetic chemicals invented starting in 1928 as refrigerants, solvents, degreasing agents, blowing agents, fire extinguishants, and fumigants. They enter the atmosphere primarily upon evaporation when they leak or when the appliances containing them are drained. Many of them (**halocarbons**) can be obtained by replacing some or all of the hydrogen (H) atoms in a methane ($CH_4$) or an ethane ($C_2H_6$) molecule with chlorine (Cl), bromine, (Br), fluorine (F), or iodine (I). The first halocarbons were obtained by replacing one or two chlorine atoms with fluorine atoms in carbon tetrachloride ($CCl_4$). Other halogens, such as sulfur hexafluoride ($SF_6$), have no carbon. Below, several types of halogens are categorized.

**Chlorofluorocarbons** (**CFCs**) are halocarbons containing Cl and F. Examples are $CFCl_3$ (CFC-11) and $CF_2Cl_2$ (CFC-12), which were used primarily as refrigerants starting in 1928 (for CFC-12). Their main applications were in ice cream coolers, whole-room coolers, refrigerators, air conditioners, and automobile cooling systems. Subsequent CFCs were used as solvents, degreasing agents, and blowing agents in spray cans and during foam production.

**Halons** are halocarbons, such as $CF_3Br$ (H-1301), that contain carbon and bromine and that are used in fire extinguishers.

**Perfluorocarbons** (**PFCs**) are halocarbons, such as $CF_4$ (PFC-14) and $C_2F_6$ (PFC-116), containing only carbon and fluorine.

**Hydrofluorocarbons** (**HFCs**) are halocarbons, such as $CHF_3$ (HFC-23), that contain carbon, fluorine, and hydrogen. PFCs and HFCs were developed primarily to replace CFCs.

Because CFCs and halons contain stratospheric ozone-destroying chlorine and bromine, most countries outlawed them through international agreement. PFCs and HFCs were developed as ozone-layer-friendly replacements. However, because many PFCs and HFCs are greenhouse gases with long lifetimes in the atmosphere, such chemicals, while not directly damaging the ozone layer, have the unintended consequence of exacerbating global warming.

### 1.2.3.4 Lifetimes and Global Warming Potentials

The contribution to global warming of each gas depends on how many molecules of the gas are in the atmosphere at a given time and how much warming each gas molecule causes. For chemicals that are directly emitted from anthropogenic sources (all the greenhouse gases, except for ozone), the number of molecules of the gas in the atmosphere at a given time depends on the emission rate of the gas and the *e*-folding lifetime of the gas against its removal by all processes. For ozone, the number of molecules depends on its production rate from chemical reaction and its lifetime against loss.

The *e***-folding lifetime** (also called the **average lifetime**) of a chemical is the time required for the amount of the chemical to decrease to $1/e$ (36.79 percent) its original amount due to chemical reaction, biological removal, removal by rain, or removal by hitting and sticking to a surface, such as a building or the ground. After two *e*-folding lifetimes, the chemical has decayed to $1/e^2$ (13.53 percent) its original amount. The time required for the amount of a chemical to reach 1 percent of its initial value is 4.61 *e*-folding lifetimes.

Table 1.2 provides the *e*-folding lifetimes of several anthropogenic greenhouse gases. It indicates that most gases have lifetimes that range from a few years to 50,000 years. Tropospheric ozone, on the other hand, has an average lifetime of only 23 days. If a chemical has a short lifetime and is an important greenhouse gas, it either has an extremely high emission rate or chemical production rate or causes significant warming per molecule. Ozone is produced rapidly in the troposphere by chemical reaction.

The strength of warming of an emitted chemical is often quantified by its **global warming potential** (**GWP**). GWP measures the time-integrated heat trapped by a chemical in the atmosphere per unit mass emission of the chemical relative to that of $CO_2$. Alternatively, it is the number of grams of $CO_2$ emissions that gives the same time-integrated atmospheric heating as 1 g emission of the substance of interest. Thus, the GWP of chemical $x$ has units of g-$CO_2$/g-$x$. The GWP of $CO_2$ is 1.

A chemical's GWP accounts for the chemical's lifetime. GWPs are usually determined by integrating over either 20 years (when examining short-term impacts) or 100 years (when examining long-term impacts). A 20-year GWP of 100 means that 1 g emission of the gas results in 100 times the heat trapping (warming) as 1 g emission of $CO_2$, averaged over 20 years. A 100-year GWP of 5 means that 1 g emissions of the gas results in 5 times the warming as 1 g emissions of $CO_2$, averaged over 100 years.

## Example 1.1  Calculating global warming potential

Calculate the 20- and 100-year global warming potentials of chemical A assuming (unrealistically) that (a) 1 kg of chemical A in the atmosphere at a given moment causes 100 times the instantaneous global warming as 1 kg of $CO_2$ and (b) 1 kg of A emitted into the atmosphere stays in the atmosphere exactly 20 years, at which point, it is suddenly removed.

### Solution:

Since 1 kg of A emitted into the air stays there for 20 years, and the warming per unit mass of A in the air is 100 times that of $CO_2$, the 20-year GWP of A is 100. Over 100 years, the GWP of A is 100 for the first 20 years and 0 for the last 80 years, or a weighted average of 20.

Table 1.2 *E-folding lifetimes, 20-year GWPs, and 100-year GWPs of several global warming agents.*

| Chemical | E-folding lifetime | 20-year GWP | 100-year GWP |
|---|---|---|---|
| [a]$CO_2$ | 50–90 years | 1 | 1 |
| [b]BC+POC in fossil-fuel soot | 3–7 days | 2,400–3,800 | 1,200–1,900 |
| [b]BC+POC in biofuel soot | 3–7 days | 2,100–4,000 | 1,060–2,020 |
| [c]$CH_4$ | 12.4 years | 86 | 34 |
| [c]$N_2O$ | 121 years | 268 | 298 |
| [c]$CFCl_3$ (CFC-11) | 45 years | 7,020 | 5,350 |
| [d]$CF_2Cl_2$ (CFC-12) | 100 years | 10,200 | 10,800 |
| [c]$CF_4$ (PFC-14) | 50,000 years | 4,950 | 7,350 |
| [d]$C_2F_6$ (PFC-116) | 10,000 years | 8,210 | 11,100 |
| [e]Tropospheric $O_3$ | 23 days | – | – |
| [f]$NO_x$-N | <2 weeks | −560 | −159 |
| [g]$SO_x$-S | <2 weeks | −1,400 | −394 |

GWP = global warming potential.

[a] The low-lifetime of $CO_2$ is the data-constrained lifetime upon increasing $CO_2$ emissions from Jacobson (2012a); the high-lifetime of $CO_2$ is calculated from Figure 9.6, which shows $CO_2$ decreasing by 65 parts-per-million-volume (ppmv) (from 400 to 335 ppmv) over 65 years upon elimination of anthropogenic $CO_2$ emissions. Since the natural $CO_2$ is 275 ppmv, the anthropogenic $CO_2$ = 400 − 275 = 125 ppmv, and the lifetime of anthropogenic $CO_2$ is ~65 y / −ln((125 − 65) ppmv/125 ppmv) = ~90 years. The GWP of $CO_2$ = 1 by definition.

Chemicals that have a greater 20-year GWP than 100-year GWP have shorter lifetimes than $CO_2$. Conversely, chemicals that have a greater 100-year GWP than 20-year GWP have longer lifetimes than $CO_2$.

Table 1.2 indicates that the *e*-folding lifetime of PFC-14, for example, is 50,000 years. Its GWP is 4,950 over 20 years, increasing to 7,350 over 100 years. The increase in GWP from 20 to 100 years indicates that the same pulse mass emission of PFC-14 and $CO_2$ results in a greater increase in the atmospheric mixing ratio of PFC-14 than of $CO_2$ over 100 years. Not only does PFC-14 last much longer than does $CO_2$, each gram of PFC-14 in the atmosphere causes more warming than does each gram of $CO_2$, resulting in a GWP of PFC-14 that exceeds 1. The only reason PFC-14 in the air today does not cause much more global warming than $CO_2$ is because the emission rate and mixing ratio of $CO_2$ are much larger than are those of PFC-14.

[b] POC is primary organic carbon co-emitted with black carbon from combustion sources. In the case of diesel exhaust, it is mostly lubricating oil and unburned fuel oil. In all cases, POC includes both absorbing organic (brown) carbon (BrC) and less absorbing organic carbon. Soot particles contain both BC and POC. The lifetime is from Jacobson (2012b) and the GWP is from Jacobson (2010a, Table 4), which accounts for direct effects, optical focusing effects, semi-direct effects, indirect effects, cloud absorption effects, and snow-albedo effects. The GWPs here are the surface temperature response after 20 or 100 years per unit continuous mass emissions (STRE) of BC+POC relative to the same for $CO_2$. STREs are analogous to GWPs (Jacobson, 2010a, Table 4 footnote).
[c] Source: Myhre et al. (2013), Table 8.7. Results from Etminan et al. (2016) suggest that the 20-y GWP of $CH_4$ may be up to 98.
[d] Source: Myhre et al. (2013), Table 8.A.1.
[e] Source: Myhre et al. (2013), Section 8.2.3.1. Tropospheric ozone is not emitted, so does not have a GWP.
[f] Source: Myhre et al. (2013), Table 8.A.3, including aerosol direct and indirect effects. Values are on a per kg nitrogen basis.
[g] Source: Streets et al. (2001) and Jacobson (2002), including aerosol direct and indirect effects. Values are on a per kg sulfur basis. These numbers are STREs, which are analogous to GWPs (Footnote b).

For $CH_4$, the 20- and 100-year GWPs also exceed 1, but methane 100-year GWP is less than its 20-year GWP. This is due to the fact that $CH_4$ has a shorter lifetime than $CO_2$. As time marches on far past 100 years, $CO_2$'s integrated warming per unit mass emission approaches that of methane. However, a danger arises from using a 100-year or longer GWP. Such GWPs hide the fact that some chemicals, such as $CH_4$, cause much more severe heating per unit mass emission over 20 years than does $CO_2$. For example, $CH_4$ directly leaked to the air from natural gas mining, pipelines, and end-use equipment causes much more climate damage over 20 years than does burning the same amount of $CH_4$ in a power plant, where it converts to $CO_2$ that is released to the air.

The 20-year heating due to methane causes impacts, including melting of the Arctic sea ice, that may not easily be reversed and that may trigger additional positive feedbacks, causing more warming. As such, implementing policies using a 20-year GWP may help delay irreversible damage, giving time for other solutions to be implemented. On the other hand, using a 100-year GWP may result in irreversible damage.

### 1.2.3.5 Carbon Dioxide Equivalent Emissions

One way to compare the global warming impacts of two different energy sources is to compare the **carbon dioxide equivalent ($CO_2e$) emission rate** of each source. The $CO_2e$ of an energy source equals the product of the emission rate (e.g., g/kilowatt hour [kWh] for electric power sources or g/km for vehicles) of each chemical from the source and the GWP of the chemical, summed over all chemicals from the source. $CO_2e$ emission rates can be defined over either a 20- or 100-year time frame. For a 20-year time frame, 20-year GWPs are used; for a 100-year time frame, 100-year GWPs are used (see Table 1.2). Since the GWP of $CO_2$ itself is always 1, the $CO_2e$ of an energy source that emits only $CO_2$ is simply the emission rate of $CO_2$.

### Example 1.2  $CO_2e$ emissions of natural gas versus coal

Calculate the methane leakage rate (fraction of total mined methane that leaks) during its mining and transport needed for the carbon dioxide equivalent emissions ($CO_2e$) of a natural gas electric power plant that uses the methane to equal that of a coal power plant over (a) a 20-year time frame and (b) a 100-year time frame. Assume the coal plant emits 1,000 g-$CO_2$/kWh and nothing else; the natural gas plant emits leaked methane and 500 g-$CO_2$/kWh of combusted methane; the molecular weight of $CO_2$ is 44.0095 g/mol; and the molecular weight of $CH_4$ is 16.04246 g/mol. Ignore any other emissions associated with either natural gas or coal.

### Solution:

Over both the 20- and 100-year time frames, the coal plant emits 1,000 g-$CO_2$/kWh, which is the $CO_2e$ emission rate of the coal plant. The $CO_2e$ emission rate of the natural gas plant is 500 g-$CO_2$/kWh + $L \times$ GWP, where $L$ is the leakage rate of methane (g-$CH_4$/kWh) and the GWP is that of methane (g-$CO_2$/g-$CH_4$). The GWPs of methane from Table 1.2 are 86 and 34 over 20 and 100 years, respectively. Equating the $CO_2e$ from the gas and coal plants and solving for the leakage rate gives $L$ = 5.81 g-$CH_4$/kWh over a 20-year time frame and $L$ = 14.7 g-$CH_4$/kWh over a 100-year time frame.

$L$ is the absolute leakage rate of methane. Next, it is necessary to calculate the leakage rate as a fraction of total methane (leaked methane plus methane combusted in the natural gas plant). This is calculated as $L / (L + B)$, where $B$ is the mass of methane that is burned per unit electricity produced in the natural gas plant. Since 1 mole of $CH_4$ is burned for every 1 mole of $CO_2$ emitted, $B$ = 500 g-$CO_2$/kWh $\times$ 16.04246 g-$CH_4$/mol / 44.0095 g-$CO_2$/mol = 182.26 g-$CH_4$/kWh.

The fraction of total methane that leaks is therefore $L / (L + B)$ = 0.031 over a 20-year time frame and 0.075 over a 100-year time frame. In other words, if the leak rate of methane equals or exceeds 3.1 percent, natural gas causes more warming than coal over a 20-year time frame, if only $CO_2$ and $CH_4$ impacts are considered. A higher GWP of $CH_4$ than what is provided in Table 1.2 would reduce this leak rate break-even point. Over a 100-year time frame, the break-even point is a 7.5 percent leak rate. However, as discussed in Chapter 3, additional chemicals must be considered when comparing the $CO_2e$ of natural gas versus coal.

### 1.2.3.6 Anthropogenic Water Vapor

**Anthropogenic water vapor** comes from two main sources: (a) evaporation of water that is used to cool power plants and industrial facilities that run on coal, natural gas, oil, nuclear power, or biofuels; and (b) emission of water vapor during the combustion of fossil fuels, biofuels, and biomass. Water vapor emitted annually from these sources is only around 1/8,800th of the ~500 million teragrams (Tg)-$H_2O$/y emitted from natural sources (Jacobson, 2014). Nevertheless, this relatively small anthropogenic emission of water vapor contributes to a modest 0.23 percent of global warming (Jacobson, 2014).

## 1.2.2 Anthropogenic Absorbing Aerosol Particle Components

Absorbing aerosol particle components, namely, black carbon and brown carbon, together cause more global warming than any other chemical aside from $CO_2$ (Jacobson, 2000, 2001a, 2002, 2010a; Bond et al., 2013). Such particle components simultaneously cause significant air pollution mortality and morbidity.

A **particle** is an aggregate of atoms or molecules bonded together as a liquid, solid, or mixture of both. An **aerosol** is an ensemble of solid, liquid, or mixed-phase particles suspended in the air. An **aerosol particle** is a single liquid, solid, or mixed-phase particle among an ensemble of suspended particles.

**Black carbon** (**BC**) is a solid agglomerate of pure carbon spherules attached together in an amorphous shape. Its source is incomplete combustion of diesel, gasoline, jet fuel, bunker fuel, kerosene, natural gas, biogas, biomass, and liquid biofuels. Black carbon is often visible to the eye and appears black because it absorbs all wavelengths of sunlight, transmitting none to the eye. Black carbon particles convert the absorbed light to heat, raising the temperature of the particles and causing them to re-radiate some of the heat to the surrounding air.

Black carbon and greenhouse gases warm the air in different ways from each other. Greenhouse gases are mostly transparent to solar radiation. They heat the air by absorbing thermal-infrared radiation emitted by the surface of the Earth and re-emitting some of the heat back toward the surface. Black carbon particles, on the other hand, heat the air primarily by absorbing solar radiation, converting the solar radiation to thermal-infrared radiation, and re-emitting the TIR back to the air. Black carbon particles also absorb and re-emit TIR, but that process is less important for them unless BC concentrations are high.

Black carbon not only warms the air directly; when it enters clouds, it can evaporate them. When BC falls on snow or sea ice, it can melt the ice or snow faster. In addition, when other aerosol material, such as sulfuric acid, nitric acid, water, or brown carbon, coats the outside of a black carbon particle, the black carbon can heat the air up to 2 to 3 times faster than it can without a coating because more light hits the larger particle, thus bends (refracts) into the particle. Inside the particle, this light bounces around until it hits and is absorbed by the black carbon core.

**Brown carbon** (**BrC**) is also an aerosol particle component that causes health problems. Major sources of brown carbon are the same as those of black carbon, except that the BrC-to-BC ratio increases with decreasing temperature of combustion. Smoldering biomass, for example, is a large source of brown carbon. BrC contains carbon, like black carbon, but also contains hydrogen (H) and possibly atomic oxygen (O), nitrogen (N), and/or other atoms. In other words, it is **organic carbon** (which is defined as containing C, H, and possibly other atoms). However, not all organic carbon is brown carbon. Brown carbon is the subset of organic carbon that selectively absorbs short (blue) and maybe some medium (green) and, even less frequently, red wavelengths of visible light. The remaining long wavelengths (red) and some of the green are transmitted to the viewer's eye, making the particle haze appear brown. The more green light that is transmitted (the less that is absorbed), the more yellow the particles appear.

All organic aerosol components and gases absorb the sun's ultraviolet (UV) wavelengths to some degree. UV wavelengths are shorter and more energetic than visible wavelengths. However, only a few organic aerosol components absorb visible wavelengths. Short-wavelength visible-light-absorbing organic particle or gas components include nitrated aromatics, benzaldehydes, benzoic acids, aromatic polycarboxylic acids, phenols, polycyclic aromatic hydrocarbons, and nitrated inorganics (Jacobson, 1999). In addition, tar balls are organic particles that form during the cooling of a biomass burning plume and strongly absorb visible radiation even in the red part of the visible spectrum.

Whereas the mass ratio of brown carbon to black carbon from diesel exhaust is around 1:1, that in biomass burning is around 8:1. Higher-temperature flames (e.g., in diesel engines) produce the lower ratio; lower-temperature flames produce the higher ratio. Although BrC concentrations are generally higher than are BC concentrations in the air, BC is a much stronger overall absorber, and so it almost always dominates the impact of the two on climate. An exception is in the case of biomass burning, where the impacts of BrC may dominate those of BC.

Table 1.2 indicates that although BC plus BrC from fossil-fuel and biofuel sources has an *e*-folding lifetime of less than a week, it has an extremely high GWP, even over 100 years. The reason is that BC in particular is about one million times more powerful per unit mass at warming the air than is $CO_2$ (Jacobson, 2002). Although a lot more $CO_2$ than BC exists in the air at a given time, the strong warming per unit mass allows BC plus BrC to be the second leading cause of global warming after $CO_2$ (see Table 1.1). Methane is third.

### 1.2.3 Anthropogenic Heat Emissions

**Anthropogenic heat emissions** include the heat from the dissipation of electricity; the heat from dissipation of motive energy by friction; the heat from combustion of fossil fuels, biofuels, and biomass for energy; the heat from nuclear reaction; and the heat from anthropogenic biomass burning. Such heat emissions warm the air directly. Much of the hot air eventually rises, expands, and cools, converting heat energy (**sensible heat**) to **gravitational potential energy** (Section 3.2.2.3), which is energy embodied in air lifted to a certain height against gravity. Differences in gravitational potential energy between one horizontal location and another result in the conversion of gravitational potential energy to **kinetic energy**, or winds. The increase in temperature and wind speed both evaporate water vapor, converting sensible heat and kinetic energy, respectively, to **latent heat**, which is energy added to liquid molecules to evaporate them into gas molecules. Since water vapor is a greenhouse gas, the production of water vapor enhances the impact of the original anthropogenic heat emissions.

In sum, much of the heat from anthropogenic heat emissions is converted to other forms of energy. Since energy is conserved, the different forms of energy persist in the atmosphere (although some of the energy is transferred to the oceans and land) and cause impacts that persist for a long time. Overall, though, the impacts of anthropogenic heat emissions are less than those of greenhouse gases, which also persist for decades to centuries but cause greater warming than do anthropogenic heat emissions. Nevertheless, anthropogenic heat may contribute to about 0.7 percent of global warming to date (see Figure 1.2 and Jacobson, 2014). So, although human actions that heat the air (e.g., fires and heat that escapes buildings and cars) contribute to global warming, the effect is relatively small. Nevertheless, eliminating combustion and nuclear reaction, which a WWS system does, eliminates such heat along with air pollutants and greenhouse gases.

### 1.2.4 The Urban Heat Island Effect

The **urban heat island (UHI) effect** is the temperature increase in urban areas due to covering soil and replacing vegetation with impervious surfaces. Covering surfaces reduces evaporation of water from soil and evapotranspiration from plants. Because evaporation and evapotranspiration are cooling processes, eliminating them warms the surface. Built-up areas also have sufficiently different properties of construction materials (e.g., heat capacities, thermal conductivities, albedos, emissivities) to enhance urban warming relative to surrounding vegetated areas. Worldwide, the UHI effect may be responsible for about 3 percent of gross global warming (warming before cooling is subtracted out) (see Figure 1.2 and Jacobson and Ten Hoeve, 2012).

### 1.2.5 Impacts of Global Warming

Global warming has already caused the world significant financial losses, and the cost is expected to grow to $25 to $30 trillion per year by 2050 (Jacobson et al., 2017). Losses arise due to coastline erosion (from sea level rise); fishery and coral reef losses; species extinction losses; agricultural losses; increased heat stress mortality and morbidity; increased migration, famine, drought, wildfires, and air pollution; and more severe storms and weather (hurricanes, tornados, hot spells).

Higher temperatures not only increase wildfire risk and damage, increasing air pollution and infrastructure loss directly, but they also increase air pollution in cities where the pollution is already severe (Jacobson, 2008a, 2010b). During November 2018, for example, three major wildfires in California, enhanced by drought and unusually high November temperatures, killed dozens of people, displaced hundreds of thousands of others, rendered several thousand people homeless, and produced dangerous levels of air pollution throughout the state for over two weeks.

Similarly, global warming increases the intensity of hurricanes, increases their wind speeds, storm surge, and the resulting damage. The failure of agriculture crops has caused mass migrations in many parts of the world. Thus, global warming has already resulted in devastation and the creation of climate migrants.

## 1.3 Energy Insecurity

Energy insecurity is a third major problem that needs to be addressed on a global scale. Several types of energy insecurity are of serious concern, as outlined in the following.

### 1.3.1 Energy Insecurity due to Diminishing Availability of Fossil Fuels and Uranium

One type of energy insecurity is the economic, social, and political instability that results from the long-term depletion of energy supply. Fossil fuels and uranium are limited resources and will run out at some point. As fossil-fuel supplies dwindle, for example, their prices will rise. Such price increases will first hit people who can least afford them – those with little or no income. Such people will suffer from not being able to warm their home as much during the winter, cool their home as much during the summer, or pay for vehicle fuel so easily when they live far from their job and public transit is not an option.

Higher energy prices will also increase the cost of food, which uses energy to be produced and transported. Higher energy prices will also ultimately lead to economic, social, and political instability. The end result may be chaos and civil war.

A solution to this problem is to transition to an energy system that is sustainable – one in which energy is at less risk of being in long-term short supply. Such a system is one that consists of **clean, renewable energy**, which is energy that is replenished by the wind, the water, and the sun. Solutions that do not solve this problem are fossil-fuel power plants, with or without carbon capture, and almost all nuclear power plants, because they rely on fuels that will be depleted over time.

### 1.3.2 Energy Insecurity due to Reliance on Centralized Power Plants and Oil Refineries

A second type of energy insecurity is the risk of power loss due to the reliance on large, centralized electric power plants and oil refineries. If a city or an island relies on centralized power plants, and one or more plants or the transmission system goes down, power to a large portion of the city or island may be unavailable for an indeterminate amount of time. Such loss can result from severe weather, power plant failure, or terrorism. Similarly, an accidental fire or act of terrorism at an oil refinery or storage facility or gas storage facility can cause a significant disruption to local and regional oil and gas supply.

### Transition highlight

On September 18, 2017, Hurricane Maria hit Puerto Rico and knocked out its power grid to 1.5 million people for almost 11 months. The hurricane toppled 80 percent of the island's utility poles and transmission lines. With 10 oil-fired power plants, 2 natural gas plants, and 1 coal plant, the island's energy supply was all but wiped out by the loss of transmission. The long delay in restoring power to individual homes and businesses occurred because of the need to rebuild most of the transmission system for that power. A more distributed energy system with rooftop solar photovoltaics, distributed onshore and offshore wind turbines, and local battery storage would have allowed hospitals, fire stations, and homes to maintain at least partial power during the entire blackout period and would have reduced the time required to restore power to most customers. In fact, in early 2019, the main utility in Puerto Rico proposed dividing the island into eight connected mini-grids dominated by solar energy sources and batteries. If one mini-grid goes down, the other seven will still function. On April 11, 2019, Puerto Rico went even further and passed a law to go to 100 percent renewable electricity by 2050 (Table 9.1).

In another example, a September 14, 2019, terrorist attack on two Saudi Arabia oil processing facilities knocked out 5 million barrels per day, or 5 percent of the world's and half of Saudi Arabia's daily oil production. Oil and gas refineries and storage facilities worldwide are continuously at risk of such attacks, and many are targets during conflict. Whereas risk does not go to zero with decentralized power generation and storage, which is what WWS largely provides, the risk significantly decreases due to the difficulty in taking down hundreds to thousands of units rather than one or two.

Another problem with large, centralized power plants is that they don't serve the 1.3 billion people who are currently without access to electricity, and they poorly serve another 1 billion with only unreliable access to electricity (Worldwatch Institute, 2019). Similarly, centralized power plants cannot provide power to remote military bases. Those bases obtain their power from diesel transported and used in diesel generators. For example, in 2009, seven liters of diesel fuel were burned during the transport of each liter of diesel used in a generator to produce electricity in U.S. military bases in Afghanistan (Vavrin, 2010). Many soldiers died during the transport of the fuel.

Because WWS technologies are largely **distributed** (decentralized), it is possible to use them in microgrids to reduce this lack of access to electricity. A **microgrid** is an isolated grid (not connected to a larger transmission network) that provides power to an individual building, hospital complex, community, or military base. A WWS microgrid consists of any combination of solar PV panels, wind turbines, batteries, hydrogen fuel cells for electricity, pumped hydropower storage facilities, or other WWS technology. When used in a microgrid,

WWS can bring electricity to many people who either are without it or have poor access to it because the fossil-fuel and nuclear systems have not been able to supply electricity to them.

In sum, a transition to WWS facilitates the creation of interconnected mini-grids and results in the use of more distributed energy generators. Both factors reduce the chance that severe weather, power plant failure, or terrorism will bring a large portion of the grid down. Fossil-fuel power plants, with or without carbon capture, and nuclear power plants do not solve this insecurity problem because these power plants, as commercialized today, are large, centralized power plants.

### 1.3.3 Energy Insecurity due to Reliance on Energy from Outside a Country

A third type of energy in security is the risk associated with one country relying on other countries to supply its energy. For example, many countries, particularly island countries, must import coal, oil, and/or natural gas to run their energy system. Importing fuel not only results in higher fuel prices but also creates a reliance of one country on others. This reliance may be tested in times of international conflict when countries that control energy may withhold it or may not be able to supply it anymore.

A clean, renewable energy system that is built within a country and that supplies most, if not all, of the country's all-purpose energy avoids this type of energy insecurity. This does not mean that countries adjacent to each other should not trade electric power. In fact, such trading is likely to reduce the cost of energy and improve reliability of the overall energy system. It means that most of a country's power can be "home grown." This reduces the risk of energy insecurity due to international conflict. The reduced risk also translates into avoided costs of war and lower costs of energy.

Fossil-fuel power plants with or without carbon capture and nuclear power plants do not avoid this problem because they both rely on fuels that must be supplied continuously from across country borders for most countries of the world. In many cases, especially for island countries, the fuels must be transported long distance.

### 1.3.4 Energy Insecurity due to Fuels That Have Mining, Pollution, Waste, Meltdown, and/or Weapons Risk

A fourth type of energy insecurity is the risk associated with byproducts of energy use. For example, the continuous mining of fossil fuels and uranium causes health damage to miners and major environmental degradation. For instance, coal mining results in black lung disease for many miners. Uranium mining results in high cancer rates from the decay products of radon (Section 3.3.2.4). In addition, energy production at fossil-fuel plants produces air pollution waste that kills millions of people each year (Section 7.6.2). Nuclear power plants similarly produce radioactive waste that must be stored for hundreds of thousands of years (Section 3.3.2.3). Nuclear plants also run the risk of a reactor core meltdown (Section 3.3.2.2). The spread of nuclear plants to dozens of countries has also contributed to the proliferation of nuclear weapons in several of these countries (Section 3.3.2.1).

A transition to clean, renewable energy avoids these risks to health, environment, and safety. The use of fossil fuels, with or without carbon capture, and nuclear power continues these energy security problems.

## 1.4 Summary

Three of the most significant world problems associated with energy today are air pollution, global warming, and energy insecurity. These problems need to be solved together because of the serious nature of all three. None should be solved in isolation. As such, a solution to any one of the problems must be a solution to all three. This is why some proposed solutions that partly address some problems but not others should not be advanced (Chapter 3). A solution that does address all three problems at the same time is to transition the world's all-purpose energy to electricity and heat that are provided by clean, renewable energy and storage. As discussed in this text, clean, renewable energy is energy in the wind, the water, and the sun (WWS), supplemented by storage. The rest of this book explores the details of how to address this energy solution and how to address emissions from non-energy sources.

## Further Reading

Burnett, R., 2018. Global estimates of mortality associated with long-term exposure to outdoor fine particulate matter, *Proc. Natl. Acad. Sci.*, **115**, 9592–9597.

Fuhrmann, M., 2009. Spreading temptation: proliferation and peaceful nuclear cooperation agreements. *Int. Secur.*, **34**(1), https://ssrn.com/abstract=1356091 (accessed September 9, 2019).

GBD (Global Burden of Disease 2013 Risk Factors Collaborators), 2015. Global, regional, and national comparative risk assessment of 79 behavioral, environmental and occupational, and metabolic risks or clusters of risks in 188 countries, 1990–2013: a systematic analysis for the Global Burden of Disease Study 2013, *Lancet*, **386**, 2287–2323.

IPCC (Intergovernmental Panel on Climate Change), 2018. Special report: Global warming of 1.5 °C, www.ipcc.ch/sr15/ (accessed June 26, 2019).

Jacobson, M. Z., 2001. Strong radiative heating due to the mixing state of black carbon in atmospheric aerosols, *Nature*, **409**, 695–697.

Jacobson, M. Z., 2009. Review of solutions to global warming, air pollution, and energy security, *Energy Environ. Sci.*, **2**, 148–173, doi:10.1039/b809990c.

Jacobson, M. Z., 2012. *Air Pollution and Global Warming: History, Science, and Solutions*, 2nd ed., Cambridge: Cambridge University Press, 375 pp.

Jacobson, M. Z., 2014. Effects of biomass burning on climate, accounting for heat and moisture fluxes, black and brown carbon, and cloud absorption effects, *J. Geophys. Res.*, **119**, 8980–9002, doi:10.1002/2014JD021861.

Jonsson, D. K., B. Johansson, A. Mansson, L. J. Nilsson, M. Nilsson, and H. Sonnsjo, 2015. Energy security matters in the EU energy roadmap, *Energy Strategy Rev.*, **6**, 48–56.

## 1.5 Problems and Exercises

1.1.  From Table 1.1 and Figure 1.2, by what percent will net global warming decrease if all anthropogenic aerosol black carbon and brown carbon were removed instantaneously from the atmosphere?

1.2.  If the ozone layer peaks in the middle of the stratosphere, why are the highest stratospheric temperatures at the top of the stratosphere?

1.3.  Give two reasons why water vapor has such a large natural greenhouse effect but such a small global warming effect in comparison.

1.4.  List two types of energy security problems faced by each coal, gasoline, and nuclear power.

1.5.  Explain how electricity from solar photovoltaics reduces all three problems of air pollution, global warming, and energy security compared with electricity from coal.

1.6.  What are the energy insecurity risks associated with nuclear power?

1.7.  Explain why global warming of the lower and mid-troposphere is causing the stratosphere to cool.

1.8.  Explain why a cooling stratosphere is inconsistent with the theory that an increase in sunspot activity (greater solar radiation) is the cause of tropospheric global warming.

1.9.  Based on Figure 1.2, what is the best overall strategy for eliminating both air pollution and global warming simultaneously? Why does eliminating greenhouse gases in isolation not solve the whole problem? Why does eliminating all aerosol particles in isolation not solve the whole problem? What is the impact of eliminating only warming particles?

1.10. Given the global warming potentials (GWPs) of methane from Table 1.2, what causes more warming over (a) a 20-year time frame and (b) a 100-year time frame, a coal plant that emits 1,000 g-$CO_2$/kWh and no methane or a gas plant that emits 500 g-$CO_2$/kWh, all from methane combustion, but leaks 4 percent of its methane before combustion? Assume the molecular weight of $CO_2$ is 44.0095 g/mol and that of $CH_4$ is 16.04246 g/mol.

# CHAPTER

# 2 Wind-Water-Solar (WWS) and Storage Solution

The solution to air pollution, global warming, and energy insecurity is, in theory, simple and straightforward: electrify or provide direct heat for everything; obtain the electricity and heat from only wind, water, and solar power; store energy; and reduce energy use.

What is meant by electrifying or providing direct heat for everything? Almost all energy worldwide is used for electricity, transportation, building heating and cooling, and industry. We would convert all modes of transportation to either battery-electric (BE) vehicles, hydrogen fuel cell (HFC) vehicles (where the hydrogen is produced from electricity), or BE-HFC hybrids. For heating/cooling, we would provide direct heat (such as from hot geothermal reservoirs or direct solar radiation) to heat some air and water. We would use electric heat pumps for remaining air and water heating, and all air conditioning. A portion of the heating and cooling would be performed in centralized facilities, and the heat and cold would be distributed through water pipes to buildings. The remainder would be produced in buildings.

For high-temperature industrial processes, we would use existing electricity-based technologies (e.g., arc furnaces, induction furnaces, dielectric heaters, and resistance furnaces) to create high-temperature heat. We would also reduce energy use by capturing and recycling more waste heat and cold, improve insulation to reduce heat and cold losses, use more energy-efficient appliances, and create more pedestrian and bike-friendly cities.

All the electricity and direct heat in this new paradigm would be powered by 100 percent wind, water, and solar (WWS) energy combined with electricity, heat, cold, and hydrogen storage. It will also be combined with short- and long-distance electricity transmission and short-distance heat, cold, and hydrogen transmission (e.g., through cities). The WWS energy generation technologies for each city, state, and country would be a combination of onshore and offshore wind (covered in detail in Chapter 6), solar photovoltaics (PV) on rooftops and in large (power plant) arrays (covered in detail in Chapter 5), concentrated solar power (CSP) plants, geothermal plants, conventional and run-of-the-river hydroelectric power, tidal and ocean current power, and wave power.

Types of storage include electricity, heat, cold, and hydrogen storage. Major electricity storage options include pumped hydroelectric power, existing hydroelectric dams, CSP coupled with thermal energy storage, batteries, flywheels, compressed air storage, and gravitational storage with solid masses. Major heat storage options include storage in water, soil, and heat-absorbing materials. Major cold storage options include water and ice. Hydrogen, used primarily for transportation, is a form of stored electricity, and would be produced from electricity by electrolysis when excess WWS electricity is available. In some systems, storage will be combined with energy generation to reduce cost. Examples include combining batteries with a rooftop solar PV system to reduce the use of grid electricity when it costs the most and drawing electricity from the batteries of electric vehicles to help meet peaks in demand on the electric power grid.

This chapter discusses the technologies needed to eliminate all global anthropogenic emissions by transitioning energy and non-energy sources of such emissions. The chapter starts by describing the WWS technologies proposed for use to replace traditional energy sources (Section 2.1). It then moves on to discuss electric and hydrogen fuel cell transportation technologies needed

(Section 2.2), followed by building heating and cooling technologies needed (Section 2.3). Next, high-temperature industrial heat technologies (Section 2.4) and an examination of electric appliance substitutes for fossil-fuel-powered appliances (Section 2.5) are described. Methods of reducing energy use and increasing energy efficiency are then provided (Section 2.6), followed by an analysis of electricity storage technologies to be used for a transition (Section 2.7). This is naturally followed by an in-depth look at heat, cold, and hydrogen storage technologies available (Section 2.8). Finally, Section 2.9 details solutions to controlling non-energy emissions.

## 2.1 WWS Electricity-Generating Technologies

In this section, the WWS electricity-producing electric power options are described. Such technologies are generally defined in terms of their nameplate capacity. **Nameplate capacity** (also called rated capacity, generating capacity, or plant capacity) is the maximum instantaneous discharge rate of electricity from an energy-producing machine's (e.g., a wind turbine's) generator, as determined by the manufacturer of the wind turbine. Thus, a wind turbine that has a nameplate capacity of 1 kW (1,000 W) can discharge no more than 1,000 J/s of electricity from its generator, where a **W** (**watt**) is a unit of power, a **J** (**joule**) is a unit of energy, and **s** is a unit of time (seconds). Power equals energy per unit time, thus power is the rate of energy discharge from an energy-producing device. Conversely, energy equals the product of power and time.

**Energy storage** is similarly defined in terms of power and energy. The **peak charge or discharge rate of storage** is the maximum power (rate of change of energy) into or out of storage, respectively. The **peak storage capacity** is then the maximum energy that can be stored and equals the peak discharge rate multiplied by the **number of hours of storage** at the peak discharge rate. Thus, energy stored in a battery is akin to water stored in a reservoir.

### 2.1.1 Onshore and Offshore Wind

**Wind turbines** convert the kinetic energy of the wind into electricity. Generally, the slow-turning turbine blade spins a shaft connected to a **gearbox**. Gears of different sizes in the gearbox convert the slow-spinning motion (e.g., 5 to 10 rotations per minute for modern turbines) to faster-spinning motion (e.g., 1,800 rotations per minute) needed to convert mechanical energy to electrical energy in a **generator**. Some modern wind turbines are gearless, with the shaft connected directly to the generator. To compensate for the slow spin rate in the generator in gearless turbines, the radius of rotation is expanded, increasing the speed at which magnets move around a coil in the generator to produce electricity. The power output of a wind turbine increases with increasing turbine hub height because wind speeds generally increase with increasing height in the lower atmosphere. As such, taller turbines capture faster winds.

Wind farms are often located on flat, open land (Figure 2.1a), within mountain passes, on ridges, and offshore (Figure 2.1b). Individual turbines to date have ranged in nameplate capacity from 1 kW to 15 MW (million watts).

Figure 2.1 (a) Onshore wind farm coexisting with agriculture in China. The wind turbines increase the flow of carbon dioxide to the crops and require little footprint on the ground. © Envision. (b) Offshore wind farm. © glimpseofSweden/AdobeStock.

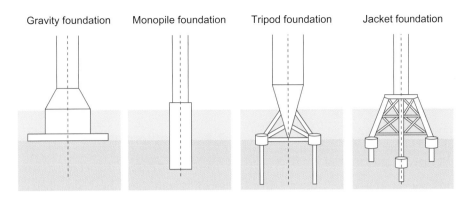

Figure 2.2 **Illustration of types of bottom-fixed foundations used to support offshore wind turbines in shallow water.** Adapted from Miceli (2012).

The smallest turbines (e.g., with nameplate capacities of 1 to 10 kW) are convenient for producing local electricity in the backyards of individual homes or city street canyons if winds are fast. **Onshore wind** farms usually contain a few to dozens of mid-sized turbines (1 to 8 MW) to power parts of towns and cities.

**Offshore wind** farms usually contain a few to dozens of mid- to large-sized (3- to 15-MW) turbines. One particular 12-MW turbine, for example, designed for offshore use, has a 150-m hub height above the ocean surface. Its rotor diameter is 220 m; thus, the height of its furthest vertical extent is 260 m, or 80 percent of the height of the Eiffel Tower. It may be able to provide electricity for up to 16,000 households (GE, 2018).

Offshore wind turbines have either bottom-fixed foundations or floating foundations. **Bottom-fixed foundations** include gravity, monopile, tripod, and jacket foundations (Figure 2.2). **Gravity foundations** are made of precast concrete, with sand, gravel, or stones added to give them stability. They are used in water that is no more than 30 to 50 m deep. **Monopile foundations** are used in water that is no more than 20 to 30 m deep. They are made of steel and hammered 30 m into the sea floor. **Tripod foundations** have three steel piles welded together, with their feet drilled into the sea floor. They are used in water depths no more than 30 to 40 m. **Jacket foundations** are made of over 500 tonnes of steel beams welded together, with their feet secured to the sea floor. They are used in water depths of up to 50 to 60 m.

**Floating wind turbines** can be placed in any depth of water. They have a floating platform secured to the sea floor by wires. Three types of floating platforms are the tension-leg platform (TLP), the semi-submersible platform, and the spar-buoy platform.

A **tension-leg platform** is a floating structure historically used to secure offshore oil and gas platforms. Stiff tethers connect the four corners of a floating platform, which has a deck above water, straight down to the sea floor. The tension in the tethers virtually eliminates vertical movement of the platform.

A **semi-submersible platform** has ballasted, watertight pontoons under the ocean surface that are attached to the platform deck by structural columns. The pontoons provide buoyancy for the platform. The base of the undersea structural columns is connected to the ocean floor by wires to prevent the platform from drifting.

A **spar-buoy platform** has a single, long, large-diameter cylinder extending vertically underwater from the surface deck. The cylinder is filled at its base with a material denser than water. This provides stability and decreases the center of gravity of the platform. The cylinder is connected to the ocean floor by wires to prevent the platform from drifting.

Chapter 6 discusses onshore (ground-based) and offshore (water-based) wind energy in more depth. It discusses both horizontal and vertical axis turbines and their characteristics along with wind energy resources. **High-altitude wind** energy capture has also been pursued, although several barriers to its implementation have prevented its commercialization to date.

### 2.1.2 Wave

Winds passing over water create surface waves. The faster the average wind speed, the longer a wave is sustained, the greater the distance the wave travels, and the greater its height. Although wave power output varies over time, it is less variable than is offshore wind power output, even at the same location (Figure 8.3). The reason is that waves form from winds dragging the water over a long distance. Over a long distance, the variability in wind speed at a given location has little impact on the flow of a wave that already has momentum due to upstream winds. As such, waves at a given point represent the impact of winds accumulated over a long distance, whereas winds at the same point are instantaneous and variable.

**Wave power devices** capture energy from ocean surface waves to produce electricity.

One type of device, shown in Figure 2.3a, is a free-floating device that bobs up and down with a wave, creating mechanical energy that is converted to electricity in a generator. The electricity is sent through an underwater transmission cable to shore. Most of the body of such a device is submerged underwater.

Figure 2.3b shows another type of wave device. This type of device has floaters, which rise and fall with the motion of a wave. Arms connect the floaters to a structure, such as a pier, jetty, breakwater, floating platform, or fixed platform. The pumping motion is transmitted, through fluid pressure, to a power station on the structure. The fluid pressure is used to spin a hydraulic turbine, whose rotating motion is converted to electricity in a generator. Because the turbine and generator are on land, this type of wave device is easier to maintain than one completely offshore. However, a shore-based wave device can be placed in fewer locations than a free-floating device, which can be placed anywhere over the ocean where waves occur. Despite the development of a variety of wave energy technologies, the wave energy industry is still less mature, in 2020, than are other WWS industries. As a result, wave energy costs are still higher than costs of other WWS technologies. Nevertheless, like with solar and wind costs, wave energy costs are expected to decline as some prototypes are deployed in increasing quantities.

### 2.1.3 Geothermal

**Geothermal energy** is energy extracted from hot water or steam (Figure 2.4a) that resides below the Earth's surface. In both cases, the heat originates from hot rocks or soil. Rocks and soil are both heated by the **conductive** (molecule-by-molecule) energy transfer from the center of the Earth (which has a temperature of 4,000 °C); the decay of radioactive elements in the Earth's crust; the conduction of heat from volcanos; and the downward conduction of solar radiation that hits the Earth's surface. Most high-temperature rocks are found near volcanic activity, which generally occurs near tectonic plates. Low-temperature rocks and soil can be found anywhere underground. Even in locations where the ground surface is frozen, the soil temperature 6 m deep is warm enough that the soil's heat can be used to warm buildings.

Geothermal energy is used to provide direct heat or, at high temperature, electricity. Low-temperature heat (0 °C to 120 °C) is used to heat buildings. High-temperature heat (120 °C to 400 °C) is used most often to generate electricity. Low-temperature heat is extracted by ground-source heat pumps (Section 2.3.3); direct capture in pipes of **native** (naturally occurring) steam or hot water from natural hot springs; and the absorption of soil heat by pipes circulating injected water or air.

(a)

(b)

**Figure 2.3** (a) Free-floating wave device offshore of the Orkney Archipelago, Scotland. © CorPower. (b) Wave device connected to a land-based structure in Gibraltar. The photo shows four of eight Eco Wave Power floaters that together produce a peak of 100 kW of power. © Chris Wood/New Atlas.

Figure 2.4 (a) Steam from Tatio Geysers, Chile. © Natalia/AdobeStock. (b) 7-MW binary geothermal power plant in Canakkale, Turkey, 2016. © Verkis.

Prior to the 1900s, steam and hot water from the Earth were used only to provide heat for buildings, industrial processes, and domestic water. The first use of geothermal energy to produce electricity was by **Prince Piero Conti** of **Larderello**, Tuscany, Italy, in 1904. He lit four light bulbs by using steam from a geothermal field near his palace to drive a steam engine attached to a generator. A **steam engine** uses steam to drive an up-and-down motion of a piston in a cylinder, which in turn drives a rotating motion. The rotating motion is converted to electricity by the generator.

In 1911, Conti installed the first geothermal power plant, which had a nameplate capacity of 250 kW. This plant grew to 405 MW by 1975. The second electricity-producing plant was built at the Geysers Resort Hotel, California, in 1922. This plant was originally used only to generate electricity for the resort, but it has since been developed to produce electricity for California.

Today, the three major types of geothermal plants for electricity production are dry steam, flash steam, and binary. Dry and flash steam geothermal plants operate where geothermal reservoir temperatures are 180 °C to 370 °C or higher. In both cases, two boreholes are drilled – one for steam alone (in the case of dry steam) or liquid water plus steam (in the case of flash steam) to flow up, and the second for condensed water to return after it passes through the plant.

In a **dry steam plant**, the pressure of the steam rising up the first borehole powers a turbine, which drives a generator to produce electricity. About 70 percent of the steam re-condenses after it passes through a condenser, and the rest is released to the air. Because $CO_2$, NO, $SO_2$, and $H_2S$ in the reservoir steam do not re-condense along with water vapor, these gases are emitted to the air. Theoretically, they can be captured, but this has not been done to date in dry steam plants.

In a **flash steam plant**, the liquid water plus steam from the reservoir enters a flash tank held at low pressure, causing some of the water to vaporize ("flash"). The vapor then drives a turbine. About 70 percent of this vapor is re-condensed. The remainder escapes with $CO_2$ and other gases (e.g., Figure 2.4b). The liquid water is injected back to the ground.

**Binary geothermal plants** are developed when the reservoir temperature is 120 °C to 180 °C. Water rising up a borehole is enclosed in a pipe and heats, through a heat exchanger, a low-boiling-point organic fluid, such as isobutene or isopentane. The evaporated organic turns a turbine that powers a generator to produce electricity. Because the water from the reservoir remains in an enclosed pipe when it passes through the power plant, and is re-injected to the reservoir, binary systems emit virtually no $CO_2$, NO, $SO_2$, or $H_2S$. About 15 percent of geothermal plants today are binary plants.

### 2.1.4 Hydroelectric

**Hydroelectricity** (**hydropower**) is produced by water flowing gravitationally through a turbine connected to a generator. Most hydropower is produced from water held in a reservoir behind a large dam. This type of hydro-power is referred to as **large, or conventional, hydropower**. Hydropower dams require a reservoir behind them (Figure 2.5(a)), which results in the flooding of large areas of land. The largest conventional hydropower plant in the world is the 22.5-GW (gigawatt) Three Gorges Dam on the Yangtze River in China. Some other large hydropower plants include the 14-GW Itaipu plant on the Parana River bordering Brazil and Paraguay and the 10.2-GW Guri plant on the Caroni River in Venezuela.

A growing portion of hydroelectricity is produced by water flowing down a river directly through a turbine (Figure 2.5b) or by water that is diverted through pipes and a turbine near the edge of a river before returning to the river. This type of hydroelectricity is referred to as **run-of-the-river hydropower**. The advantage of run-of-the-river hydropower over conventional hydropower is that large amounts of land are not flooded behind a dam with the former. As a result, though, the former is less useful for storage. On the other hand, run-of-the river hydro with a modest storage pond behind it and conventional hydropower can both provide electricity within 15 to 30 seconds of a need (Table 2.1). The largest run-of-the-river hydropower dam is the 11.2-GW Belo Monte Dam in Brazil. A conventional hydropower plant consists of a dam, a water storage reservoir behind the dam, penstocks, sluice gates, a **powerhouse** containing water turbines and generators, power transmission cables, and a downstream water outlet (Figure 2.6). A **penstock** is a pipe, channel, or tunnel through which water flows from the storage reservoir to a water turbine. A **sluice gate** is a

**Table 2.1** Comparison of startup times and maximum ramp rates among several types of electric power and electric storage plants.

| Type of power or storage plant | Startup time | Maximum ramp rate |
| --- | --- | --- |
| Concentrated solar power (CSP) w/storage | 60 minutes | 10 percent/minute |
| Hydropower | 15–30 seconds | 100 percent/minute |
| Pumped hydropower storage (PHS) | 15–30 seconds | 100 percent/minute |
| Batteries | <0.1 second | 100 percent/second |
| Flywheels | <0.1 second | 100 percent/second |
| Compressed air energy storage (CAES) | <0.1 second | 100 percent/second |
| Gravitational storage with solid masses | <0.1 second | 100 percent/second |
| Open cycle gas turbine (OCGT) | 10 to 20 minutes | 20 percent/minute |
| Combined cycle gas turbine (CCGT) | 30 to 60 minutes | 5 to 10 percent/minute |
| Coal plant | 1 to 10 hours | 1 to 5 percent/minute |
| Nuclear plant | 2 hours to 2 days | 1 to 5 percent/minute |

Source: Denholm et al. (2014) for CSP; from Nonbol (2013) for OCGT, CCGT, coal, and nuclear.

(a)

(b)

**Figure 2.5** (a) Bonneville Dam in the Columbia River Gorge bordering Oregon and Washington State. It has a 1.242-GW nameplate capacity. © Cascoly2/AdobeStock. (b) Run-of-the-river hydroelectric plant along the Rhine River, Germany. © HappyAlex/AdobeStock.

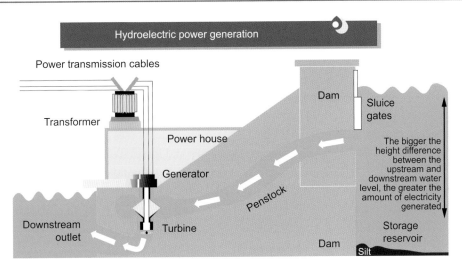

Figure 2.6 **Diagram of a hydropower electricity generation facility.** Adapted from Environment Canada.

gate to stop or control water flow between the storage reservoir and a penstock. When water passes through a **water turbine**, the turbine's blades spin, rotating a metal shaft connected to the **generator**, which converts the mechanical rotating motion into electricity. A run-of-the-river hydropower facility consists of a turbine and generator, but no reservoir, except for a small holding pond in some cases, and no pipes, except when the water is diverted from the river through the turbine and back to the river.

Only a fraction of the many dams with reservoirs worldwide have hydroelectric equipment associated with them. For example, the United States has about 84,000 dams, but only 2,500 (about 3 percent) of these have hydroelectric power associated with them (Jacobson et al., 2015a). The nameplate capacities of U.S. hydroelectric facilities vary from a few kW to 6.8 GW for the Grand Coulee Dam on the Columbia River in Washington State.

A hydropower plant can be run as a baseload plant, a load-following plant, or a peaking power plant. A **load** is a demand for electricity. When a light bulb connected by wire to the electric power grid is turned on, a load, or demand for electricity, arises on the grid. A **baseload** power plant produces a constant supply of electricity for an extended period of time to meet loads on the grid. The constant supply provided by a baseload plant is typically the smaller of the nameplate capacity of the turbines in the plant and the minimum demand for electricity over the year.

A **load-following plant** (Section 8.1.6.1) is an electricity-producing plant that runs continuously but ramps its power production up and down to meet 5- to

15-minute average changes in load on the grid. Hydropower plants, when run as load-following plants, run continuously and adjust their output every 5 to 15 minutes to meet such changes.

A **peaking power plant** generates electricity only to meet specific peaks in electricity demand that other electricity generators (e.g., wind, solar) cannot meet immediately. The rest of the time, peaking power plants produce no electricity. Hydropower plants can be run as peaking plants as well (Section 8.1.6.2). Section 2.7.3 discusses the load-following and peaking characteristics of hydropower plants in more detail.

In a hydropower plant, neither the powerhouse nor the penstock needs to be located inside the dam. The penstocks can be built to channel water around the dam to a powerhouse in front of or to the side of the dam. This may be desirable in cases where turbines are added to a hydropower plant, years after the dam was built, to increase the plant's peak discharge rate (nameplate capacity) while keeping the annual average water stored in the reservoir constant. Adding turbines is useful when a plant is run as a load-following or peaking plant. Such additions are called **uprating** a hydropower facility. Uprating a hydropower plant may be "one of the most immediate, cost-effective, and environmentally acceptable means of developing additional electric power" (U.S. DOI, 2005).

The average power output from a hydropower facility is limited not only by the nameplate capacity of its generators but also by the annual average amount of water available in the reservoir to run through the turbines. A measure of the practical average power output of a hydropower facility that accounts for both the water availability and the turbine nameplate capacity is the installed capacity.

**Installed capacity** for hydropower is the smaller of the average power produced by available water in a hydropower reservoir and the nameplate capacity of turbines in the hydropower plant itself (Rahi and Kumar, 2016; Business Dictionary, 2019; Free Dictionary, 2019). For other types of power-producing technologies (e.g., wind turbines and solar panels), the installed capacity equals the nameplate capacity of the technology.

For example, the power output (W) of a hydropower turbine, ignoring losses in pipes, can be approximated as

$$P_g = \eta \gamma Q H_g \tag{2.1}$$

where $\eta$ is the combined turbine, generator, and drive system efficiency (usually around 0.8); $\gamma$ is the **specific weight** of water (9,807 N/m$^3$ = 9,807 J/m$^4$), which is water density multiplied by gravity; Q is the **volumetric flow rate** (m$^3$/s) of water through the turbine; and $H_g$ is the **gross head**, or elevation difference (m) between the storage reservoir surface and the water surface just beyond the turbine. The water level just beyond the turbine may be lower than, the same as, or higher than the turbine elevation.

To account for pipe losses, a more accurate equation for power output from the turbine is

$$P_{net} = \eta \gamma Q H_{net} \tag{2.2}$$

where $H_{net}$ is the **net (dynamic) head** (m) available to the turbine when water is flowing. It is defined as the difference between the gross head and all hydraulic head losses in the waterway between the storage reservoir and the turbine. Such losses are primarily due to penstock friction losses and losses due to water bending around corners in pipes. When the flow rate and net head used in Equation 2.2 are averaged over a year, the installed capacity of a hydropower facility is the smaller of Equation 2.2 and the facility's nameplate capacity (Rahi and Kumar, 2016).

A simple estimate of $H_{net} = 0.9\ H_g$. However, a more physical estimate of the net head that accounts for both the pressure at the turbine and the velocity of water going through the turbine is

$$H_{net} = p/\gamma - H_d + v^2/2g \tag{2.3}$$

where p is pressure (N/m$^2$ = J/m$^3$) in the pipe measured at the turbine, $H_d$ is the elevation difference between the water surface downstream of the turbine and the turbine itself, v (m/s) is the speed of the moving water in the penstock, and g is **gravity** (9.807 m/s$^2$). For still water, $p/\gamma - H_d = H_g$ and v = 0, so $H_{net} = H_g$.

---

### Example 2.1 Power delivered by a hydropower turbine

Suppose a penstock with diameter D = 10.2 cm delivers a flow rate to a turbine of Q = 0.568 m$^3$/min of water through an elevation change (gross head) of $H_g$ = 30.5 m. The pressure in the pipe at the turbine is p = 186.2 kN/m$^3$ and the elevation difference between the water surface downstream of the turbine and the turbine itself is $H_d$ = 0. What fraction of available head is lost in the pipe and what power is delivered by the turbine, assuming a combined turbine, generator, and drive system efficiency of $\eta$ = 0.8?

### Solution:

The pressure head (p/$\gamma$) = 186.2 kN/m$^2$/9.807 kN/m$^3$ = 19.0 m

The speed of water in the penstock is v = Q/A = (0.568 m$^3$/min) × (1 min/60 s)/($\pi$ × (0.102/2)$^2$ m$^2$) = 1.16 m/s

The kinetic energy head is v$^2$/2g = (1.16 m/s)$^2$/(2 × 9.81 m/s$^2$) = 0.069 m

The net head to the turbine is $H_{net}$ = pressure head + kinetic energy head = 19.069 m

The head lost $H_L$ = $H_g$-$H_{net}$ = 30.5 m – 19.069 m = 11.43 m

Fraction of head lost in pipe $H_L$/$H_g$ = 11.43 m/30.5 m = 37.5 percent

Power to turbine = $\gamma Q H_{net}$ = 9.807 kN/m$^3$ × (0.568 m$^3$/min) × (1 min/60 s) × 19.07 m = 1.77 kW

Power output from turbine $P_{net}$ = $\eta$ × power to turbine = 0.8 × 1.77 kW = 1.42 kW

## 2.1.5 Tidal and Ocean Currents

**Tidal currents (tides)** are oscillating (back and forth) currents in the ocean caused by the rise and fall of the ocean surface due to the gravitational attraction among the Earth, moon, and sun. The rising and sinking motion of the ocean surface forces water below the surface to move horizontally as a current. Because tides run about six hours in one direction before switching directions for another six hours, they are fairly predictable, so tidal turbines are usually used to supply baseload energy.

An **ocean current** is a continuous flow (in one direction) of seawater driven primarily by winds or by temperature and salinity gradients in the ocean. The Gulf Stream current, for example, is a warm, fast-flowing current driven by winds. It runs from the Gulf of Mexico, past the tip of Florida, up the U.S. East Coast and Newfoundland coast, to the North Atlantic Ocean, where it splits into two other currents. Ocean currents, in fact, run along all major ocean coastlines in the North and South Pacific and Atlantic Oceans and the Indian Ocean.

A **tidal turbine** captures the kinetic energy of an ebbing and flowing tidal current or the continuous flow of an ocean current, just as a wind turbine captures the kinetic energy of the wind. Tidal turbines can be mounted on the sea floor (Figure 2.7) or hang from under a floating platform. They can also be placed in rivers to capture the kinetic energy of continuously flowing river water (similarly to run-of-the-river hydro) or of tidal water that ebbs and flows in a river. For example, tides from the Atlantic Ocean flow up and down the Potomac River, all the way up to Harpers

Ferry, West Virginia, a land-locked U.S. state. In this text, the term "tidal turbine" is used to refer to a turbine that captures either tidal, ocean current, or river tidal energy.

Like a wind turbine, a tidal turbine consists of a blade that spins a turbine, which provides rotational kinetic energy to a generator. The generator converts the rotational energy to electrical energy that is transmitted to shore. A tidal turbine's rotor, which lies underwater, may be fully exposed to the water or placed within a narrowing duct that directs water toward it.

## 2.1.6 Solar Photovoltaics

**Solar photovoltaics (PVs)** are arrays of cells containing a material that converts solar radiation into **direct current (DC)** electricity. An inverter is then used to convert DC electricity into **alternating current (AC) electricity**. Photovoltaics are often mounted above the ground in large (utility-scale) power plants. They are also mounted on or built into the roofs or walls of commercial, industrial, or residential buildings. They are additionally placed above carports, parking lots, and parking structures (Figure 2.8) and are built into the sides of cars, trucks, and buses. In utility-scale plants, PV panels are either mounted at a fixed tilt or on trackers that rotate to follow the sun. For most other applications, they are mounted at a fixed tilt.

Solar PV systems are also built to float on lake or ocean surfaces or to rest over canals (Figure 2.8). A 3.6-km stretch of the Narmada irrigation canal in Vadodara,

(a)                                          (b)

**Figure 2.7 (a) Tidal turbine with a similar structure as a wind turbine.** © ANDRITZ HYDRO Hammerfest. **(b) Different design of a tidal turbine.** © Oceana Energy Co.

India, for example, is covered with 33,816 solar PV panels generating 10 MW of peak power. The system not only saves cropland from being covered with solar, but it also reduces water evaporation from the canal (Kougias et al., 2016).

The presence of clouds, shadow-casting buildings or trees, dust or snow that accumulates on solar panels, and extremely high temperatures can reduce the efficiency of a solar PV panel. However, even when the sun is behind a tree, the presence of thin clouds can enhance solar

Figure 2.8 (a) Utility-scale solar PV farm in the desert. © Netterstock/AdobeStock. (b) 1 MW solar PV system on an industrial facility in West Jordan, Utah. © Nikola Motor Company. (c) PV system over an irrigation canal in India (Kougias et al., 2016). (d) 70 MW floating PV farm, Anhui Province, China. © Ciel & Terre International. (e) Rooftop PV, Nottingham, England. © Ingo Bartusek/AdobeStock. (f) Solar PV over a parking lot. © fovivaphoto/AdobeStock.

radiation incident upon a panel by scattering light from multiple locations in the cloud onto the panel. Similarly, snow on the ground can enhance the scattering of light to a panel. Low temperatures in high-latitude locations can also increase panel efficiency compared with high temperatures in deserts.

Chapter 5 discusses how PV cells, panels, and arrays work. It also discusses solar resources and radiative transfer through the atmosphere.

### 2.1.7 Concentrated Solar Power

With **concentrated solar power (CSP)**, mirrors or reflective lenses focus (concentrate) sunlight onto a collector containing a fluid to heat the fluid to a high temperature. The heated fluid (e.g., pressurized steam, synthetic oil, molten salt, or a phase change material)

flows from the collector to a heat engine, where a portion of the heat (up to 30 percent) is converted to electricity. Some types of CSP allow the heat to be stored for many hours so that electricity can be produced during nighttime or when heavy clouds are present.

One type of collector is a set of **long parabolic trough (U-shaped) mirror reflectors**, which focus light onto a pipe containing oil. The hot oil flows to a chamber, where the heat is transferred to water, which boils to produce steam that is sent to a steam turbine connected to a generator to produce electricity.

A second type of collector is a **central tower receiver** with a field of mirrors surrounding it (Figure 2.9). In the central tower, the focused light heats a circulating thermal storage fluid, such as a **molten salt mix**, to 500 to 600 °C. The molten mix often consists of sodium nitrate and potassium nitrate. The hot salt flows to a heat

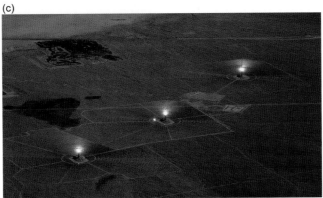

Figure 2.9 (a), (b) Gemasolar solar thermal (CSP) plant, owned by Terresol Energy. © SENER. The plant has a 19.9 MW-nameplate capacity and peak storage capacity of 299 MWh (thus 15 hours of storage at the maximum discharge rate). It has provided electricity 24 hours per day for up to 36 days in a row. (c) 392 MW-Ivanpah solar thermal (CSP) plant with three central tower receivers (but no storage), in the Mojave Desert, California. © jllm06/Wikimedia Commons.

exchanger, where the heat is transferred to water, which boils to produce steam for a steam turbine connected to a generator to produce electricity. After passing through the steam turbine, the steam is piped to a condenser, where it reforms a liquid. From there, the liquid water is piped back to the heat exchanger to be boiled again. Thus, the water circulates in a closed loop (Figure 2.21). Meanwhile, after exchanging its heat with water to boil it in the heat exchanger, the molten salt temperature is now down to about 260 °C. The salt is then moved to a "cold" holding tank until it is re-circulated to the central tower. Thus, the molten salt also circulates in a closed loop (Figure 2.21).

In many parabolic trough and central tower CSP plants, some of the hot fluid (oil or molten salt mix) is stored in an insulated thermal storage tank before it is used to boil water in the heat exchanger (Figure 2.21). The purpose is to delay the heating of water, and thus electricity production, until nighttime, when no immediate solar electricity from the plant is available, or until a time of peak electricity demand (see Section 2.7.1).

A third type of CSP technology is a **parabolic dish-shaped** (e.g., satellite dish) **reflector** that reflects light onto a receiver while rotating to track the sun. The receiver transfers the energy to hydrogen in a closed loop. The expansion of hydrogen against a piston or turbine produces mechanical power used to run a generator to produce electricity. The power conversion unit is air cooled; thus, water cooling is not needed. Parabolic dish CSP is not coupled with thermal energy storage.

CSP plant condensers require either air or water cooling. The use of air cooling, which is desirable in water-constrained locations, reduces overall CSP plant water requirements by 90 percent at a cost of only about 1 to 5 percent less electric power production (DOE, 2008).

Because the components of CSP plants are made of abundant raw materials, material shortages are not expected to limit the mass production of CSP plants. For example, CSP plants consist primarily of mirrors, receivers, and thermal storage fluid. Mirrors are made mostly of glass with a reflective silver layer on the back of the glass. Receivers are stainless steel tubes with an outer surface that selectively absorbs solar radiation and that is surrounded by an outer anti-reflective glass tube. The receiver heats the thermal storage fluid circulating through the receiver inner tube. Many alternate mirror types, receivers, and thermal storage fluids are possible.

## 2.2 WWS Transportation Technologies

**Transportation** is the conveyance of people, animals, or goods from one place to another. Types of transportation include skateboards, bicycles, motorcycles, passenger vehicles, sport utility vehicles (SUVs), small trucks, large trucks, semi-trucks, buses, forklifts, cranes, tractors, bulldozers, asphalt pavers, backhoe loaders, cold planers, compactors, excavators, harvesters, graders, off-road vehicles, trains, ferries, motorboats, yachts, ships, helicopters, aircraft, battle tanks, infantry fighting vehicles, armored personnel carriers, armored combat support vehicles, mine-protected vehicles, light armored vehicles, light utility vehicles, amphibious vehicles, and more. All of these, except skateboards and bicycles, currently run primarily by burning gasoline, diesel, biodiesel, methanol, ethanol, liquefied natural gas, bunker fuel, or jet fuel in an internal combustion engine (ICE). These fuels are all derivatives of fossil fuels or biofuels.

The cleanest and most efficient method of replacing these fossil-fuel and biofuel vehicles is to convert them to **battery-electric (BE)** vehicles, **hydrogen fuel cell (HFC)** vehicles, or **BE-HFC hybrid** vehicles.

### 2.2.1 Battery-Electric Vehicles

BE vehicles (Figure 2.10) store electricity in and draw power from batteries to run an electric motor that drives the vehicle. Because BE vehicles have zero tailpipe emissions, their only emissions beyond those from producing and decommissioning the vehicles and their batteries are associated with the electricity used for battery charging. So long as such electricity is from a WWS source, the air pollution and global warming-relevant gas and particle emissions from such vehicles over their lifetimes are nearly zero.

Another advantage of a BE vehicle compared with an ICE vehicle is that a BE vehicle can travel 3 to 5 times further per unit of energy input at the plug outlet than can an ICE vehicle travel per unit of energy in gasoline. For example, whereas only 17 to 20 percent of the energy in gasoline moves a gasoline vehicle (because the rest of the energy is lost as waste heat), 64 to 89 percent of the electricity from the plug outlet moves a BE vehicle, and the rest is waste heat. As such, a BE vehicle requires less energy than does an ICE vehicle, and a conversion from ICE to BE vehicles reduces energy demand for transportation fuel by a factor of 3 to 5 (Section 7.3.1.1).

A BE vehicle can increase its range with the use of regenerative braking. An electric vehicle uses a motor to convert electricity to rotational motion to rotate the wheels of the vehicle. In an ICE vehicle, the car is slowed with brake pads that squeeze against a turning tire disk, reducing tire rotation and slowing the vehicle. The friction that results produces substantial heat. With **regenerative braking**, the motor is run in reverse as an electricity generator. Rotational kinetic energy from the rotating wheels is fed into the generator to "regenerate" electricity. In this way, rotational kinetic energy is converted to electricity, slowing the car. The electricity is then stored in the vehicle's battery for subsequent use.

## Transition highlight

Regenerative braking can not only extend BE vehicle range substantially, but it can also eliminate, in at least one special case, the recharging of an electric vehicle entirely. The largest stand-alone electric vehicle in the world in 2019 was an enormous electric dump truck operating in a quarry in Biel, Switzerland. The empty truck used electricity to climb from the bottom to the top of the quarry. At the top, it was filled with ore, more than doubling the truck's weight compared with its uphill trip. The additional weight created so much kinetic energy that the electricity produced from regenerative braking during the downhill trip exceeded the electricity consumed during the uphill trip. As such, the truck did not need recharging during its daily operation (Evarts, 2019).

In a WWS world, battery electricity will be used not only to power light-duty vehicles, but also short- and medium-range semi-trucks, short-range aircraft, some military vehicles, and short-distance boats and ships (Figure 2.10).

Contemporary BE vehicles use lithium-ion batteries. Lithium itself is not a toxic element. However, lithium-ion batteries also contain nickel, cobalt, aluminum, manganese, titanium, and/or phosphorus (e.g., Harvey, 2018), some of which are toxic if breathed in high enough concentrations. However, these chemicals are never breathed in, as they are confined to the batteries that contain them. Further, the quantities of those toxic chemicals are less than the quantities of lead, nickel, or cadmium in lead acid batteries and nickel cadmium batteries.

Nevertheless, mining for lithium, as with mining for other elements, causes environmental degradation. On the other hand, damage from the one-time mining of lithium for a battery is minor in comparison with damage from the continuous, forever, mining of oil that is used in gasoline and diesel vehicles.

Other types of batteries, aside from lithium ion, nickel cadmium, and lead acid, are currently being developed. Section 2.7.4 discusses these types of batteries, how batteries work, and limits to lithium supplies.

If BE vehicles are produced on a massive scale, the two materials that are most limiting are lithium for batteries and neodymium for electric motors. As discussed in Section 2.7.4, sufficient lithium resources exist in the world to power all vehicles needed in 2050 several times over. However, recycling may be needed to ensure short-term shortages in lithium and other elements do not occur (e.g., Jacobson and Delucchi, 2011; Harvey, 2018; Earthworks, 2019).

Another important element used in many electric vehicles is neodymium. Neodymium is used in permanent magnet alternating current motors. However, the amount of neodymium required per vehicle is about an order of magnitude less than that required per wind turbine, and neodymium is not a limiting factor in wind turbine production (Section 6.5). In addition, one common type of electric motor, an induction motor, does not use neodymium.

## 2.2.2 Hydrogen Fuel Cell Vehicles

**Hydrogen gas**, also known as **molecular hydrogen** ($H_2$), has been present in the Earth's atmosphere since the formation of the Earth 4.6 billion years ago. Today, it is a well-mixed gas in the lower atmosphere whose main natural atmospheric source is metabolism by bacteria in the oceans and soil. Humans contribute to hydrogen's atmospheric burden too, as hydrogen is a byproduct of fossil-fuel combustion. Hydrogen is removed from the air primarily by soil bacteria.

**Paracelsus** (1493–1541), a Swiss physician and alchemist born Theophrastus von Hohenheim, may have been the first to isolate $H_2$. He discovered that pouring sulfuric acid over the metals iron (Fe), zinc (Zn), or tin (Sn) gave off a highly flammable vapor. In 1766, **Henry Cavendish** also made this determination and isolated its properties.

Figure 2.10  (a) Battery-electric 2010 Tesla Roadster, which uses thousands of lithium-ion laptop batteries. When new in 2009, the vehicle had a range of 244 miles (395 km). © Mark Z. Jacobson. (b) Battery-electric Tesla semi-truck, which has a range of about 850 km, in Rocklin, California. © Korbitr/Wikimedia Commons. (c) Battery-electric Nikola all-terrain military vehicle. © Nikola Motor Company. (d) Battery-electric Ampaire 9-seat aircraft, with an intended fully loaded range of 550 km. © Ampaire. (e) Battery-electric Nikola jet-ski style watercraft with 150-km range. © Nikola Motor Company. (f) Battery-electric ferry in Norway with a designed range of 12.5 km. © Fjellstrand.

Because hydrogen gas (molecular weight of 2.0159 g/mol) is much lighter than air (molecular weight of 28.966 g/mol), hydrogen has been used to lift passenger balloons and airships for flight as well as to lift party balloons. Hydrogen has also been burned to propel rockets and space shuttles into space. Its use in the **Hindenburg** airship, which caught on fire and crashed, killing 36 people on May 6, 1937, set back its use for transportation for many decades. However, due to its lightness, a hydrogen flame shoots straight up, minimizing damage to a vehicle carrying it, whereas gasoline explodes outward, destroying the vehicle.

### 2.2.2.1 Mechanisms of Hydrogen Production

Hydrogen can be produced synthetically by steam reforming of methane, coal gasification, or electrolysis (Jacobson et al., 2005). **Steam reforming of methane** produces hydrogen by

$$CH_4 + H_2O \text{ (steam)} + heat \rightarrow CO + 3H_2 \qquad (2.4)$$

where the reaction occurs at a high temperature (700 to 1,000 $^\circ$C) and pressure (3 to 25 bars), and the water is in the form of steam. In a second reaction, $CO_2$ and additional hydrogen are produced by the water-shift reaction,

$$CO + H_2O \text{ (steam)} \rightarrow CO_2 + H_2 + \text{small amount of heat} \quad (2.5)$$

As such, steam reforming, while producing hydrogen, also produces one mole of carbon dioxide for each mole of methane processed. Because steam reforming usually involves processing natural gas, whose main component is methane, but which also contains other chemicals, steam reforming results in emissions of hydrocarbons, nitrogen oxides, and CO along with some unprocessed methane (Colella et al., 2005, Table 4). However, such emissions are relatively small. On the other hand, additional emissions and land degradation occur continuously during the mining and transporting of natural gas. Energy is also needed to produce the high temperature and high pressure, and producing this energy results in pollution as well.

With **coal gasification**, coal reacts with molecular oxygen and steam under high temperature and pressure to form a mixture of $H_2$, CO, $CO_2$, and other chemicals. After impurities are removed, the CO reacts with steam through the water-shift reaction (Equation 2.5) to form additional hydrogen. Coal gasification results in more $CO_2$ and other pollutant emissions than does steam

reforming of methane. Additional emissions and land degradation occur during the continuous mining and transport of coal.

Because both steam reforming of methane and coal gasification emit $CO_2$ and other pollutants and continuously degrade land, they are not suitable candidates for producing $H_2$ in a 100 percent WWS world. In a WWS world, all carbon and pollution emissions and continuous fuel mining are eliminated. Instead, in a WWS world, electrolysis is the more appropriate method of producing hydrogen.

**Electrolysis** involves combining electricity with water in an **electrolyzer** to split the water into hydrogen and oxygen by

$$H_2O + electricity \rightarrow H_2 + 0.5\,O_2 \qquad (2.6)$$

This process has no emissions if the electricity originates from a WWS source. The resulting hydrogen is usually compressed and stored in a storage tank for later use in a fuel cell. Three types of electrolyzers are the **polymer electrolyte membrane (PEM) electrolyzer** (Figure 2.11), the **alkaline water electrolyzer**, and the **solid oxide**

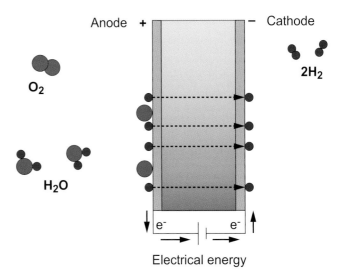

**Figure 2.11** Drawing of polymer electrolyte membrane (PEM) electrolyzer. Some water molecules supplied to the anode side come in contact with a catalyst and are split into oxygen atoms (blue dots), hydrogen ions (red dots), and electrons ($e^-$). The hydrogen ions diffuse along with additional water molecules to the cathode side. The oxygen atoms can't diffuse through. Instead, they combine to form oxygen molecules that bubble up and are released to the air. The electrons travel through an external circuit, driven by a power source, to the cathode side, where they recombine on a platinum catalyst with hydrogen ions to form molecular hydrogen gas, which is captured.

**electrolyzer**. The PEM, which is the most commonly used electrolyzer, is discussed here.

A PEM electrolyzer consists of a cell divided into two compartments by a thin (100 to 200 μm), solid proton-conducting plastic membrane coated by a catalyst on each side. Each compartment has an electrode, and the two electrodes are connected by wire with a power source in between. The membrane is a polymer electrolyte. An **electrolyte** is a material that conducts protons from one electrode to another. Whereas hydrogen ions ($H^+$) and liquid water can permeate the membrane, neutral gases and electrons cannot. The only inputs into the cell are water and electricity. The outputs are molecular hydrogen gas and molecular oxygen gas.

One electrode in the electrolyzer, the **anode**, is positively charged. The other, the **cathode**, is negatively charged. When a voltage from the power supply is applied, liquid water supplied to the anode side reacts with the iridium catalyst coating the membrane. Some of the water splits into dissolved oxygen, dissolved hydrogen ions, and electrons by

$$2H_2O \rightarrow O_2 + 4H^+ + 4e^- \tag{2.7}$$

This is referred to as the **oxygen evolution reaction**. The rest of the water remains unaffected. The electrons then flow through the external circuit from the anode side to the cathode side. The hydrogen ions and liquid water selectively diffuse across the membrane to the cathode side. The oxygen, which cannot permeate the membrane, bubbles up through the water and escapes to the air or is captured. On the cathode side, the hydrogen ions that permeated the membrane recombine on a **platinum** catalyst with electrons from the external circuit to form molecular hydrogen. The reaction is

$$4H^+ + 4e^- \rightarrow 2H_2 \tag{2.8}$$

This is referred to as the **hydrogen evolution reaction**. The hydrogen bubbles to the top, where it is captured. Equation 2.6 is the net reaction.

Once hydrogen is produced, it is usually compressed or liquefied and stored before use. Section 2.8.6 discusses methods of storing hydrogen.

### 2.2.2.2 Hydrogen Fuel Cells

In a 100 percent WWS world, hydrogen will be used in fuel cells, primarily to provide electricity for electric motors in hydrogen fuel cell transportation vehicles. HFC vehicles use a fuel cell to convert hydrogen from an onboard storage tank plus oxygen from the air to electricity that is used to run an electric motor.

Some hydrogen may be used in stationary fuel cells to provide electricity for buildings or industry. However, using electricity to produce hydrogen and then using the hydrogen to reproduce electricity is inefficient compared with storing electricity in a battery, then discharging the battery to reproduce electricity. The overall efficiency of using hydrogen for normal electric power use can be improved if heat released during a fuel cell use is captured and used.

One useful application of a hydrogen fuel cell aside from powering vehicles is for producing electricity and heat in a microgrid. A **microgrid** is an isolated grid that provides power for a single home, a few homes, a few buildings, a hospital or university campus, a remote community, or a remote military base. In 2016, for example, an apartment complex was built in Brutten, Switzerland, that included 127 kW of solar PV cells, an electrolyzer, two hydrogen storage tanks, a fuel cell, and a ground-source heat pump. Excess electricity from the solar PV was used to electrolyze water into hydrogen, which was stored until needed. During winter, hydrogen was run through the fuel cell to generate electricity and heat. The heat was used to heat water. Ground-source heat pumps that ran on electricity from either the fuel cell or the solar PV were used to heat and cool the air.

A **fuel cell** is a device that converts chemical energy from a fuel into electricity. A single fuel cell consists of two porous gas diffusion electrodes separated by an electrolyte. The type of electrolyte defines the different types of fuel cells. The major types of fuel cells are proton exchange membrane (PEM) fuel cells, alkaline fuel cells (AFCs), solid oxide fuel cells (SOFCs), and molten carbonate fuel cells (MCFCs). PEM cells and AFCs are used primarily in vehicles for transportation, SOFCs and MCFCs are used primarily for stationary electric power and heat (DOE, 2015a).

In a PEM fuel cell, the membrane is thin and made of a proton-conducting electrolyte, usually **nafion**, which is a synthetic polymer invented in the 1960s. Pores in nafion allow movement of positively charged ions (**cations**), such as the hydrogen ion ($H^+$), through the membrane, but not negatively charged ions (**anions**), electrons, or neutral gases. In a fuel cell, the membrane is sandwiched between two platinum catalyst-coated carbon papers.

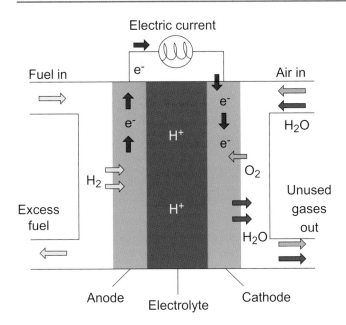

Electric current

Fuel in

Air in

$H_2O$

$e^-$

$H^+$

$H_2$

$O_2$

Excess fuel

$H^+$

Unused gases out

$H_2O$

Anode    Electrolyte    Cathode

**Figure 2.12 Drawing of a fuel cell. The text describes its operation.**

Hydrogen is introduced on one side (anode side) of the fuel cell and oxygen is introduced on the cathode side (Figure 2.12). On the anode side, the hydrogen encounters the platinum catalyst and dissociates slightly by the reaction,

$$H_2 \rightarrow 2H^+ + 2e^- \qquad (2.9)$$

$H^+$ is released near the anode. The concentration gradient causes $H^+$ to diffuse through the membrane toward the cathode, leaving electrons, which cannot pass through the membrane, behind.

Due to the transfer of hydrogen ions to the cathode side of the fuel cell, the cathode takes on a positive charge relative to the anode. Electrons left behind on the anode side are then drawn toward the positively charged cathode. However, since they can't pass through the membrane, they must take another route. Their path of least resistance is through an external circuit created between electrodes. The flow of electrons from the anode to the cathode delivers DC electricity to the end-use load (e.g., the electric motor in the fuel cell vehicle) and to the cathode. At the cathode, the electrons combine with $O_2$ from the air and $H^+$ on the surface of the platinum catalyst, which coats the membrane, to produce water by

$$0.5O_2 + 2H^+ + 2e^- \rightarrow H_2O \qquad (2.10)$$

The net effect of Equations 2.9 and 2.10 is

$$H_2 + 0.5O_2 \rightarrow H_2O \qquad (2.11)$$

In sum, a fuel cell converts molecular hydrogen plus oxygen to water, 0.5 to 1 volt of DC electricity, and heat. The total energy available in the chemical bonds of hydrogen is 141.8 MJ/kg-$H_2$. This is the **higher heating value (HHV)** of hydrogen. When hydrogen is used in a fuel cell, 21.84 MJ/kg-$H_2$ of the HHV are used to evaporate the water produced by Equation 2.11. The remaining 119.96 MJ/kg-$H_2$ can be used to generate electricity. This is called the **lower heating value (LHV)** of hydrogen. However, because a fuel cell is much less than 100 percent efficient, only a portion of the LHV is converted to electricity. The rest is waste heat released to the air (unless it is captured). The fraction of the LHV that is converted to electricity is called the **fuel cell efficiency**.

The **maximum theoretical fuel cell efficiency** is about 83 percent. In other words, of the lower heating value of hydrogen, at most 83 percent can be converted to electricity; the rest goes to heat. Actual fuel cell efficiencies are about 60 (50 to 70) percent for PEM fuel cells; 60 percent for AFCs; 50 percent for MCFCs; and 60 percent for SOFCs (DOE, 2015a).

The overall plug-to-wheel efficiency of an HFC vehicle depends not only on fuel cell efficiency but also on the efficiencies of the electrolyzer and compressor (to compress hydrogen for storage), the leakage rate of hydrogen to the air from leaky pipes, electrical losses between the fuel cell and the motor, and the motor efficiency. The overall **plug-to-wheel efficiency for an HFC passenger vehicle** ranges from 23 to 37 percent (Table 7.2), which is much lower than that of an electric passenger vehicle but higher than that of an internal combustion engine vehicle.

Because the voltage produced by a single fuel cell is small, fuel cells are connected in series to form a **fuel cell stack** to increase voltage. Stacks contain a few to hundreds of fuel cells. Fuel cell stacks of rated power 50 to 125 kW are used in vehicles. Stacks of 1 to 200 MW are used in power plants.

Additional advantages of hydrogen fuel cell transportation over fossil-fuel transportation are (1) the only emissions from a hydrogen fuel cell system is leaked hydrogen (usually during transport and storage) and water vapor, and (2) hydrogen can be produced from entirely clean and renewable electricity generation. Because the likely emission rate of leaked hydrogen is less than the emission rate of hydrogen from fossil-fuel vehicle combustion, a replacement of fossil-fuel vehicles with hydrogen fuel cell vehicles reduces $H_2$ emissions (Jacobson, 2008b).

Because HFC passenger vehicles are much less efficient than are BE vehicles, HFC passenger vehicles require 1.7 to 3.9 times the number of wind turbines or solar panels to go the same distance as does an equivalent BE vehicle (Table 7.2). However, aside from the leaked hydrogen and water vapor for the HFC vehicles, the tailpipe emissions from both are zero. HFC vehicles become advantageous over BE vehicles for extra-long-distance and heavy truck, train, ship, and aircraft transport for the reasons discussed in Section 7.3.1.2. For example, in 2019, a hydrogen fuel cell semi-truck (Figure 2.13) had a range of up to 1,200 km, whereas an electric semi (see Figure 2.10) had a range of about 850 km.

### 2.2.2.3 Is Platinum a Limitation If Hydrogen Fuel Cells Are Adopted on a Large Scale?

One concern with the large-scale production of HFC vehicles is potential shortages of the scarce metal catalyst, platinum. However, because HFC vehicles are proposed to replace gasoline and diesel vehicles, which use platinum in their catalytic converters, the increase in platinum demand for HFC vehicles will be partly offset by a decrease in platinum demand for use in catalytic converters.

Further, because BE passenger vehicles will comprise the largest share of vehicles in a WWS world due to their greater efficiency, the number of HFC passenger vehicles required will be limited. If even 10 percent of the world's vehicle fleet in 2050 were HFC vehicles, this might mean about 150 million 50-kW HFC vehicles. Since each 50-kW fuel cell stack requires about 12.5 g Pt (Jacobson and Delucchi, 2011), 150 million such vehicles require about 1.875 million kg Pt. This is much less than the 17 million kg Pt that can be extracted from platinum group metals worldwide.

**Platinum group metals** (PGMs) are six elements – ruthenium (Ru), rhodium (Rh), palladium (Pd), osmium (Os), iridium (Ir), and platinum (Pt) – that have similar physical and chemical properties and are often found in the same deposits. The estimated resource of platinum

**Figure 2.13** Nikola Tre hydrogen fuel cell semi, which has a range of up to 1,200 km and a refueling time of 15 to 20 minutes. © Nikola Motor Company.

group metals in deposits worldwide in 2017 is more than 100 million kg (USGS, 2018a). The largest known deposit of PGMs is in the Bushveld Igneous Complex of the Transvaal Basin, South Africa. Other large deposits are near Norilsk, Russia, and Sudbury, Canada. Platinum comprises at least one-sixth, or about 17 million kg Pt, of total PGM deposits.

In addition, platinum is already recycled, and additional Pt will be available due to the reduced need for it in catalytic converters upon conversion to WWS. As such, platinum is not considered a limiting element in a WWS.

### 2.2.3 Comparing Masses and Volumes among BE, HFC, and ICE Vehicles

Some of the main considerations in designing BE and HFC vehicles to replace internal combustion engine (ICE) vehicles are whether the BE or HFC vehicles can have the same or lower mass and volume while maintaining or exceeding the range and power-to-weight ratio (PWR) or thrust-to-weight ratio (TWR) as the ICE vehicles. A vehicle's mass and volume affect not just its capabilities, but also how it interacts with supporting infrastructure, like roads, bridges, tunnels, runways, locks, canals, and ports. The PWR defines how fast a vehicle can accelerate to a top speed. For an airplane, the TWR defines its ability to take off within a runway's length. Those considering adopting new BE or HFC vehicles are more likely to do so if their performance and range are either similar to or better than existing ICE vehicle options.

The major components in an ICE vehicle that can be removed when transitioning to a BE or HFC vehicle include the vehicle's engine, transmission, oil, coolant, automatic transmission fluid, fuel tank, and fuel. The major components a BE vehicle replaces them with include a battery pack, inverter, electric motor, gearbox (optional), and wiring. An HFC vehicle replaces them with a hydrogen storage tank, compressed or liquefied hydrogen fuel, a fuel cell stack, an electric motor, a gearbox, and wiring.

Figure 2.14 illustrates the feasibility of transitioning nine different military and civilian ICE vehicle types in terms of mass, volume, PWR (or TWR), and range to either BE or HFC vehicles using technology that is commercially available today (solid bars) or feasible in the near future based on published data (hatched bars). The bars represent ranges of possible characteristics. Only solutions where the overall PWR or TWR can be equaled or improved with BE or HFC vehicles are shown, since otherwise, helicopters and airplanes could not take off.

The figure indicates that no solution exists for long-haul jet airliners using today's commercial HFC technology; however, solutions exist using future technology. For example, the black bars within the subset shown for long-haul jet airliners illustrate one possible design solution using future feasible HFC technology. A future HFC airliner could have the same TWR and range as today's ICE airliner while simultaneously having 22 percent lower mass, but it would need a 21 percent larger volume. Vehicle designers can seek solutions by defining what is acceptable for each characteristic. For example, if it is acceptable for a vehicle to become slightly heavier and/or larger, then better results can be achieved in terms of PWR and range.

Example 2.2 provides a calculation of the mass and volume ratios of a hydrogen fuel cell 747-8 commercial aircraft versus a jet fuel aircraft. The hydrogen used in the HFC aircraft is **low-temperature liquid (cryogenic) hydrogen**. Using cryogenic hydrogen in aircraft is not a novel idea. Such hydrogen was combusted in the space shuttles for decades, most rockets lifted into space to date, and in demonstration flights for the Soviet Union's commercial aircraft Tupolev Tu-154B, beginning April 15, 1988. In the last case, the aircraft was fitted with a thermally insulated fuel tank behind the passenger cabin that contained liquid hydrogen at a temperature of 20 K. The hydrogen powered a third engine on the aircraft.

## Example 2.2 Simplified estimate of HFC:ICE mass and volume ratios for a 747-8 commercial aircraft powered with hydrogen fuel cells compared with jet fuel. The estimate does not consider the changes in mass, volume, or thrust from using electric motors versus jet engines. It treats only the change in mass and volume due to using hydrogen, a hydrogen storage tank, and a fuel cell stack instead of jet fuel and a jet fuel storage tank assuming the power required and delivered is the same in both cases.

1. **Mass and volume of $H_2$ versus jet fuel fuselage**
a. Lower heating value of jet fuel[i]    42.8 MJ/kg-jet fuel
b. Overall energy conversion efficiency jet fuel[ii]    0.25–0.40
c. Jet fuel density[i]    810.53 kg/m$^3$
d. Lower heating value of hydrogen[i]    119.96 MJ/kg-$H_2$
e. Fuel cell system efficiency (Section 7.3.1.2)    0.56–0.34
f. Cryogenic hydrogen density[i]    70.9 kg/m$^3$
g. Volume ratio $H_2$:jet fuel [abc/(def)]    1.8–4.8 m$^3$/m$^3$
h. Fuel mass ratio $H_2$:jet fuel [ab/(de)]    0.16–0.42 kg/kg
2. **Size of fuel stack needed**
a. Power for full 747-8 to takeoff[iii]    90–145 MW
b. Ultralight fuel cell stack power (60 cells)[iv]    1–1.3 kW
c. Number of fuel cell stacks needed (a/b)    69,000–145,000
d. Volume of one fuel cell stack[iv]    0.0046 m$^3$

e. Mass of one stack[iv]                                                    1.8 kg
f. Volume of all $H_2$ fuel cell stacks needed (cd)                          316–661 $m^3$
g. Mass of all $H_2$ fuel cell stacks needed (ce)                            125,000–261,000 kg

3. **Mass and volume of 747-8 aircraft**
a. Mass of 747-8 aircraft with no fuel[iii]                                  295,289 kg
b. 747-8 jet fuel mass at takeoff[iii]                                        153,800 kg
c. Jet fuel volume at takeoff 747-8 (b/810.5 kg/$m^3$)                        189.7 $m^3$
d. Mass ratio of jet fuel to jet fuel + spherical storage tank[i]            0.75
e. Mass of jet fuel cylindrical storage tank [b(1 – d)/d]                     51,256 kg
f. Volume of cylindrical jet fuel storage tank (e/2,825 kg/$m^3$)            18.1 $m^3$

4. **Cryogenic $H_2$ volume and storage tank size**
a. 747-8 jet fuel mass at takeoff (3b)                                        153,800 kg
b. Fuel mass ratio $H_2$: jet fuel (1h)                                       0.16–0.42 kg/kg
c. Mass of $H_2$ fuel needed (ab)                                            24,500–64,500 kg
d. Volume of $H_2$ fuel needed (c/70.9 kg/$m^3$)                             345–910 $m^3$
e. Mass ratio of $H_2$ fuel to $H_2$ fuel + storage tank[i]                  0.64
f. Mass of cylindrical $H_2$ storage tank [c(1 – e)/e]                        13,800–36,300 kg
g. Volume of cylindrical $H_2$ storage tank (f/2,567 kg/$m^3$)               5–13 $m^3$

5. **Mass ratio of HFC: ICE 747-8 aircraft**
a. Mass of 747-8 aircraft with no fuel (3a)                                  295,289 kg
b. Mass of $H_2$ fuel needed (4c)                                            24,500–64,500 kg
c. Mass of all $H_2$ fuel cell stacks needed (2g)                            125,000–261,000 kg
d. Mass of spherical $H_2$ storage tank (4f)                                 13,800–36,300 kg
e. 747-8 jet fuel mass at takeoff (3b)                                        153,800 kg
f. Mass of jet fuel storage tank (3e)                                         51,256 kg
g. Mass of $H_2$ aircraft (a+b+c+d-f)                                        407,000–606,000 kg
h. Mass of jet fuel aircraft (a+e)                                            449,100 kg
i. Overall mass ratio of $H_2$ to jet fuel aircraft (g/h)                    0.91–1.35

6. **Volume ratio of HFC: ICE 747-8 aircraft**
a. Volume of 747 aircraft[iii]                                               4,000 $m^3$
b. Volume of $H_2$ fuel needed (4d)                                          345–910 $m^3$
c. Volume of $H_2$ storage tank (4g)                                         5–13 $m^3$
d. Volume of $H_2$ fuel cell stacks needed (2f)                             316–661 $m^3$
e. Jet fuel volume at takeoff 747-8 (3c)                                     189.7 $m^3$
f. Volume of jet fuel storage tank (3f)                                      18.1 $m^3$
g. Volume of $H_2$ fuel cell aircraft (a+b+c+d-e-f)                         4,460–5,380 $m^3$
h. Volume ratio $H_2$ to jet fuel+stack+storage (b+c+d)/(e+f)              3.2–7.6
i. Overall volume ratio of $H_2$ to jet fuel aircraft (g/a)                 1.1–1.3

[i]Winnefeld et al. (2018)
[ii]Spakovsky (2008)
[iii]Boeing (2012)
[iv]Ultralight fuel cell stack.

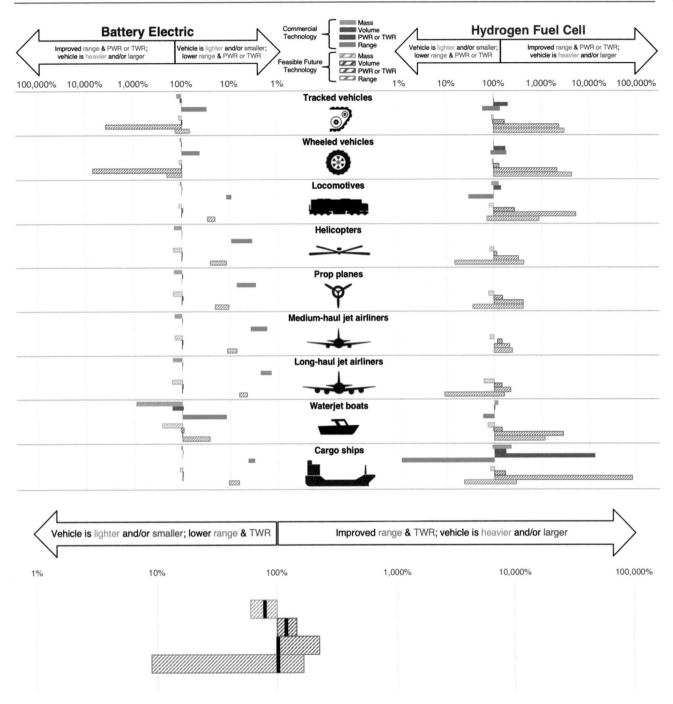

Figure 2.14 Percent increase or decrease in mass, volume, and range upon converting from an internal combustion engine (ICE) military or civilian vehicle to a battery-electric (BE) or hydrogen fuel cell (HFC) vehicle assuming the power-to-weight ratio (PWR) or thrust-to-weight ratio (TWR) is the same for all vehicles. Solid bars represent technology that is commercially available today with the following characteristics: an electric motor PWR of 3 kW/kg; a battery pack-level specific energy of 157 Wh/kg (watt-hours per kilogram) and energy density of 260 Wh/L; a hydrogen storage system specific energy of 1,465 Wh/kg and energy density of 833 Wh/L; and a fuel cell stack specific power of 659 W/kg and power density of 640 W/L. Hatched bars represent future feasible technology with the following published characteristics: an electric motor PWR of 15 kW/kg; a battery pack-level specific energy of 335 Wh/kg and energy density of 670 Wh/L (assumed using solid state batteries); a hydrogen storage system specific energy of 21 kWh/kg and energy density of 1.7 kWh/L; and a fuel cell stack specific power of 8 kW/kg and power density of 850 W/L. The bottom of the figure shows an enlarged rendering of the solution shown for HFC long-haul jet airliners using feasible future technology. Black bars identify one feasible design for such aircraft. From Katalenich and Jacobson (2020).

Liquid hydrogen requires about 1.8 to 4.8 times the volume of jet fuel to move an aircraft the same distance (Example 2.2, line 1g). As such, a liquid hydrogen aircraft requires a larger fuel tank than does a jet-fueled aircraft, increasing drag. However, jet fuel weighs 2.4 to 6.2 times what liquid hydrogen weighs to go the same distance (Example 2.2, line 1h), so the fuel in a liquid HFC aircraft weighs much less than does that in a jet-fueled aircraft. On the other hand, the fuel cells themselves required for an HFC aircraft are heavy. Nevertheless, accounting for all these factors, but ignoring changes in the power difference between using motors instead of jet engines, suggests that the overall HFC-to-jet fuel aircraft mass and volume ratios are 0.91 to 1.35 (Example 2.2, line 5i) and 1.1 to 1.3 (line 6i), respectively. This result supports the technical potential, based on the assumptions made, of transitioning a large commercial aircraft to renewable energy without compromising distance or payload. When all differences between a 747-8 and an equivalent HFC aircraft (same mass, power-to-weight ratio, and range) are considered, the HFC aircraft is about 21 percent larger assuming technology determined as feasible based on the published literature (Katalenich and Jacobson, 2020).

The main advantage of an HFC aircraft is that it eliminates all aircraft exhaust emissions except water vapor. Thus, HFC aircraft eliminate exhaust emissions of particles containing black carbon, brown carbon, and sulfate and gases such as carbon dioxide, carbon monoxide, nitrogen oxides, and sulfur oxides, along with a soup of organic gases. Due to the water vapor emissions, contrails still form from an HFC aircraft under the right temperature and humidity conditions in the background atmosphere. However, such contrails form only on background aerosol particles rather than on exhaust plus background aerosol particles. As such, contrails from HFC aircraft are about 70 percent thinner, thus dissipate faster, than contrails from jet fuel aircraft. Pure battery-electric aircraft do not even emit water vapor so eliminate all contrails.

## 2.3 WWS Building Heating and Cooling Technologies

A significant portion of the world's energy is used to provide air and water heating, air conditioning, and refrigeration in buildings. Heat is also used for cooking, which is discussed in Section 2.5.1. Away from the tropics, demand for air heating is usually greatest during winter, whereas demand for air conditioning is usually greatest during summer. Hot water and refrigeration are needed year-round, although energy demand for hot water generally peaks during cold months and demand for refrigeration peaks during warm months.

In a 100 percent WWS world, building air and water heating and air conditioning will be provided either by district heating and cooling systems or individual building heating and cooling systems. In both cases, heating and cooling will be provided primarily by electric heat pumps (Section 2.3.3), where the heat or cold is extracted from the air, ground, water, or waste heat/cold. The heat and cold may be stored or used immediately. Heating may be augmented by direct geothermal and solar heat. Both district heating/cooling and individual building heating/cooling are discussed here.

### 2.3.1 District Heating and Cooling

**District heating** is the production of hot water or steam at a centralized location and the distribution of that water or steam by pipes to buildings to provide hot air and water to them (Section 2.8.2). District cooling is the production of cold water at a centralized location (usually the same location as district heating) and the distribution of that water by pipes to buildings to provide cold air and/or refrigeration to them (Section 2.8.2).

In a WWS world, the heat for district heating will be provided by electric heat pumps or direct solar or geothermal heat. The electricity for the heat pumps will come from WWS sources. The source of heat for the heat pumps will either be the air, ground, water, or recycled waste heat. For example, waste heat from industry or buildings can be captured and piped back to a district-heating center. Over 60 percent of Denmark, Latvia, and Estonia's collective building heat today is from district heating (Section 2.8.2). The heat produced for district heating will either be used immediately or stored in water tanks (Section 2.8.1) or underground (Section 2.8.3) for later use.

The cold for district cooling will be provided mostly by electric heat pumps, where the electricity comes from WWS. The source of cold will either be the air, ground, water, or recycled wasted cold. The cold produced for district cooling will either be used immediately or stored in water tanks (Section 2.8.1), ice (Section 2.8.5), or in aquifers (Section 2.8.3.3) for later use. In some cases, the cold for district cooling can be drawn directly from cold lake water and passed through a heat exchanger, where the cold water absorbs heat from a building's air to cool the building (e.g., Cornell University, 2019).

An advantage of district heating/cooling is that it avoids the need for individual buildings to have their own heat pumps. It also allows heat to be stored for later use in large water tanks or underground and cold to be stored in water tanks, ice, or underground. Such storage helps to keep the electric grid stable in a WWS world (Section 8.1.3). A disadvantage of district heating/cooling is that it requires dedicated water piping to all buildings in a town or city. This can be costly if the buildings are far apart. As a result, district heating is most effective in densely populated cities.

### 2.3.2 Rooftop Solar Water Heaters

**Rooftop solar water heaters** (or domestic hot water systems) have been used worldwide for decades. Solar water heating systems include a solar collector and a storage tank. There are two types of solar hot water systems, active and passive. Active systems can have direct or indirect circulation.

An **active direct-circulation system** has a pump that circulates water through a collector. The water either goes to a storage tank or is used immediately in the building. Direct systems work well so long as the temperature does not drop below the freezing point of water.

An **active indirect-circulation system** passes a non-freezing heat-transfer fluid through a collector and transfers heat from the fluid to water with a heat exchanger. The water is then stored in the storage tank or used directly.

**Passive systems** generally use city water pressure to push water directly from a city water source through a rooftop collector that holds a fair amount of water. Sunlight heats the water in the collector. As water is used in the building, the water in the collector moves to the household storage tank, where a heat pump water heater warms it further. Thus, the solar collector serves to preheat the water, reducing the energy required by the heat pump. An evaluation should be performed in each case to determine whether the additional labor and capital cost of installing a passive solar hot water system in addition to a heat pump/solar PV system may be more costly than merely adding a few more PV panels to run the heat pump/PV system.

### 2.3.3 Heat Pumps

Heat pumps are efficient devices that extract heat or cold from the air, ground, water, or a waste stream of heat/cold and move it to where it is needed. A heat pump that extracts heat or cold from the ground is a **ground-source heat pump**, whereas one that extracts heat or cold from the ambient air or from a waste heat or cold stream of air is an **air-source heat pump** (Figure 2.15). One that extracts heat or cold from water, such as a swimming pool or lake, is a **water-source heat pump**.

The advantage of a ground- or water-source heat pump over an air-source heat pump is that temperatures under the ground and in the water are relatively stable, even when the air outside is very cold or very hot; thus, ground- and water-source heat pumps are more efficient under extremely cold or hot air conditions than are air-source heat pumps. The advantage of an air-source heat pump is that it is easier to install and maintain because none of its parts is buried underground or in the water.

Traditionally, heat and cold for ground-source heat pumps have been obtained from coils buried in a shallow backyard pit spread over a large area and covered with soil. More recently, the development of a technology to vertically drill two narrow but deep holes has reduced the cost of installing ground-source heat pumps significantly (Dandelion, 2018).

When heat pumps are used for cooling, they operate just like air conditioners or refrigerators, which move heat from a room or refrigerator, respectively, to the outside of the house or to outside the refrigerator, respectively. The difference between a heat pump and an air conditioner or refrigerator is that the heat pump operates in reverse to provide heating when needed. An air conditioner and refrigerator provide only cooling.

One type of heat pump is the **ductless mini-split heat pump**. As its name implies, it does not require ducts to move heat or cold air around a building. Instead, an indoor air handling unit (e.g., Figure 2.15) is placed in each of several rooms or zones of the building. The indoor units are connected to an outdoor compressor/condenser unit by a thin pipe containing refrigerant, a power cable, suction tubing, and a condensate drain line.

For air conditioning (cooling of warm indoor air), a fan in the indoor unit blows the warm air from the room over evaporator coils. Simultaneously, a thin pipe brings

(a)

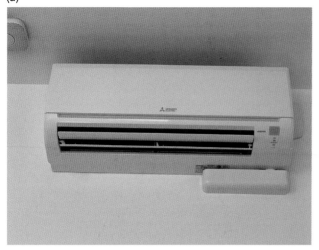

(b)

Figure 2.15  (a) Ductless mini-split heat pump for air heating and air conditioning. This is an indoor unit, several of which are placed throughout a building, each in a different zone. © Mark Z. Jacobson. (b) Outdoor compressor/condenser unit, used for drawing in heat from or releasing heat to the outdoor air. © Mark Z. Jacobson.

liquid refrigerant to the evaporator. Inside the evaporator, the liquid pressure is dropped, cooling the refrigerant. Conduction transfers heat from the room air to the cold refrigerant, bringing the refrigerant to its boiling point, evaporating the refrigerant. The room heat is therefore converted to latent heat stored in the refrigerant molecules. The room air that passed over the coils is now cold and flows back to the room, cooling the room. The gaseous refrigerant then travels by tubing to the outside unit (Figure 2.15), where it is heated by compression. In a condenser, the hot gas then transfers some of its heat by conduction to the cooler (relative to the hot refrigerant) outside air, causing the refrigerant's temperature to drop below its boiling point and condense back to a liquid. The condensation releases the stored latent heat, which conducts to the outside air as well. In sum, a ductless heat pump moves heat from a room to the outside, cooling the room, using only electricity as a power source.

For heating cold indoor air, the heat pump runs in reverse. Gaseous refrigerant in the outdoor unit is compressed and liquefied, with both processes heating the refrigerant. The hot liquid is then piped to the evaporator coils. Simultaneously, cold air from the room is blown over the coils. At the coils, heat from the refrigerant transfers by conduction to the cold air blowing over the coils, warming the air, which then flows back to the room. The refrigerant is now cold and transfers back to the outside unit. Dropping the pressure cools the liquid refrigerant even more. At the outside unit, heat from the outside air transfers by conduction to the cold, liquid refrigerant, raising its temperature above its boiling point, evaporating the refrigerant while cooling the outside air. The cycle then starts again. In sum, heat from the outside air is transferred to the room to warm it.

Air-source heat pumps become inefficient for heating at very low outdoor temperatures and inefficient for cooling at very high outdoor temperatures. For very low temperatures, sometimes a resistance-heating element is used as a backup to ensure a room stays warm. Ground- and water-source heat pumps largely solve this problem because ground and water temperatures rarely become extremely low or high. Thus, ground- and water-source heat pumps are recommended for locations that experience extreme seasonal heat or cold or both. Air-source heat pumps are recommended for more mild climates.

Whereas heat pump air heaters and air conditioners move heat between a building and the outside, **heat pump water heaters** often extract heat from the air within a room they sit in. For example, the heat pump water heater in Figure 2.16 extracts heat from the mechanical room it sits in. This can reduce the temperature of the air in the room by 1 to 3 °C. In the summer, this is advantageous, because the door to the room can then be opened, providing additional cooling to the house, reducing energy requirements for air heating. In the winter, the extra cooling slightly increases the air heating requirement.

An air-source heat pump water heater heats water as follows: First, liquid refrigerant is brought to an evaporator (Figure 2.16). At the evaporator, the refrigerant's pressure is dropped, cooling the refrigerant. A fan inhales warm (relative to the cool refrigerant) room air and blows it over the evaporator's coils. Heat from the room air transfers by conduction to the refrigerant, raising the refrigerant's temperature to its boiling point, evaporating the refrigerant. During evaporation, heat from the room is stored as latent heat in the gas molecules of the refrigerant. As such, the inhaled room air is now cooler and is expelled back to the room.

The gaseous coolant is heated further by compression in a compressor (Figure 2.16). The gas then goes to a condenser (Figure 2.16), where it condenses back to a liquid, releasing its latent heat and compressional heat to water in the tank, heating the water. The net result is hot water and a slightly cooler mechanical room.

Because heat pumps move, rather than create, heat and cold, they reduce the energy required to heat or cool by a factor of 3 to 5 compared with conventional gas heaters or electric resistance heaters. The efficiency of a heat pump is defined by its **coefficient of performance (COP)**, which is the ratio of useful heating or cooling required to work (electricity input) required. The COP of a heat pump exceeds 1, which means that the heat pump moves more than one unit of heat or cold for every one unit of electricity required to run it.

Air-source heat pumps have a COP of 3.2 to 4.5, whereas ground-source heat pumps have a COP of 4.2 to 5.2 (Fischer and Madani, 2017). This compares with **electric resistance heaters**, which have a COP of about 0.97 and fossil-fuel-powered boilers, which have a typical COP of about 0.8. Since only one joule (J) of electricity is therefore needed to move 3.2 to 5.2 J of hot or cold air

(a)

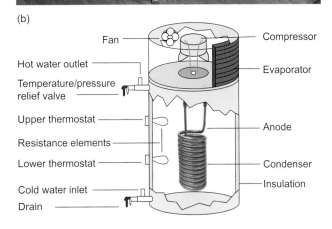

(b)

Figure 2.16 (a) Air-source heat pump water heater in a mechanical room of a house. © Mark Z. Jacobson. (b) Diagram of such a water heater.

with a heat pump, heat pumps reduce power demand compared with natural gas boilers by 75 to 85 percent (2.4 to 4.4 J). The use of heat pumps for all air and water heating worldwide is estimated to reduce world all-purpose end-use power demand compared with a business-as-usual case by about 13.2 percent (Section 7.3.6).

## 2.4 WWS High-Temperature Industrial Heat Technologies

The industrial sector creates products made of metal, plastic, rubber, concrete, glass, and ceramics, among other materials. Energy is needed in industry for heating, drying, curing, and producing phase changes. In the United States in 2010, energy for creating industrial heat comprised about 69.6 percent of industrial energy demand. Energy for cooling and refrigeration comprised 3.4 percent; for machine drive, 19.4 percent; for electro-chemical processes, 2.6 percent; and for other processes, 5.0 percent (DOE, 2015b). Globally, energy for heat is around one-fifth of world energy demand (Bellevrat and West, 2018). Industrial heat ranges from low to high temperature. About half of industrial heat is high-temperature heat (above 400 °C) and the other half, low-temperature heat (30 to 200 °C) and medium-temperature heat (200 to 400 °C) (Bellevrat and West, 2018).

High-temperature heat is used for plastics and rubber manufacturing, casting, steel production, other metal production, glass production, lime calcining in cement manufacturing, metal heat treating and reheating, and ironmaking. Low- and medium-temperature heat is used for drying and washing in the food industry, chemical manufacturing, distilling, cracking, pulp and paper manufacturing, and petroleum refining, for example.

Energy for industrial heat is currently obtained from fuels, electricity, and steam.

**Fuels** producing industrial heat include natural gas, coal, fuel oils, liquefied gases, and biomass, which are burned in ovens, fired heaters, kilns, and melters. These fuels are burned, and the resulting heat is transferred either directly or indirectly to the material being melted. With **direct heating** (**convection**), combusted gases come in direct contact with the material. With **indirect heating** (**radiant heating**), the hot gases pass through radiant burner tubes or panels separated from the material, and the heat is transferred radiantly.

**Electricity-based heating** technologies include electric arc furnaces, induction furnaces, resistance furnaces, dielectric heaters (including radio frequency driers and microwave processors), and electron beam heaters. The first three produce high-temperature heat (greater than 400 °C). The remaining technologies produce primarily medium- and low-temperature heat. All technologies use either direct or indirect heating. With direct heating, an electric current is sent directly to a material, heating the material by resistance heating. With indirect heating, high-frequency energy is inductively coupled with a specific material to heat it.

**Steam-based heating** technologies directly heat materials with steam or indirectly heat materials through a heat-transfer mechanism. Almost all steam heating is for low-temperature (<200 °C) processes, such as pulp and paper manufacturing, chemical manufacturing, and petroleum refining. The steam for these processes is often generated simultaneously (co-generated) with electricity produced by the burning of a fossil fuel or biomass.

In the United States in 2010, fuels comprised 48.5 percent of energy supplied to industry; electricity supplied 22.7 percent, and steam provided 28.8 percent (DOE, 2015b). Of all the energy consumed, only 58 percent was applied usefully, and the rest was lost.

Transitioning industrial-sector energy to 100 percent clean, renewable WWS energy and storage requires moving high-, medium-, and low-temperature fuel-based combustion to electric alternatives. Such alternatives include primarily electric arc furnaces, induction furnaces, resistance heaters, dielectric heaters, electron beam heaters, heat pump steam, and CSP steam.

In addition, it is necessary to eliminate non-energy $CO_2$ emissions produced chemically during steel manufacturing and concrete production. This requires either alternative methods of producing steel and concrete or methods of capturing the $CO_2$ upon their emissions from these two processes. Industrial-sector electrification and alternatives to steel and concrete manufacturing are discussed next.

### 2.4.1 Electric Arc Furnaces

An **electric arc furnace** is a spherical-shaped furnace for melting metal (Figure 2.17). It has a retractable roof and contains graphite electric rods. It has a hearth coated by a heat-resistant material that is used for collecting the molten metal. Scrap steel or iron is fed into the furnace, the roof is closed, and the electric rods (cathodes) are lowered onto the metal. An electric current that passes between the rods and an anode mounted at the bottom of the furnace creates an arc (extremely hot, bright light). The current that passes from the negatively charged cathode rods to the positively charged anode base melts the metal, as does the radiant heat emitted by the arc. When the metal is melted, the metal alloy, slag (stony waste

Figure 2.17 **Electric arc furnace.** © Scanrail/AdobeStock.

Figure 2.18 **Electric induction furnace.** © The Gund Company.

matter separated from metals during smelting), and oxygen formed by the process are removed through side doors.

The arc furnace derives from the carbon arc lamp. In 1800, Sir Humphrey Davy invented the **carbon arc lamp**, which consists of two carbon rods, acting as electrodes, in free air, with an electric current running through both rods. When the rods first contact each other, a spark is ignited. The rods are then pulled apart slowly. The current forms an extremely bright arc of light across the air gap. The bright arc forms because the hot (3,600 to 6,300 °C) carbon rod tips vaporize and ionize, creating a **plasma** that contains positive carbon ions and free electrons at high temperature. The electrons turn the gas into a good conductor that can be maintained at the high temperature. When the current strikes the ionized carbon vapor, the result is a bright light. The carbon rods slowly burn away as they gasify, requiring the distance between them to be adjusted as well. Carbon arc lamps were the only form of light used for street lighting and industrial indoor lighting from 1801 to 1901. Disadvantages are that they produce UV-B and UV-C radiation, which are both dangerous to humans; they create a buzzing sound; and they produce flickering light and sparks, which can set fires.

In 1878, Carl W. Siemens extended the idea of the arc lamp to build and patent an electric arc furnace. James B. Readman subsequently invented (1888) and patented (1889) an arc furnace in Edinburgh, Scotland, to create phosphorus. Paul Heroult of France developed a commercial arc furnace for steel production in 1900. In 1905, he went to the United States to work with several steel companies, including the Sanderson Brothers Steel

Company in Syracuse, New York, which installed the first commercial arc furnace worldwide in 1906.

Arc furnaces are currently used primarily in foundries (workshops for casting metals) to produce cast iron products, and in steel mills. In steel mills, they are used to produce steel from scrap metal, reducing the energy needed compared with making steel directly from ores. Because arc furnaces require a lot of power, they are often used when electricity prices are low, thus their use responds well to electricity pricing, which is a central feature of **demand response** management of the grid (Section 8.1).

In a 100 percent clean, renewable energy world, arc furnaces are one technology that will replace fossil fuels and biomass to produce heat for making steel and for casting metals.

### 2.4.2 Induction Furnaces

Another method of melting metals, such as iron, steel, copper, aluminum, and precious metals, is with an **induction furnace**, which also runs on electricity. Induction furnaces can melt as little as a kilogram to a hundred tonnes of metal.

An induction furnace consists of a non-conductive crucible surrounded by a coil of copper wire (Figure 2.18). Metal is placed inside the crucible to be melted. A strong alternating current is sent through the wire coil, creating a rapidly reversing magnetic field that induces circular electric currents (called eddy currents) inside the metal. The metal heats by resistance heating as the eddy currents pass through it. Thus, the heating is by electromagnetic induction.

Because metals are heated by induction, the temperature of the metal rises to no higher than the temperature required to melt the metal. This prevents the loss of some alloying elements. Thus, an induction furnace differs from an arc furnace, where the temperature rises above that required to melt the metal.

In a 100 percent clean, renewable WWS world, induction furnaces are another technology that will replace fossil fuels and biomass for producing high temperatures to melt metals.

### 2.4.3 Electric Resistance Furnaces

A third method of obtaining high-temperature heat is with electric resistance furnaces. With this technology, a DC current is passed between a negative electrode (cathode) and a positive electrode (anode), through a conducting material, which heats up due to resistance within the material. In **direct resistance heating (DRH)**, the material targeted for heating or melting is itself the conductor. In **indirect resistance heating (IRH)**, a separate heating element or conductor is heated up and transfers its heat by a combination of conduction, convection, and radiation to the material targeted for heating or melting.

The resistance of a conducting material is proportional to its length (l) and inversely proportional to its area (A):

$$R = \rho l/A \qquad (2.12)$$

where $\rho$ is the resistivity of the material. For efficient DRH, the resistance, thus the length-to-area ratio and/or the resistivity of the material, needs to be high. For typical metals, the length-to-width ratio should be about 6:1 and the cross section should be uniform. As such, the most common applications of DRH are the heating of long rods, billets of ferrous metals prior to forging, and continuous **annealing** (heating followed by slow cooling) of ferrous (iron-containing) and non-ferrous wire (Barber, 1982).

IRH is used to heat solids, liquids, and gases in many industries by a combination of direct contact heating, convection, and radiation. IRH is used in the heating and metals industries for melting, hot working, plasma-heating processes, stress relieving, and preheating. It is also used in the food industry; paper, print, and textile industries; rubber and plastic industries; and glass and ceramic industries.

A variety of metals (e.g., nickel-based alloys, iron-based alloys, and refractory metals, including platinum, molybdenum, tungsten, tantalum, and rhodium) and non-metals (silicon carbide, graphite) can serve as the heating element for IRH. Depending on the material, these materials heat from 400 to 3,000 °C. Silicon carbide, for example, which is heated to an operating temperature of 1,600 °C, is used in the heat treatment of metals and glass. Molybdenum, which is heated to an operating temperature of 1,750 °C, is used in hydrogen muffle furnaces to sinter metallic and ceramic materials. Tungsten and tantalum, which operate at 1,800 °C and 2,500 °C, respectively, are used in vacuum furnaces (Barber, 1982).

### 2.4.4 Dielectric Heaters

A fourth method of electric heating is **dielectric heating**, which is what microwave ovens use to cook food. Dielectric heating is used for much lower temperature applications than arc furnaces and induction furnaces.

Dielectric heating, also referred to as electronic heating, uses electromagnetic radiation in the frequency range of 300 kilohertz (kHz) to 300 gigahertz (GHz). Thus, it covers the radio frequency range (300 kHz to 300 megahertz [MHz]) and the microwave frequency range (300 MHz to 300 GHz). The former type of dielectric heating is referred to as **radio frequency (RF) heating** and the latter, **microwave (MW) heating**.

In a dielectric heater, the radio wave or microwave is used to heat a **dielectric material**, which is a poor conductor of electric current. RF heating is often applied to heat large materials because the heating is more uniform than with MW heating. As such, radio frequency heating is used for most dielectric industrial heating applications, including gluing, welding, plastic production, preheating, bread baking, textile drying, adhesive and paper drying, microwave preheating, and vulcanizing rubber. Microwave heating is generally used for the tempering of meat and other applications of food processing (Hulls, 2016).

### 2.4.5 Electron Beam Heaters

**Electron beam heating** is a method of melting metals or modifying materials with a fine beam of electrons. When the high concentration of electrons hits a solid material, the kinetic energy of the electrons is converted to heat, which melts the material. The electrons are produced with an **electron gun**, which ejects a narrow stream of electrons from a heated cathode and accelerates them using high voltage. The beam is obtained by using electric and magnetic fields to force free electrons in a vacuum into a straight line. The power density of electron beam heating is about 100 W/cm$^2$, which is 1,000 times that of typical peak solar radiation reaching the surface of the Earth during the day (1,000 W/m$^2$). Around 50 to 80 percent of the energy in the electron beam is transferred to the material.

Electron beam heating takes place in a large vacuum furnace and is applied to mass-produce steel and to purify metals, such as titanium, vanadium, tantalum, molybdenum, tungsten, zirconium, niobium, and hafnium. The electronics industry uses many of these metals. A **vacuum furnace** is a furnace in which the material being operated on is surrounded by a vacuum, or absence of air. Temperatures in a vacuum furnace reach up to 3,000 °C. Electron beam heating is also used in vacuum chambers to weld and precisely cut materials to make machines, evaporate and deposit thin layers of metal on solar cells, and modify surface layers of metals.

### 2.4.6 Steam Production from Heat Pumps and CSP

Steam is needed for many low- and some medium-temperature processes in a 100 percent WWS world. Such steam is currently obtained as a co-product of burning fossil fuels or biomass for electricity production. These sources will be replaced by steam from electric heat pumps or steam co-generated with electricity from concentrated solar power (CSP) plants. A heat pump can produce heat up to 160 °C (e.g., Viking Heat Engines, 2019), which can be used to boil water to produce steam of a similar temperature. Parabolic trough CSP plants produce temperatures of 60 to 350 °C; parabolic dish plants produce temperatures of 100 to 500 °C; and central tower receivers produce temperatures up to 600 °C (Ramaiah and Shekhar, 2018).

### 2.4.7 Steel Manufacturing

A significant use of energy as well as a source of pollution and carbon dioxide in the industrial sector is steel manufacturing. With steel manufacturing, $CO_2$ emissions arise not only due to the fossil-fuel or biomass energy used to produce high-temperature heat but also due to chemical reaction, which releases $CO_2$ during the extraction of iron metal from iron ore.

Steel can be produced from raw iron ore or recycled metal. Steel produced from iron ore is produced in two stages. The first is called ironmaking; the second, steelmaking.

In the **ironmaking** step, molten iron metal [Fe(l)] is extracted from solid iron ore, also called hematite [$Fe_2O_3$(s)], by filling a **blast furnace (BF)** with hematite, relatively pure solid carbon [C(s)] in the form of coke (coal heated in the absence of air), and **limestone** [calcium carbonate, $CaCO_3$(s)]. Hot air containing molecular oxygen gas [$O_2$(g)] is then forced up the bottom of the blast furnace. It reacts with the coke to form carbon monoxide gas [CO(g)] and heat by

$$2C(s) + O_2(g) \rightarrow 2CO(g) + \text{heat} \qquad (2.13)$$

At high temperature in the furnace (up to 2,550 °C), the hematite reacts with the carbon monoxide and the coke itself to produce molten iron metal and carbon dioxide gas [$CO_2$(g)]:

$$Fe_2O_3(s) + 3CO(g) \rightarrow 2Fe(l) + 3CO_2(g) \qquad (2.14)$$

$$2Fe_2O_3(s) + 3C(s) \rightarrow 4Fe(l) + 3CO_2(g) \qquad (2.15)$$

The limestone simultaneously heats up and thermally decomposes to quicklime [calcium oxide, CaO(s)] and $CO_2$ by

$$CaCO_3(s) + \text{heat} \rightarrow CaO(s) + CO_2(g) \qquad (2.16)$$

The calcium oxide then reacts with and removes sandy (containing silicon dioxide) remnants of the iron ore to form slag [$CaSiO_3$(l)]

$$CaO(s) + SiO_2(s) \rightarrow CaSiO_3(l) \qquad (2.17)$$

Slag is less dense than molten iron, so floats on top of the iron. Slag is then cooled and removed for use in roads. Thus, traditional steelmaking releases $CO_2$, not only through fossil-fuel and biofuel combustion to produce high temperatures, but also from chemical reaction.

**Steelmaking** is the second step in steel production. In this step, impurities are removed from the raw iron, and carbon and other alloying elements are added to make crude steel. Impurities removed include phosphorus, sulfur, and excess carbon. Alloying elements added include chromium, nickel, vanadium, and manganese.

Steelmaking is performed in one of two ways. **Primary steelmaking** is the use of new iron from ironmaking. **Secondary steelmaking** is the recycling of scrap steel and melting it in an electric arc furnace to produce steel.

The main method of primary steelmaking is the **basic oxygen steelmaking (BOS)** method. With this method, the molten iron and impurities from the blast furnace are mixed with scrap steel and placed in a **basic oxygen furnace (BOF)**. Oxygen is then blown through the furnace and reacts with carbon in the molten mix to form $CO_2(g)$, which is released to the air. Calcium oxide in the molten mix also reacts with phosphorous and sulfur, the products of which rise to the top as slag and are removed. Finally, alloys are mixed in, and the molten steel is poured into pre-shaped molds, where it cools and hardens.

With secondary steelmaking, scrap metal is melted in an arc furnace. Oxygen is blown through the metal to help remove the carbon and speed the meltdown of the metal by increasing combustion. Calcium, phosphorous, and sulfur are removed, and alloys are mixed in, similar to the process in the basic oxygen furnace.

In sum, the sources of $CO_2$ during the two-step steel formation process are (a) its emissions during fossil-fuel combustion to produce heat in the blast furnace and in the basic oxygen furnace, (b) its chemical release during reaction of carbon with hematite, (c) its release upon the thermal decomposition of limestone during ironmaking, and (d) its release during chemical reaction of carbon with oxygen during steelmaking. In addition, in an arc furnace, there is a small amount of $CO_2$ released due to the vaporization of graphite and its reaction with oxygen. The overall carbon emissions in the ironmaking plus steelmaking process using a blast furnace and basic oxygen furnace are about 1,870 kg-$CO_2$ per tonne-steel (Vogl et al., 2018). Of this, ironmaking produces about 70 to 80 percent of the $CO_2$ emissions.

### 2.4.7.1 Reducing Carbon Emissions with Hydrogen Direct Reduction

An alternative to extracting molten iron from iron oxide with coke during ironmaking is to extract it with hydrogen gas [$H_2(g)$], where the hydrogen is produced with 100 percent WWS (Vogl et al., 2018). This process is called the **hydrogen direct reduction (HDR)** process. The main extraction reaction is

$$Fe_2O_3(s) + 3H_2(g) \rightarrow 2Fe(l) + 3H_2O(g) \qquad (2.18)$$

This reaction occurs optimally at a temperature of around 800 °C, which is lower than the temperature needed in a blast furnace (Vogl et al., 2018). The reaction eliminates chemical $CO_2$ produced from the hematite reaction with carbon during ironmaking. However, an injection of carbon into the molten iron during the steelmaking process is still needed to create an iron-carbon alloy (0.002 to 2.14 percent C) to strengthen the steel. In addition, some carbon is still emitted from the thermal decomposition of limestone and subsequent emission of $CO_2$.

If the heat required for the HDR ironmaking process is obtained with an electric resistance furnace instead of with fossil fuels, if the hydrogen for the HDR process is produced by electrolysis (passing of electricity through water), if an electric arc furnace is used for the steelmaking process, and if all electricity is provided by 100 percent WWS, the HDR process emits only 53 kg-$CO_2$ per tonne of steel, or only 2.8 percent of the emissions of the blast furnace/basic oxygen furnace process (1870 kg-$CO_2$ per tonne of steel) (Vogl et al., 2018). The only emissions in the HDR process are some oxidation of injected carbon in the arc furnace, some $CO_2$ from the thermal decomposition of limestone, and some oxidation of the vaporized carbon in the arc furnace electrodes.

The system just described is a 100 percent clean, renewable energy system, but still results in a residual of 53 kg-$CO_2$ per tonne of steel (2.8 percent of the original emissions). The remainder will either be released to the air or captured. However, capturing $CO_2$ requires energy, and regardless of whether the electricity is from WWS or natural gas, it is better to use that electricity to displace a fossil-fuel electricity source than to capture $CO_2$, since displacing a fossil-fuel source eliminates not only $CO_2$ from the source but also air pollution and upstream mining and emissions (Section 3.6).

### 2.4.7.2 Reducing Carbon Emissions with Molten Oxide Electrolysis

A second alternative to extracting molten iron from iron oxide with coke during ironmaking is with **molten oxide electrolysis (MOE)** (Allanore et al., 2013; Wiencke et al., 2018). With MOE, solid iron ore, $Fe_2O_3(s)$, is first heated in a molten electrolyte soup above 1,688 K, where it decomposes to magnetite [$Fe_3O_4(s)$] and oxygen by

$$6Fe_2O_3(s) + heat \rightarrow 4Fe_3O_4(s) + O_2(g) \qquad (2.19)$$

The molten electrolyte soup contains oxides, such as silica ($SiO_2$), alumina ($Al_2O_3$), and magnesia (MgO), and has the purpose of helping electricity flow. The soup is heated further, past the melting point of pure iron, which is 1,811 K. Above this temperature, electricity is passed through the soup, producing pure molten iron by

$$0.5Fe_3O_4(s) + electricity \rightarrow 1.5Fe(l) + O_2(g) \qquad (2.20)$$

The pure liquid iron sinks to the bottom of the cauldron, where it is drained. As such, the MOE process produces pure iron without emitting chemically produced carbon dioxide.

## 2.4.8 Concrete Manufacturing

**Concrete** is a mixture of **aggregate** (sand, gravel, and crushed stone) and a paste (water and **Portland cement**). The paste binds the aggregate together, making a hard surface. Concrete is used for roads, foundations, buildings, runways, sidewalks, driveways, and a variety of other purposes.

**Joseph Aspdin** (1788–1855) of Leeds, England, invented Portland cement in the early nineteenth century. He formed it by burning powdered limestone and clay on his kitchen stove. Today, cement contains limestone, shells, or chalk, all of which contain $CaCO_3(s)$, mixed with clay, shale, slate, blast furnace slag, silica sand, or iron ore. These ingredients are heated to 1,500 °C to form a hard substance that is ground into a fine, powdery cement.

The concrete industry produces about 5 percent of the world's $CO_2$ emissions. About half of the emissions during concrete production are from energy use and the other half are from chemical reaction during cement manufacturing. The chemical emissions rate of $CO_2$ from cement manufacturing is 900 kg-$CO_2$ per tonne-cement. These emissions arise due to the reaction,

$$CaCO_3(s) + clay + heat \rightarrow clinker(SiO_2, Fe_2O_3, Al_2O_3,$$
$$CaO) + CO_2 \qquad (2.21)$$

where the $CO_2$ originates from calcium carbonate. The clinker is then mixed with gypsum ($CaSO_4$–$2H_2O$) to form cement by

$$Clinker + gypsum \rightarrow cement \qquad (2.22)$$

The cement is subsequently mixed with water to form a paste, which is combined with the aggregate to form concrete.

Three ways of reducing emissions from concrete manufacturing are to (1) use an alternative to concrete that doesn't emit $CO_2$ as part of the chemical process,

(2) make concrete that traps $CO_2$, and (3) recycle concrete.

### 2.4.8.1 A Type of Concrete That Emits No CO$_2$

One commercialized alternative to concrete is **Ferrock**, or iron carbonate ($FeCO_3$) (Build Abroad, 2016; Stone, 2017). Ferrock is derived by first mixing waste steel dust containing iron (FeO) with crushed glass containing silicon dioxide ($SiO_2$), limestone ($CaCO_3$), kaolinite or another clay, stabilizers, promoters, and a catalyst into a mixer at room temperature. The mixture is then poured into a mold containing seawater. The filled mold is put into a curing chamber, where $CO_2$ from a furnace is injected. The iron, $CO_2$, and saltwater react together to form Ferrock and molecular hydrogen ($H_2$). When the final product dries, it is about five times harder and more flexible than cement. The production of Ferrock not only avoids the chemical $CO_2$ emissions and most of the energy emissions of concrete production, but it also sequesters $CO_2$ and produces hydrogen, which can be used for other applications.

### 2.4.8.2 Sequestering CO$_2$ in Concrete

Trapping $CO_2$ from combustion emissions, as done in Ferrock, is a method of offsetting chemically produced $CO_2$ emissions from the concrete production process. Trapping $CO_2$ within concrete itself is another option (e.g., Carbon Cure, 2018). The clinker in cement contains CaO. When the cement is mixed with water to form a paste, the CaO reacts with water to form calcium hydroxide, $Ca(OH)_2$, by

$$CaO + H_2O \rightarrow Ca(OH)_2 \qquad (2.23)$$

If $CO_2$ captured from any source is mixed with the clinker, it will react to form $CaCO_3 + H_2O$ within the cement. Upon drying, the solid $CaCO_3$ strengthens the cement. Even if the cement breaks, the $CO_2$ will not break free because it is a solid bound to the cement.

Cement also contains **calcium silicate hydrate (3CaO–SiO$_2$–4H$_2$O)**. $CO_2$ can react with this chemical to form calcium carbonate, which is bound in the cement, by

$$CO_2 + 3CaO\text{–}SiO_2\text{–}4H_2O \rightarrow CaCO_3 + 2CaO$$
$$\text{–}SiO_2\text{–}4H_2O \qquad (2.24)$$

### 2.4.8.3 Concrete Recycling

A third method of reducing $CO_2$ emissions from concrete manufacturing is **concrete recycling**. Concrete structures or roads are often demolished. Historically, such concrete has been sent to a landfill. However, if the concrete is uncontaminated (free of trash, wood, and paper), it can be recycled. Rebar in concrete can also be recycled, as magnets can remove it. The rebar can then be melted and used for other purposes. The broken concrete is crushed. Crushed concrete is often used as gravel in new construction projects or as aggregate in new concrete.

Remaining $CO_2$ emitted chemically during the cement formation process could be captured chemically upon emissions. However, as stated previously, capturing $CO_2$ requires energy, and regardless of whether the electricity is from WWS or natural gas, using that electricity to displace a fossil-fuel electricity source is better than using it to capture $CO_2$, since displacing a fossil-fuel source eliminates not only $CO_2$ from the source but also air pollution and upstream mining and emissions (Section 3.6).

## 2.5 WWS Electric Substitutes for Fossil-Fuel Appliances and Machines

A key step in the electrification of all energy sectors is to ensure electric appliances and technologies that replace fossil-fuel ones perform at least as well and at a similar or lower cost. So far, electricity, transportation, heating/cooling, and industrial technologies have been discussed. Here, some additional electric replacement technologies are described.

### 2.5.1 Electric Induction Cookers

In a 100 percent WWS world, **electric induction cookers** (also referred to as induction burners) should replace cookers that run on gas, wood, waste, dung, or coal combustion. Induction cookers also represent an improvement over electric resistance cookers. An induction cooker consists of a ceramic plate with an electromagnetic coil beneath it. When the cooker is turned on, an electric current runs through the coil, generating a fluctuating magnetic field but not heat in the cooker itself. This is why the cooker does not feel hot when it is touched.

Induction cookers work only with pots and pans that stick to a magnet. The base of such cookware must be made of a high-resistance material, such as iron or stainless steel, not a conducting material, such as aluminum or copper. Once an iron or stainless steel pot or pan is placed on the cooker, the magnetic field induces many smaller electric currents in the cookware's metal base. Because iron does not conduct well, much of the electricity in the small currents turns to heat, heating the metal. The heat is produced in the metal base of the pot or pan, not in the cooker. As such, the cooker itself feels only warm due to conduction of heat from the base of the pot or pan to it. Once the pot or pan is removed from the cooker, the cooker cools off relatively quickly.

Water boils with an induction cooker in about half the time as it does with a gas cooker. Induction cookers also result in less scorching of food because they have fewer hot spots within a pot or pan than do gas cookers.

Electric induction cookers have the potential to help eliminate millions of premature deaths per year. This is because about 2.2 million children and adults die each year, primarily in developing countries, due to breathing air pollution from the indoor burning of wood, waste, dung, and coal during home cooking and heating (Table 7.13). Induction cookers can also save time for hundreds of millions of people who collect their own wood or waste. Figure 2.19 shows the simplicity of a single induction cooker and multiple cookers on a cooktop, respectively. All that is needed is an electricity source and an iron or stainless steel pot or pan. While bringing electricity to remote villages in developing countries is a challenge, the advent of low-cost solar PV and batteries now permits the widespread adoption in remote communities of clean, renewable electricity and electric appliances, such as induction cooktops.

### 2.5.2 Electric Fireplaces

A fire risk and source of carbon monoxide, nitrogen oxide, organic gas, and particle air pollution inside and outside a home are wood and natural gas fireplaces. Such fires are cozy and warm and give a home character, but they also produce indoor and outdoor air pollution. Wood fires can also cause smoke damage and unpredictable impacts from their embers that burst from a fireplace into a living room. Fortunately, warm and visually appealing electric fireplaces are now available to replace wood and gas fireplaces (Figure 2.20). These eliminate the air pollution and smell of wood or gas burning while providing the same warmth and coziness.

(a)

(b)

Figure 2.19 (a) Stand-alone electric induction cooker. © Della Liner/AdobeStock. (b) Five electric induction cookers on a cooktop. © Mark Z. Jacobson.

Figure 2.20 Electric fireplace. © Andriana_cz/AdobeStock.

### 2.5.3 Electric Leaf Blowers

Possibly the most annoying fossil-fuel machine today is the gasoline **leaf blower**. It not only smells and creates noxious air pollution, but it is also noisy, since it has no muffler or other means of noise suppression. Both corded and battery-powered electric leaf blowers are available. They perform the same task as a gasoline leaf blower but do not release a smell and are quiet. Some battery-powered leaf blowers run for 1 to 3 hours at low speed or 10 to 30 minutes at full speed and take 30 to 100 minutes to charge, depending on battery size. Most models allow the battery to be swapped, so an electric leaf blower can be used continuously if multiple batteries are charged and swapped when needed.

### 2.5.4 Electric Lawnmowers

Another noisy machine that creates a smell and uncontrolled, unhealthy air pollution is the gasoline **lawnmower**. Fortunately, cost-competitive (over the lifetime of the lawnmower) battery-powered lawnmowers are available to replace fossil-fuel lawnmowers. Not only do electric lawnmowers eliminate the exhaust-related air pollution (unburned hydrocarbons, oxides of nitrogen, carbon monoxide, and particulate matter) from a gasoline lawnmower, but they also start instantly, are quiet, and require less maintenance (since they have no engine and fewer parts) than does a fossil-fuel mower. Although the upfront cost of an electric lawnmower may be more than that of a gasoline mower, electric mowers have no gasoline cost and need less equivalent electricity than do gasoline mowers because the work output to energy input ratio of electricity powering a motor is about four times that of gasoline powering an engine. As such, the lifetime (10 years or more) energy and overall cost to run an electric mower is less than that of a gasoline mower even if the upfront cost of the electric mower is higher (Hope, 2017). Runtimes of electric lawnmowers are currently 30 to 60 minutes before recharging or battery swapping is needed.

### 2.5.5 Other Appliances and Technologies

All other home, business, and industrial technologies that run on fossil fuels (natural gas, gasoline, diesel, etc.) or biofuels have an electric counterpart that are needed as part of a 100 percent clean, renewable energy economy. Two examples are chainsaws and clothes dryers.

The first **electric chainsaw** for woodcutting was invented in 1926 in Germany by Andreas Stihl. Stihl invented a gas-powered chainsaw as well, but three years later. Today, both electric and gas chainsaws are available. Advantages of electric chainsaws are that they have minimal noise, emit no fumes, require no mixing of oil and gasoline, require minimal maintenance, are lighter, and cost less. Cordless versions have traditionally lasted about an hour before recharging and have been used for smaller jobs than gas chainsaws. However, with the improved storage capacity of batteries and battery swapping capabilities, both limits are no longer an issue.

## Transition highlight

In 1938, J. Ross Moore of North Dakota commercialized the first **electric clothes dryer** after tinkering with the technology since 1915. The first **electric heat pump clothes dryer** was commercialized in Europe in 1997. By 2007, heat pump dryers were being manufactured in Japan as well. By 2009, about 4 percent of European dryers were heat pump dryers (Meyers et al., 2010). Today, heat pump dryers are available worldwide. Heat pump dryers require about half the electricity as conventional electric dryers. However, heat pump dryers themselves cost more upfront. This upfront cost difference can be erased for new construction homes if a ventless heat pump dryer is purchased because the construction cost of vents is usually more than the dryer cost itself. As such, a ventless heat pump dryer in a new construction home will almost always have a lower lifecycle cost than a conventional electric dryer due to both the energy cost savings and the construction cost savings, despite the higher dryer cost. A heat pump dryer will also have a lower lifecycle cost if it is installed in an existing building if it is used often (Meyers et al., 2010).

## 2.6 Reducing Energy Use and Increasing Energy Efficiency

Two additional components of a 100 percent WWS world are **reducing energy use** and **increasing energy efficiency** of existing electric technologies or buildings. Both of these methods are referred to as **demand-side energy conservation** measures. Such measures reduce the need for energy.

Some examples of reducing energy use include the following:

- Using public transit or telecommuting instead of driving;
- Carpooling and minimizing the number of driving trips required;
- Teleconferencing instead of flying in an airplane to a conference;
- Designing cities to facilitate the use of public transit and bicycles and to improve traffic flow;
- Building high-rise apartments instead of single-family homes to reduce construction materials and heat loss;
- Eating foods that require less energy to produce;
- Eating locally sourced food instead of food transported a long distance; and
- Ensuring electronic equipment is shut off when it is not in use.

Some example methods of increasing energy efficiency include the following:

- Performing building energy audits to identify where energy is wasted;
- Incorporating **green building standards** in the construction of new buildings to minimize their energy use;
- Replacing incandescent and compact fluorescent light bulbs with light-emitting diode (LED) light bulbs;
- Adding advanced lighting controls;
- Replacing electric resistance air and water heaters with heat pump air and water heaters;
- Using ductless heating and air conditioning to eliminate duct losses of heat or cold;
- Improving wall, floor, ceiling, and pipe insulation;
- **Weatherizing** (sealing) windows, doors, and fireplaces;
- Using triple-paned windows to reduce heat loss from windows;
- Using thermal mass in walls and floors to modulate temperature changes of homes (Section 2.8.4.1);
- Using a ventilated façade to shield a building from extreme heat or cold (Section 2.8.4.2);
- Using window blinds to control sunlight into buildings (Section 2.8.4.3);
- Adding window film to reduce heat loss and sunlight intake (Section 2.8.4.4);
- Using night ventilation cooling (Section 2.8.4.5);
- Improving air flow management in buildings;

- Using passive solar heating;
- Purchasing more energy-efficient appliances;
- Improving data center design; and
- Improving the efficiencies of solar cells, wind turbines, batteries, and electric cars.

Several of these techniques are discussed in Section 2.8. As discussed in Section 7.3, transitioning to 100 percent WWS results in over a 50 percent reduction in end-use energy demand due to four factors aside from improving energy efficiency and reducing energy use: (1) the efficiency of electricity and electrolytic hydrogen over combustion for transportation; (2) the efficiency of electricity over combustion for high-temperature industrial heat; (3) the efficiency of moving low-temperature building air and water heat with heat pumps instead of creating heat with combustion; and (4) eliminating the energy needed to mine, transport, and process fossil fuels, biofuels, bioenergy, and uranium.

Increasing energy efficiency and reducing energy use avoid the use of fossil fuels, thus avoiding pollution. As such, they have the same effect as, but are usually less expensive than, replacing fossil fuels with WWS. Because they are less expensive, increasing efficiency and reducing energy use should be prioritized first in any strategy to address global warming, air pollution, and energy security. However, increasing efficiency and reducing energy use cannot solve the problems on their own. As such, strategies to solve these problems must include not only increasing efficiency and reducing energy use, but also transitioning to 100 percent clean, renewable energy and storage and eliminating non-energy emissions (Section 2.9).

## 2.7 WWS Electricity Storage Technologies

For any energy system, matching energy demand with supply continuously is essential. In the 100 percent WWS system discussed here, the types of demand for energy include electricity demand, high-temperature heat demand, low-temperature heat demand, cold demand, and hydrogen demand.

**Electricity demand** will be met by a combination of (a) electricity currently available from local WWS electricity generators, (b) electricity that is stored, and (c) electricity that is transmitted in from distant WWS energy

generators. **High-temperature heat demand** for industrial processes will be met primarily by furnaces and other equipment running on electricity and by CSP heat (Section 2.4). **Low-temperature air and water heat demand** will be met either by heat currently available from WWS heat sources (geothermal or solar heat), heat that has been stored, or heat produced from heat pumps run on electricity. **Cold demand** for air conditioning will be met by cold produced from heat pumps run on electricity or cold that is stored in water or ice. **Hydrogen demand** for transportation will be met by hydrogen produced immediately from electrolysis or hydrogen that has been stored.

This section discusses different types of electricity storage in a 100 percent WWS system. Heat, cold, and hydrogen storage are discussed in subsequent sections.

### 2.7.1 Concentrated Solar Power with Storage

As discussed in Section 2.1.7, parabolic trough and central tower concentrated solar power (CSP) plants work by heating a fluid (either oil for a parabolic trough or a molten salt mix for a central tower), which passes through a heat exchanger to boil water to create steam to run a steam turbine to generate electricity. For both types of plants, the hot fluid can first be stored in an insulated thermal storage tank before the fluid is sent to the heat exchanger (Figure 2.21). This allows electricity production to be delayed until nighttime, until heavy clouds are present, until a time of high electricity demand, or until no other WWS electricity source is available (Section 8.1). After the steam passes through the turbine, it is routed to a condenser, which liquefies the water and sends it back to the heat exchanger in a closed loop (Figure 2.21). After the molten salt exchanges its heat with water in the heat exchanger, the molten salt is cold and is sent to a holding tank before it is passed through either the parabolic trough or the central tower receiver again.

While molten salt can itself be stored, its heat can also be transferred to a **phase change material** (PCM), which requires less volume, thus costs less, to store heat than does molten salt (Nithyanandam and Pitchumani, 2014).

The storage associated with CSP is usually sized to last up to 15 hours at its peak discharge rate before storage is

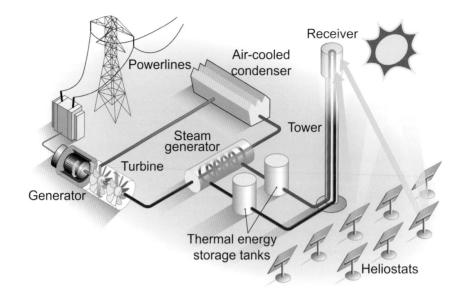

**Figure 2.21 Diagram of central tower CSP plant with storage. See text for explanation.** From DOE (2014).

depleted. The peak discharge rate of a CSP plant is limited by the nameplate capacity of the steam turbines used in the plant. Storage of up to 15 hours allows for 24 hours per day of electricity production from CSP plus storage. The Gemasolar CSP plant in Seville, Spain (see Figure 2.9) first demonstrated 24 hours per day of CSP-with-storage electricity production during July 2011. It has provided electricity continuously (24 hours per day) for stretches of up to 36 days. During cloudy and winter days, a CSP-with-storage plant also produces electricity at night, but for fewer hours.

The **capacity factor** of a CSP plant is the actual annually averaged power produced by the steam turbines of the plant divided by the maximum possible power produced by the turbines (their summed nameplate capacities). With no storage, the capacity factor of a CSP plant is limited by the instantaneous solar power available and collected and sent through the steam turbine, averaged over a year. Excess heat not used to produce steam is wasted because the turbine can produce only a limited amount of power. A typical capacity factor of a CSP plant without storage in a sunny location is around 25 percent. With storage, the capacity factor increases to around 65 percent or higher. This is accomplished first by adding more mirrors and molten salt (for a central tower CSP

plant) and storing the excess salt and its heat. During the night, or when solar radiation is weak, the stored hot salt is used to boil water. The additional steam increases the power generated by the turbines, increasing the plant's electricity output and thus annual average capacity factor.

The thermal efficiency of a CSP plant is the heat available for the steam turbine after storage to the heat from the solar collector that goes into storage. If the fluid loops of a CSP plant are well insulated, the overall thermal efficiency of the plant is around 99 percent. However, 45 percent of incident solar energy reaching the CSP mirrors is lost due to reflection and absorption by the mirrors. Also, of the heat going to the steam turbines, only 28.7 percent goes to electricity production and 71.3 percent is wasted (Jacobson et al., 2018a). However, the waste heat can be captured and used for industrial heat processes (Section 2.4.6) or as a source of heat for heat pumps (Section 2.3.3).

Finally, CSP plants are excellent for helping to match power demand on the electric grid with intermittent solar supply. With storage, CSP plants can ramp their power production up and down more quickly than coal or nuclear plants and similarly to natural gas plants (Table 2.1), allowing them to help match demand with CSP supply.

## 2.7.2 Hydroelectric Power Dam Storage

Conventional hydroelectric power (hydropower) plants have built-in storage, since the water that generates the electricity for them is stored in a reservoir behind a dam. Whereas the water can be drained through a turbine to create electricity as needed, the reservoir can only be charged naturally through rainfall and runoff, except if the reservoir and turbines are turned into a pumped hydropower storage (PHS) station (Section 2.7.3). Thus, the electricity storage associated with a conventional hydropower plant differs from that of battery storage. A battery can be charged when excess electricity is available and discharged when electricity is needed. A hydropower plant can only produce electricity upon demand, but excess electricity on the grid can't be used to recharge the hydropower storage unless the facility is turned into a PHS facility.

As stated in Section 2.4, hydropower plants can be run as baseload plants, load-following plants, or peaking power plants. In the case of load-following and peaking plants, hydropower can meet the minute-by-minute load fluctuations faster than can thermal power plants. Table 2.1 compares the startup time and ramp rate of hydropower with those of other electric power sources. It indicates that both the startup time and ramp rate of a hydropower plant are much faster than those of a gas, coal, or nuclear plant. The reason is that a hydropower turbine can generate full power starting from zero power within 15 to 30 seconds after the sluice gate between a hydropower storage reservoir and penstock opens. That is the time required for the water moving from the sluice gate to reach the turbine.

Because hydropower plants can produce electricity within 15 to 30 seconds of a need, they are often used to provide peaking power when no other electricity is available to match power demand with supply on the electric power grid. At a high continuous electricity discharge rate, hydropower plants can meet a high continuous demand for hours to days, depending on the size of the reservoir and the nameplate capacity of installed turbines in the plant. When used only to meet gaps in supply, such plants may be idle for days to weeks then used several days in a row for 2 to 3 hours a day.

### 2.7.3 Pumped Hydropower Storage

Another type of storage that can ramp its supply up and down quickly is **pumped hydropower storage** (**PHS**). Like with CSP, PHS is used primarily to fill in short-term gaps in supply (on the order of minutes to a day).

PHS consists of two reservoirs – an upper and a lower one (Figure 2.22). The lower one can be the ocean, a natural lake, a human-made lake or reservoir, or a continuously running river. The upper one can be a natural lake or a human-made lake or reservoir. One newly proposed PHS system consists of Lake Mead, the reservoir behind Hoover Dam, and a pumping station 32 km downstream along the Colorado River.

When excess electricity is available or when electricity prices are low, water is pumped through pipes from a

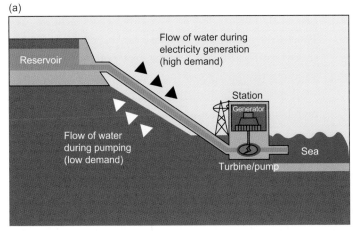

**Figure 2.22** (a) Diagram of a pumped hydropower storage system. (b) Srinigaring, Thailand, combined hydropower and pumped hydropower system. The hydropower system is 720 MW, of which 360 MW is pumped hydropower. © Terrapun Fuangtong/AdobeStock.

pumping station in the lower reservoir to the upper reservoir. When electricity is needed, water drains downhill through a **water turbine** connected to a generator to produce electricity.

The overall efficiency of a PHS system (ratio of electricity delivered to the sum of electricity delivered and electricity used to pump the water) is around 80 percent (Jacobson et al., 2018a). The ramp rate for PHS is the same as for hydropower (see Table 2.1). As such, PHS ramps faster than do thermal power plants, including natural gas plants.

In a PHS facility, the electricity available from the upper reservoir, after losses are accounted for, to meet a community's energy demand is

$$E = \gamma\,A\,h\,H\,\varepsilon_p\,\varepsilon_{tp} \qquad\qquad (2.25)$$

where $\gamma = 9{,}807$ N/m$^3$ is the specific weight (density multiplied by gravity) of water, A is the surface area (m$^2$) of the upper reservoir, h is the water depth (m) of the upper reservoir, H is the elevation difference (m) between the upper and lower reservoir, $\varepsilon_p$ is the penstock efficiency, and $\varepsilon_{tp}$ is the combined turbine-generator-pump efficiency. A penstock efficiency below 1 accounts for frictional losses of energy in the penstock. A combined turbine-generator efficiency below 1 indicates that not all the water going through the turbine generates power. A pump efficiency less than 1 indicates that a portion of the gravitational energy in the water pumped up a hill is used for the pumping.

---

### Example 2.3  Calculating the size of a reservoir for pumped hydropower storage

Suppose the average electric power demand in a town is P = 100 kW. What surface area of an upper storage pond is needed for t = 2 days of storage if the average difference in elevation between the upper and lower reservoir is H = 250 m, the allowable change in depth of the upper reservoir is h = 2 m, the penstock efficiency is $\varepsilon_p = 0.9$, and the combined turbine, generator, and pump efficiency is $\varepsilon_{tp} = 0.85$? Note that 1 kWh = 3,600,000 N-m.

#### Solution:
Combining Equation 2.25 with the fact that E = power (P) × time (t) gives the area required as

$$A = P\,t\,/\,(\gamma\,h\,H\,\varepsilon_p\,\varepsilon_{tp})$$
$$= 100 \text{ kW} \times 48 \text{ h} \times 3{,}600{,}000 \text{ N-m/kWh} \,/\, (9{,}807 \text{ N/m}^3 \times 2 \text{ m} \times 250 \text{ m} \times 0.9 \times 0.85) = 4{,}607 \text{ m}^2$$

---

Aside from the energy stored behind hydropower dams, 97 percent of all energy stored for electricity use is currently in the form of pumped hydropower storage. Worldwide, about 530,000 potential pumped hydro sites exist. These sites can store an estimated 22 million gigawatt hours (GWh) of energy for the electricity grid. This represents 100 times the electricity needed to support a 100 percent worldwide electricity system (Blakers et al., 2019).

### 2.7.4 Stationary Batteries

Batteries in a 100 percent WWS world will be used for transportation, equipment (e.g., leaf blowers, lawnmowers), and stationary electric power storage. Batteries are useful because they generate electricity almost instantaneously on demand (see Table 2.1). As such, an electric vehicle accelerates more quickly than does an internal combustion engine vehicle. Similarly, batteries can provide needed electricity to the grid much faster than can gas, coal, nuclear, or even hydroelectric facilities (see Table 2.1). Stationary batteries can be used to provide short-term load-following power or peaking power for a large electric power grid or a small, isolated microgrid. Wall-mounted or floor-mounted battery packs are commercially available for use in a home, business, microgrid, or city electric grid.

A **battery** is an electrochemical cell that converts chemical energy into electricity. A **battery pack** is a collection of such cells. The electrochemical cell in a battery is referred to as a **galvanic cell** or, alternatively, a **voltaic cell**, named after **Luigi Galvani** (1737–1798) and **Alessandro Volta** (1745–1827), respectively. The cell consists of two half-cells. Each half-cell consists of an electrode, a current collector, and an electrolyte solution. A separator separates the half-cells.

The **electrode** in one half-cell is a positively charged terminal, called a **cathode**. The electrode in the other half-cell is a negatively charged terminal, called an **anode**. The electrodes don't touch each other. The electrolyte solution and separator separate the electrodes.

The **current collector** is a metal cap in each half-cell through which electrons move either forward or backward. The current collector at the anode is the negative current collector. That at the cathode is the positive current collector.

The **electrolyte solution** contains positively charged **cations** and negatively charged **anions**. The solution conducts the passage of lithium ions between the two half cells. A **separator** between the half-cells is permeable to cations, such as $Li^+$, thus allows them to move between the two half-cells, but it is impermeable to electrons, thus prevents them from moving between the half-cells. The movement of cations from one half-cell to the other creates a positive charge in one half-cell and a negative charge in the other.

In a **lithium-ion battery**, the anode is usually made from carbon (graphite). The cathode is usually made from a metal oxide (e.g., lithium cobalt oxide, lithium iron phosphate, or lithium manganese oxide). The electrolyte is usually a lithium salt (e.g., $LiPF_6$, $LiAsF_6$, $LiClO_4$, $LiBF_4$, or $LiCF_3SO_3$) dissolved in an organic solvent (e.g., ethylene carbonate or diethyl carbonate).

During battery charging, an external charger is connected to the positive and negative current collectors by a wire. A battery charger provides electrons to a battery. Electrons first flow from the charger to the negative current collector at the anode (Figure 2.23a). The buildup of negative charges at the anode induces a flow of positively charged lithium ions ($Li^+$) from the cathode through the separator, to meet the electrons, neutralizing charge at the anode. The lithium ions originate from, for example, lithium cobalt oxide ($LiCoO_2$) in the cathode, which separates into $Li^+$, an electron, and $CoO_2$.

Electrons left alone by the lithium ions at the cathode can't pass through the separator. So, they take the path of least resistance by flowing from the positive current collector at the cathode, through the wire, to the charging source. There, they join electrons added by the charging source and continue by wire to the negative current collector at the anode, completing the circuit. At the anode, the electrons recombine with lithium ions and anode material (e.g., graphite, $C_6$) to form $LiC_6$, neutralizing the positive charges of the lithium cations. When no more lithium cations flow from the cathode, through the separator, to the anode, the battery is fully charged (Figure 2.23b).

During battery discharging to the power electronic device, the device draws electrons from the negative current collector at the anode through a wire to the device. The device consumes some of the electrons (Figure 2.24a). Remaining electrons continue to flow past the device, through a wire, to the positive current collector at the cathode. The negative charge buildup at the cathode due to the buildup of electrons there induces positively charged lithium ions to flow from the anode,

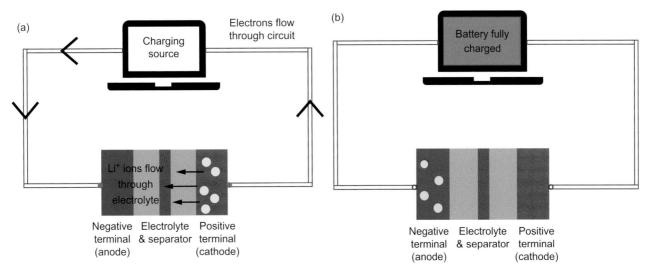

Figure 2.23 **Illustration of battery charging, as described in the text.** Adapted from Woodford (2019).

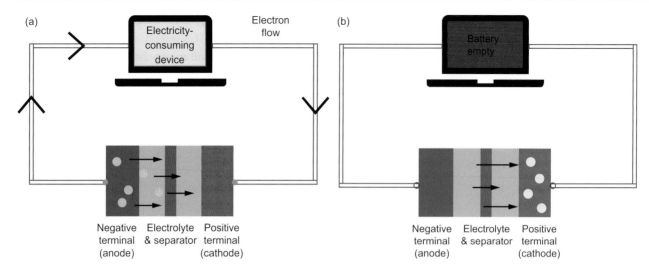

Figure 2.24 **Illustration of battery discharging, as described in the text.** Adapted from Woodford (2019).

through the separator, to the cathode, where their charge is neutralized by the abundance of electrons there. When all ions have moved from the anode to the cathode, the battery is depleted and needs recharging (Figure 2.24b).

Three parameters important for determining the use of batteries in different applications are their **energy density** (Wh-electricity output/kg-battery), **power density** (W-power charge or discharge rate/kg-battery), and **round-trip efficiency** (ratio of electricity delivered to energy put into the battery).

Energy density is a measure of the total amount of energy that can be drawn. Power density is a measure of how fast the battery can be charged or discharged. The range of energy densities for modern lithium-ion batteries in 2019 was 250 to 305 Wh/kg (Kane, 2019). The range of power densities is 250 to 350 W/kg. Battery round-trip efficiencies are 80 to 90 percent.

Another characteristic of a lithium-ion battery is the number of times it can **cycle** (fully discharge plus charge) before the battery is no longer useful. A typical car battery goes through 500 to 1,000 cycles before needing to be replaced.

An issue related to lithium-ion batteries is whether sufficient resource exists in the world to provide enough lithium to power all the electric vehicles, equipment, and stationary batteries needed for a 100 percent WWS world. Table 2.2 gives an estimate of known lithium resources worldwide.

---

### Example 2.4 Estimating the maximum possible number of lithium-ion battery-electric vehicles worldwide

Estimate the number of battery-electric vehicles that can be built worldwide with lithium-ion batteries given the known world resources of lithium from Table 2.2. Assume a typical power density of lithium itself within a battery is 6.67 kWh/kg-Li and a typical vehicle (e.g., Tesla Model S) stores 70 kWh.

#### Solution:

Dividing 70 kWh by 6.67 kWh/kg-Li gives 10.5 kg as the mass of lithium required per vehicle. With a known resource of 53.35 Tg-Li from Table 2.2, the total number of battery-electric vehicles possible worldwide is 5.08 billion. In 2018, the world had ~1.1 billion ICE vehicles.

---

Example 2.4 illustrates that plenty of lithium resource exists to provide power for over 5 billion electric vehicles of substantial range if lithium batteries were not needed for any other purpose. Currently the world has only 22 percent of this number of vehicles. It is also possible that more resources will be discovered in the next decade or two. However, because batteries will also be used for stationary electricity storage and for some equipment in a

Table 2.2 **Known lithium resources available worldwide.**

| Country | Li resource (Tg) |
|---|---|
| Argentina | 9.8 |
| Bolivia | 9.0 |
| Chile | 8.4 |
| China | 7.0 |
| United States | 6.8 |
| Australia | 5.0 |
| Canada | 1.9 |
| D.R. Congo | 1.0 |
| Russia | 1.0 |
| Serbia | 1.0 |
| Czech Republic | 0.84 |
| Zimbabwe | 0.5 |
| Spain | 0.4 |
| Mali | 0.2 |
| Brazil | 0.18 |
| Mexico | 0.18 |
| Portugal | 0.1 |
| Austria | 0.05 |
| World land | 53.35 |
| Oceans | 240 |

Tg = teragram (1 Tg = $10^{12}$ g)
Source: USGS (2018b).

100 percent WWS world and because international conflict may make it difficult to obtain lithium from some countries, it is possible that lithium shortages could arise. For that reason, the future world infrastructure may also need other types of batteries.

Different types of batteries and other variations of lithium batteries have been developed. These are discussed next.

**Sodium sulfur batteries.** Sodium sulfur batteries contain a liquid solution of sodium and sulfur. They have high energy densities, high charge and discharge efficiencies, and long cycle lives. Because they are made of inexpensive material, they are increasingly economical at large sizes, which makes them ideal for stationary electricity storage. However, they can also be used for vehicles. In fact, one of the first applications of the sodium sulfur battery was in the 1991 Ford Ecostar demonstration electric vehicle, which never went into commercial production. Sodium sulfur batteries operate at high temperatures (300 to 350 °C). Substantial research is currently being carried out to allow the batteries to be used at room temperature (Boukhalf and Kaul, 2019).

**Aluminum-ion batteries.** These batteries use aluminum instead of lithium to flow from the cathode to the anode. An aluminum ion, $Al^{3+}$, carried three times the charge of a lithium ion, $Li^+$. This allows aluminum-ion batteries to be smaller than lithium-ion batteries. However, the higher charge also results in greater interactions with other chemicals in the battery, reducing efficiency. Aluminum-ion batteries are less flammable than lithium-ion batteries.

**Saltwater batteries.** In these batteries, a concentrated saline solution (e.g., saltwater) is used as the electrolyte to conduct ions. The voltages are one-third those in lithium-ion batteries, but these batteries are suitable for stationary power storage and can be run for 10,000 cycles or more, versus 500 to 1,000 cycles for a lithium-ion battery. In addition, saltwater batteries are nonflammable and have no toxic chemicals, so they are easily recycled.

**Graphite dual ion batteries.** As their name implies, both the cation and anion in these batteries are involved in the battery reaction. The batteries use a graphite cathode and potassium anode as one option. The electrolyte may be a stable organic or an ionic liquid without lithium. Graphite dual ion batteries are used for stationary electricity storage.

**Flow (redox) batteries.** These batteries convert chemical energy directly to electricity. They are much larger than a lithium-ion battery (because of their low energy density and power density). As such, they are used only for stationary electric power storage, not for transportation or movable equipment. An advantage of a flow battery is that it lasts for more charge/discharge cycles and can operate at a higher current and power density than does a lithium-ion battery. The most common type of flow battery is the **vanadium redox flow battery**, which uses vanadium at both electrodes.

**Silicon-based lithium batteries.** These are lithium batteries that use silicon instead of graphite in the anode. Silicon anodes can bind lithium 25 times more strongly than can graphite anodes. However, silicon-based anodes alone have low electrical conductivity, low diffusion rates, and large volume fluctuations. These problems may be overcome by combining silicon with carbon in the anode.

## 2.7.5 Flywheels

A **flywheel** is a spinning wheel or disk, usually made of steel or carbon fiber, that rotates around an axis. The axis is perpendicular to the ground, like with a spinning top, so that gravity acts symmetrically about the spinning wheel.

A flywheel stores energy as rotational kinetic energy and later converts that energy to electricity. A flywheel is an electric motor, an energy storage device, and a generator, all at the same time. When excess electricity is available, it powers an electric motor to rotate the flywheel up to a high speed. Of the energy added to a flywheel, a small portion is used to keep the flywheel rotating. The rest is stored energy. If the flywheel is run in a vacuum, air resistance is zero, so frictional losses are minimized. An alternate way to minimize frictional losses is to use an electromagnetic bearing or permanent magnet. This allows the spinning rotor to float.

When electricity is needed, a flywheel is turned into a generator, which converts rotational kinetic energy to electricity, with built-in electronics. When electricity is produced, the flywheel slows down but does not stop.

A flywheel can store more and more rotational energy until its rotor shatters. Steel flywheels are limited to around 3,000 rotations per minute (rpm). High-energy-density carbon fiber flywheels can rotate up to 60,000 rpm. Aside from possible breakage, flywheels require little maintenance and have a long lifetime (about 20 years) with no impact on the environment past their use of materials to build them.

A flywheel accumulates and stores kinetic energy produced by electricity intermittently over any period of time and releases energy as electricity at a fast rate over a short period. As such, flywheels are ideal for storing excess electricity from intermittent solar and wind energy on the electric power grid. However, flywheels developed to date do not store large amounts of energy (e.g., 3 to 25 kWh) and they have relatively high loss rates (e.g., 3 percent per hour), so the stored electricity must be used quickly. On the other hand, it can begin to discharge quickly (e.g., within 4 milliseconds) and discharge at a high rate (e.g., 10 to 100 kW). Thus, two useful applications of a flywheel are to provide short-term peaking power for the electric grid and to charge electric vehicles quickly (e.g., Chakratek, 2019). For example, a flywheel that stores 25 kWh and has a discharge rate of 100 kW can add 20 kWh to an electric vehicle in 12 minutes.

## 2.7.6 Compressed Air Energy Storage

Using **compressed air energy storage (CAES)** is another way to accumulate intermittent renewable electricity in storage over a long time, then to re-supply that electricity to the grid in short but powerful spurts when needed. With CAES, excess intermittent electricity is used to compress air. When electricity is needed, the compressed air is allowed to expand in a turbine. The turbine's rotating shaft is connected to a generator, which converts the rotational energy to electricity.

The storage location for CAES can be an underground cavern, a salt dome, an aquifer, or a closed vessel. Whereas CAES has been studied extensively, only a few large-scale CAES facilities have been built worldwide, including one in Germany and two in the United States (Alabama and Texas). The locations for large-scale underground CAES are limited geographically.

However, a small CAES system can be connected to an electricity-producing device, such as a wind turbine. The wind turbine operates normally to produce electricity in a generator. Electricity from the generator at hub height is then used to power a motor that compresses air that is stored in a storage tank. When electricity is needed for the grid, the compressed air is expanded and converted back to electricity in an expander-driven generator. The motor, storage vessel, and second generator are all housed at ground level, below the turbine (Crossley, 2018).

### Transition highlight

Another variation of CAES is to use off-peak grid electricity or excess renewable electricity to compress and cool air until it condenses as a liquid. The high-pressure, cold liquid is then stored in a container until electricity is needed. At that point, the air is warmed until it re-evaporates. The expanding air then drives a turbine to generate electricity. The process is made efficient by storing the heat released during condensation and using that heat to help re-evaporate the air. The process does not require special materials, such as lithium used in batteries. A 50-MW storage facility based on this concept, which was developed by inventor Peter Dearman, is being built in the north of England (Harrabin, 2019).

## 2.7.7 Gravitational Storage with Solid Masses

A form of electricity storage similar to pumped hydro-power storage is **gravitational storage with solid masses**. In one version of this storage type, excess electricity from the grid is used to power an electric motor in a crane to lift cement blocks against the force of gravity and stack them, one at a time, on a tower. When electricity is needed, the crane grabs each block and slowly drops the block toward the surface. The downward motion uncoils the hoist chain holding the block. The rotating motion during the uncoiling is translated to a rotating motion inside the same electric motor that lifted the block, turning the motor into an electric generator (e.g., Allain, 2018). The electricity produced by the generator is sent to the grid. Using a motor in reverse as a generator to produce electricity is similar to using a motor in a vehicle to produce electricity during regenerative braking (Section 2.2.1).

### Transition highlight

One company that is building a gravitational storage system will use a large crane with multiple arms so that multiple blocks can be lifted or dropped simultaneously (Hanley, 2018). They state that the efficiencies for charging and discharging storage are each 90.5 percent, resulting in a round-trip efficiency (electricity consumed per unit electricity input) of 82 percent. The one-way efficiencies break down into a crane/pulley efficiency of 96 percent, a gear efficiency of 98 percent, a motor/generator efficiency of 98.2 percent, a variable frequency drive efficiency of 98.7 percent, and a transformer efficiency of 99.3 percent (Gross on Twitter, 2019). So long as the crane is holding a block, electricity can be produced almost instantaneously when needed (Table 2.1). The company suggests that the cost should be less than the cost of batteries (Hanley, 2018).

In another version of gravitational storage, excess grid electricity is used in an electric motor to move a train carrying rocks or concrete blocks up a hill. When electricity is needed, the train rolls down the hill, producing electricity in a generator almost instantaneously. The company producing this storage technology says that the round-trip efficiency is about 86 percent, similar to that from moving mass vertically (Roberts, 2016).

In a third version, excess grid electricity is run through a motor to lift sand or gravel in a container up a mountain slope via a cable. Containers of sand or gravel are stored at the top of the mountain. When electricity is needed, the containers are lowered down the slope via the cable. The rotating motion of the coil at the top of the mountain produces electricity in the generator (which is the motor used to lift the containers, run in reverse). This type of storage is referred to as **mountain gravity energy storage** (Hunt et al., 2020). It can store electricity for use on the timescale of seconds, minutes, days, weeks, or months. It is most efficient in locations with tall, steep mountain slopes.

## 2.8 WWS Heat, Cold, and Hydrogen Storage Technologies

In a 100 percent WWS world, low-temperature heat storage, cold storage, and hydrogen storage are needed along with electricity storage. Whereas most heat for air and water heating in buildings and most cold for air conditioning will be obtained directly from heat pumps, which run on electricity, some heat and cold will also come from hot and cold storage through domestic hot water heaters or district heating and cooling systems. The rest will come from solar or geothermal direct heat. In a 100 percent WWS world, excess electricity will be converted to heat, cold, and hydrogen, all of which will be stored or used immediately. Excess electricity occurs if current electricity demand has been met and electricity storage is already full.

This section discusses heat, cold, and hydrogen storage. It first examines short-term heat and cold storage in water tanks and their application to district heating. It then moves on to analyze seasonal heat and cold underground storage in boreholes, water pits, and aquifers. It subsequently summarizes heat and cold storage in building materials and cold storage in ice. Hydrogen storage is discussed last.

### 2.8.1 Heat and Cold Storage in Water Tanks

**Tank thermal energy storage (TTES)** is the most common type of heat and cold storage worldwide. It involves heating or chilling water as it sits in a storage tank. Water tanks are used primarily as part of small or large district heating and/or cooling systems. Domestic hot water tank heaters for individual homes or buildings are also a form of tank storage, but such tanks are much smaller than are those used for district heating.

District heating tanks are made of concrete, steel, or fiber-reinforced plastic. Concrete tanks usually contain an interior polymer or stainless steel liner to minimize water vapor and heat diffusion out of the tank. They are also insulated on the outside. Steel tanks are generally insulated as well.

Hot water tanks are also called **boilers**. The stored hot water is used to heat building air or as potable hot tap water for showering, bathing, cooking, hand washing, and drinking. Cold water tanks are also called **chillers**. The cold water stored in a chiller is used for air conditioning and, sometimes, refrigeration.

Although community-scale boilers and chillers can be large, they are generally used to store heat and cold for only days to weeks. Underground thermal energy storage (Section 2.8.3) is less costly per unit energy; thus, it is now the main type of seasonal (between summer and winter) heat and cold storage. However, an advantage of above-ground boilers and chillers is that they can be built almost anywhere, whereas borehole and aquifer underground storage can be built only where soil and/or groundwater conditions are sufficient.

Water stored in tanks can be heated directly with solar or geothermal heat or with heat from a heat pump powered by electricity. Similarly, water in a tank can be cooled with a heat pump. Ideally, electricity used to heat water is excess electricity that is not needed on the grid. Using excess electricity to produce heat is an ideal way of decreasing the cost of electricity because it reduces the **curtailment** (**shedding** or wasting) of excess renewable electricity.

Using heat pumps in conjunction with boilers and chillers also reduces winter energy demand by 68 to 81 percent because heat pumps have a much higher coefficient of performance (COP) than do gas heaters or electric resistance heaters (Section 2.3). Thus, boilers and chillers powered by heat pumps not only help to lower the cost of stabilizing the electric power grid by absorbing excess electricity instead of wasting it; they also reduce energy requirements for heating.

Hot water stored in a water tank heats building air when the water passes through a radiator. **Radiators** are

heat exchangers. **Heat exchangers** are devices that allow a heated liquid or gas to transfer its heat to another liquid or gas without the two fluids mixing together or otherwise coming into direct contact. The transfer occurs by conduction, where heat energy is passed from one molecule to the next in the metal or other material that separates the two fluids. In the case of a radiator, water flows through pipes, often with fins to increase the surface area exposed to the air. Heat from the water in a radiator conducts through the metal in the radiator to the air, where convection by heated air molecules carries the heat through the room. Radiators can be placed in or against a wall, under a floor (**radiant floor heating**), or in a ceiling. They may or may not be accompanied by a fan. After hot water passes through a radiator, it is piped to subsequent radiators until it returns, now cold, to the original source of the hot water for reheating.

Similarly, cold water piped to a building is used to cool air in the building when the water passes through a heat exchanger. The heat exchanger moves heat from the warm air in the building to the cool water in the pipe, which warms up. The warm water is then circulated back to the chiller, where it is cooled down again, ideally with a heat pump.

## 2.8.2 District Heating Systems

**District heating** is the name given to a heating system whereby hot water in centralized boilers is distributed in a closed loop through insulated pipes to multiple residential or commercial buildings for either air heating (through radiators), potable water heating, or both. Once the heat is distributed, the cooler water is returned to the centralized boilers for reheating.

District heating originated in the United States. On March 3, 1882, the New York Steam Company began sending high-temperature (less than 200 °C) steam through pipes to heat buildings in Lower Manhattan. Burning coal to boil water created the steam. Heat losses from coal combustion and through poorly insulated and leaky pipes were significant, but coal was plentiful. Steam district heating still exists today in Manhattan, with Consolidated Edison providing steam to over 1,700 buildings. From 1882 to 1930, steam-based district heating using coal combustion (**first-generation district heating**) was common in many cities in Europe and the United States.

From 1930 to 1970, **second-generation district heating** emerged. This system was based on burning coal and oil to heat water and sending the hot water through pressurized pipes to buildings. Water temperatures exceeded 100 °C and system losses were less than with steam, partly due to the use of concrete ducts around pipes. These systems were installed worldwide but particularly in Eastern Europe after World War II.

**Third-generation district heating** evolved from 1970 to the present. It was developed and used first in Scandinavian countries. Today, 50 to 65 percent of building air and potable water heat in Denmark, Iceland, Sweden, Finland, Estonia, Latvia, Lithuania, Poland, Russia, and northern China is from district heating. About 12.5 percent of all heat delivered in the European Union and about 8 percent delivered worldwide is district heating (Werner, 2017).

Third-generation district heating uses prefabricated, insulated water pipes buried underground to reduce heat loss. The insulation allows water temperatures to stay below 100 °C, which in turn reduces the energy needed to heat the water. The heat for water in these systems was originally provided by a combination of coal, natural gas, and biomass combustion and by recycling the heat of combustion from electricity generated by these fuels. More recently, heat has also been recycled from data centers and created from excess electricity produced by wind, solar, and geothermal plants. In some district heating systems, the heat is stored underground (Section 2.8.3) instead of in aboveground boilers.

**Fourth-generation district heating** emerged in 2015 and is currently evolving. It combines district cooling with district heating. District cooling involves a centralized chiller that stores water that has been cooled. As with district heating, the cold water is sent through an insulated piping network to buildings to cool air in the buildings. With fourth-generation district heating, heat pumps are used for heating water in boilers and cooling water in chillers. The heat pumps, which run on electricity, extract heat or cold from the air, water, or the ground. If waste heat or cold from buildings, data centers, or manufacturing processes is the source of heat for each heat pump, the heat pump requires less electricity and thus runs more efficiently. Ideally, in a fourth-generation

system, all electricity for a heat pump is obtained from 100 percent WWS, and the heat pump uses hot and cold water circulated from buildings to help heat or chill water in the boiler or chiller, respectively.

An example of a fourth-generation district heating system is the Stanford Energy System Innovations (SESI) project (Stagner, 2016, Section 2.8.7). The round-trip efficiency of the heating portion of this system (ratio of the energy returned as heating after storage to the energy in the electricity used to heat the water for storage) is around 83 percent. That of the cooling portion is around 84.7 percent (Stagner, 2017). Hot water temperatures in a fourth-generation system are below 70 °C to minimize losses.

In a 100 percent WWS world, fourth-generation district heating and cooling would be used as much as possible in densely populated areas, such as cities, where such systems are most advantageous. District heating can also be used on college campuses, military bases, and remote communities. Rural and many suburban homes and buildings are less ideal for district heating and cooling due to the length of trenches and pipes required and pipe losses. Homes and buildings that are not on district heating and cooling loops will have their own domestic hot water tanks and use heat pumps for all building and water heating and air conditioning.

## 2.8.3 Underground Thermal Energy Storage

Whereas boilers are useful forms of district or home heat storage for periods of days to weeks, the winter heat for boilers can be obtained from larger heat storage reservoirs filled primarily during summer. Such seasonal heat storage takes the form of **underground thermal energy storage (UTES)**. Three main types of UTES have been developed: borehole, pit, and aquifer thermal energy storage. Table 2.3 summarizes some of the similarities and differences among these UTES systems and tank thermal energy storage (TTES) (Section 2.8.1). The different UTES options are discussed, next, in turn.

### 2.8.3.1 Borehole Thermal Energy Storage

A **borehole thermal energy storage (BTES)** system is effectively a large, underground heat exchanger that stores solar heat collected in summer for later use during

Table 2.3 **Features of four different types of thermal energy storage systems.**

| Parameter | TTES | BTES | PTES | ATES |
|---|---|---|---|---|
| Storage medium | Water | Soil/rock | Water; gravel/water | Sand/water; Rock/water |
| Heat capacity $(kWh/m^3)$ | 60–80 | 15–30 | 60–80; 30–50 | 30–40 |
| Storage temperature $(^\circ C)$ | 5–95 | –5 to 95 | 5–95 | Shallow: 2–20; Deep: 2–80 |
| Depth (m) | 5–15 | 30–100 | 5–15 | 20–100 |
| Storage volume $(m^3)$ for 1 $m^3$ water equivalent | 1 | 3–5 | 1; 1.3–2 | 2–3 |

Source: IEA (2018a).
TTES is tank thermal energy storage, BTES is borehole thermal energy storage; PTES is pit thermal energy storage; and ATES is aquifer thermal energy storage.

winter. It consists of an array of boreholes drilled into soil. A plastic pipe with a U-bend at the bottom is inserted down each borehole. The space around each pipe in each borehole is then filled with highly conducting grouting material to increase the molecule-to-molecule conduction of heat from hot water that passes through the pipe to the soil surrounding the grouting material. After water in the pipe transfers its heat to the soil, the water is cold and sent back to solar collectors or heat pumps to collect more heat, particularly during summer. When hot water is needed during winter or another time, cold water is sent down the pipe; heat stored in the surrounding soil conducts through the grouting material to the water through the pipe wall and the hot water is sent to a boiler for subsequent distribution to buildings.

A good example of a BTES storage system is the **Drake Landing Solar Community** (Figure 2.25) in Okotoks, Alberta, Canada, which is one hour's drive south of Calgary. Starting in 2004, 52 homes were constructed. On the garage roof of each home, solar collectors were installed. The collectors contain a glycol solution (mix of water and non-toxic glycol) that absorbs solar heat, particularly during long summer days. The heated solution is transferred through underground, insulated pipes to a building (Figure 2.25), where the heat from the solution is transferred through a heat exchanger to water stored in a short-term hot water storage tank (boiler). The water temperature in the boiler is maintained at 40 to 50 °C. This tank is the connection source of all water and air heat for the 52 homes all year.

During summer and autumn, all home heating needs are met through water supplied to the boiler by the solar collectors without the need of heat stored in the nearby borehole field. In fact, the solar collectors produce a surplus of heat, and the excess heat is sent to the field to be stored until winter. The excess heat is piped by water to the

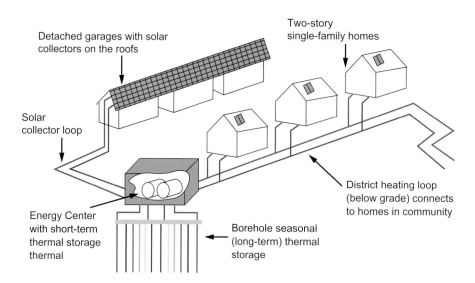

Figure 2.25 **Diagram of Drake Landing Solar Community district heating system.** Adapted from Drake Landing (2016).

borehole field, which is about 35 m in diameter. In this field, 144 holes were drilled to a 35-m depth, and each was filled with a pipe with a U-bend at the bottom. Insulation was placed on top, and a grassy field, used as a community park, was grown on top of that (Figure 2.26).

The hot water collected from the homes and sent to the boiler is then delivered through separate pipes to the borehole field. Each pipe containing hot water extends to the bottom of each borehole. The heat from the pipes is conducted to the surrounding soil, raising the soil temperature up to 80 °C by the end of summer. Each pipe returns upward with cooler water and is sent back to the solar collectors of each home to collect more heat.

During winter or other times of the year, when the hot water tank alone cannot satisfy all building heat demand, cold water is piped through the boreholes to collect heat

to bring back to the hot water tank. The hot water from the water tank is then distributed to the 52 homes for air and domestic water heating. With this system, up to 100 percent of winter heat is satisfied by summer heat collection.

The Drake Landing system has been operational since 2007. The efficiency of the UTES system, which is defined as the fraction of heated fluid entering underground storage that is ultimately returned during the year for air or water heat, is about 56 percent (Sibbitt et al., 2012).

In sum, Drake Landing is a district heating system with a community hot water tank but where much of the wintertime hot water is stored underground between summer and winter. The technology is repeatable for almost any district heating system and requires little land. The borehole field is not visible and can be used

(a)

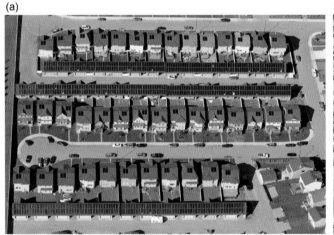

(b)

(c)

Figure 2.26 Homes of the Drake Landing Solar Community during (a) summer (September 9, 2007) and (b) winter (January 12, 2010). © Drake Landing. (c) Energy center housing the hot water tank and heat exchangers, and the borehole field covered with grass in 2015. © Mark Z. Jacobson.

simultaneously as grazing land, parkland, open space, or for the solar collectors used to provide heat to be stored in the field.

## Transition highlight

Another borehole thermal energy storage facility serving a district heating system resides in Braedstrup, Denmark. The Braedstrup district heating system services 1,500 customers. The borehole field, installed in 2012, is honeycomb shaped and has 48 boreholes, each 45 m deep and 3 m apart. U-bend pipes are used, just as in Drake Landing. The storage field, which has a volume of 19,000 m$^3$ for heating and storage capacity of 400 MWh, is heated up to 55 to 60 °C during summer and cooled down to 12 to 15 °C during winter, when heat is drawn from the field. The round-trip efficiency is about 63 percent, slightly higher than that of Drake Landing, and the investment cost is 0.65 euros/kWh ($0.72/kWh in 2020) (Sorensen and Schmidt, 2018). The boreholes, along with two additional steel water tanks totaling 7,700 m$^3$, are fed by 18,600 m$^2$ of solar collectors.

### 2.8.3.2 Pit Thermal Energy Storage

**Pit thermal energy storage (PTES)** consists of a lined pit dug into the ground and filled with a storage material, such as water or a mix of gravel and water, then covered with insulation and soil. The pit can be small or large. The material in the pit is supplied with heat primarily during the summer. The heat is extracted primarily during the winter and sent to a boiler, where it is distributed via district heating to homes for air and water heating.

As of 2018, the largest pit worldwide for PTES was in Vojens, Denmark, completed in 2015 (Figure 2.27). It is 13 m deep, 610 m in circumference, and filled with 200,000 m$^3$ of water. The pit is lined underneath and on the sides with welded plastic to ensure water does not leak through it. A few meters thick of soil outside the lining become warm, along with the water, insulating the storage pit. The surface of the water is also lined with a floating plastic cover topped with 60 cm of insulating expanding clay and a draining system to remove rainwater. The water temperature during summer can reach 95 °C, but is maintained at a maximum of 80 °C to extend the life of the liner (Ramboll, 2016).

Figure 2.27 Pit thermal energy storage system in Vojens, Denmark. The pit, empty in the figure, is on the left. When filled, it is then covered with floating plastic. The solar thermal collectors are to the right of the pit. © Ramboll.

There are 5,439 solar thermal panels (70,000 m²) that heat the water. The pit is connected to the district heating system of Vojens, which has 2,000 customers. The solar collectors plus PTES facility provide 45 to 50 percent, or 28 GWh/y, of the annual heat to the city. The peak heat discharge rate is 49 MW (Arcon/Sunmark, 2017).

---

### Example 2.5  Energy storage in the Vojens water pit

Estimate the maximum energy (GWh) that the Vojens water pit can store at a given time assuming the water temperature without storage is 15 °C and the water temperature with storage is (a) 80 °C, (b) 95 °C.

a. The specific heat of liquid water is 4,186 J/kg-K. Multiplying this by the density of liquid water, 1,000 kg/m³, the volume of water in the pit, 200,000 m³, the temperature difference with and without storage, 80 – 15 = 65 °C = 65 K and by 1W-s/J, $10^{-9}$ GW/W, and 1 h/3600 s gives 15.1 GWh maximum energy storage.
b. The calculation is the same, except the temperature difference is 80 K, resulting in a maximum storage of 22.1 GWh.

---

The cost of wintertime heat from the solar heating combined with this PTES interseasonal heat storage in Vojens is competitive, even without subsidy, with the cost of heat from gas boilers due to economies of scale. In fact, heating bills declined by 10 to 15 percent with the seasonal heat storage system (Ramboll, 2016).

### Transition highlight

A modified version of the Vojens PTES plant was built for the district heating system in Gram, Denmark, in 2015. Gram is a town with 2,500 residents, 99 percent of whom are connected to the district heating system. The storage pit in this case is 15 m deep, with 10 m below ground and 5 m above ground. A 5-m-tall, sloped dam was built to raise the height of the pit. The water volume in the pit is 122,000 m³. The facility also has 44,800 m² of solar collectors, which directly

and through PTES provide 61 percent, or 18.3 GWh/y, of the town's heating. The peak discharge rate of heat is 31 MW (Damkjaer, 2016).

Two other PTES storage systems coupled with solar collectors and district heating are in Marstal and Dronninglund, Denmark.

The Marstal system, which serves 2,200 residents, consists of 33,365 m² of solar collector, and 75,000 m³ of pit water storage with a thermal capacity of 6 GWh-th (gigawatt-hours-thermal; completed in 2012). It also has 2,100 m³ of steel tank water storage, a pit maximum charge and discharge rate of 10 MW, a maximum measured temperature of 88 °C, a minimum measured temperature of 17 °C, an efficiency of 52 percent, and an investment cost of 0.44 euros/kWh ($0.48/kWh in 2020) (Sorensen and Schmidt, 2018).

The Dronninglund system, which serves 3,300 residents, consists of 37,573 m² of solar collectors and 62,000 m³ of pit water storage. It began operating during March 2014. The storage has a capacity of 5.4 GWh-th, a maximum charge and discharge rate of 27 MW, a maximum measured temperature of 89 °C, a minimum measured temperature of 12 °C, an efficiency of 81 percent, and an investment cost of 0.43 euros/kWh ($0.47/kWh in 2020) (Sorensen and Schmidt, 2018).

#### 2.8.3.3 Aquifer Thermal Energy Storage

**Aquifer thermal energy storage** (ATES) is similar to PTES, except that the water used for storing heat is naturally occurring water in underground layers of groundwater or aquifers. Aquifers can also be used to store cold water for cooling applications. Aquifers contain permeable sand, gravel, sandstone, or limestone layers with high hydraulic conductivity (IEA, 2018a). Aquifers can be used for UTES if they are encapsulated above and below them by impervious layers and if natural groundwater flow is slow.

With ATES, two wells, or several pairs of wells, are drilled into an aquifer. During summer, cold water is extracted from one part of the well, heated with building heat (cooling the building), then returned to another part of the well, where it mixes with and heats that part of the well (Figure 2.28). During winter, warm water from the warm part of the aquifer is extracted, the heat is removed to heat the building, and the cold water is returned to the cold part of the aquifer (Figure 2.28).

Summer                                              Winter

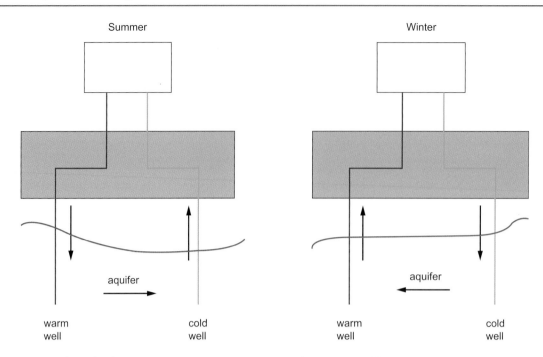

aquifer                                             aquifer

warm              cold                   warm              cold
well              well                   well              well

Figure 2.28 Diagram of aquifer thermal energy storage system. During the summer, cold water is drawn up. Heat is transferred from building air to the water, warming the water and cooling the building air. The warm water is sent down to another part of the aquifer. During winter, heat is drawn from the warm part of the aquifer and released to heat the building air. The residual cold water is sent back down to the cold part of the aquifer and stored until summer. Adapted from Socaciu (2011).

Because aquifers cannot be insulated, heat storage at temperatures greater than 50 °C is efficient only for large, deep aquifers with volumes greater than 50,000 m$^3$ and a low surface-to-volume ratio (IEA, 2018a). As such, the maximum temperature in a shallow aquifer for heating is generally limited to 20 °C. For cooling, shallow aquifers are generally used.

Some ATES systems built to date include a 1.7-million-cubic-meter storage reservoir under Eindhoven University of Technology, the Netherlands, for 20 university buildings (2001); a 1-million-cubic-meter storage reservoir under Arlanda Airport in Stockholm for that airport (2011); and a 180,000-cubic-meter storage reservoir under the Riverlight Project in London for a new residential and mixed-use complex (2014).

The **Eindhoven ATES system** provides both direct cooling during summer and low-temperature heating during winter for heat pumps. Sixteen wells for cooling and 16 wells for heating are maintained. The aquifer lies between 28 and 80 m below the surface and has a natural temperature of 11.8 °C. The temperature in the aquifer varies from 6 to 16 °C during the year due to extraction or addition of heat. The maximum heating and cooling energy delivered is 15 to 30 GWh per year. The

maximum charge and discharge rate of the reservoir for both heating and cooling is 17 MW (IEA, 2018a).

The **Stockholm airport ATES system** is also used for both heating and cooling. During winter, heat stored in the aquifer from the summer is used to preheat air used for ventilation in the terminal buildings and to melt snow at the gates. The extraction of heat cools the aquifer during winter. As such, by summer, the aquifer water is cold. The cold water is extracted during summer and used to provide air conditioning for the terminals. The aquifer well lies 15 to 30 m below the surface. The maximum heating and cooling energy delivered is 20 GWh per year. The maximum charge and discharge rates of the reservoir for both heating and cooling are both 10 MW (IEA, 2018a). The temperature in the aquifer varies from 2 to 25 °C during the year due to extraction or addition of heat.

## Transition highlight

The **Riverlight Project ATES system** is along the river Thames in London. It serves 806 residential apartments in addition to several commercial businesses. The ATES system is coupled with heat pumps and provides cooling during the summer and heating during the winter. The system consists of four warm wells and four cold wells drilled into the London chalk aquifer. The well depths are about 100 m each. The thickness of the aquifer in this layer is about 25 m. The maximum heating and cooling energy delivered is 1.4 GWh per year. The maximum charge and discharge rate is 3.7 MW for cooling and 1.6 MW for heating (IEA, 2018a). The temperature in the aquifer varies from 8 to 24 °C during the year due to the extraction or addition of heat.

### 2.8.4 Passive Heating and Cooling in Buildings

The energy required to heat and cool buildings can be reduced significantly with **passive heating and cooling techniques**. These techniques include installing thermal mass, ventilated façades, window blinds, films on windows, and night ventilation (de Gracia and Cabeza, 2015). Each is discussed, in turn.

#### 2.8.4.1 Thermal Mass

**Thermal mass** is a building material used to absorb, store, and release heat in a building in such a way as to keep building temperature relatively constant during day and night. At least three types of thermal mass materials store energy. These include sensible-heat storage materials, latent heat storage materials, and thermochemical storage materials.

**Sensible-heat storage materials** are materials that do not change temperature much during day or night, thus helping to keep buildings near a constant temperature. Good sensible-heat storage materials are materials with a combination of a high volumetric heat capacity and a moderate thermal conductivity. **Volumetric heat capacity** is the product of specific heat capacity (or just specific heat) and density. **Specific heat** is the energy (J) required to increase the temperature of 1 kg of a substance 1 K (or 1 °C). Substances with a high volumetric heat capacity absorb solar radiation during the day without their temperature increasing much and release thermal-infrared (heat) radiation at night without their temperature decreasing much. On the other hand, substances with a low volumetric heat capacity warm rapidly during the day upon absorbing sunlight and cool rapidly at night upon releasing heat radiation.

The volumetric heat capacity of liquid water is 3.5 times that of loose sand (Table 2.4). As such, the same amount of sunlight heats up a given volume of loose sand 3.5 times as much as it does water. This is why ocean beach water is relatively cool during the day, whereas sand feels hot. Conversely, the loss of the same amount of heat radiation at night decreases the temperature of sand 3.5 times the temperature decrease of water.

**Conduction** is the transfer of energy from molecule to molecule. The **thermal conductivity** of a substance is a measure of its ability to conduct heat in the presence of a temperature gradient. Units of thermal conductivity are W/m-K, which is the same as J/m-K-s. A thermal conductivity of 1 W/m-K means that a 1-m-long material will transfer 1 J per second from one end of the material to the other if the temperature difference between one end and the other is 1 K. Steel has a thermal conductivity of 45 W/m-K, thus it is extremely conductive. Pinewood, on the other hand, has a conductivity of 0.15 W/m-K, which is hardly conductive.

A good thermal mass material should have a moderate thermal conductivity. If the conductivity is too high, such as with steel, the material will transfer its heat (or cold) too fast to the air or other materials, even if the material has a high volumetric heat capacity. If the conductivity is too low, such as with wood, the material won't conduct its heat (or cold) to the air when the heat or cold is needed.

Table 2.4 identifies materials that are **effective thermal mass materials**, namely those with high volumetric heat capacities and moderate thermal conductivities. Table 2.4 indicates that the use of water concrete, marble, granite, or brick in building walls or floors modulates the temperature of the building in comparison with using wood or steel. Another way to think of it is that these materials store heat during the day without raising the temperature of the building and slowly release this heat to the building air at night, keeping the building warm at night. Water can be used as a thermal mass material when it is placed in tubes in a wall or under a floor. A disadvantage of sensible-heat storage in comparison with, for example, phase change material storage, is that sensible-heat storage requires a larger volume of storage mass to have the same impact on building temperature as does phase change material storage.

**Latent heat storage materials** are materials that modulate building temperature by changing phase near room temperature. As such, they are also called **phase change materials** (PCMs). PCMs used in buildings

Table 2.4 **Specific heats, densities, volumetric heat capacities, and thermal conductivities of several substances. The last column indicates the effectiveness of the substance as a thermal mass.**

| Substance | Specific heat (J/kg-K) | Density (kg/m³) | Volumetric heat capacity (kJ/m³-K) | Thermal conductivity (W/m-K) | Effectiveness of thermal mass materials |
|---|---|---|---|---|---|
| Water | 4,186 | 1,000 | 4,186 | 0.6 | High |
| Steel | 7,800 | 480 | 3,744 | 45 | Low |
| Dense concrete | 1,100 | 2,400 | 2,640 | 1.63 | High |
| Marble | 880 | 2,700 | 2,376 | 1.8 | High |
| Granite | 790 | 2,750 | 2,173 | 1.7 | High |
| Brick | 800 | 1,700 | 1,360 | 0.73 | High |
| Polyvinyl chloride (PVC) | 1,000 | 1,380 | 1,380 | 0.19 | Low |
| Oak | 2,390 | 545 | 1,303 | 0.16 | Low |
| Loose sand | 830 | 1,442 | 1,197 | 0.2 | Low |
| Pine | 2,300 | 400 | 920 | 0.15 | Low |

include paraffin wax, fatty acids, and salt hydrates. Paraffin wax consists of hydrocarbon molecules with the chemical formula $C_nH_{2n+2}$, where n is the number of carbon atoms, typically numbering 20 to 40.

Much less mass of a phase change material than of a high volumetric heat capacity material is needed to modulate building temperature. When a PCM absorbs heat, its temperature quickly reaches its melting point, at which point the material absorbs more heat in order to melt without raising its temperature.

For example, if paraffin wax is distributed within south-facing walls of a building in the Northern Hemisphere, the wax absorbs heat, increasing its temperature until the melting point (37 to 65 °C, depending on which wax is used) is reached. During melting, the wax absorbs more heat, but that heat is used to melt the wax, so the wax temperature doesn't rise further until all the wax is melted. By absorbing heat without increasing in temperature, the wax keeps the building cool. After the sun goes down, the wax loses heat due to thermal-infrared radiation cooling. Once it cools to its freezing point (similar to its melting point), the melted wax freezes (solidifies), releasing latent heat to the air, keeping the air warm without the wax dropping its temperature further. Thus, PCMs keep the temperature of a wall relatively constant during the day and at night. A PCM can be encapsulated within a wallboard, mixed in concrete, or mixed with insulation material.

With **thermochemical heat storage**, a chemical absorbs heat (from sunlight) and decomposes into two chemical products, which store the heat. The two products are separated until the heat is needed again (e.g., at night), at which time the products are brought together to chemically react to reproduce the original chemical and release the heat back to the air. Thus, the reaction is reversible. Thermochemical heat storage has not been used much in buildings to date.

### 2.8.4.2 Ventilated Façades
A **ventilated façade** is a weather-protective wall, separated by air from the wall of a building. The air between the façade and the wall can flow, providing ventilation and exchange of the warm or cold cavity air with ambient air. The cavity provides acoustic as well as thermal benefits to the building. The façade can also be made of a thermal mass material to modulate temperatures received by the building wall.

### 2.8.4.3 Window Blinds
**Window blinds** are used to control sunlight into buildings. Lowering window blinds reduces direct and diffuse ultraviolet, visible, and solar-infrared radiation into a building during hot, sunny days. Raising the blinds increases direct and diffuse sunlight penetrating into the building during cloudy days.

#### 2.8.4.4 Window Film

Similar to window blinds, putting a transparent or translucent **window film** on a window reduces the penetration of UV, visible, and solar-infrared light into a building by increasing the window's solar reflectivity. Window film also reduces the outgoing thermal-infrared heat radiation from the building. An advantage of window film is that it can significantly reduce the penetration of UV radiation into a building, reducing damage to furniture and people's skin and eyes. During the winter, though, the reduction in solar radiation penetrating into a room may be greater than the reduced heat loss from the room, resulting in a cooler room than desired. However, in the annual average, film is usually beneficial in terms of heat and always beneficial in terms of reducing UV radiation.

#### 2.8.4.5 Night Ventilation

**Night ventilation** is the use of natural ventilation to remove heat from a building during the night. During the day, thermal mass in a building heats up due to absorption of solar and thermal-infrared radiation. During the night, thermal mass releases its heat to the building, keeping the building warm. However, in hot climates, building heat can accumulate during the day and night over periods of days to weeks, even with a thermal mass. To shed some of this heat, night ventilation is useful. Night ventilation takes advantage of either nighttime wind or thermally generated pressure to push heat added to the air inside the building out of a stack to the outside air. In all cases, cool outside air flushes the warm inside air. Since the inside air is now cooler, the thermal mass in the building cools down further as well. The next day, the thermal mass does not heat as much as it would with no nighttime ventilation.

### 2.8.5 Cold Storage in Ice

A type of thermal energy storage that is effectively a type of electricity storage is cold storage in ice. **Ice storage** has been used for decades in universities, hospitals, stadiums, and other large facilities. It involves freezing liquid water to produce ice when excess electricity is available or when electricity prices are low, then running water through coils in the ice when cooling is needed (Figure 2.29). Water passing through coils embedded in ice is chilled by conduction, and the cold water is piped to a building. In the building, the cold water is run through a radiator

Figure 2.29 Example of ice storage. Ice is formed when excess electricity is available or when the electricity price is low. When cooling is needed, such as during a hot afternoon, liquid water flows through the coils in the ice. The cold water is then piped to a building to cool that building, thereby reducing air conditioning electricity demand. As such, ice storage shifts the time of electricity use, just as a battery does. © Ice Energy (Ice Bear 40 energy storage copper ice coils).

to air condition the building. In this way, ice storage avoids the need for daytime electricity use for air conditioning. Since summer electricity demand peaks during the late afternoon in many places, ice storage is a method of reducing peak electricity demand in these places. The main advantage of ice storage over battery storage is the lower cost of ice storage. Ice storage costs \$35 to \$40/kWh-thermal storage. The round-trip efficiency (cooling energy output to electricity input) of ice storage is around 82.5 percent (IRENA, 2013).

### 2.8.6 Hydrogen Storage

In a 100 percent WWS world, hydrogen will be used primarily in fuel cells to provide electricity to run an electric motor in transportation vehicles (Section 2.2.2). The modes of transport that will benefit most from hydrogen use will be long-distance, heavy transport – namely, long-distance ships, aircraft, trucks, and trains. Electrolysis will produce the hydrogen for fuel cells. Electrolysis couples electricity with water in an electrolyzer to split the water into hydrogen plus oxygen (Section 2.2.2.1). In a WWS world, all electricity will be produced with WWS sources.

Hydrogen for fuel cell vehicles will either be used immediately or stored in a storage tank. An advantage of using hydrogen in a 100 percent WWS economy is that

WWS electric power generators can produce hydrogen when excess WWS electricity is available on the grid. Currently, excess WWS electricity is curtailed or shed, resulting in wasted electricity and a higher cost of WWS generators than necessary. If excess WWS electricity is used to produce hydrogen, and if the hydrogen is stored, the cost of WWS generation will be lower than if the WWS electricity is shed. In a similar manner, excess WWS electricity can power heat pumps to produce heat or cold for storage and later use. As such, electricity, heat, cold, and hydrogen can work together to help power a 100 percent WWS economy at low cost.

At 1 bar of pressure and 0 °C, the density of hydrogen gas is 0.08988 kg/m$^3$. For comparison, gasoline's density at 1 bar is 750 kg/m$^3$. Since hydrogen gas has a low density, it requires a large storage volume. Alternatively, hydrogen can be compressed into a small volume or its temperature can be dropped until it liquefies.

Compressed hydrogen is usually used in hydrogen fuel cell ground vehicles or stored in tanks for later use. Liquefied (**cryogenic**) hydrogen is usually used in aircraft, rockets, and space shuttles to minimize storage volumes. However, cryogenic hydrogen can be used in ground vehicles as well.

For ground transport, hydrogen is usually compressed to 350 bars or 700 bars. At 350 bars and 15 °C, hydrogen's density is about 26 kg/m$^3$. At 700 bars and 15 °C, hydrogen's density is about 40 kg/m$^3$. Hydrogen storage containers for use in vehicles at these pressures are made of either low-alloy steel or a composite material or both. Long-term exposure to hydrogen can make the steel brittle, which is why composites are useful. Stationary compressed hydrogen storage containers are often made of these materials, but in some cases, they are made of both stainless steel and carbon steel encapsulated within pre-stressed concrete (e.g., Feng, 2018).

Liquefied hydrogen is obtained by reducing hydrogen's temperature below its boiling point, which is – 252.882 °C (20.268 K) at an ambient pressure of 1 bar. At 1 bar of pressure, liquefied hydrogen's density is 70.9 kg/m$^3$, thus higher than compressed hydrogen's density. Cryogenic hydrogen is stored in cryogenic tanks below the boiling point temperature of hydrogen. If the tank is closed, even a small increase in temperature will evaporate hydrogen, building up the pressure from 1 bar to 10$^4$ bars. As such, cryogenic hydrogen tanks require additional insulation to prevent hydrogen from evaporating (boiling).

About 12.8 percent of the energy stored in cryogenic hydrogen (15.3 MJ/kg out of 119.96 MJ/kg) is needed to cool the hydrogen enough to liquefy it. Cryogenic hydrogen has one-fourth the energy density as jet fuel. But, because the overall energy conversion efficiency of jet fuel is 0.25 to 0.4, whereas that of a fuel cell system is 0.34 to 0.56, the volume ratio of hydrogen fuel to jet fuel in an aircraft is only 1.8 to 4.8 (see Example 2.2). As such, a cryogenic hydrogen storage tank needs to be more voluminous than does a jet fuel storage tank.

## 2.8.7 Stanford University 100 Percent Renewable Electricity, Heat, and Cold Energy System

Stanford University in California has developed a district heating and cooling system coupled with solar PV. The university purchased most of the PV at off-campus locations through power purchase agreements and installed the rest of the PV on campus buildings. This combined system will provide 100 percent of the campus annual electricity demand and almost 100 percent of the heating and cooling demands with renewable energy by 2021. The electricity, heat, and cold demands were about 208 GWh/y, 158 GWh/y, and 225 GWh/y, respectively, in 2015. For an average California home that uses 7 MWh/y of electricity, this corresponds to the demand from roughly 32,000 homes.

The **Stanford Energy System Innovations** (SESI) project (Figure 2.30) is a fourth-generation district heating and cooling system (Section 2.8.2). It was first conceived in 2009 and became operational in March 2015, replacing a natural gas co-generation plant that supplied 80 percent of the electricity and heat for the university campus.

The SESI project consists of a 2.3-million-gallon insulated steel hot water tank, two 4.75-million-gallon insulated steel cold water tanks (Figure 2.30), three waste-heat-driven 2,500-tonne heat pumps (heat recovery chillers) that raise the temperature of the hot water tank and expel cold to the cold water tanks, three hot water generators (boilers) to provide additional heat to the hot water tank, four 3,000-tonne cold water generators (chillers) to provide additional cold to the cold water tanks, cooling towers (14,500 tonnes total) to cool water further for the cold water tanks and provide additional waste heat for the heat pumps, an elaborate 35.4-km hot water pipeline system made of pre-insulated welded steel, a 42.2-km welded steel and polyvinyl chloride (PVC) cold

Figure 2.30 **Stanford Energy System Innovations (SESI) project.** © Steve Proehl.

water pipeline system, a heat recovery and distribution system that takes advantage of the fact that different parts of the university have both heating and cooling needs at the same time, a planning model for optimizing the distribution of heating and cooling around campus every 15 minutes, and a high voltage substation that receives electricity from the grid (Stagner, 2016).

The heat recovery system collects waste heat from buildings and moves it to a hot water loop, where it is fed to the heat pumps. The recycling of waste heat eliminates the discharge of waste heat to the air.

The SESI project requires electricity, as do all buildings on campus. To provide this electricity, the university first built 5 MW of solar PV on university rooftops, then signed a power purchase agreement over 25 years to generate 67 MW of solar PV from a new PV farm in Rosamond, California, dedicated to the campus. The farm came online during December 2016. The university then signed another power purchase agreement for dedicated power from a 72-MW solar PV farm near Lemoore, California, which will be completed in 2021. The combination of the SESI project and the 144 MW of PV will result in 100 percent of the Stanford University campus's

annual electricity demand supplied by clean, renewable energy.

The 100 percent renewable electricity portfolio of the campus will reduce its overall greenhouse gas footprint by 80 percent compared with 1990 levels and its water use by 18 percent compared with 2014 due to eliminating water needed to cool the gas turbines. The remaining greenhouse gas emissions are from university vehicles, natural gas steam heating units in some buildings that have not yet been tied to the SESI hot and chilled water system, and emergency diesel generators. Plans exist to transition these remaining sources by 2025.

## 2.8.8 Electrified Home with Battery Storage and Heat Pumps

Part of the process of transitioning society to 100 percent clean, renewable energy is to transition individual buildings, including residences. In this section, a case study for a new-construction home, my own, completed in 2017, is provided.

The home is two stories, with approximately 278 m$^2$ (3,000 ft$^2$) of floor space and 46.5 m$^2$ (500 ft$^2$) of garage space. The structure itself was made of prefabricated steel, 80 percent of it recycled (Figure 2.31). The advantages of a prefabricated steel structure are that it eliminates wood waste on the property during assembly of the structure, it eliminates mold and termites associated with the structure itself (the exterior of the house and floors are still wood), and it produces walls that are perfectly flat, with corners that are at exact 90° angles. The precise construction reduces the risk of air leaks and mistakes in constructing the rest of the house.

The home uses double-glazed windows that include a krypton-gas-filled cavity in lieu of a third glazing to suppress conduction and convection between glazes and to reduce 99.5 percent of incoming ultraviolet radiation. The low-conductivity fiberglass window frames have insulating foam within them and triple weather stripping around them. The R-values of the windows, including the frames, are up to 1.04 $m^2$-K/W (5.9 $ft^2$-°F-h/BTU), and the U-factors are down to 0.962 W/$m^2$-K (0.17 BTU/h-$ft^2$-°F).

An **R-value** measures how well a window resists conduction per unit area. The greater the R-value is, the greater the resistance will be, and the better the insulation will be. The temperature difference between the two sides of a window multiplied by the surface area of the window plus its frame, all divided by the R-value, gives the heat flow rate (W or BTU/h) through the window. For walls, R-values of different components in the wall can be added to obtain a total R-value for the wall.

A **U-factor** is the thermal conductivity of heat through a window assembly. In the case of windows, it is exactly the inverse of the R-value and has units of thermal conductivity (W/$m^2$-K or BTU/h-$ft^2$-°F). The lower the U-factor is, the less conductive the window will be and the greater the resistance of the material against heat loss will be. Good double-pane windows have a U-factor of around 1.70 W/$m^2$-K (0.3 BTU/h-$ft^2$-°F). Triple-paned windows have U-factors closer to 0.849 W/$m^2$-K (0.15 BTU/h-$ft^2$-°F). As such, the windows in this house are insulated to a similar degree as triple-paned windows.

No gas pipes are connected to the property. Instead, the house generates more than its annual average electric power consumption from 43 320-W roof-mounted panels, for a total nameplate capacity of 13.76 kW (Figure 2.31). However, each panel is rated at 298.7 W, so the expected peak output is 12.84 kW. The panels are mounted at a 15° tilt, facing south-by-southwest, in four rows, with at least 0.914 m (3 ft) between rows. The house is also connected to the electric power grid because, although the house produces about 120 percent of its

(a)

(b)

Figure 2.31 (a) Prefabricated, 80 percent recycled steel structure, from Bone Structure, of an all-electric, energy-efficient home. © Mark Z. Jacobson. (b) Solar PV panel array on finished home. © Bone Structure.

annual electricity consumption, including that required for charging electric cars, enough solar PV electricity is not always available at the exact times it is needed to meet instantaneous electricity demand.

In order to improve the matching of power demand with supply over time, the house has four wall-mounted batteries (Figure 2.32). Each battery can hold 6.4 kWh of energy for a total of 25.6 kWh. The peak charge and discharge rate of the batteries is 3.3 kW, for a total of 13.2 kW. Two batteries each are connected to each of two 7.6 kW inverters (15.2 kW for both inverters) that have a 97.5 percent efficiency. The solar PV system is also connected to the inverters. Although four batteries are available, the utility permitted only two batteries to operate. As such, the other two batteries, while still physically connected to an inverter, are idle, and the battery system really stores only 12.8 kWh of energy and charges and discharges at a maximum rate of 6.6 kW. During the year, up to three electric cars are charged with one of two electric car chargers (e.g., Figure 2.32).

---

### Example 2.6 Do more batteries reduce utility bills?

Consider two cases.

In Case A, Bob charges two batteries that store 6 kWh each with his solar system, from 6 AM to 9 AM. He discharges the two batteries for use in his home between 6 PM and 11 PM. He then pays for using 12 kWh of grid electricity from 11 PM until 6 AM the next morning.

In Case B, Mary charges four batteries that store 6 kWh each with her solar system, from 6 AM to 10 AM. She then discharges two of the batteries between 6 PM and 11 PM and the other two batteries between 11 PM and 6 AM.

Assume the cost of grid electricity is $0.3/kWh between 6 AM and 11 PM and $0.1/kWh between 11 PM and 6 AM. Also assume that any solar produced by either Bob or Mary can be sold to the utility at the same price the utility would charge for electricity at that time of day.

Who loses more money and by how much?

### Solution:

Bob's only payment to the utility is for the 12 kWh he used between 11 PM to 6 AM, and the total cost of that (at $0.1/kWh) was $1.20. If he had sold his morning solar electricity to the utility instead of charging his battery, he would have made 2 batteries × 6 kWh/battery × $0.3/kWh = $3.60 in the morning. However, he would have spent that money buying 12 kWh of electricity from the utility at $0.3/kWh from 6 PM to 11 PM.

Mary pays nothing to the utility in this case. However, by charging her batteries with 24 kWh when the rate is $0.3/kWh, she loses a payment from the utility of $7.20 in the morning. On the other hand, by using two batteries from 6 PM to 11 PM, she avoids paying 12 kWh × $0.3/kWh = $3.60; by using two from 11 PM to 6 AM, she avoids paying 12 kWh × $0.1/kWh = $1.20.

In sum, whereas Mary avoids paying $3.60 + $1.20 = $4.80 to the utility, she also avoids receiving $7.20, so loses a total of $2.40, whereas Bob paid only $1.20 to the utility. So, in this case, the use of two batteries cost Bob less than the use of four batteries cost Mary.

(a)                                                                    (b)

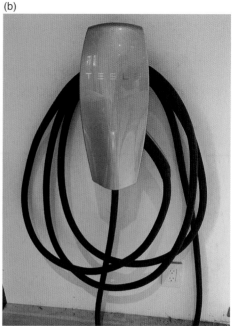

Figure 2.32 Four wall-mounted batteries, two inverters, and an electric car charger. © Mark Z. Jacobson.

## Example 2.7  Cost of grid electricity versus home battery storage

Assume a home battery starts full, stores 8 kWh of energy, and has a peak discharge rate of 3 kW. Assume also that, if either limit is exceeded, grid electricity, at $0.2/kWh, is used to provide the remaining energy or power. Suppose Sue charges her electric car, which draws 10 kW, for 2 hours. How much will she need to pay the utility for grid electricity and how much did she save during the car charging from using her battery?

### Solution:

The total energy required to charge Sue's car is 10 kW × 2 h = 20 kWh. Sue can draw no more than 3 kW from her battery. She does this for 2 hours, drawing a total of 6 kWh from her battery. Thus, the total drawn from the grid is 20 – 6 = 14 kWh. The cost of this, at $0.2/kWh, is $2.80. The cost savings of using her battery was 6 kWh × $0.2/kWh = $1.20.

Because the home has no gas, all heating, cooling, and cooking must be done with electric appliances. For air heating and air conditioning, ductless mini-split heat pump air heaters and air conditioners are used (Section 2.3). Nine indoor units are placed in different rooms or zones of the house, and two compressor/condensers are placed outside either to release heat to or extract heat from the outside air. The indoor units are connected to the outdoor units by pipes containing a refrigerant, thus the system is ductless.

For water heating, an air-source heat pump water heater is used (see Figure 2.15, Section 2.3). The source of the heat in this case is the mechanical room in which the water heater sits. When the heat pump is used, it cools the room slightly. Thus, in summer, the mechanical room door can be opened to provide additional cooling to the home. Hot water is circulated around the house through pipes controlled by a timer. The timer is manually set to eliminate its use during times of peak electricity cost (usually 2 to 9 PM) and to use electricity only sporadically during other times of the day.

For cooking, an electric induction cooktop is used (see Figure 2.19, Section 2.5.1). Induction cookers boil water in half the time as gas cookers, yet don't feel hot when

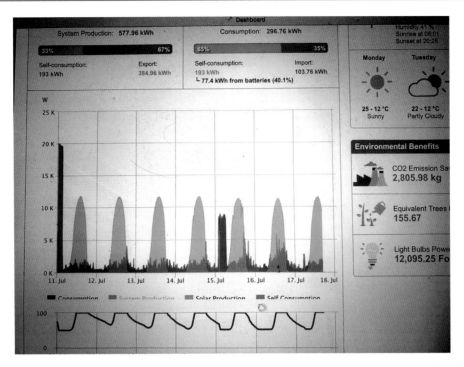

**Figure 2.33** One week of electricity use in an electrified home. The bottom graph shows the charge level of the four batteries. Since two batteries are inactive, the minimum charge level is 50 percent and the maximum level is 100 percent. Green is solar PV production; light blue is electricity used for the home directly from the solar PV, including for charging the batteries or electric cars; dark blue at night is for electricity used for the home from the batteries; and red is electricity used from the grid. The two spikes are for car charging. See the text for details. © Mark Z. Jacobson.

they are touched because they induce resistance heating in the pot rather than create heat themselves. As such, the cooking pot gets hot, but the stove stays warm, heating only by conduction from the pot.

Other major electric appliances in the house include a washer, drier, refrigerator, microwave oven, convection oven, toaster, garbage disposal, LED lights, televisions, computers, garage door, and an air filtration system.

Figure 2.33 shows one week of summer electricity use in the home. During the night, the battery is drained to supply any needed electricity in the home. Thus, by the time the sun rises, the battery level is usually low, but not empty. Because two batteries are inactive, the minimum charge level of the four batteries is 50 percent and the maximum is 100 percent (Figure 2.33). As the sun rises farther into the sky, the rooftop PV begins to generate electric power, which is consumed first by household electric needs. Excess electricity after that is used to charge the batteries at a maximum charge rate of 6.6 kW (since only two batteries are active). Excess electricity after that is sent back to the grid. As the day proceeds, the batteries fill up completely, then sit idle until after the PV can no longer sustain the household electricity demands.

At that point, the batteries automatically provide household power and continue doing so all night unless they are drained. When the batteries are drained, grid electricity is used (red in Figure 2.33). If the batteries do not drain during the night, then the house runs 24 hours per day on solar PV (days and nights with no red in Figure 2.33).

The electric cars are charged at night for cost reasons, as discussed shortly. The two electric cars charged during the week shown in Figure 2.33 both draw power quickly (either 20 kW or 8 kW) and draw a large amount of energy (up to a maximum 85 kWh and 53 kWh, respectively). The two batteries alone cannot normally supply either the full charge rate or the total energy requirement of either car. However, the charge rate of both cars can be modified to limit it to the maximum discharge rate of the batteries. Even then, though, grid electricity is still needed to provide the total energy drawn during the night for the cars (red bars in Figure 2.33).

Figure 2.33 indicates that, during the summer week, about 384.96 kWh was exported back to the grid and 103.76 kWh was imported for a net export of 281.2 kWh. At the same time, 577.96 kWh was generated. As such,

48.7 percent (281.2/577.96) of all electricity generated during the week was exported back to the grid and the rest, 51.3 percent, was consumed. Another way to look it is that the solar PV system produced 195 percent (577.96/296.76) of the home's weekly electricity needs, including car charging.

For the first year of household electricity use, including winter, the PV system produced 120 percent of its household electricity use. In other words, there was no electric bill (aside from a $10 per month hookup fee), no natural gas heating bill (because no gas was connected to the property), and no gasoline bill (because the cars were powered by the rooftop PV). In fact, excess electricity was sent back to the grid, and the utility paid $530 for it.

In California and several other states, many **Community Choice Aggregation (CCA)** utilities have sprung up. These utilities procure clean, renewable electricity by signing power purchase agreements with wind, solar, geothermal, and hydroelectric energy farms. The CCAs provide the electricity at competitive rates to customers who sign up with the CCA. The CCAs also procure electricity from homeowners who have solar PV on their roof or a small wind turbine in their backyard. The CCAs pay for the solar PV at the same rate a ratepayer would pay for electricity at the same time of day. Thus, if a ratepayer pays $0.25/kWh for electricity at 4 PM, a homeowner who sends electricity back to the grid at that time will also receive $0.25/kWh. The transmission and distribution (T&D) portion of a ratepayer's bill is handled by a different utility, but T&D costs are applied only when more electricity is drawn from the grid than sent to the grid.

Paying a different electricity rate at different times of the day is called **time-of-use pricing**. Often, there is a time period during the day of **peak** electricity price, such as 2 PM to 9 PM, a time period of minimum (**off-peak**) electricity price (e.g., 11 PM to 7 AM), and time periods of intermediate (**partial-peak**) electricity price (all other times). Time-of-use pricing may differ between weekends and weekdays. For example, one weekend distribution would be to charge peak prices between 3 PM to 7 PM and off-peak prices for all other hours.

An alternative to time-of-use pricing is to use constant pricing all day and to charge increasingly higher rates per kWh of energy used the more that energy is used. The advantage of time-of-use pricing is that it is a built-in method of **demand response** management of the grid. In other words, it helps utility operators balance supply of

**Table 2.5.** Range in cost savings in a new-construction home due to eliminating natural gas, electrifying everything, and using solar PV plus batteries to power the electricity while staying on the grid. Also shown is the payback time without and with subsidies. The payback time without subsidy is calculated as the cost of the solar-plus-battery system minus the upfront avoided cost, all divided by the annual avoided cost. The payback time with subsidy is the same, except the subsidy is also subtracted from the cost of the solar-plus-battery system upfront cost.

| Item | Avoided cost |
|---|---|
| Gas hookup fee | $3,000–$8,000 upfront |
| Gas pipes | $1,000–$7,000 upfront |
| Electric bill | $1,000–$3,000/y |
| Natural gas bill | $1,000–$3,000/y |
| Vehicle fuel bill | $1,000–$4,000/y |
| Total avoided cost | $4,000–$15,000 upfront + $3,000–$10,000/y |
| Cost of solar-plus-battery system | $30,000–$115,000 |
| Subsidies (35%) | $10,500–$40,250 |
| Payback time no subsidies | 8.6–10 y |
| Payback time with subsidies | 5.2–6 y |

electricity with its demand. With demand response, utility operators give individuals and businesses financial incentives not to use electricity at certain times of the day when electricity demand is high by increasing electricity prices during those times or by paying customers not to use electricity during those times. Higher prices between 2 PM and 9 PM, for example, shift electricity use from that period to another period, such as between 11 PM and 7 AM, when electricity prices are low. In fact, the nighttime car charging in Figure 2.33 was encouraged directly by low nighttime electricity prices.

The building of a superefficient all-electric home with heat pumps, electric appliances, and electric cars and powered by solar PV and batteries not only eliminated electricity, natural gas, and gasoline bills, and resulted in a payment for excess electricity, but it also eliminated installation costs related to natural gas. For example, it eliminated the cost of a gas hookup fee and the cost of gas pipes.

## Transition highlight

Table 2.5 summarizes the range in upfront and annual cost savings for a new construction all-electric home without gas. The table suggests that eliminating gas and electrifying a new home can save $4,000 to $15,000 in upfront costs and $3,000 to $10,000 per year in electricity and fuel bills. These savings offset part of the capital cost of installing a solar-battery system that powers all household and vehicle needs. Such a system has an upfront capital cost ranging from $30,000 to $115,000. The capital and maintenance cost of heat pumps for air and water heating and air conditioning and other electric appliances and vehicles roughly cancel those of fossil-fuel equivalent appliances and vehicles used. As such, a solar-plus-battery system powering the home can pay itself off in 5.2 to 6 years with government incentives and 8.6 to 10 years without (Table 2.5). This payback time is decreasing yearly, as solar and battery costs are declining yearly.

## 2.9 Controlling Non-Energy Air Pollution and Climate-Relevant Emissions

WWS technologies address energy-related emissions. However, some emissions that affect human health and climate are not energy-related emissions but are still necessary to eliminate or reduce their impacts on human health and climate. These include open biomass burning; methane emissions from agriculture and waste; halogen emissions from leakage and draining; and nitrous oxide emissions from fertilizers, industry, and wastewater treatment. These sources of emissions and methods to control them are discussed briefly.

### 2.9.1 Open Biomass Burning and Waste Burning

Open biomass burning is the burning of evergreen forests, deciduous forests, woodland, grassland, and agricultural land, either to clear land for other use, to stimulate grass growth, to manage forest growth, to satisfy a ritual, by accident (campfires, debris burning, cigarettes), or by arson. About 17 percent of all global $CO_2$ and air pollution emissions worldwide are from open biomass burning (Jacobson, 2014). Although such burning may be natural or anthropogenic in origin, humans cause around 93 percent of it today, and nature causes the rest (Jacobson, 2014).

Agricultural fields are often burned after harvest to remove old straw to clear the field for a new harvest during the next spring. Sugarcane fields are usually burned before harvest to remove the outer leaves around the sugarcane stalks to facilitate the sugarcane extraction during harvesting.

Waste burning is the burning of trash, such as in a landfill (Figure 2.34), open pit, garbage can, or backyard incinerator. Such waste burning is illegal in many countries but still occurs in many others.

(a)

(b)

Figure 2.34 (a) Landfill plastic waste burning in India. © cgdeaw/AdobeStock. (b) United Kingdom landfill site. Smaller pipes are carrying methane captured from under the landfill to a common larger pipe. © Angella Streluk/Wikimedia Commons.

Biomass burning produces not only gases that warm the climate, including $CO_2$, $CH_4$, and CO, but also climate-warming particles, including black carbon (BC) and brown carbon (BrC). BC and BrC, along with other particle components emitted during biomass burning (ash, other organic carbon, and sulfate), cause substantial health impacts to people and animals who breathe them in. In addition, the oxides of nitrogen and organic gases from biomass burning result in elevated levels of ozone, formaldehyde, and other gaseous pollutants that affect human health. Waste burning emits the same chemicals as biomass burning but also emits toxic chemicals from the burning of plastics, paints, varnishes, pesticides, medical waste, and chemical byproducts.

While some argue that biomass burning followed by regrowth of vegetation results in no net increase in $CO_2$ to the air, that contention is incorrect. It is incorrect because, even though $CO_2$ released upon burning is offset by $CO_2$ used to regrow the vegetation in the first place, the time lag between burning and regrowth (from 1 to 10 years for savannah and 80 years for a forest, for example) mathematically always results in elevated $CO_2$ in the air (Jacobson, 2004). The contention is also misleading because biomass burning emits black carbon, brown carbon, water vapor, methane, carbon monoxide, ozone precursors (organic gases and oxides of nitrogen), and heat, all of which increase global temperatures and are not recycled as $CO_2$ is. As such, biomass burning causes net global warming (Jacobson, 2014).

The only solution to biomass burning is to stop it. No technology exists to control its emissions. An alternative to burning agricultural waste straw is to till it into the soil. An alternative to sugarcane burning is to cut away the leafy parts of the sugarcane before harvest and mix them into the soil. Since humans cause more than 90 percent of all open biomass fires worldwide, biomass burning is largely (but not completely) preventable through government policies restricting burning and discouraging the conversion of forest land to agricultural land or another land use type.

Similarly, the only method of reducing the impacts of landfill waste burning is to stop it because its emissions cannot be trapped. If waste is burned in an incinerator, many of its emissions can theoretically be controlled with emission control technologies; however, no technology eliminates all the emissions, including all the $CO_2$. Thus, even incinerators with emission controls result in significant pollution. The best control of waste burning is to eliminate it by recycling the waste or keeping it in a landfill. Waste should not be dumped into the oceans, since the plastics and many other materials do not degrade for centuries. The accumulation of plastics in the oceans has caused an environmental catastrophe, resulting in, for example, the **Great Pacific Garbage Patch**, a plastic wasteland in the middle of the Pacific Ocean between California and Hawaii that is three times the size of France.

## 2.9.2 Methane from Agriculture and Waste

**Methane** is a long-lived greenhouse gas present in the atmosphere that selectively absorbs certain thermal-infrared wavelengths of radiation, trapping some of that radiation near the surface of the Earth. It causes about 12 percent of gross global warming (see Table 1.1).

Anthropogenic sources of methane include not only open biomass burning and energy-related natural gas leaks and fossil-fuel combustion, but also biological sources enabled by human activity. A 100 percent WWS energy infrastructure will eliminate methane emissions from energy production. Controls of biomass burning (Section 2.9.1) will reduce that source of methane. This section focuses on controlling methane from human-enabled biological sources.

The root biological source of most methane is **methanogenic bacteria**, which live in **anaerobic** (without oxygen) environments. Methanogenic bacteria consume organic material and excrete methane. Ripe anaerobic environments include the digestive tracts of cattle, sheep, and termites; manure from cows, sheep, pigs, and chickens; rice paddies; landfills; and wetlands. Of these, emissions from cattle, sheep, manure, rice paddies, and landfills are human-enabled emissions.

The main method of reducing methane from bacteria in digestive tracts and manure is to reduce human consumption of meat and poultry. This requires people changing their diets, which can have additional health benefits.

Methane released from **manure** can also be captured with a methane digester. A **methane digester**, or anaerobic manure digester, is an airtight tank in which manure is placed after water is separated from it. The manure is heated and stirred to simulate the inside of a cow's stomach. Methane rises from the manure to the top of the tank, where it is captured in a bag or piped out of the digester to a storage tank.

In a 100 percent WWS world, the use of captured methane should be used only for steam reforming of methane (Section 2.2.2.1) to produce hydrogen for use in a fuel cell, not for methane combustion. Whereas $CO_2$ emissions from steam reforming are similar to those of methane combustion, emissions of all other pollutants from steam reforming are small (Colella et al., 2005, Table 4) in comparison with those from natural gas combustion.

Methane gas from **landfills** can also be captured. **Landfill gas** is often a mixture of nearly 50 percent methane, nearly 50 percent carbon dioxide, and trace amounts of non-methane organic gases. Landfill gas is extracted by drilling multiple half-meter-wide boreholes up to 30 m deep into a section of the landfill. Trash is then removed so that a well, which consists of a perforated or slotted siding with a cap on the bottom, can be installed. Most of the top of the well is sealed to prevent gas escape. A wellhead is installed at the top through which gas is piped to its end destination. In some cases, a network of multiple vertical and horizontal pipes can capture the landfill gas (Figure 2.34).

Once captured, the landfill gas is usually filtered to separate out the methane from the other gases. Once methane is relatively isolated, it can be used to produce hydrogen by steam reforming (Section 2.2.2.1).

## Transition highlight

**Rice paddies** release a significant amount of methane globally, both through the leaves of rice plants and in the anaerobic environment of the flooded soil in which rice plants usually grow. On average, rice paddy soil is flooded for four months of the year. Direct seeding of rice plants, instead of transplanting them into already-flooded paddies, can reduce the time needed for flooding down to one month. This reduces methane emissions from rice paddies by 15 to 90 percent. A system of pipes can also be used to capture rice paddy methane just as it does with methane from landfill gas. The methane should be used only to produce hydrogen by steam reforming.

### 2.9.3 Halogen Emissions

Table 1.1 indicates that halogens are responsible for about 9 percent of gross global warming. Halogens are still used today as refrigerants, solvents, blowing agents, fire extinguishants, and fumigants. They enter the atmosphere primarily upon evaporation when they leak or when the appliances containing them are drained in a way that exposes them to the air. Their persistence in the atmosphere and impacts on warming depend on the specific global warming potential of each halogen (see Table 1.2).

The main methods of reducing halogens and their impacts in the atmosphere are (a) substituting lower GWP halogens or non-halogen compounds for higher GWP halogens to perform the same function, (b) requiring more stringent standards for sealing halogens in the equipment or appliance they are used in to reduce leakage, and (c) requiring tougher standards for disposing of halogens at the end of life of the equipment or appliance they are used in. For these suggestions to be effective, they need to be implemented and enforced worldwide.

### 2.9.4 Nitrous Oxide and Ammonia Emissions from Fertilizers

Nitrous oxide is a strong greenhouse gas with a high global warming potential and long lifetime (see Table 1.2). $N_2O$ is produced by bacteria and contributes to about 4.3 percent of gross anthropogenic global warming (see Table 1.1); 67 to 80 percent of anthropogenic nitrous oxide originates from agriculture (Ussiri and Lal, 2012). In particular, nitrogen-containing fertilizers emit $N_2O$. In addition, the cultivation of legumes (plants in the pea family) results in the conversion of atmospheric **molecular nitrogen** ($N_2$) to $N_2O$, which is released to the air, but to a lesser extent than with fertilized crops. A third agricultural source of $N_2O$ is the solid waste of domesticated animals.

Bacterial metabolism in nitrogen-containing fertilizers also results in the emission of **ammonia** ($NH_3$). $NH_3$ is a gas that dissolves in liquid aerosol particles to form the **ammonium ion** ($NH_4^+$), which reacts with other chemicals inside the particles. $NH_4^+$ inside of liquid particles also causes the particles to swell by encouraging water vapor to condense. Swollen particles reduce visibility and attract more toxic chemicals, increasing health problems for those who breathe in the aerosol particles.

Some methods of reducing nitrous oxide and ammonia emissions from fertilizers are (a) using less nitrogen-based fertilizer, (b) cultivating leguminous crops that don't require fertilizer in the crop rotation, and (c) reducing tillage to reduce the breakdown of organic fertilizer, thereby reducing reaction and release of chemicals.

The remaining anthropogenic sources of $N_2O$ are fossil-fuel combustion, open biomass burning, industrial processes, and treatment of wastewater. Transitioning fossil energy to 100 percent WWS will eliminate fossil combustion sources of $N_2O$. Section 2.9.1 describes how open biomass burning sources will be controlled.

The two main industrial sources of $N_2O$ are the production of **nitric acid** for use in fertilizers and the production of **adipic acid** for use in the production of nylon fibers and plastics. To date, $N_2O$ emissions from adipic acid production have been reduced effectively with emission control technologies in several production plants, so the expansion of such technologies to adipic acid plants worldwide will help reduce $N_2O$ emissions. Similarly, $N_2O$ emission control technologies for nitric acid production plants are available (NACAG, 2014) and can be implemented worldwide with stringent policies.

The source of $N_2O$ in wastewater is organic material from human or animal waste. Modulating the dissolved oxygen content in the wastewater treatment process can control the $N_2O$ content of wastewater (e.g., Boiocchi et al., 2016; Santin et al., 2017).

## 2.10 Summary

This chapter described the technologies needed to address air pollution, global warming, and energy insecurity. Solving these problems requires electrifying or providing direct heat for all energy sectors (electricity, transportation, heating/cooling, industry, agriculture/forestry/fishing, and the military), and using WWS to produce that electricity and heat. Some electricity will be converted to hydrogen. In addition, electricity, heat, cold, and hydrogen storage and transmission will be used. Other technologies that will be used in a WWS world include electric and hydrogen fuel cell vehicles, heat pumps, electric appliances, and electric industrial machines. About 90 to 95 percent of all technologies to transition to 100 percent WWS are commercially available as of 2020. The ones that are not are expected to be available within the next 5 to 20 years. Energy efficiency and reducing energy use will also be important in a transition. In addition to transitioning energy, it will be necessary to reduce or eliminate non-energy sources of emissions. Such sources include open biomass burning, methane from agriculture and waste, halogen emissions, nitrous oxide and ammonia emissions from fertilizers, and carbon dioxide emissions from steel manufacturing and cement production.

## Further Reading

Alva, G., Y. Lin, and G. Fang, 2018. An overview of thermal energy storage systems, *Energy*, **144**, 341–378.

Cano, Z. P., D. Banham, S. Ye, A. Hintennach, J. Lu, M. Fowler, and Z. Chen, 2018. Batteries and fuel cells for emerging electric vehicle markets, *Nat. Energy*, **3**, 279–289.

Cerri, C. E. P., C. C. Cerri, S. M. F. Maia, M. R. Cherubin, B. J. Feigl, and R. Lal, 2018. Reducing Amazon deforestation through agricultural intensification in the Cerrado for advancing food security and mitigating climate change, *Sustainability*, **10**, 989, doi:10.3390/su10040989.

Hames, Y., K. Kaya, E. Baltacioglu, and A. Turksoy, 2018. Analysis of the control strategies for the fuel savings in the hydrogen fuel cell vehicles, *Int. J. Hydrog. Energy*, **43**, 10810–10821.

Masters, G., 2013. *Renewable and Efficient Electric Power Systems*, 2nd ed., Hoboken, NJ: Wiley, 712 pp.

Vogl, V., M. Ahman, and L. J. Nilsson, 2018. Assessment of hydrogen direct reduction for fossil-free steelmaking, *J. Clean. Prod.*, **203**, 736–745.

Zhang, H., J. Baeyens, G. Caceres, J. Degreve, and Y. Lv, 2016. Thermal energy storage: recent developments and practical aspects, *Prog. Energy Combust. Sci.*, **53**, 1–40.

# 2.11 Problems and Exercises

2.1. Identify a non-combustion alternative technology for each of the following fossil-fuel technologies:

a. Natural gas stove

b. Natural gas water heater

c. Coal-fired furnace for high-temperature industrial heat

d. Gasoline automobile

e. Bunker-fuel ferry

f. Diesel tractor

2.2. Identify three WWS electricity-generating technologies, three electricity storage technologies, and three heat storage technologies.

2.3. How much power can be generated over 3 days from a pumped hydropower storage facility if the difference in elevation between the upper and lower reservoir is 300 m? Assume the maximum change in depth of the upper reservoir, which has surface area of 3,000 $m^2$, is 3 m; the penstock efficiency is 0.9; and the combined turbine, generator, and pump efficiency is 0.85. Note that 1 kWh = 3,600,000 N-m.

2.4. Suppose a penstock of a hydropower facility with diameter D = 7 cm delivers a flow rate to a turbine of Q = 0.4 $m^3$/min of water through an elevation change (gross head) of $H_g$ = 20 m. The pressure in the pipe at the turbine is p = 180 $kN/m^3$ and the elevation difference between the water surface downstream of the turbine and the turbine itself is $H_d$ = 0. What fraction of available head is lost in the pipe and what power is delivered by the turbine, assuming a turbine plus generator turbine plus drive system efficiency of $\eta$ = 0.8?

2.5. Given a lower heating value of jet fuel of 42.8 MJ/kg-jet fuel, an overall energy conversion efficiency of jet fuel of 0.25, a jet fuel density of 810.53 $kg/m^3$, a lower heating value of hydrogen of 119.96 MJ/kg-$H_2$, a fuel cell system efficiency of 0.55, and a cryogenic hydrogen density of 70.9 $kg/m^3$, calculate the volume ratio and mass ratio of $H_2$-to-jet fuel needed to provide the exact same energy, assuming these are the only assumptions that matter.

2.6. Suppose a 747-8 aircraft weighs 295,000 kg with no jet fuel (dry mass) and a volume of 4,000 $m^3$. Suppose also that it requires 100 MW of power to take off. If a fuel cell stack of mass 1.8 kg and volume 0.0046 $m^3$ provides 1.2 kW of power, what percent of the aircraft's (a) dry mass plus fuel cell stack and (b) total volume (assuming the stack fits within the existing volume) is required for the fuel cell stacks alone?

2.7. Identify one advantage and one disadvantage of a district heating/cooling system over a home heat pump heating/cooling system.

2.8. If a single electric vehicle requires 10.5 kg of lithium for its batteries, how much energy is stored in 100 million vehicles if they are fully charged and each vehicle stores up to 6.67 kWh/kg-Li?

2.9. Assume a home battery starts full, stores 6 kWh of energy, and has a peak discharge rate of 3 kW. Assume also that, if either limit is exceeded, grid electricity, at $0.2/kWh, is used to provide the remaining energy or power. Suppose Sue charges her electric car, which draws 10 kW, for 3 hours. How much will she need to pay the utility for grid electricity?

2.10. Consider two cases.

In Case A, Gwen uses her rooftop solar PV to charge one battery that stores 8 kWh, from 6 AM to 8 AM. She discharges the battery for use in her home between 6 PM and 11 PM. She then pays for using 8 kWh of grid electricity from 11 PM until 6 AM the next morning.

In Case B, Tim uses his rooftop PV to charge two batteries that store 8 kWh, each from 6 AM to 9 AM. He then discharges one battery between 6 PM and 11 PM and the other between 11 PM and 6 AM.

Assume the cost of grid electricity is $0.2/kWh between 6 AM and 11 PM and $0.1/kWh between 11 PM and 6 AM. Also assume that any solar produced by either Gwen or Tim can be sold to the utility at the same price the utility would charge for electricity at that time of day.

Who, if either, loses more money and by how much? Consider only the costs and benefits of using grid electricity and battery electricity during the hours specified.

2.11. If a water pit contains 150,000 $m^3$ of water, and its natural temperature without heat storage is 15 °C, to what temperature must the pit temperature be raised to store 10 GWh of heat? Assume the specific heat of liquid water is 4,186 J/kg-K and the density of liquid water is 1,000 $kg/m^3$.

2.12. Identify four electric alternatives to obtaining high-temperature industrial heat. Why can't a heat pump be used to produce the heat needed during steel production?

2.13. Briefly identify one alternative method of producing (a) steel and (b) cement that significantly reduces chemically produced carbon emissions during each process.

2.14. Identify one policy or technological solution that can help reduce emissions from each of these non-energy sources of carbon:

a. Open biomass burning

b. Methane from agriculture and waste

c. Halogens

d. Nitrous oxide from fertilizers

# CHAPTER

## 3 Why Some Technologies Are Not Included

Prior to the Industrial Revolution of the mid-1700s, the world relied primarily on biomass but also on some coal for its energy, which was primarily heat. During the Industrial Revolution, manufacturing processes transitioned from artisan shops to factories following expanded use of the steam engine. The use of coal, which powered the steam engine, has grown ever since. Although natural gas was discovered accidentally in China around 500 BC, it was not used on a large scale until after the first natural gas well was constructed in Fredonia, New York, in 1821. On August 27, 1859, oil was discovered in Titusville, Pennsylvania. Worldwide, oil consumption has increased since then.

Although biomass, coal, natural gas, and oil have provided energy necessary for societies to function until today, such energy has come at a cost – namely, increased air pollution, global warming, and energy insecurity. While such a cost was accepted in the past, scientists and policymakers no longer believe societies can function properly in the future while allowing the emissions associated with these fuels to continue. As such, researchers have proposed a variety of technologies to address the emissions.

Aside from the WWS and storage technologies described in Chapter 2, the main suggestions for reducing or eliminating emissions have included using natural gas for electricity instead of coal, using natural gas or coal with carbon capture and storage or use (CCS/U), using nuclear power instead of fossil fuels for electricity, using biofuels instead of gasoline or diesel for transportation, and using biomass with or without carbon capture for electricity. The justification for these technologies is that they reduce carbon relative to the current carbon intensity of energy worldwide.

The main issue with all of these technologies going forward is that they result in greater global warming, air pollution, and land degradation or all three per unit energy compared with WWS resources. Second, some increase energy insecurity. As such, they are opportunity costs compared with investing in WWS technologies and are not recommended.

In addition, some methods have been proposed to remediate global warming, including synthetic direct air carbon capture and storage (SDACCS) and geoengineering. SDACCS involves removing carbon dioxide directly from the air by chemical reaction and storing it. Geoengineering consists of a variety of measures to increase the reflectivity (albedo) of the Earth in order to cool the Earth's surface. These technologies also have several serious issues, which render them incapable of helping to reduce air pollution, global warming, or energy insecurity, and are also not recommended.

This chapter discusses the above technologies and clarifies the reasons why they are not needed or helpful for solving global warming, air pollution, and energy security problems. The chapter starts with a discussion about why natural gas is not useful as a bridge fuel between coal and renewables (Section 3.1). It then describes why coal and natural gas with carbon capture do not help to solve air pollution, climate, or energy security problems (Section 3.2). Section 3.3 analyzes why new and some existing nuclear power plants are unhelpful. Sections 3.4 and 3.5 then outline problems with biomass with or without carbon capture and liquid biofuels, respectively. Section 3.6 focuses on why synthetic direct air capture causes more harm to the solution than good. Finally, Section 3.7 details the problems with geoengineering.

## 3.1 Why Not Use Natural Gas as a Bridge Fuel?

**Natural gas** is a colorless, flammable gas containing a mass (mole) fraction of about 88.5 (93.9) percent methane plus smaller amounts of ethane, propane, butane, pentane, hexane, nitrogen, carbon dioxide, and oxygen (Union Gas, 2018). It is often found near petroleum deposits. It is usually either combusted in a gas turbine that is coupled with a generator to produce electricity or combusted in a burner to produce either building heat or high-temperature industrial heat.

Because natural gas is not very dense, it can be stored on its own only in a large container. As such, natural gas is often compressed or liquefied for transport and storage. **Compressed natural gas (CNG)** is natural gas compressed to less than 1 percent of its gas volume at room temperature. **Liquefied natural gas (LNG)** is natural gas that has been cooled to –162 °C, the temperature at which it condenses to a liquid at ambient pressure. LNG has a volume that is 1/600th the volume of the original gas. Both CNG and LNG can be sent through pipelines, although different pipelines are needed for each. CNG and LNG can also be stored and used directly in automobiles that are designed to run on them. CNG and LNG can further be transported by truck or bus with a special fuel tank and can be stored at a power plant for backup use when pipeline gas is not available. In addition, pipeline CNG is often converted to LNG at a marine export terminal, put on a tanker ship with supercooled cryogenic tanks, and shipped overseas. At the import terminal, it is re-gasified and piped to its final destination – either a power plant, industrial company, or company that transmits and distributes it to buildings for heating or other purposes.

Natural gas is obtained from underground conventional wells containing both oil and natural gas or by hydraulic fracturing. **Hydraulic fracturing (fracking)** is the process by which natural gas is extracted from shale rock formations instead of wells. **Shale** is sedimentary rock composed of a muddy mix of clay mineral flakes and small fragments of quartz and calcite. Large shale formations containing natural gas can be found in eastern North America, close to population centers, among many other locations worldwide. In the United States, about 67 percent of natural gas in 2015 was extracted from shale rock (EIA, 2016b). Extraction of natural gas from shale requires large volumes of water, laced with chemicals, forced under pressure to fracture and re-fracture the rock to increase the flow of natural gas. As the water returns to the surface over days to weeks, it is accompanied by methane that escapes to the air. As such, more methane leaks occur during fracking than during the drilling of conventional gas wells (Howarth et al., 2011, 2012; Howarth, 2019). Methane also leaks during the transmission, distribution, and processing of natural gas.

For electricity production, natural gas is usually used in either an **open cycle gas turbine (OCGT)** or a **combined cycle gas turbine (CCGT)**. In an OCGT, air is sent to a compressor, and the compressed air and natural gas are both sent to a combustion chamber, where the mixture is burned. The hot gas expands quickly, flowing through a turbine to perform work by spinning the turbine's blades. The rotating blades turn a shaft connected to a generator, which converts a portion of the rotating mechanical energy to electricity.

The main disadvantage of an OCGT is that the exhaust contains a lot of waste heat that could otherwise be used to generate more electricity. A CCGT routes that heat to a heat recovery steam generator, which boils water with the heat to create steam. The steam is then sent to a steam turbine connected to the generator to generate 50 percent more electricity than the OCGT alone. Thus, a CCGT produces about 150 percent of the electricity of an OCGT with the same input mass of natural gas, thus carbon dioxide emissions, in each case.

On the other hand, the ramp rate of an OCGT is 20 percent per minute, which is 2 to 4 times that of a CCGT (5 to 10 percent per minute) (see Table 2.1). Thus, the less efficient OCGT, which also releases more $CO_2$ per unit electricity generated (Table 3.1), is more useful for filling in short-term gaps in supply on the grid than is a CCGT.

It has long been suggested that natural gas could be used as a **bridge fuel** between coal and renewables (e.g., MIT, 2011). The two main arguments for this suggestion are (1) natural gas emits less carbon dioxide equivalent emissions per unit energy produced ($CO_2e$ – Section 1.2.3.5) than coal and (2) natural gas electric power plants are better suited to be used with intermittent renewables than coal.

However, the justifications for using gas as a bridge fuel are incorrect and insufficient. Natural gas is not recommended for use together with WWS technologies for multiple reasons. These are discussed in the following sections.

Table 3.1 Comparison of 20- and 100-year lifecycle global $CO_2$ equivalent ($CO_2e$) emissions from coal versus natural gas used in either an open cycle gas turbine (OCGT) or a combined cycle gas turbine (CCGT) for electricity generation.

| Chemical (X) | 20-y GWP | 100-y GWP | Coal | | | Natural Gas Open Cycle Gas Turbine | | | Natural Gas Combined Cycle Gas Turbine | | |
|---|---|---|---|---|---|---|---|---|---|---|---|
| | | | Emis. factor (g-X/kWh) | 20-y $CO_2e$ (g-$CO_2e$/kWh) | 100-y $CO_2e$ (g-$CO_2e$/kWh) | Emis. factor (g-X/kWh) | 20-y $CO_2e$ (g-$CO_2e$/kWh) | 100-y $CO_2e$ (g-$CO_2e$/kWh) | Emis. factor (g-X/kWh) | 20-y $CO_2e$ (g-$CO_2e$/kWh) | 100-y $CO_2e$ (g-$CO_2e$/kWh) |
| [a]$CO_2$-upstream | | | | 97.2 | 97.2 | | 54 | 54 | | 54 | 54 |
| [b]$CH_4$-leak | 86 | 34 | 4.1 | 353 | 140 | 4.35 | 374 | 148 | 3.26 | 280 | 111 |
| [c]$CO_2$-plant | 1 | 1 | 905 | 905 | 905 | 540 | 540 | 540 | 404 | 404 | 404 |
| [d]BC+OM-plant | 3,100 | 1,550 | 0.045 | 141 | 70 | 0.0003 | 0.93 | 0.47 | 0.0003 | 0.93 | 0.47 |
| [c]$NO_x$-N-plant | −560 | −159 | 0.23 | −129 | −37 | 0.15 | −84 | −24 | 0.015 | −8.4 | −2.4 |
| [c]$SO_2$-S-plant | −1,400 | −394 | 0.75 | −1,050 | −393 | 0.005 | −7 | −2 | 0.0015 | −2.1 | −2 |
| **Total** | | | | 317 | 782 | | 878 | 716 | | 728 | 565 |

All 20- and 100-year GWPs are from Table 1.2. Each $CO_2e$, except for upstream values, is the product of an emission factor and a GWP.

[a] Upstream $CO_2$ emissions from coal mining, processing, and transport are 27 g-$CO_2$/MJ, or 97.2 g-$CO_2$/kWh and from natural gas are 15 g-$CO_2$/MJ = 54 g-$CO_2$/kWh (Howarth, 2014).

[b] For coal, the 100-year $CO_2e$ from $CH_4$ leaks is estimated from Skone (2015), Slide 17. The emission factor in the present table is then derived from that number and the 100-year GWP of $CH_4$ in the present table. The 20-year $CO_2e$ is then calculated from the emission factor and the 20-year GWP. For natural gas, the $CH_4$-leak emission factors in the present table are obtained by multiplying the mass of $CH_4$ required per kWh of electricity by L/(1-L), where L is the fractional leakage rate of methane between mining and use in a power plant. The mass of $CH_4$ required per kWh for a natural gas turbine is the $CO_2$ combustion emissions from the turbine (present table) multiplied by the ratio of the molecular weights of $CH_4$:$CO_2$ (16.04276 g/mol:44.0098 g/mol) and by the mole fraction of natural gas that consists of methane (0.939) (Union Gas, 2018). The results are 184.8 g-$CH_4$/kWh-electricity for open cycle and 138.3 g-$CH_4$/kWh-electricity for combined cycle. The overall U.S. methane leakage rate, from shale gas, which includes leaks from drilling and from pipe transmission and distribution to electric power plants, industrial facilities, homes, and other buildings, may be ~3.5 percent (Howarth, 2019). Shale gas was about 2/3 of the U.S. natural gas production in 2015 (EIA, 2016b). The leakage rate of all natural gas (shale gas plus conventional gas) for only drilling and transmission to large facilities may be ~2.3 percent (Alvarez et al., 2018). This number is used in this table, which is for electric power plant generation.

[c] Emission factors from Figure 4 of de Gouw et al. (2014) for 2012 U.S. plants; emission factors for $NO_x$-$NO_2$ were multiplied by the ratio of the molecular weight of N to that of $NO_2$. For $SO_2$-S, emission factors for $SO_2$ were multiplied by the ratio of the molecular weight of S to that of $SO_2$.

[d] The emission factors of BC+OM for coal and natural gas were obtained from Bond et al. (2004) assuming, for coal, pulverized coal and a mix between hard and lignite coal.

## 3.1.1 Climate Impacts of Natural Gas versus Other Fossil Fuels

When used in an electric power plant, natural gas substantially increases, rather than decreases, global warming (by increasing $CO_2e$) compared with coal over a 20-year time frame. The difference over 100 years, while more favorable to gas, is relatively small (Table 3.1). Regardless, $CO_2e$ emissions (and health-affecting air pollutant emissions) from both gas and coal are much larger than are those from WWS technologies, so spending money on natural gas or coal represents an opportunity cost relative to spending the same money on WWS.

Over a 20-year time frame, the $CO_2e$ from using natural gas with a CCGT or an OCGT is **2.3 and 2.8 times**, respectively, that of using coal (Table 3.1). Over a 100-year time frame, the $CO_2e$ from a natural gas OCGT is only **8 percent less** than that of coal, and the $CO_2e$ from a natural gas CCGT is only **28 percent less** than that of coal.

The fact that natural gas causes far more global warming than coal over a 20-year time frame is a concern because of the severe damage global warming is already causing that will only be made worse over the next two decades, including the triggering of some difficult-to-reverse impacts, such as the complete melting of the Arctic ice.

The reasons that the $CO_2e$ of natural gas exceeds that of coal over 20 years and is close to that of coal over 100 years are as follows.

First, although natural gas combustion in an OCGT or CCGT emits only 60 or 45 percent, respectively, of the $CO_2$ per kilowatt-hour (kWh) of coal combustion, natural gas leaks during the mining and transport of natural gas emit similar or more $CH_4$ than do $CH_4$ leaks during coal mining. This is an issue, because $CH_4$ has a high, positive 20- and 100-year GWP (see Table 1.2). As such, the leaked $CH_4$ from natural gas mining and transport contributes substantially to the $CO_2e$ of natural gas.

Second, and more important, coal combustion emits more $NO_x$ and $SO_2$ per kWh than does natural gas combustion (Table 3.1), and $NO_x$ and $SO_2$ both produce cooling aerosol particles, which offset or mask much of global warming, particularly over a 20-year time frame (see Figure 1.2). The cooling impacts of these particles are through their direct reflection of sunlight back to space and their enhancement of cloud thickness. Thicker clouds reflect more sunlight back to space. As such, $NO_x$ and $SO_2$, which are both short-lived, have high negative GWPs over both 20 years and 100 years (Table 3.1).

Howarth et al. (2011, 2012) identified the importance of methane leaks, particularly natural gas fracking of shale gas on the $CO_2e$ emissions of natural gas versus coal on a 20- versus 100-year lifetime. Wigley (2011), for one, estimated the cooling impact of $SO_2$, but not $NO_x$, when comparing $CO_2e$ from coal versus natural gas power plants.

Neither natural gas nor coal is recommended in a 100 percent WWS world because, among other reasons, the natural gas lifecycle of 100-year $CO_2e$ for electricity generation (565 to 716 g-$CO_2e$/kWh) (see Table 3.1) is on the order of 56 to 72 times that of wind (~10 g-$CO_2e$/kWh) (Table 3.5), and the 100-year coal $CO_2e$ (~780 g-$CO_2e$/kWh) is about 78 times that of wind. Similarly, both coal and gas produce much more air pollution than do WWS sources (Section 3.1.2).

The $CO_2e$ emissions from natural gas versus other fossil fuels are higher for heating and transportation than for electricity. For building heat and industrial process heat, for example, natural gas offers less efficiency advantage over oil or coal than it does for electricity generation. As such, after accounting for all chemical emissions and their respective global warming potentials, natural gas may cause greater long-term global warming than does oil or coal for heating.

With respect to transportation fuels, the carbon dioxide equivalent emissions of natural gas may also exceed that of oil, since the efficiency of natural gas used in transportation is similar to that of oil. Thus, when methane leaks are added in, natural gas causes more overall warming than oil (Alvarez et al., 2012). In sum, in terms of climate, natural gas causes greater global warming than other fossil fuels over 20 years across all applications. Over a 100-year time frame, natural gas causes similar or less warming than coal used for electricity generation and greater warming than oil for heating and transportation over 100 years. All fossil fuels emit 1.5 to 2 orders of magnitude the $CO_2e$ as WWS sources.

### 3.1.2 Air Pollution Impacts of Natural Gas versus Coal and Renewables

Whereas natural gas causes more $CO_2e$ emissions than coal over 20 years and a similar or slightly less level over 100 years, coal emits more health-affecting air pollutants than does natural gas, which is the main reason it has a lower $CO_2e$ over 20 years than does natural gas. Nevertheless, both natural gas and coal are much worse for human health than are WWS technologies, which emit no air pollutants during their operation, only during their manufacture and decommissioning. Such WWS emissions will disappear to zero as all energy transitions to WWS since even manufacturing will be powered by WWS at that point.

Table 3.2 provides U.S. emissions from all natural gas and coal uses in the United States in 2008. The table indicates that natural gas production and use in the United States emitted more CO, volatile organic carbon (VOC), $CH_4$, and ammonia ($NH_3$) than coal production and use, whereas coal emitted more $NO_x$, $SO_2$, and particulate matter smaller than 2.5 and 10 μm in diameter ($PM_{2.5}$, $PM_{10}$). Thus, both fuels resulted in significant air pollution, although the higher $SO_2$, $NO_x$, and particulate

Table 3.2 **2008 U.S. emissions from natural gas and coal (metric tonnes/y). Bold indicates higher overall emissions between coal and natural gas (NG).**

|              | Coal all uses | NG all uses |
|--------------|---------------|-------------|
| CO           | 680           | **900**     |
| VOC          | 40            | **1,130**   |
| $CH_4$       | 5             | **310**     |
| $NH_3$       | 11            | **54**      |
| $NO_x$       | **2,800**     | 1,540       |
| $SO_2$       | **7,600**     | 123         |
| $PM_{2.5}$   | **290**       | 61          |
| $PM_{10}$    | **420**       | 71          |

Source: U.S. EPA (2011). VOCs exclude methane. The methane emissions from the EPA inventory are likely underestimated (e.g., Alvarez et al., 2018).

matter emissions from coal resulted in overall greater air pollution health problems from coal than natural gas.

Most $SO_2$ and $NO_x$ emissions evolve to sulfate and nitrate aerosol particles, respectively. Natural gas also emits $NO_x$, but less so than does coal (see Tables 2.6 and 2.7). Natural gas, on the other hand, emits much less $SO_2$ than does coal (Tables 2.6 and 2.7). Aerosol particles, including those containing sulfate and nitrate formed from gases in the atmosphere, and those emitted directly, cause 90 percent of the 4 to 9 million air pollution deaths that occur annually worldwide (Section 1.1.1). As such, coal in particular, but also natural gas, causes significant health damage.

Model simulations over the United States with the emission data from Table 3.2 suggests that emissions from all natural gas sources may cause 5,000 to 10,000 premature mortalities each year in the United States from air pollution (Jacobson et al., 2015a). Coal-related emissions may cause 20,000 to 50,000 premature mortalities in the United States. Many of the remaining premature mortalities are due to pollution associated with oil (e.g., traffic exhaust, oil refinery evaporation), biofuels for transportation, and wood smoke emissions from open fires, fireplaces, and cooking.

In sum, coal causes more mortalities than does natural gas, but both coal and gas cause far more mortalities than do WWS technologies. The combination of the much higher $CO_2e$ emissions and premature mortalities due to natural gas than WWS technologies renders natural gas not an option as a bridge fuel.

### 3.1.3 Using Natural Gas for Peaking or Load Following

Another argument for using natural gas as a bridge fuel is that it can be used in a load-following or peaking plant (defined in Section 2.4), and WWS technologies will need load-following or peaking plants that use natural gas to back them up when not enough wind or solar is available.

Whereas natural gas plants can help with peaking and load following, they are not needed (Section 8.2.1). Other types of WWS electric power storage options available include CSP with storage, hydroelectric dam storage, pumped hydropower storage, stationary batteries, flywheels, compressed air energy storage, and gravitational storage with solid masses (Section 2.7). By 2019, the cost of a system consisting of wind and solar plus batteries decreased to below that of using natural gas. For example, a Florida utility is replacing two natural gas plants with a combined solar-battery system due the lower cost of the latter (Geuss, 2019).

More important, a 100 percent WWS world involves electrifying or providing direct heat for all energy sectors, where the electricity or heat comes from WWS. Such a transition allows heat, cold, and hydrogen storage to work together with demand response to facilitate matching electric power demand with supply on the grid while also satisfying heat, cold, and hydrogen demands minute by minute at low cost. Chapter 8 discusses this issue in detail.

### 3.1.4 Land Required for Natural Gas Infrastructure

The continuous use of natural gas for electricity and heat results in the continuous and cumulative degradation of land for as long as the gas use goes on. Wells must be dug, and pipes laid every year to supply a world thirsty for gas. When gas wells become depleted, new wells much be drilled. Allred et al. (2015) estimate that 50,000 new natural gas wells are drilled each year in North America alone to satisfy gas demand. The land area required for the well pads, roads, and storage facilities of these new wells amounts to 2,500 $km^2$ of additional land consumed per year (Allred et al., 2015). Once a gas well is depleted, it is sealed and abandoned, and a portion of the abandoned land cannot be used for any other purpose. The natural gas infrastructure also requires land for underground and aboveground pipes, power plants, fueling

Table 3.3 **Estimated land areas required for the fossil-fuel and nuclear infrastructure in California and the United States circa 2018.**

| | Area per installation (km²) | California | | United States | |
|---|---|---|---|---|---|
| | | Number | Area (km²) | Number | Area (km²) |
| [a]Active oil and gas wells | 0.05 | 105,000 | 3,327 | 1.3 million | 65,000 |
| [b]Abandoned oil wells | 0.00005 | 225,000 | 6.6 | 2.6 million | 128.5 |
| [b]Abandoned gas wells | 0.000025 | 48,000 | 0.7 | 550,000 | 13.8 |
| [c]Coal mines | 50 | 0 | 0 | 680 | 34,000 |
| [d]Oil refineries | 7.28 | 17 | 124 | 135 | 983 |
| [e]Kilometers of oil pipeline | 0.006 | 4,800 | 29 | 258,000 | 1,550 |
| [e]Kilometers of gas pipeline | 0.006 | 180,000 | 1,080 | 2.62 million | 15,700 |
| [f]Coal power plants | 1.74 | 1 | 1.74 | 359 | 626 |
| [f]Gas power plants | 0.12 | 37 | 4.5 | 1,820 | 221 |
| [f]Petroleum power plants | 0.93 | 0 | 0 | 1,080 | 1,007 |
| [f]Nuclear power plants | 14.9 | 1 | 14.9 | 61 | 911 |
| [f]Other power plants | 0.93 | 0 | 0 | 41 | 41 |
| [g]Fueling stations | 0.0018 | 10,200 | 18 | 156,000 | 275 |
| [h]Gas storage facilities | 12.95 | 10 | 130 | 394 | 5,102 |
| **Total** | | | **4,736** | | **126,000** |
| **Percent of CA or United States** | | | **1.2** | | **1.3** |

[a] Number of active oil and gas wells, compressors, and processors from Oil and Gas (2018). The area of each is calculated from the 3 million ha of well pads, roads, and storage facilities required for 600,000 new wells from 2000 to 2012 (Allred et al., 2015).

[b] The number of abandoned U.S. oil and gas wells is from U.S. EPA (2017), Slide 11. The California number is calculated as the U.S. number multiplied by the California to U.S. ratio of active wells. The area of each abandoned oil well is estimated as 50 m², and of each gas well, 25 m² from K. Jepsen (pers. comm., 2018).

[c] The number of coal mines is from EIA (2018a). The area per mine is estimated from the total area among all mines from Sourcewatch (2011) divided by number of mines here.

[d] The number of oil refineries is from EIA (2018b). The area of each refinery is based on the area of the Richmond, California, refinery.

[e] Kilometers of oil and gas pipeline for the United States were from BTS (2018); for California, they were estimated. The area needed for each 1 km of pipeline is estimated to be 6 m (3 m on each side of the pipe) multiplied by 1 km.

[f] The numbers of coal, gas, petroleum, nuclear, and other power plants are from EIA (2018c). The areas for each coal, gas, and nuclear plant is derived from Strata (2017). For coal, the area includes those for the plant and waste disposal (mining is a separate line in this table). For gas, the area is just for the plant. For nuclear, the area includes the areas required for uranium mining, the plant itself, and waste disposal. The areas required for petroleum and other power plants are an average of that for a coal and gas plant.

[g] The number of retail fueling stations in the United States is from AFDC (2014) for 2012 and in California, from Statistica (2017) for 2016. The area of a fueling station is estimated from the area of a typical gas station.

[h] The number of gas storage facilities is from FERC (2004). The area of a gas storage facility is estimated as that of the Aliso Canyon storage facility.

stations, and underground storage facilities. The flammability of natural gas further results in explosions in homes and urban areas that have had fatal consequences.

Table 3.3 shows the estimated land required for the entire fossil-fuel and nuclear infrastructure in California and the United States. The table indicates that the fossil-fuel infrastructure currently takes up about 1.3 percent of the U.S. land area and 1.2 percent of California's land area. Whereas all fossil fuels contribute to this land area

degradation, natural gas's share is growing due to the phaseout of coal and the growth of natural gas, particularly of hydraulically fracked gas. The damage due to fracking includes damage not only to the landscape but also to groundwater, in which fracked natural gas often leaks. Additional damage occurs to roads, which must carry heavy trucks associated with natural gas development. Gas flaring is another form of local environmental degradation, as the flaring emits soot (containing black carbon), which warms the air, evaporates clouds, and melts snow.

A transition to 100 percent WWS, on the other hand, eliminates the energy needed continuously to mine, transport, and process fossil fuels and uranium, which amounts to 12.1 percent of all energy worldwide (Table 7.1). Wind, on the other hand, comes right to the turbine, and sunlight comes right to the solar panel. In other words, eliminating all fossil fuels and uranium by switching to WWS will eliminate immediately 12.1 percent of all energy needed worldwide and will prevent the future degradation of land due to the continuous mining of fossil fuels and uranium.

## 3.2 Why Not Use Natural Gas or Coal with Carbon Capture?

Another proposal to help solve the climate problem is to capture the $CO_2$ emitted from a coal or natural gas power plant before the $CO_2$ is released from the exhaust stack. This would be accomplished with carbon capture and storage (CCS) equipment added to the plant. However, this solution is poor for four reasons: it increases emissions and health problems of all gases and particles aside from $CO_2$ compared with no CCS; it only marginally reduces $CO_2$; it increases the land degradation from the mining of fossil fuels compared with no CCS; and its high cost prevents more effective climate and pollution mitigation with lower-cost renewables.

**Carbon capture and storage** (CCS) is the separation of $CO_2$ from other exhaust gases after fossil-fuel or biofuel combustion, followed by the transfer of the $CO_2$ to an underground geological formation (e.g., saline aquifer, depleted oil and gas field, or un-minable coal seam). The remaining combustion gases are emitted into the air or filtered further. Geological formations worldwide may theoretically store up to 2,000 Gt-$CO_2$, which compares with a global fossil-fuel emission rate in 2017 of about 37 Gt-$CO_2$/y.

Another proposed CCS method is to inject the $CO_2$ into the deep ocean. The addition of $CO_2$ to the ocean, however, results in ocean acidification. Dissolved $CO_2$ in the deep ocean eventually equilibrates with $CO_2$ in the surface ocean, reducing ocean pH and simultaneously supersaturating the surface ocean with $CO_2$, forcing some of it back into the air.

A third type of sequestration method is to mix captured $CO_2$ with concrete material, trapping the $CO_2$ inside the concrete (Section 2.4.8.2).

Carbon capture and use (CCU) is the same as CCS, except that the isolated $CO_2$ with CCU is sold to reduce the cost of the carbon capture equipment. To date, the major application of CCU has been **enhanced oil recovery**. With this process, $CO_2$ is pumped underground into an oil field. It binds with oil, reducing its density and allowing it to rise to the surface more readily. Once the oil rises up, the $CO_2$ is separated from it and sent back into the reservoir. About two additional barrels of oil can be extracted for every tonne of $CO_2$ injected into the ground.

Another proposed use has been to create carbon-based fuels to replace gasoline and diesel. The problem with this proposal is that it allows combustion to continue in vehicles. Combustion creates air pollution, only some of which can be stopped by emission control technologies.

### 3.2.1 Air Pollution Increases and Only Modest Lifecycle $CO_2$e Decreases due to Carbon Capture

Whereas carbon capture equipment is nominally expected to capture 85 to 90 percent of the $CO_2$ from a fossil-fuel exhaust stream, several factors cause the overall $CO_2$ and $CO_2$e savings due to carbon capture to be much smaller than this but also cause an increase in emissions of health-affecting air pollutants relative to no carbon capture. The reasons for these impacts are summarized as follows:

1. A fossil fuel with carbon capture power plant needs to produce 25 to 50 percent more energy, thus requires 25 to 50 percent more fuel, to run the carbon capture equipment than does a plant without the equipment (IPCC, 2005; EIA, 2017).
2. Carbon capture equipment does not capture the upstream $CO_2$e emissions resulting from mining, transporting, or processing the fossil fuel used in the plant. Instead, such emissions increase 25 to 50 percent because 25 to 50 percent more fuel is needed. These emissions offset a portion of the captured $CO_2$ from the plant exhaust and increase the air pollution associated with the mining, transporting, and processing of the fuel.
3. The carbon capture equipment does not capture any of the non-$CO_2$ air pollutants from the fossil-fuel exhaust. Such pollutants include CO, $NO_x$, $SO_2$, organic gases, mercury, toxins, BC, BrC, fly ash, and other aerosol components, all of which affect health. Instead, those pollutants increase 25 to 50 percent

Table 3.4 Theoretical lifecycle 20-year and 100-year $CO_2$e emissions from average U.S. coal power plants, a supercritical pulverized coal (SCPC) power plant, and a natural gas combined cycle gas turbine (CCGT) plant with and without carbon capture. Compare with data from a real carbon capture facility in Table 3.6.

| | Average U.S. Coal Plant | | | Coal SCPC Plant | | | Natural Gas CCGT Plant | | |
|---|---|---|---|---|---|---|---|---|---|
| | No carbon capture | With carbon capture | $CO_2$e remaining (%) | No carbon capture | With carbon capture | $CO_2$e remaining (%) | No carbon capture | With carbon capture | $CO_2$e remaining (%) |
| 20-y $CO_2$e/kWh | 1,316 | 664 | 50.4 | 1,188 | 599 | 50.4 | 896 | 305 | 34.0 |
| 100-y $CO_2$e/kWh | 1,205 | 346 | 28.7 | 965 | 277 | 28.7 | 506 | 179 | 35.4 |

Source: all values are from Skone (2015), except the percent remaining for average U.S. coal was assumed the same as from coal SCPC, and the $CO_2$e values with carbon capture for average U.S. coal were calculated from the percent remaining and the no carbon capture values.

because much more fossil fuel from the plant is needed to run the CCS equipment.

4. The chance that $CO_2$ sequestered underground leaks increases over time and varies with geological formation.

One way to estimate the climate impact of carbon capture equipment attached to a fossil-fuel plant is to examine the plant's lifecycle emissions before and after the equipment is added. **Lifecycle emissions** are carbon-equivalent ($CO_2$e) emissions of a technology per unit electric power generation (kWh or MWh), averaged over a 20- or 100-year time frame. The emissions accounted for include those during the construction, operation, and decommissioning of the plant. For a fossil-fuel (or nuclear) plant, the operation phase includes mining, transporting, and processing the fuel as well as running the plant equipment, repairing the plant over its life, and disposing of waste (e.g., coal residue or nuclear waste) over its life. Lifecycle $CO_2$e is calculated as the lifecycle emission of $CO_2$ plus the lifecycle emission of each other gas or particle pollutant from the technology multiplied by its respective 20- or 100-year GWP (see Table 1.2).

Table 3.4 shows theoretical estimates of 20- and 100-year lifecycle $CO_2$e emissions from an average U.S. coal plant, a modern **supercritical pulverized coal (SCPC)** plant, and a natural gas combined cycle gas turbine (CCGT) plant, each with and without carbon capture. An SCPC plant operates at a higher temperature and pressure than a normal coal plant. As such, the efficiency of combustion (electricity production per mass of coal) is higher. Table 3.4 indicates that, even after carbon capture, the coal SCPC plant still emits 50.4 percent of its $CO_2$e

over 20 years and 28.7 percent over 100 years compared with no carbon capture. A natural gas CCGT emits 34 percent of its $CO_2$e over 20 years and 35.4 percent over 100 years compared with no capture. These results reflect the fact that the carbon capture equipment increases the upstream emissions of $CO_2$e due to increasing the fuel needed to be burned in the power plant. The results also reflect the fact that the carbon capture equipment lets 10 to 15 percent of the $CO_2$ emitted by the stack escape. Whereas these estimates are theoretical, data from a real coal plant before and after carbon capture (Table 3.6) show much less removal of $CO_2$e than indicated in Table 3.4.

The results in Table 3.4 suggest that carbon capture does not come close to eliminating $CO_2$e emissions from coal or gas power plants. Data from real-world projects (Section 3.2.3) indicate even less reduction in $CO_2$e emissions due to carbon capture than Table 3.4 suggests. Further, the lifecycle $CO_2$e emissions from a natural gas or coal plant with carbon capture are not the only emissions associated with a fossil-fuel plant. Other important emissions include those that affect human and animal health. Lifecycle emissions can be placed in context only when all relevant climate-affecting emissions or avoided emissions associated with a plant are accounted for and compared with the same from other energy technologies, as discussed next.

## 3.2.2 Total $CO_2$e Emissions of Energy Technologies

Lifecycle emissions are one component of total carbon equivalent ($CO_2$e) emissions. Additional components relevant to fossil fuels with carbon capture include opportunity cost emissions, anthropogenic heat emissions, anthropogenic water vapor emissions, emissions risk due to $CO_2$ leakage, and emissions due to covering or clearing land for energy development. These are discussed next, in turn.

### 3.2.2.1 Opportunity Cost Emissions

**Opportunity cost emissions** are emissions from the background electric power grid, averaged over a defined period of time (e.g., either 20 years or 100 years), due to two factors. The first factor is the longer time lag between planning and operation of one energy technology relative to another. The second factor is the longer downtime needed to refurbish one technology at the end of its useful life when its useful life is shorter than that of another technology (Jacobson, 2009).

For example, if Plant A takes 4 years and Plant B takes 10 years between planning and operation, the background grid will emit pollution for 6 more years out of 100 years with Plant B than with Plant A. The emissions during those additional 6 years are opportunity cost emissions. Such additional emissions include emissions of both health- and climate-affecting air pollutants.

Similarly, if Plants A and B have the same planning-to-operation time but Plant A has a useful life of 20 years and requires 2 years of refurbishing to last another 20 years and Plant B has a useful life of 30 years but takes

only 1 year of refurbishing, then Plant A is down 2 y / 22 y = 9.1 percent of the time for refurbishing and Plant B is down 1 y / 31 y = 3.2 percent of the time for refurbishing. As such, Plant B is down an additional $(0.091 - 0.032) \times 100$ y = 5.9 years out of every 100 years for refurbishing. During those additional years, the background grid will emit pollution with Plant B.

Mathematically, opportunity cost emissions ($E_{OC}$, in g-$CO_2$e/kWh) are calculated as

$$E_{OC} = E_{BR,H} - E_{BR,L} \qquad (3.1)$$

where $E_{BR,H}$ is the background grid emissions over a specified number of years due to delays between planning and operation and due to refurbishing the technology with the longer delays. $E_{BR,L}$ is the same but for the technology, for the technology with the shorter delays. Background emissions (for either technology) over the number of years of interest, Y, are

$$E_{BR} = E_G \times [T_{PO} + (Y - T_{PO}) \times T_R/(L + T_R)]/Y \qquad (3.2)$$

where $E_G$ is the emissions intensity of the background grid (g-$CO_2$e/kWh for analyses of climate impacts and g-pollutant/kWh for analyses of health-affecting air pollutants), $T_{PO}$ is the time lag (in years) between planning and operation of the technology, $T_R$ is the time (years) to refurbish the technology, and L is the operating life (years) of the technology before it needs to be refurbished.

---

## Example 3.1 Opportunity cost emissions

What are the opportunity cost emissions (g-$CO_2$e/kWh) over 100 years resulting from Plant B if its planning-to-operation time is 15 years, its lifetime is 40 years, and its refurbishing time is 3 years, whereas these values for Plant A are 3 years, 30 years, and 1 year, respectively? Assume both plants produce the same number of kWh/y once operating, and the background grid emits 550 g-$CO_2$e/kWh.

### Solution:

The opportunity cost emissions are calculated as the emissions from the background grid over 100 years of the plant with the higher background emissions (Plant B in this case) minus those from the plant with the lower background emissions (Plant A).

The background emissions from Plant B are calculated from Equation 3.2 with $E_G$ = 550 g-$CO_2$e/kWh, Y = 100 y, $T_{PO}$ = 15 y, L = 40 y, and $T_R$ = 3 y as $E_{BR,H}$ = 550 g-$CO_2$e/kWh $\times$ [15 y + (100 y – 15 y) $\times$ 3 y / 43 y)] / 100 y = 115 g-$CO_2$e/kWh.

Similarly, the background emissions from Plant A averaged over 100 years are $E_{BR,L}$ = 550 g-$CO_2$e/kWh $\times$ [3 y + (100 y – 3 y) $\times$ 1 y / 31 y)] / 100 y = 33.7 g-$CO_2$e/kWh. The difference between the two from Equation 3.1, $E_{OC} = E_{BR,H} - E_{BR,L}$ = 81.3 g-$CO_2$e/kWh, is the opportunity cost emissions of Plant B over 100 years.

The time lag between planning and operation of a technology includes a development time and construction time. The development time is the time required to identify a site, obtain a site permit, purchase or lease the land, obtain a construction permit, obtain financing and insurance for construction, install transmission, negotiate a power purchase agreement, and obtain permits. The construction period is the period of building the plant, connecting it to transmission, and obtaining a final operating license.

The development phase of a coal-fired power plant without carbon capture equipment is generally 1 to 3 years, and the construction phase is another 5 to 8 years, for a total of 6 to 11 years between planning and operation (Jacobson, 2009). No coal plant has been built from scratch with carbon capture, so this could add to the planning-to-operation time. However, for a new plant, it is assumed that the carbon capture equipment can be added during the long planning-to-operation time of the coal plant itself. As such, Table 3.5 assumes the planning-to-operation time of a coal plant without carbon capture is the same as that with carbon capture. The typical lifetime of a coal plant before it needs to be refurbished is 30 to 35 years. The refurbishing time is an estimated 2 to 3 years.

No natural gas plant with carbon capture exists. The estimated planning-to-operation time of a natural gas plant without carbon capture is less than that of a coal plant. However, because of the shorter time, the addition of carbon capture equipment to a new natural gas plant is likely to extend its planning-to-operation time to that of a coal plant with or without carbon capture (6 to 11 years).

For comparison, the planning-to-operation time of a utility-scale wind or solar farm is generally 3 to 5 years, with a development period of 1 to 3 years and a construction period of 1 to 2 years (Jacobson, 2009). This time applies to both onshore and offshore wind. For example, the 407-MW (49 turbine) Horns Rev 3 offshore wind farm, located in the North Sea off the west coast of Denmark, required 1 year and 10 months to build (Frangoul, 2019). Wind turbines often last 30 years before refurbishing, and the refurbishing time is 0.25 to 1 year.

Table 3.5 provides the estimated opportunity cost emissions of coal and natural gas with carbon capture due to the time lag between planning and operation of those plants relative to wind or solar farms. The table indicates an investment in fossil fuels with carbon capture instead of wind and solar results in an additional 46 to 62 g-$CO_2$e/kWh in opportunity cost emissions from the background grid.

### 3.2.2.2 Anthropogenic Heat Emissions

Anthropogenic heat emissions were defined in Section 1.2.3 to include the heat released to the air from the dissipation of electricity; from the dissipation of motive energy by friction; from the combustion of fossil fuels, biofuels, and biomass for energy; from nuclear reaction; and from anthropogenic biomass burning. Jacobson (2014) provides the relative contributions of different energy-generating technologies to worldwide anthropogenic heat emissions.

Table 3.5 includes the g-$CO_2$e/kWh emissions from heat of combustion (for biomass, natural gas, and coal) and from nuclear reaction. However, because the dissipation of the resulting electricity back to heat is due to the consumption rather than production of electricity, that heat release term is not included in the table. In any case, the heat released per unit electricity produced is the same for all technologies.

Solar PV and CSP convert solar radiation to electricity, thereby reducing the flux of heat to the ground or to rooftops below PV panels. This is reflected in Table 3.5 as a negative heat flux. Wind turbines also cause a negative heat flux, discussed in Section 3.2.2.3.

The $CO_2$e emissions (g-$CO_2$e/kWh) due to the anthropogenic heat flux is calculated for all technologies (including the negative heat fluxes due to solar and wind) as follows:

$$H = E_{CO2} \times A_h/(F_{CO2} \times G_{elec}) \tag{3.3}$$

where $E_{CO2}$ is the equilibrium global anthropogenic emission rate of $CO_2$ (g-$CO_2$/y) that gives a specified anthropogenic mixing ratio of $CO_2$ in the atmosphere, $F_{CO2}$ is the direct radiative forcing (W/m$^2$) of $CO_2$ at the specified mixing ratio, $A_h$ is the anthropogenic heat flux (W/m$^2$) due to a specific electric-power-producing technology, and $G_{elec}$ is the annual global energy output of the technology (kWh/y). Radiative forcing is the net increase (+) or decrease (–) in downward solar plus thermal infrared radiation at the top of the atmosphere when a pollutant is present versus absent in the atmosphere. A positive radiative forcing indicates the pollutant warms the air.

The idea behind this equation is that the current radiative forcing (W/m$^2$) in the atmosphere due to $CO_2$ can be maintained at an equilibrium $CO_2$ emission rate,

$$E_{CO2} = \chi_{CO2}C/\tau_{CO2} \tag{3.4}$$

Table 3.5 Total 100-year CO$_2$e emissions from several different energy technologies. The total includes lifecycle emissions, opportunity cost emissions, anthropogenic heat and water vapor emissions, weapons and leakage risk emissions, and emissions from loss of carbon storage in land and vegetation. All units are g-CO$_2$e/kWh-electricity, except the last column, which gives the ratio of total emissions of a technology to the emissions from onshore wind. CCS/U is carbon capture and storage or use.

| Technology | [a] Lifecycle emissions | [b] Opportunity cost emissions due to delays | [c] Anthropogenic heat emissions | [d] Anthropogenic water vapor emissions | [e] Nuclear weapons risk or 100-year CCS/U leakage risk | [f] Loss of CO$_2$ due to covering land or clearing vegetation | [g] Total 100-year CO$_2$e | Ratio of 100-year CO$_2$e to that of wind-onshore |
|---|---|---|---|---|---|---|---|---|
| Solar PV-rooftop | 15–34 | –12 to –16 | –2.2 | 0 | 0 | 0 | 0.8–15.8 | 0.1–3.3 |
| Solar PV-utility | 10–29 | 0 | –2.2 | 0 | 0 | 0.054–0.11 | 7.85–26.9 | 0.91–5.6 |
| CSP | 8.5–24.3 | 0 | –2.2 | 0 to 2.8 | 0 | 0.13–0.34 | 6.43–25.2 | 0.75–5.3 |
| Wind-onshore | 7.0–10.8 | 0 | –1.7 to –0.7 | –0.5 to –1.5 | 0 | 0.0002–0.0004 | 4.8–8.6 | 1 |
| Wind-offshore | 9–17 | 0 | –1.7 to –0.7 | –0.5 to –1.5 | 0 | 0 | 6.8–14.8 | 0.79–3.1 |
| Geothermal | 15.1–55 | 14–21 | 0 | 0–2.8 | 0 | 0.088–0.093 | 29–79 | 3.4–16 |
| Hydroelectric | 17–22 | 41–61 | 0 | 2.7–26 | 0 | 0 | 61–109 | 7.1–22.7 |
| Wave | 21.7 | 4–16 | 0 | 0 | 0 | 0 | 26–38 | 3.0–7.9 |
| Tidal | 10–20 | 4–16 | 0 | 0 | 0 | 0 | 14–36 | 1.6–7.5 |
| Nuclear | 9–70 | 64–102 | 1.6 | 2.8 | 0–1.4 | 0.17–0.28 | 78–178 | 9.0–37 |
| Biomass | 43–1,730 | 36–51 | 3.4 | 3.2 | 0 | 0.09–0.5 | 86–1,788 | 10–373 |
| Natural gas-CCS/U | 179–405 | 46–62 | 0.61 | 3.7 | 0.36–8.6 | 0.41–0.69 | 230–481 | 27–100 |
| Coal-CCS/U | 230–935 | 46–62 | 1.5 | 3.6 | 0.36–8.6 | 0.41–0.69 | 282–1,011 | 33–211 |

[a] Lifecycle emissions are 100-year carbon equivalent (CO$_2$e) emissions that result from the construction, operation, and decommissioning of a plant. They are determined as follows:

Solar PV-rooftop: The range is assumed to be the same as the solar PV-utility range, but with 5 g-CO$_2$/kWh added to both the low and high ends to account for the use of fixed tilt for all rooftop PV versus the use of some tracking for utility PV.

Solar PV-utility: The range is derived from Fthenakis and Raugei (2017). It is inclusive of the 17 g-CO$_2$/kWh mean for cadmium telluride (CdTe) panels at 11 percent efficiency, the 27 g-CO$_2$e/kWh mean for multi-crystalline silicon panels at 13.2 percent efficiency, and the 29 g-CO$_2$e/kWh mean for mono-crystalline silicon panels at 14 percent efficiency. The upper limit of the range is held at the mean for multi-crystalline silicon, since panel efficiencies are now much higher than 13.2 percent. The lower limit is calculated by scaling the CdTe mean to 18.5 percent efficiency, its maximum in 2018.

CSP: The lower limit CSP lifecycle emission rate is from Jacobson (2009). The upper limit is from Ko et al. (2018).

Wind-onshore and wind-offshore: The range is derived from Kaldelis and Apostolou (2017).

Geothermal: The range is from Jacobson (2009) and consistent with the review of Tomasini-Montenegro et al. (2017).

Hydroelectric and wave: From Jacobson (2009).

Tidal: From Douglas et al. (2008).

Nuclear: The range of 9–70 g-$CO_2$e/kWh is from Jacobson (2009), which is within the Intergovernmental Panel on Climate Change's (IPCC) range of 4–110 g-$CO_2$e/kWh (Bruckner et al., 2014), and conservative relative to the 68 (10–130) g-$CO_2$e/kWh from the review of Lenzen (2008) and the 66 (1.4–288) g-$CO_2$e/kWh from the review of Sovacool (2008).

Biomass: The range provided is for biomass electricity generated by forestry residues (43 g-$CO_2$e/kWh), industry residues (46), energy crops (208), agriculture residues (291), and municipal solid waste (1,730) (Kadiyala et al., 2016).

Natural gas-CCS/U: The lower bound is for the CCGT with carbon capture plant from Skone (2015), also provided in Table 3.4. The upper bound is CCGT value without carbon capture, 506 g-$CO_2$e/kWh from Table 3.4, multiplied by 80 percent, which is the percent of $CO_2$e emissions expected to be captured from the Petra Nova facility that will remain in the air over 100 years (Table 3.6).

Coal-CCS/U: The lower bound is for IGCC with carbon capture from Skone (2015). The upper bound is the coal value without carbon capture, 1,168 g-$CO_2$e/kWh from Table 3.6, multiplied by 80 percent, which is the percent of coal lifecycle $CO_2$e emissions from the Petra Nova facility that will remain in the air over 100 years (Table 3.6).

[b] Opportunity cost emissions are emissions per kWh over 100 years from the background electric power grid, calculated from Equations 3.1 and 3.2 due to (i) the longer time lag between planning and operation of one energy technology relative to another and (ii) additional downtime to refurbish a technology at the end of its useful life compared with the other technology. The planning-to-operation times of the technologies in this table are 0.5–2 years for solar PV-rooftop; 2–5 years for solar PV-utility, CSP, wind-onshore, wind-offshore, tidal, and wave; 3–6 years for geothermal; 8–16 years for hydroelectric; 10–19 years for nuclear; 4–9 years for biomass (without CCS/U); and 6–11 years for natural gas-CCS/U and coal-CCS/U (Jacobson, 2009, except rooftop PV and natural gas-CCS/U values are added and solar PV-rooftop is updated here). The refurbishment times are 0.05–1 year for solar PV-rooftop; 0.25–1 year for solar PV-utility, CSP, wind-onshore, wind-offshore, wave, and tidal; 1–2 years for geothermal and hydroelectric; 2–4 years for nuclear; and 2–3 years for biomass, coal-CCS/U, and natural gas-CCS/U. The lifetimes before refurbishment are 15 years for tidal and wave; 30 years for solar PV-rooftop, solar PV-utility, CSP, wind-onshore, and wind-offshore; 40 years for hydroelectric; 30–40 years for nuclear; and 80 years for hydroelectric (Jacobson, 2009). The opportunity cost emissions are calculated here relative to the utility-scale technologies with the shortest time between planning and operation (solar-PV-utility, CSP, wind-onshore, and wind-offshore). The opportunity cost emissions of all other technologies are calculated as in Example 3.1 while assuming a background U.S. grid emission intensity equal to 557.3 g-$CO_2$e/kWh in 2017. This is derived from an electricity mix from EIA (2018d) and emissions, weighted by their 100-year GWPs, of $CO_2$, $CH_4$, and $N_2O$ from mining, transporting, processing, and using fossil fuels, biomass, or uranium. The reason tidal power has opportunity cost emissions although its planning-to-operation time is the same as onshore wind is the shorter lifetime of tidal turbines than wind turbines. Thus, tidal has more downtime over 100 years than do other technologies. See Section 3.2.2.1. The opportunity cost emissions of offshore and onshore wind are assumed to be the same because new projects suggest offshore wind, particularly with faster assembly techniques and with floating turbines, are easier to permit and install now than a decade ago. Although natural gas plants don't take as long as coal plants between planning and operation, natural gas combined with CCS/U is assumed to take the same time as coal with CCS/U.

[c] Anthropogenic heat emissions here include the heat released to the air from combustion (for coal and natural gas) or nuclear reaction, converted to $CO_2$e (see Section 3.2.2.2). For solar PV and CSP, heat emissions are negative because these three technologies reduce sunlight to the surface by converting it to electricity. The lower flux to the surface cools the ground or a building below the PV panels. For wind turbines, heat emissions are negative because turbines extract energy from wind to convert it to electricity (Section 3.2.2.3 and Example 3.6). For binary geothermal plants (low end), it is assumed all heat is re-injected back into the well. For non-binary plants, it is assumed that some heat is used to evaporate water vapor (thus the anthropogenic water vapor flux is positive), but remaining heat is injected back into the well. The electricity from all electric power generation also dissipates to heat, but this is due to the consumption rather than production of power and is the same amount per kWh for all technologies so is not included in this table.

[d] Anthropogenic water vapor emissions here include the water vapor released to the air from combustion (for coal and natural gas) or from evaporation (water-cooled CSP, water-cooled geothermal, hydroelectric, nuclear natural gas, and coal), converted to $CO_2$e (see Section 3.2.2.3). Air-cooled CSP and geothermal plants have zero water vapor flux, representing the low end of these technologies. The high end is assumed to be the same as for nuclear, which also uses water for cooling. The low end for hydroelectric power assumes 1.75 kg-$H_2O$/kWh evaporated from reservoirs at mid- to high latitudes (Flury and Frischknecht, 2012). The upper end is 17.0 kg-$H_2O$/kWh from Jacobson (2009) for lower latitude reservoirs and assumes reservoirs serve multiple purposes. For biomass, the number is based only on the water emitted from the plant due to evaporation or combustion, not water to irrigate some energy crops. Thus, the upper estimate is low. The negative water vapor flux for onshore and offshore wind is due to the reduced water evaporation caused by wind turbines (Section 3.2.2.3 and Example 3.6).

[e] Nuclear weapons risk is the risk of emissions due to nuclear weapons use resulting from weapons proliferation caused by the spread of nuclear energy. The risk ranges from zero (no use of weapons over 100 years) to 1.4 g-$CO_2$e/kWh (one nuclear exchange in 100 years) (Section 3.3.2.1). The 100-year CCS/U leakage risk is the estimated rate, averaged over 100 years, that $CO_2$ sequestered underground leaks back to the atmosphere. Section 3.2.2.4 contains a derivation. The leakage rate from natural gas-CCS/U is assumed to be the same as for coal-CCS/U.

[f] Loss of carbon, averaged over 100 years, due to covering land or clearing vegetation is the loss of carbon sequestered in soil or in vegetation due to the covering or clearing of land by an energy facility; by a mine where the fuel is extracted from (in the case of fossil fuels and uranium); by roads, railways, or pipelines needed to transport the fuel; and by waste disposal sites. No loss of carbon occurs for solar PV-rooftop, wind-offshore, wave, or tidal power. In all remaining cases, except for solar PV-utility and CSP, the energy facility is assumed to replace grassland with the organic carbon content as described in the text. For solar PV-utility and CSP, it is assumed that the organic content of both the vegetation and soil are 7 percent that of grassland because (i) almost all CSP and many PV arrays are located in deserts with low carbon storage and (ii) most utility PV arrays and CSP mirrors are elevated above the ground. For biomass, the low value assumes the source of biomass is industry residues or contaminated wastes. The high value assumes energy crops, agricultural residues, or forestry residues. See Section 3.2.2.5.

[g] The total column is the sum of the previous six columns.

where $\chi_{CO2}$ (ppmv) is the specified anthropogenic mixing ratio that gives the current $CO_2$ radiative forcing, C is a conversion factor ($8.0055 \times 10^{15}$ g-$CO_2$/ppmv-$CO_2$), and $\tau_{CO2}$ is the data-constrained *e*-folding lifetime of $CO_2$ against loss by all processes. As of 2020, $\tau_{CO2}$ is ~50 years but increasing over time (e.g., Jacobson, 2012a).

Equation 3.4 is derived by noting that the time rate of change of the atmospheric mixing ratio of a well-mixed gas, such as $CO_2$, is simply $d\chi/dt = E - \chi C/\tau$. In steady state, this simplifies to $E = \chi C/\tau$. Scaling the ratio of this equilibrium $CO_2$ emission rate to the radiative forcing of $CO_2$ by the ratio of the anthropogenic heat flux to the electricity generation per year producing that heat flux, gives Equation 3.3, the $CO_2$e emission rate of the heat flux.

Thus, Equation 3.3 accounts for the emission rate of $CO_2$ needed to maintain a mixing ratio of $CO_2$ in the air that gives a specific radiative forcing. It does not use the present-day emission rate because that results in a much higher $CO_2$ mixing ratio than is currently in the atmosphere, since $CO_2$ emissions are not in equilibrium with the $CO_2$ atmospheric mixing ratio. Equation 3.3 requires a constant emission rate that gives the observed mixing ratio of $CO_2$ for which the current direct radiative forcing applies. Similarly, the energy production rate in Equation 3.3 gives a consistent anthropogenic heat flux.

Whereas radiative forcing is a top-of-the-atmosphere value (and represents changes in heat integrated over the whole atmosphere) and heat flux is added to the bottom of the atmosphere, they both represent the same amount of heat added to the atmosphere. In fact, because the anthropogenic heat flux adds heat to near-surface air, it has a slightly greater impact on surface air temperature per unit radiative forcing than does $CO_2$. For example, the globally averaged temperature change per unit direct radiative forcing for $CO_2$ is ~0.6 K/(W/m$^2$) (Jacobson, 2002), whereas the temperature change per unit anthropogenic heat plus water vapor flux is ~0.83 K/(W/m$^2$) (Jacobson, 2014). As such, the estimated $CO_2$e values for heat fluxes in particular in Table 3.5 may be slightly underestimated.

---

### Example 3.2 Calculate the carbon equivalent heat emissions for coal and nuclear power worldwide

In 2005, the anthropogenic flux of heat (aside from heat used to evaporate water) from all anthropogenic heat sources worldwide was $A_h = 0.027$ W/m$^2$ (Jacobson, 2014). Assume the percent of all heat from coal combustion was 4.87 percent and from nuclear reaction was 1.55 percent.

Estimate the $CO_2$e emissions corresponding to the coal and nuclear heat fluxes given the energy generation of $G_{elec} = 8.622 \times 10^{12}$ kWh/y from coal combustion and $2.64 \times 10^{12}$ kWh/y from nuclear reaction.

Assume an anthropogenic $CO_2$ direct radiative forcing of $F_{CO2} = 1.82$ W/m$^2$, which corresponds to an anthropogenic mixing ratio of $CO_2$ of $\chi_{CO2} = 113$ ppmv (Myhre et al., 2013). Also assume a $CO_2$ *e*-folding lifetime of $\tau_{CO2} = 50$ years.

#### Solution:

From Equation 3.4, the equilibrium emission rate of $CO_2$ giving the anthropogenic mixing ratio is

$$E_{CO2} = 1.809 \times 10^{16} \text{g} - CO_2/\text{y}.$$

Multiplying the total anthropogenic heat flux by the respective fractions of heat from coal combustion and nuclear reaction gives $A_h = 0.00132$ W/m$^2$ for coal and 0.00042 W/m$^2$ for nuclear. Substituting these and the other given values into Equation 3.3 gives H = 1.52 g-$CO_2$e/kWh for coal and 1.57 g-$CO_2$e/kWh for nuclear.

---

## Example 3.3 Calculate the carbon-equivalent negative heat emissions of a solar PV panel

Solar panels convert about 20 percent of the sun's energy to electricity, thereby reducing the flux of sunlight to the ground. What is the reduction in heat flux ($W/m^2$) per kWh/y of electricity generated by a solar panel and what is the corresponding $CO_2e$ emission reduction? The surface area of the Earth is $5.092 \times 10^{14}$ $m^2$.

### Solution:

If a solar panel produces $G_{elec}$=1 kWh/y of electricity, the panel prevents exactly that much solar radiation from converting to heat compared with the sunlight otherwise hitting an equally reflective surface. Eventually, the electricity converts to heat as well (as does the electricity from all electric power generators). However, other electric power generators do not remove energy from the sun on the same timescale as solar panels do.

Multiplying the avoided heat (−1 kWh/y) by 1,000 W/kW and dividing by 8,760 h/y and by the area of the Earth gives $A_h = -2.24 \times 10^{-16}$ $W/m^2$. Substituting this, $G_{elec}$ = 1 kWh/y, and $E_{CO2}$ and $F_{CO2}$ from Example 3.2 into Equation 3.3 gives H = −2.23 g-$CO_2e$/kWh.

---

Finally, for hydropower, evaporation of water vapor at the surface of a reservoir by the sun increases anthropogenic water vapor emissions (Section 3.2.2.3). Because evaporation requires energy, it cools the surface of the reservoir. The energy used to evaporate the water becomes embodied in latent heat carried by the water vapor. However, the water vapor eventually condenses in the air (forming clouds), releasing the heat back to the air. As a result, warming of the air offsets cooling at the surface, so hydropower causes no net anthropogenic heat flux. On the other hand, water vapor is a greenhouse gas, resulting in a net warming of the air due to evaporation. This warming is accounted for in the next section.

### 3.2.2.3 Anthropogenic Water Vapor Emissions

Fossil-fuel, biofuel, and biomass burning releases not only heat, but also water vapor. The water vapor results from chemical reaction between the hydrogen in the fuel and oxygen in the air. In addition, coal, natural gas, and nuclear plants require cool liquid water to re-condense the hot steam as it leaves a steam turbine. This process results in significant water evaporating out of a cooling tower to the sky. Many CSP plants also use water cooling, although some use air cooling. Similarly, whereas non-binary geothermal plants and some binary plants use water cooling, thus emitting water vapor, binary plants that use air cooling do not emit any water vapor. Further, water evaporates from reservoirs behind hydroelectric power plant dams. Table 1.1 indicates that anthropogenic water vapor from all anthropogenic sources causes about 0.23 percent of gross global warming.

On the other hand, wind turbines reduce water vapor, a greenhouse gas, by reducing wind speeds (Chapter 7)

(Jacobson and Archer, 2012; Jacobson et al., 2018a). Water evaporation is a function of wind speed (and temperature).

In this section, the positive or negative $CO_2e$ emissions per unit energy (M, g-$CO_2e$/kWh) due to increases or decreases in water vapor fluxes resulting from an electric power source are quantified. The emissions are estimated with an equation similar to Equation 3.3, except the anthropogenic moisture energy flux ($A_m$, $W/m^2$) is substituted for the heat flux:

$$M = E_{CO2} \times A_m / (F_{CO2} \times G_{elec}) \tag{3.5}$$

In this equation, the globally averaged moisture energy flux can be obtained from the water vapor flux per unit energy (V, kg-$H_2O$/kWh) by

$$A_m = V \times L_e \times G_{elec} / (S \times A_e) \tag{3.6}$$

where $L_e = 2.465 \times 10^6$ J/kg-$H_2O$ is the latent heat of evaporation, $S = 3.1536 \times 10^7$ seconds per year, and $A_e = 5.092 \times 10^{14}$ $m^2$ is the surface area of the Earth. For water evaporating from a hydropower reservoir, V = 1.75 to 17 kg-$H_2O$/kWh (Table 3.5, Footnote c).

Combining Equations 3.5 and 3.6 gives the globally averaged $CO_2e$ emissions per unit energy due to a positive or negative water vapor flux resulting from an energy generator as

$$M = E_{CO2} \times V \times L_e / (F_{CO2} \times S \times A_e) \tag{3.7}$$

This equation is independent of the total annual energy production ($G_{elec}$). Examples 3.4 to 3.6 provide calculations of anthropogenic water vapor fluxes for several of the generators in Table 3.5.

## Example 3.4 Calculate the carbon-equivalent anthropogenic water vapor emissions from natural gas and nuclear plants

The global anthropogenic water vapor flux from natural gas power plants in 2005 was $A_m = 0.00268$ W/m$^2$ and from nuclear power plants it was $A_m = 0.000746$ W/m$^2$ (Jacobson, 2014). The total energy generation from natural gas use was $G_{elec} = 7.208 \times 10^{12}$ kWh/y and from nuclear was $2.64 \times 10^{12}$ kWh/y. Calculate the $CO_2e$ emissions associated with these fluxes.

### Solution:

Substituting $E_{CO2}$ and $F_{CO2}$ from Example 3.2 and $A_m$ and $G_{elec}$ provided in the problem into Equation 3.5 gives M = 3.69 g-$CO_2e$/kWh for natural gas and 2.81 g-$CO_2e$/kWh for nuclear.

## Example 3.5 Calculate the carbon-equivalent anthropogenic water vapor emissions from a hydropower reservoir

If the evaporation rate of water from a hydropower reservoir is V = 1.75 kg-$H_2O$/kWh (Flury and Frischknecht, 2012), determine the $CO_2e$ emissions of water vapor from the reservoir.

### Solution:

Substituting V into Equation 3.7 and using $E_{CO2}$ and $F_{CO2}$ from Example 3.2 gives the carbon equivalent emissions due to hydropower reservoir evaporation as M = 2.66 g-$CO_2e$/kWh.

Wind turbines extract kinetic energy from the wind and convert it to electricity. **Kinetic energy** is the energy embodied in air due to its motion. For every 1 kWh of electricity produced, 1 kWh of kinetic energy is extracted. As with all electric power generation, the 1 kWh of electricity eventually converts back to heat that is added back to the air. However, for purposes of assigning $CO_2e$ emissions or savings, the conversion of electricity back to heat is not assigned to any particular electric power generator in Table 3.5. However, the addition or extraction of heat and water vapor by the energy technology is.

When electricity dissipates to heat, some of that heat returns to kinetic energy. Heat is **internal energy**, which is the energy associated with the random, disordered motion of molecules. Higher-temperature molecules move faster than lower-temperature molecules. Some of the internal energy in the air causes air to rise since warm, low-density air rises when it is surrounded by cool, high-density air. To raise the air, internal energy is converted to **gravitational potential energy (GPE)**, which is the energy required to lift an object of a given mass against gravity a certain distance. The lifted parcel is now cooler

as a result of giving away some of its internal energy to GPE. Differences in GPE over horizontal distance create a pressure gradient, which recreates some kinetic energy in the form of wind (Section 6.8).

In sum, wind turbines convert kinetic energy to electricity, which dissipates to heat. Some of that heat converts to GPE, some of which converts back to kinetic energy. If a wind turbine did not extract kinetic energy from the wind, that energy would otherwise still dissipate to heat due to the wind bashing into rough surfaces, which are sources of friction. But such dissipation would occur over a longer time.

However, **wind turbines have an additional effect, which is to reduce water vapor, a greenhouse gas.** When wind from dry land blows over a lake, for example, the dry wind sweeps water vapor molecules away from the surface of the lake. More water vapor molecules must then evaporate from the lake to maintain saturation of water over the lake surface. In this way, winds increase the evaporation of water over not only lakes, but also over oceans, rivers, streams, and soils. Because a wind turbine extracts energy from the wind, it slows the wind, reducing evaporation of water.

By reducing evaporation, wind turbines warm the water or soil near the turbine because evaporation is a cooling process, so less evaporation causes warming. However, because the air now contains less water vapor, less condensation occurs in the air. Since condensation releases heat, less of it means the air cools.

In addition, because a wind turbine slows the wind in its wake, it drops the air pressure in its wake as well (Section 6.4). Lower pressure decreases temperature, as evidenced by the increased fog thickness in the wake of wind turbines at the Horns Rev offshore wind farm (Hasager et al., 2013). The increase in fog thickness results from a slight increase in the relative humidity in a turbine's wake upon a slight drop in temperature, which is due to the drop in pressure.

Thus, the surface warming due to wind turbines reducing evaporation is cancelled by the air cooling due to both lesser atmospheric condensation and lower temperatures in the turbines wake.

However, because water vapor is a greenhouse gas, less of it in the air means that more heat radiation from the Earth's surface escapes to space, cooling the ground, reducing internal energy. Since water vapor stays in the air for days to weeks, its absence due to a wind turbine reduces heat trapped near the surface over that time more than the one-time dissipation of electricity, created by the wind turbine, increases heat to the air.

In sum, wind turbines allow a net escape of energy to space by reducing water vapor. A portion of the lost energy comes from the air's internal energy, resulting in lower air temperatures. The rest comes from kinetic energy, reducing wind speeds, and from gravitational potential energy, reducing air heights. As such, a new equilibrium is reached in the atmosphere. Section 6.9.1 quantifies the impacts of different numbers of turbines worldwide on temperatures and water vapor.

Thus, wind turbines reduce temperatures in the global average by reducing both heat fluxes and water vapor fluxes. Wind turbines do increase temperatures on the ground downwind of a wind farm because they reduce evaporation, but in the global average, this warming is more than offset by atmospheric cooling due to less condensation plus the loss of more heat radiation to space due to the reduction in water vapor caused by wind turbines.

The energy taken out of the atmosphere temporarily (because it is returned later as heat from dissipation of electricity) by wind turbines is 1 kWh per 1 kWh of electricity production. The maximum reduction in water vapor, based on global computer model calculations (Chapter 7), due to wind turbines ranges from $-0.3$ to $-1$ kg-$H_2O$/kWh, where the variation depends on the number and location of wind turbines. Example 3.6 provides an estimate of the $CO_2e$ savings due to wind turbines from these two factors.

---

## Example 3.6 Estimate the globally averaged $CO_2e$ water vapor and heat emission reductions due to wind turbines

Assuming that wind turbines extract 1 kWh of the wind's kinetic energy for each 1 kWh of electricity produced, estimate the $CO_2e$ savings per unit energy from reduced heat and water vapor fluxes due to wind turbines considering that, when the turbine is not operating, every 1 kWh of kinetic energy in the wind evaporates 0.3 to 1 kg-$H_2O$/kWh and the rest of the energy remains in the atmosphere. Assume the equilibrium emission rate and resulting radiative forcing of $CO_2$ from Example 3.2.

### Solution:

Multiplying the latent heat of evaporation ($L_e = 2.465 \times 10^6$ J/kg) and 1 kWh/3.6 $\times 10^6$ J by $-0.3$ to $-1$ kg-$H_2O$/kWh gives the reduction in energy available to evaporate water as $-0.21$ to $-0.69$ kWh per kWh of electricity produced. Multiplying 1,000 W/kW and dividing by 8,760 h/y and by the area of the Earth, $5.092 \times 10^{14}$ $m^2$, gives $A_m/G_{elec} = -4.6 \times 10^{-17}$ to $-1.53 \times 10^{-16}$ (W/$m^2$)/(kWh/y). Substituting this and $E_{CO2}$ and $F_{CO2}$ from Example 3.2 into Equation 3.5 gives the anthropogenic water vapor energy flux from wind turbines as $-0.46$ to $-1.53$ g-$CO_2e$/kWh.

The heat flux is the difference between $-1$ kWh/kWh-electricity and $-0.21$ to $-0.69$ kWh/kWh-electricity, which is $-0.79$ to $-0.31$ kWh/kWh-electricity. Performing the same calculation as above gives the anthropogenic heat flux from wind turbines as $-1.77$ to $-0.70$ g-$CO_2e$/kWh. The total heat plus water vapor energy flux savings due to wind turbines is thus $-2.23$ g-$CO_2e$/kWh, the same as for solar panels (Example 3.3).

### 3.2.2.4 Leaks of $CO_2$ Sequestered Underground

The sequestration of carbon underground due to CCS or CCU (e.g., from injecting $CO_2$ during enhanced oil recovery) runs the risk of $CO_2$ leaking back to the atmosphere through existing fractured rock or overly porous soil or through new fractures in rock or soil resulting from an earthquake. Here, a range in the potential emission rate due to $CO_2$ leakage from the ground is estimated.

The ability of a geological formation to sequester $CO_2$ for decades to centuries varies with location and tectonic activity. IPCC (2005, p. 216) references $CO_2$ leakage rates for an enhanced oil recovery operation of 0.00076 percent per year, or 1 percent over 1,000 years, and $CH_4$ leakage from historical natural gas storage systems of 0.1 to 10 percent per 1,000 years. Thus, while some well-selected sites could theoretically sequester 99 percent of $CO_2$ for 1,000 years, there is no certainty of this since tectonic activity or natural leakage over 1,000 years is not possible to predict. Because liquefied $CO_2$ injected underground will be under high pressure, it will take advantage of horizontal and vertical fractures in rocks to escape as a gas back to the air. Because $CO_2$ is an acid, its low pH will also cause it to weather rocks over time. If a leak from an underground formation to the atmosphere occurs, it may or may not be detected. If a leak is detected, it may or may not be sealed, particularly if it occurs over a large area.

The time-averaged leakage rate of $CO_2$ from a reservoir can be calculated by first estimating how the stored mass of $CO_2$ changes over time. The stored mass ($S$) of $CO_2$ at any time t in a reservoir, resulting from a constant injection at rate $I$ (mass/y) and $e$-folding lifetime against leakage, T (years), is

$$S(t) = S(0)e^{-t/T} + TI(1 - e^{-t/T}) \qquad (3.8)$$

where $S(0)$ is the stored mass at time t = 0. The average leakage rate over t years is then simply the injection rate minus the remaining stored mass at time t divided by t years,

$$L(t) = I - S(t)/t \qquad (3.9)$$

Once an injection rate and lifetime against leakage are known, the average leakage rate of $CO_2$ from an underground storage reservoir over a specified period can be calculated from Equations 3.8 and 3.9.

---

### Example 3.7 Estimating average leakage rates from underground storage reservoirs

Assume a coal-fired power plant has a $CO_2$ emission rate before carbon capture and sequestration ranging from 790 to 1,017 g-$CO_2$/kWh. Assume also that carbon capture equipment added to the plant captures 90 and 80 percent, respectively, of the $CO_2$ (giving a low and high, respectively, emission rate of remaining $CO_2$ to the air). If the captured $CO_2$ is injected underground into a geological formation that has no initial $CO_2$ in it, calculate a low and high $CO_2$ emission rate from leakage averaged over 100 years, 500 years, and 1,000 years. Assume a low and high $e$-folding lifetime against leakage of 5,000 years and 100,000 years, respectively. The low value corresponds to an 18 percent leakage over 1,000 years, close to that of some observed methane leakage rates. The high value corresponds to a 1 percent loss of $CO_2$ over 1,000 years (e.g., IPCC, 2005).

#### Solution:

The low and high injection rates are $790 \times 0.9 = 711$ g-$CO_2$/kWh and $1,017 \times 0.85 = 864.5$ g-$CO_2$/kWh, respectively. Substituting these injection rates into Equation 3.8 (using the high lifetime with the low injection rate and the low lifetime with the high injection rate) and the result into Equation 3.9 gives a leakage rate range of 0.36 to 8.6 g-$CO_2$/kWh over 100 years; 1.8 to 42 g-$CO_2$/kWh over 500 years, and 3.5 to 81 g-$CO_2$/kWh over 1,000 years.

Thus, the longer the averaging period is, the greater the average emission rate over the period due to $CO_2$ leakage will be.

### 3.2.2.5 Emissions from Covering Land or Clearing Vegetation

**Emissions from covering land or clearing vegetation** are emissions of $CO_2$ itself due to (a) reducing the carbon stored in soil and in the vegetation above the soil by covering land with impervious material or (b) reducing the carbon stored in vegetation by clearing land so less vegetation grows. When soil is covered with impervious material, such as concrete or asphalt, vegetation can't grow or decay, or its remains become part of the soil. Similarly, when land is cleared of vegetation, less carbon is stored in the vegetation and below ground. Energy facilities cover land and reduce vegetation.

Estimates of the organic carbon stored in grassland and the soil under grassland are 1.15 kg-C/m$^2$ and 13.2 kg-C/m$^2$, respectively (Ni, 2002). Normally, when grass dies, the dead grass contributes to the soil organic carbon. The grass then regrows, removing carbon from the air by photosynthesis. If the soil is instead covered with concrete, the grass no longer exists to remove carbon from the air or store carbon in the soil. However, existing carbon stored underground remains. Some of this is oxidized, though, over time and carried away by groundwater.

The carbon emissions due to developing land for an energy facility can be estimated simplistically by first summing the land areas covered by the facility; the mine where the fuel is extracted from (in the case of fossil fuels and uranium); the roads, railways, or pipelines needed to transport the fuel; and the waste disposal site associated with the facility. This summed area is then multiplied by the organic carbon content normally stored in vegetation per unit area that is lost plus the organic carbon content normally stored in soil under the vegetation per unit area that is lost. The latter value can be estimated as approximately one-third the original organic carbon content of the soil. The loss of carbon is then converted to a loss of carbon per unit electricity produced by the energy facility over a specified period of time. For purposes of Table 3.5, this period is 100 years. Example 3.8 provides an example calculation of $CO_2$e emissions from the covering land with an energy facility.

---

### Example 3.8 Estimating the loss of carbon stored in vegetation and soil due to an energy facility

Assume a 425-MW coal facility has a 65 percent capacity factor and has a footprint of 5.2 km$^2$, including the land for the coal facility, mining, railway transport, and waste disposal. Calculate the emission rate of $CO_2$ from the soil and vegetation, averaged over 100 years, due to this facility, assuming that it replaces grass and 34 percent of the soil carbon is lost.

#### Solution:

The energy generated over 1 year from this plant is 425 MW × 8,760 h/y × 0.65 × 1,000 kW/MW = 2.42 × $10^9$ kWh/y. Over 100 years, the energy produced is 2.42 × $10^{11}$ kWh.

The carbon lost in soil is 0.34 × 13.2 kg-C/m$^2$ = 4.5 kg-C/m$^2$ and that lost from vegetation is 1.15 kg-C/m$^2$, for a total of 5.64 kg-C/m$^2$. Multiplying by 1,000 g/kg and the molecular weight of $CO_2$ (44.0095 g-$CO_2$/mol), then dividing by the molecular weight of carbon (12.0107 g-C/mol) give 20,700 g-$CO_2$/m$^2$. Multiplying this by the land area covered by the facility and dividing by the 100-year energy use gives an emission rate from lost soil and vegetation carbon as 0.44 g-$CO_2$/kWh, averaged over 100 years.

---

Because most of the carbon in soil and vegetation is lost immediately, the 100-year average loss of carbon from the soil provided in Table 3.5 underestimates the impact on climate damage of an energy facility that occupies land. Most climate impacts from the loss of carbon will begin when the emissions occur. Thus, for example, the impacts over 10 years of carbon loss in soil are 10 times those in Table 3.5. However, to maintain consistency with the other types of carbon-equivalent emissions in the table, that from soil carbon loss are also averaged over 100 years.

### 3.2.2.6 Comparison of Coal and Natural Gas with Carbon Capture with Other Energy Technologies

Table 3.5 compares the overall 100-year $CO_2$e emissions from coal and natural gas power plants that have carbon capture (CCS or CCU) with emissions from other electricity-generating technologies. The table indicates that coal-CCS/U results in 33 to 211 times the $CO_2$e emissions as onshore wind per unit electricity generated. Natural gas-CCS/U results in 27 to 100 times the emissions as onshore wind.

The reasons for the high $CO_2$e emissions of coal and natural gas with carbon capture are (1) coal and gas plants need 25 to 50 percent more energy to run the carbon capture equipment, and this increases the upstream emissions (fuel mining, transport, and processing) of coal and gas by 25 to 50 percent (Example 3.9); (2) the capture equipment allows 10 to 30 percent of the $CO_2$ in the power plant exhaust to escape (Example 3.9); (3) $CO_2$e emissions from the background grid occur due to the time lag between planning and operation of a coal or gas plant with capture relative to a wind or solar farm; (4) coal and natural gas have heat and water vapor emissions associated with them; (5) some leakages of $CO_2$ occurs once $CO_2$ is sequestered; and (6) coal and gas facilities reduce the storage of carbon in the ground.

Table 3.5 provides climate-relevant emissions, but not health-relevant emissions. Air pollution emissions of coal and natural gas without carbon capture are 100 to 400 times those of onshore wind per unit energy. Adding carbon capture to a coal or gas plant increases air pollution emissions another 25 to 50 percent.

The high air pollution and climate-relevant emission rates of coal and natural gas with carbon capture suggest that spending money on them represents an opportunity cost relative to spending money on lower-emitting technologies. Another issue is that, in a future WWS system, the number of hours of fossil-fuel use at any given plant decreases, making CCS equipment, which is already costly, even more uneconomical (Lund and Mathiesen, 2012).

### 3.2.3 Carbon Capture Projects

To date, $CO_2$ has been captured and separated primarily from mined natural gas or, in one case, from gasified coal. In all such cases, the $CO_2$ has been used to enhance oil recovery.

As of 2020, only two fossil-fuel power plants have operated with carbon capture equipment. In both cases, the separated $CO_2$ was used for enhanced oil recovery, and the CCU equipment was installed at high cost. One project experienced equipment problems, resulting in much more $CO_2$ released to the air than anticipated. The other project required a natural gas plant to be built to power the CCU equipment, also resulting in much less benefit than anticipated. Future projects like these must also be in proximity to an oil and gas production field.

The first electric power plant with CCU equipment was the **Boundary Dam power station** in Estevan, Saskatchewan, Canada. It has been operating with CCU equipment on one coal boiler connected to a steam turbine since October 2014. The cost of the retrofit project was $1.5 billion ($13.6 million/MW for a 110-MW turbine). This cost included a $240 million subsidy from the Canadian government and was on top of the original coal plant cost. Whereas half the captured $CO_2$ from the CCU equipment has been sold for enhanced oil recovery, the other half has been released to the air. In addition, since 2016, the CCU equipment has been operating only 40 percent of the time due to design problems. The energy required to run the CCU equipment also results in additional uncaptured $CO_2$ emissions. Mining the coal for the plant results in even more uncaptured emissions.

The second plant with CCU equipment was the **W.A. Parish coal power plant** near Thompsons, Texas. The plant was retrofitted with CCU equipment as part of the **Petra Nova** project and began using the equipment during January 2017. The CC equipment (240 MW) receives 36.7 percent of the emissions from a 654-MW boiler at the plant. The equipment requires about 0.497 kWh of electricity to run per kWh produced by the coal plant (Table 3.6, Footnote 7). A natural gas turbine with a heat recovery boiler was installed to provide this electricity. A cooling tower and water treatment facility were also added. The retrofit cost $1 billion ($4,200/kW) beyond the coal plant cost (Scottmadden, 2017).

The captured $CO_2$ is compressed and piped to an oil field, where it is used to enhance oil recovery. $CO_2$ from the gas turbine is not captured. Natural gas production also has upstream $CO_2$e emissions, including $CH_4$ leaks, which are not captured. Upstream $CO_2$ and $CH_4$ emissions from the coal plant are also uncaptured.

Table 3.6 indicates that, when upstream emissions are excluded, the CCU equipment captures an average of only 55.4 percent of coal combustion $CO_2$ and only 33.9 percent of coal plus gas combustion $CO_2$. Table 3.6 also shows the upstream emissions from the mining and processing of coal and natural gas. When these are accounted for, the CCU equipment reduces coal and gas combustion plus upstream $CO_2$ a net of only 10.8 percent

Table 3.6 **Comparison of relative $CO_2e$ emissions, electricity use, and electricity total social costs among three scenarios related to the Petra Nova coal-CCU facility, each over a 20-yr and 100-yr time frame. The first scenario is using natural gas to power the carbon capture (CC) equipment. This is based on data from the Petra Nova facility (EIA, 2017). The second scenario is running the CC equipment with onshore wind instead of natural gas. The third is using the same quantity of wind electricity required to run the CC equipment to instead replace coal electricity from the coal plant. In all cases, the additional energy required to run the CC equipment is equivalent to 49.7 percent of the energy output of the coal plant (Footnote 7). The coal plant has a nameplate capacity of 654 MW, but only 240 MW (36.7 percent) is subject to CC. The numbers in the table are all based on the portion subject to CC. All emission units (including of natural gas emissions) are $g\text{-}CO_2e/kWh$-coal-electricity-generation.**

| | Coal with gas-powered CC 20 yr | Coal with gas-powered CC 100 yr | Coal with wind-powered CC 20 yr | Coal with wind-powered CC 100 yr | Wind used for CC replacing coal + remaining coal 20 yr | Wind used for CC replacing coal + remaining coal 100 yr |
|---|---|---|---|---|---|---|
| a) Upstream $CO_2$ from coal[1] | 97.2 | 97.2 | 97.2 | 97.2 | 48.9 | 48.9 |
| b) Upstream $CO_2e$ of leaked $CH_4$ from coal[2] | 353 | 140 | 353 | 140 | 177.6 | 70.4 |
| c) Coal stack $CO_2$ before capture[3] | 930.6 | 930.6 | 930.6 | 930.6 | 468.1 | 468.1 |
| d) Total coal $CO_2e$ before capture (a+b+c) | 1,381 | 1,168 | 1,381 | 1,168 | 695 | 587 |
| e) Remaining stack $CO_2$ after capture[4] | 414.6 | 414.6 | 414.6 | 414.6 | – | – |
| f) $CO_2$ captured from stack (c-e) | 516.0 | 516 | 516 | 516 | – | – |
| g) Percent stack $CO_2$ captured (f/c) | 55.4% | 55.4% | 55.4% | 55.4% | – | – |
| h) $CO_2$ emissions gas combustion[5] | 200.9 | 200.9 | 0 | 0 | 0 | 0 |
| i) Upstream $CO_2e$ of $CH_4$ from gas leaks[6] | 139.2 | 55.03 | 0 | 0 | 0 | 0 |
| j) Upstream $CO_2$ from gas mining, transport[7] | 26.85 | 26.85 | 0 | 0 | 0 | 0 |
| k) Total $CO_2e$ emissions (a+b+e +h+i+j) | 1,232 | 934.5 | 865 | 652 | 695 | 587 |
| l) Percent of coal $CO_2e$ re-emitted (k/d)[8] | **89.2%** | **80.0%** | **62.6%** | **55.8%** | **50.3%** | **50.3%** |
| m) Percent of coal $CO_2e$ captured (100-l) | **10.8%** | **20%** | **37.4%** | **44.2%** | **49.7%** | **49.7%** |
| n) Relative $CO_2e$ to original (l/100)[9] | 0.892 | 0.80 | 0.626 | 0.558 | 0.503 | 0.503 |

| | | | | | | |
|---|---|---|---|---|---|---|
| o) Relative air pollution to original[10] | 1.25 | 1.25 | 1.0 | 1.0 | 0.503 | 0.503 |
| p) Energy required relative to original[11] | 1.497 | 1.497 | 1.497 | 1.497 | 1 | 1 |
| q) Private energy cost/kWh relative to original[12] | 1.74 | 1.74 | 1.74 | 1.74 | 0.71 | 0.71 |
| r) Social cost before changes ($/MWh)[13] | 334 | 334 | 334 | 334 | 334 | 334 |
| s) Social cost after changes ($/MWh)[14] | 413 | 399 | 353 | 342 | 189 | 189 |
| t) Social cost ratio (s/r) | 1.24 | 1.19 | 1.06 | 1.02 | 0.57 | 0.57 |

[1] Coal upstream emissions are estimated as 27 g-$CO_2$/MJ = 97.2 g-$CO_2$/kWh (Howarth, 2014). Upstream emissions include emissions from fuel extraction, fuel processing, and fuel transport. Upstream $CO_2$ emissions (from the portion of the coal plant not replaced) for the wind-replacing-some-coal cases (last two columns) are the same as in the other cases, but multiplied by 0.503, which equals 1 minus the fraction of coal electricity used to run the carbon capture equipment, which is derived in Footnote 7. Since the electricity used to run the CC equipment is used to replace coal in this case, upstream coal emissions are reduced accordingly.

[2] For coal, the 100-year $CO_2$e from $CH_4$ leaks is estimated from (Skone, 2015, Slide 17). The emission factor is derived from that number and the 100-year GWP of $CH_4$, 34 from Myhre et al. (2013). The 20-year $CO_2$e is then derived from the resulting emission factor (4.1 g-$CH_4$/kWh) and the 20-year GWP of $CH_4$, 86. Emissions in the wind cases are reduced as described under Footnote 1.

[3] The average coal stack emission rate for the Petra Nova facility in 2016, prior to the addition of CC equipment, is from EIA (2017). In the wind-replacing-coal cases (last two columns), the emission rate is reduced as described under Footnote 1.

[4] The coal-stack $CO_2$ remaining after capture is from EIA (2017).

[5] The natural gas combustion emissions resulting from powering the CC equipment is from EIA (2017).

[6] Natural gas upstream leaks are obtained by dividing the raw emission rate of $CO_2$ from natural gas for each month January through June 2017 from EIA (2017) (in kg-$CO_2$/MWh-coal-electricity) by the molecular weight of $CO_2$ (44.0098 g-$CO_2$/mol) to give the moles of natural gas burned per MWh-coal-electricity. Multiplying the moles burned per MWh by the fractional number of moles burned that are methane (0.939) (Union Gas, 2018) and the molecular weight of methane (16.04276 g-$CH_4$/mol) gives the mass intensity of methane in the natural gas burned each month (kg-$CH_4$-burned/MWh-coal-electricity). The upstream leakage rate of methane is then the kg-$CH_4$-burned/MWh-coal-electricity multiplied by L/(1-L), where L=0.023 is the fraction of all methane produced (from conventional and shale rock sources) that leaks (Alvarez et al., 2018), giving the methane leakage rate in kg-$CH_4$/MWh-coal-electricity. This leakage rate is conservative based on a more recent full-lifecycle leakage rate estimate of methane from shale rock alone of L=0.035 (Howarth, 2019). Using the latter estimate would result in CCS/U with natural gas re-emitting even more $CO_2$e than calculated here. Multiplying the kg-$CH_4$/MWh-coal-electricity by the 20- and 100-year GWPs of $CH_4$ (86 and 34, respectively) (Myhre et al., 2013) gives the $CO_2$e emission rate of methane leaks each month. The monthly values are linearly averaged over January through June 2017.

[7] The non-$CH_4$ upstream $CO_2$e emissions rate is estimated as 15 g-$CO_2$/MJ-gas-electricity = 54 g-$CO_2$/kWh-gas-electricity (Howarth, 2014). Multiplying that by 0.497 MWh-electricity from natural gas per MWh-coal-electricity produced gives 26.8 kg-$CH_4$/MWh-coal-electricity. 0.497 MWh-electricity from natural gas per MWh-coal-electricity produced, or 49.7 percent, is calculated by dividing the average gas combustion emission from Petra Nova (200.9 g-$CO_2$/kWh-coal from the present table) by the combustion emissions per unit electricity from a combined cycle gas plant (404 g-$CO_2$/kWh-natural-gas).

[8] The percent $CO_2$ re-emitted for the wind cases (last two columns) equals Row k for the wind cases divided by Row d for either of the non-wind cases.

[9] $CO_2$e emissions relative to coal with no CC equipment.

[10] Air pollution emissions relative to coal with no CC equipment. In the natural-gas cases, all air pollution from coal emissions still occurs. Although gas is required to produce 0.497 MWh of electricity for the CC equipment per MWh of coal electricity, gas is assumed to be 50 percent cleaner than coal, so the overall air pollution in this case increases only 25 percent relative to the no-CC case. In the wind-CC cases, all upstream and combustion emissions from coal still occur.

[11] The electricity required (for end-use consumption plus to run the CC equipment) in all CC cases is 49.7 percent higher than with no CC. In the wind-replacing-coal case, no electricity is needed to run the CC equipment, but electricity is still needed for end use.

[12] The private energy cost in all CC cases is assumed to be 74 percent higher than coal with no CC because the CC equipment (including the gas plant) costs $4,200/kW, which represents about 74 percent of the mean capital cost of a new coal plant ($5,700/kW) from Lazard (2018). For simplicity, it was assumed that the cost of a wind turbine running the CC equipment was the same as that of a gas turbine running the equipment. In the wind-replacing-coal cases, the cost of coal was assumed to be a mean of c = $102/MWh and of wind, w = $42.5/MWh (Lazard, 2018). The final ratio was calculated as (0.503c+0.497w)/c.

over a 20-year time frame and 20 percent over a 100-year time frame. Twenty years is a relevant time frame to avoid 1.5 °C global warming and resulting climate feedbacks (IPCC, 2018).

When wind, instead of gas, is used to power the CC equipment, $CO_2e$ decreases by 37.4 percent over 20 years and 44.2 percent over 100 years compared with no CC (Table 3.6, Figure 3.1). The $CO_2e$ decrease exceeds that in the CCU-gas case because wind powering CC equipment case does not result in any combustion or upstream emissions from wind, as seen in Figure 3.1.

However, using the wind electricity that powers the CC equipment instead to replace coal electricity directly at the same plant reduces $CO_2e$ by 49.7 percent compared with no CC (Table 3.6, Figure 3.1). It is not 100 percent because only the wind used to run the capture equipment replaces coal. More wind would be needed to replace the whole coal plant. This third strategy is the best for reducing $CO_2e$ among the three cases. Using solar PV to replace coal directly results in a similar benefit as using wind.

But, $CO_2e$ is only part of the story. Because CCU equipment does not capture health-affecting air pollutants, air pollution emissions continue from coal and rise by about 25 percent compared with no capture from the use of natural gas to run the Petra Nova equipment (Table 3.6). Even when wind powers the CC equipment, air pollution from the coal plant continues as before (but not from using the new wind turbine). Only when wind partially replaces the use of coal itself does air pollution decrease by ~50 percent (Table 3.6).

The equipment costs of new coal and wind electricity in the United States are a mean of $102/MWh and $42.5/MWh, respectively (Lazard, 2018). The capital cost of CC equipment, $4,200/kW (Scottmadden, 2017), is about 74 percent the capital cost of a new coal plant ($5,700/kW) (Lazard, 2018), suggesting that new coal plus CCU is $1.74 \times \$102/\text{MWh} / \$42.5/\text{MWh} = 4.2$ times the equipment cost of new wind. Since CC equipment reduces only 10.8 percent of coal $CO_2e$ over 20 years and 20 percent over 100 years, the equipment for coal-CCU powered by natural gas alone costs 39 and 21 times that of wind-replacing coal per mass-$CO_2$ removed over 20 and 100 years, respectively.

Major additional social costs associated with coal electricity generation are air pollution and climate costs. The health cost of coal emissions in the United States is calculated as a mean of $80/MWh, which is much lower than the world average ($169/MWh, Table 3.6, Footnote 13). Since the use of CC equipment requires 50 percent more electricity than the coal plant produces but the health cost of natural gas emissions are about half those of coal, the use of gas to run the CC equipment increases health costs by about 25 percent compared with no capture (Table 3.6, Row o). Mean climate costs of U.S. emissions are estimated as $152/MWh, close to the world mean of $160/MWh (Table 3.6, Footnote 13). CC equipment with natural gas is estimated to reduce this cost by only 10.8 and 20 percent over 20 and 100 years, respectively (Table 3.6, Row n).

In sum, the total social cost (equipment plus health plus climate cost) of coal-CCU powered by natural gas is over twice that of wind replacing coal directly (Table 3.6, Figure 3.1). Moreover, the social cost of coal with CC powered by natural gas is 24 percent higher over 20 years and 19 percent higher over 100 years than coal without CC. Thus, no net social benefit exists of using CC equipment. In other words, from a social cost perspective, using CC equipment powered by natural gas causes more damage than does doing nothing at all.

In fact, in order for the social cost of using the CC equipment powered by natural gas to be less than that of doing nothing, the percent $CO_2e$ re-emitted by the Petra Nova plant would need to be 37 percent or less instead of 89.8 percent over 20 years. However, this is all but impossible, because 59.2 percent of the re-emissions is due to

Table 3.6 (*cont.*)

[13] The social cost before changes is the private energy cost of coal without CCU [$102/MWh from Lazard (2018)] plus air pollution mortality, morbidity, and non-health environmental costs of coal power plant emissions in the United States plus the global climate costs of U.S. emissions ($152/MWh) (Jacobson et al., 2017). U.S. coal power plant emissions health costs are estimated as $80/MWh, which is twice the background grid health cost of $40/MWh (Jacobson et al., 2019). In the worldwide average, from the same source, the health cost of background grid emissions is estimated as $169/MWh, so use of the U.S. number here is likely to underestimate the health costs of using carbon capture outside the United States.

[14] The social cost after changes is the sum of the private-energy cost multiplied by Row q, the air-pollution-health cost multiplied by Row o, and the climate cost multiplied by Row n.

Source: Jacobson (2019).

upstream coal and gas emissions and natural gas combustion emissions, so has little to do with how effective the CC equipment is at capturing carbon. In other words, even if the CC equipment captured 100 percent of the stack $CO_2$, which no one is proposing is feasible, the re-emissions would still be 59.2 percent. As such, the data indicate that **no technological improvement will result in the social cost of using CC equipment powered by natural gas being less than that of not using the equipment.**

When wind powers CC equipment, the social costs are still 6 and 2 percent higher over 20 and 100 years, respectively, than not using CC (Table 3.6, Figure 3.1). Although wind-powering-CC decreases $CO_2$e, thus climate cost, compared with coal without CC, wind-CC allows the same air pollution emissions from coal as no CC, and the cost of the wind plus CC equipment outweighs the $CO_2$e cost reduction (Figure 3.1).

Only when wind replaces coal electricity production directly does the total social cost drop 43 percent compared with no CC (Table 3.6). This is the best scenario.

A similar benefit occurs if wind replaces natural gas and no CC is used.

When CC is powered by wind, it is theoretically possible, albeit challenging, to reduce the total social cost below that of no CC. However, **it is impossible to reduce the total social cost below that of wind replacing coal electricity directly** because wind-powering-CC also incurs a CC equipment cost and never reduces air pollution or mining from coal, whereas wind replacing coal incurs no CC equipment cost and eliminates coal air pollution and mining (Jacobson, 2019).

Finally, Figure 3.1 illustrates that CCU powered by wind results in more $CO_2$e than the same wind replacing the coal plant and its upstream emissions. As such, CCU is an opportunity cost not only in terms of total social cost, but also in terms of $CO_2$e emissions. Thus, using WWS to replace fossil fuels (or bioenergy) reduces $CO_2$e and social cost over using it to power CCS/U equipment.

Example 3.9 shows, in more detail, how some of the calculations in Table 3.6 are performed using one out of the six months of data from EIA (2017).

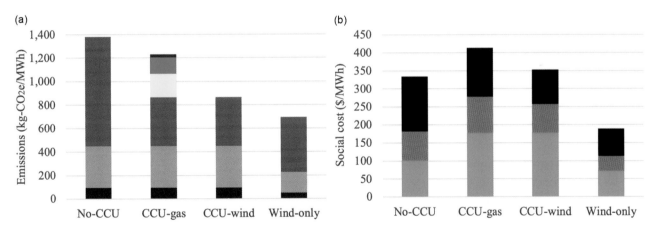

Figure 3.1 (a) $CO_2$e emissions, averaged over 20 years, from the Petra-Nova coal plant before (No-CCU) and after (CCU-gas) the addition of CCU equipment powered by natural gas. Also shown are emissions when the CCU equipment is powered by wind energy (CCU-wind) and when the portion of wind energy used to power the CCU equipment is instead used only to replace a portion of the coal power (thus some power is generated by coal and some by wind). Blue is upstream $CO_2$e from coal mining and transport aside from $CH_4$ leaks; orange is upstream $CO_2$e from coal mining $CH_4$ leaks; red is coal combustion $CO_2$; yellow is natural gas combustion $CO_2$; green is $CO_2$e from natural gas mining and transport $CH_4$ leaks; and purple is natural gas mining and transport $CO_2$e aside from $CH_4$ leaks. (b) Mean estimate of social costs per unit electricity over 20 years generated by the coal plant (in the first three cases) or the residual coal plant plus replacement wind plant (fourth case) for each of the four cases. Light blue is the cost of electricity generation plus CCU equipment; brown is air pollution health cost; and black is 20-year climate cost. All data are from Table 3.6 and Jacobson (2019).

## Example 3.9  Calculating $CO_2$ emission reduction due to carbon capture

According to EIA (2017), emissions of $CO_2$ during January 2016 from the Texas Parish coal power plant, before carbon capture, were 934.4 kg-$CO_2$/MWh. Emissions during January 2017, from the portion of the coal emissions subject to carbon capture, were 242.2 kg-$CO_2$/MWh, for a reduction of 692.2 kg-$CO_2$/MWh. However, the natural gas plant needed to run the carbon capture equipment itself emitted 271.9 kg-$CO_2$/MWh-coal-electricity. In addition, coal upstream emissions were 97.2, 353, and 140 kg-$CO_2$e/MWh-coal-electricity for $CO_2$ from coal mining, $CH_4$ leaks from coal mining over 20 years, and $CH_4$ leaks from coal mining over 100 years, respectively. The non-$CH_4$ upstream emissions from natural gas required for the capture equipment were 36.3 kg-$CO_2$/MWh-coal-electricity.

First, estimate the upstream $CO_2$e emissions from $CH_4$ leaks associated with mining, transporting, and processing the natural gas used to run the gas turbine. Then, calculate the overall 20-year and 100-year $CO_2$e of the upstream plus stack emissions from the facility after carbon capture. Finally, what percent of the $CO_2$e upstream plus stack emissions from the facility continues to be emitted over a 20-year and a 100-year time horizon after carbon capture?

Assume natural gas contains a 93.9 percent mole fraction of methane, and the upstream leakage rate of natural gas is 2.3 percent (Alvarez et al., 2018).

### Solution:

Dividing the combustion emission rate of $CO_2$ from natural gas, 271.9 kg-$CO_2$/MWh-coal-electricity, by the molecular weight of $CO_2$ (44.0098 g-$CO_2$/mol) gives the moles of natural gas burned. Multiplying the moles burned by the fractional number of moles burned that are methane (0.939) and the molecular weight of methane (16.04276 g-$CH_4$/mol) gives the mass intensity of methane in the natural gas burned, 93.1 kg-$CH_4$-burned/MWh-coal-electricity.

The upstream leakage rate of methane is then 93.1 × 0.023 / (1 − 0.023) = 2.19 kg-$CH_4$/MWh-coal-electricity. Multiplying by the 20- and 100-year GWPs of $CH_4$ (86 and 34, respectively) from Table 1.2 gives $CO_2$e emissions of the methane leaks as 188.4 and 74.5 kg-$CO_2$e/MWh-coal-electricity, respectively. Adding the upstream $CO_2$+$CH_4$ emissions to the natural gas stack emissions gives 20- and 100-year $CO_2$e emissions from the gas turbine as 497 and 382 kg-$CO_2$e/MWh-coal-electricity, respectively.

Adding these emissions to the coal upstream emissions and uncaptured coal emissions gives the 20- and 100-year $CO_2$e emissions from the gas turbine plus coal plant after capture as 1,189 and 862 kg-$CO_2$e/MWh-coal-electricity, respectively. The coal stack plus upstream emissions before capture were 1,385 and 1,172 kg-$CO_2$e/MWh-coal-electricity, respectively.

As such, averaged over 20 years, 1,189 / 1,385 = 85.8 percent of the original $CO_2$e from the coal plant and its upstream emissions remain in the air; Over 100 years, 73.6 percent remain in the air. These re-emissions are on top of downstream leaks that may occur with the captured $CO_2$.

Example 3.10 illustrates that, because coal with CCS/U is (a) expensive, (b) results in more air pollution emissions than coal without CCS/U, and (c) only modestly decreases $CO_2$ emissions, its social cost is more than 12 times that of wind energy providing the same energy. In addition, the energy cost is 4.2 times that of a wind turbine to produce the same energy.

## Example 3.10 Calculating the cost to society of using coal with CCS instead of wind

Estimate the energy plus health plus climate change cost of a new coal plant with CCS versus that of wind energy under the following assumptions. The cost of wind energy is 4.25 ¢/kWh (Table 7.9), the cost of a new coal plant is 10.2 ¢/kWh (Table 7.9), the cost of CCS equipment is 7.5 ¢/kWh, the average global health cost of coal pollution is 16.9 ¢/kWh (Table 7.11), and the average global climate cost of coal pollution is 16 ¢/kWh (Table 7.11). Also assume that the CCS equipment requires 25 percent more energy thus increases all emissions by 25 percent. Finally, assume that the CCS equipment reduces the overall $CO_2$ emission by 15 percent compared with before CCS equipment is added.

## Solution:

The social cost of the coal plant is the energy plus health plus climate cost of the plant. In this case, the energy cost of the plant plus equipment is $10.2 + 7.5 = 17.7$ ¢/kWh. The health cost is $1.25 \times 16.9$ ¢/kWh = 21.1 ¢/kWh. The climate cost is $0.85 \times 16$ ¢/kWh = 13.6 ¢/kWh. Adding these three together gives 52.4 ¢/kWh. Dividing this by the cost of wind, 4.25 ¢/kWh, gives 12.3. Thus, the social cost of coal-CCS in this case is 12.3 times that of wind. The private energy cost of coal-CCS is 4.2 times that of wind.

Tables 3.6 suggests little carbon benefit of and greater air pollution damage from CCS/U powered by natural gas before even considering the disposition of the captured $CO_2$. Two reasons for this, aside from the additional combustion emissions of the natural gas turbine, are the upstream coal and gas emissions and the air pollution resulting from the additional gas mining and use.

The results here are independent of the fate of the $CO_2$ after it leaves the CC equipment, thus apply to CCS/U with bioenergy (e.g., BECCS/U) or cement manufacturing. The CCS/U equipment always requires energy. If the energy comes from a fossil fuel, mining and combustion emissions from the fuel cancel most $CO_2$ captured. If it comes from a renewable, the renewable will not be able to displace fossil emissions, increasing those emissions. In all cases, air pollution increases.

When the fate of captured $CO_2$ is considered, the problem may deepen. If $CO_2$ is sealed underground without leaks, little added emissions occur. If the captured $CO_2$ is used to enhance oil recovery, its current major application, more oil is extracted and burned, increasing combustion $CO_2$, some leaked $CO_2$, and air pollution. If the captured $CO_2$ is used to create carbon-based fuels to replace gasoline and diesel, energy is still required to produce the fuel, the fuel is still burned in vehicles (creating pollution), and little $CO_2$ is captured to produce the fuel with. A third proposal is to use the $CO_2$ to produce carbonated drinks. However, along with the issues

previously listed, most $CO_2$ in carbonated drinks is released to the air during consumption.

In sum, CCS/U is not close to a zero-carbon technology. For the same energy cost, wind turbines and solar panels reduce much more $CO_2$ while also eliminating fossil air pollution and mining, which CCS/U increases. Using renewables also reduces pipelines, refineries, gas stations, tanker trucks, oil tankers, and coal trains, oil spills, oil fires, gas leaks, gas explosions, and international conflicts over energy. CCS/U increases these by increasing energy use. In fact, CCS/U powered by either fossils or renewables always increases total social costs relative to using renewables to eliminate fossil-fuel and bioenergy power generation directly.

## 3.3 Why Nuclear Power Represents an Opportunity Cost

In evaluating solutions to global warming, air pollution, and energy security, there are two important questions that arise: (1) Should new nuclear plants be built to help solve these problems? (2) Should existing, aged nuclear plants be kept open as long as possible to help solve these problems? To answer these questions, the main risks associated with nuclear power are first examined.

The risks associated with nuclear power can be broken down into two categories: (1) risks affecting nuclear's ability to reduce global warming and air pollution and

(2) risks affecting nuclear's ability to provide energy and environmental (aside from climate and air pollution) security. Risks under Category 1 include delays between planning and operation, emissions contributing to global warming and outdoor air pollution, and costs. Risks under Category 2 include weapons proliferation risk, reactor meltdown risk, radioactive waste risk, and mining cancer and land despoilment risks. These risks are discussed in this section.

**Nuclear fission** is the process by which tiny neutrons bombard and split certain fissile heavy elements, such as **uranium-235** ($^{235}U$) or **plutonium-239** ($^{239}Pu$) in a **nuclear reactor**. The numbers 235 and 239 refer to the isotope, or number of protons plus neutrons in the nucleus of a uranium or plutonium atom, respectively. A **fissile** element is one that can be split during fission upon neutron bombardment and whose neutrons released during splitting can split other fissile atoms in a chain reaction. Fissile elements do not spontaneously release neutrons, creating a chain reaction. Instead, they require outside neutrons bombarding them, thereby initiating a chain reaction. The nuclide $^{235}U$ is the only fissile element found in nature. The nuclide $^{239}Pu$ is a product of **uranium-238** ($^{238}U$) capturing a free neutron in a nuclear reactor. The resulting $^{239}U$ decays to $^{239}Pu$, a fissile element.

When a neutron approaches $^{235}U$ in a nuclear reactor, the neutron may be absorbed by or pass through the atom. Fast-moving neutrons have a higher probability of passing through the atom, whereas slow-moving neutrons have a higher probability of being absorbed. If the neutron is absorbed, the uranium atom's total energy is spread among the 236 protons and neutrons now present in the atom's nucleus. The nucleus is now unstable, and some of the uranium atoms fragment into two smaller elements, whereas the remaining atoms form $^{236}U$. A variety of element pairs arise from fragmentation. Two of the most common are Krypton-92 ($^{92}Kr$) and Barium-141 ($^{141}Ba$). The fragmentation, with this product pair, also produces gamma rays and three free neutrons. The overall reaction is thus

$$1 \text{ neutron} + {}^{235}U \rightarrow {}^{92}Kr + {}^{141}Ba + 3 \text{ neutrons} + \text{gamma rays} \tag{3.10}$$

The new neutrons may then collide with other $^{235}U$ atoms or with $^{239}Pu$ atoms, splitting them in a chain reaction. When the fragments and the gamma rays collide with water, the collision converts kinetic energy and electromagnetic energy, respectively, to massive amounts of heat.

In a **boiling water reactor (BWR) nuclear power plant**, the heat boils water directly. The high-pressure steam turns a turbine connected to a generator to produces electricity. The steam is then re-condensed to liquid water in a condenser, and the liquid water is returned back to the reactor core. In the condenser, heat from the steam is transferred to a separate (in an enclosed pipe) stream of cooling water that originates from a lake, river, or the coastal ocean. The warmed water is then returned to where it originated, warming the outdoor water body, creating thermal pollution. Other thermal power plants, such as those running on coal, oil, or gas, similarly warm water bodies.

In a **pressurized water reactor (PWR)** plant, the air pressure in the reactor is increased substantially, up to 155 bar (air pressure at Earth's surface is 1 bar). Because the boiling point of water increases with increasing atmospheric pressure, water in the reactor doesn't boil, even though the temperature in the reactor reaches 282 °C (at Earth's surface, water usually boils at 100 °C). The hot water in the reactor, which is radioactive, passes through a pipe and exchanges its heat with a different batch of water maintained at normal air pressure, causing the latter water to boil. The boiling water creates steam to run a steam turbine. The water batches are kept separate to ensure radioactive material in the high-pressure reactor does not pass through to the water vapor running through the steam turbine. BWR and PWR reactors are both **light water reactors (LWRs)**, which are reactors that use normal water.

Uranium in a nuclear power plant is originally stored in small ceramic pellets within a metal fuel rod, often 3.7-m long. A conventional BWR or PWR nuclear reactor will go through one rod during about six years. The rod and remaining material in it then become radioactive waste. Reactors that use rods once are referred to as **once-through** reactors. The radioactive waste in the fuel rod must be stored for several hundred thousand years.

A fuel rod that has gone through a fission reactor once still has 99 percent of its uranium left over, including slightly more $^{235}U$ than natural uranium. This remaining uranium and its fission product, plutonium, can be extracted and reprocessed for use in a **breeder reactor**, extending the life of a given mass of uranium and reducing waste significantly. However, the reprocessing increases both the cost and the production of $^{239}Pu$ by the collision of $^{238}$U with fast-moving neutrons. Breeder reactors can thus be optimized to produce $^{239}$Pu for use in nuclear weapons (Karam, 2006), so they are a concern with respect to weapons proliferation.

As of 2020, over 400 active nuclear reactors provide electric power among 31 countries. Only two of these reactors are breeder reactors. For this number of reactors, uranium mines produce about 60,000 tonnes of uranium per year (World Nuclear Association, 2019). Uranium reserves (aside from hard-to-extract uranium in seawater) as of 2015 were about 7.6 million tonnes. This suggests that about 127 years of uranium are available for current once-through fuel cycle reactors at near-current rates of uranium use. As such, even if the issues discussed below were not issues, uranium is a limited resource, and growing nuclear power will deplete uranium faster.

An alternative fuel to uranium in nuclear reactors is thorium. **Thorium**, like uranium, can be used to produce nuclear fuel in a breeder reactor. The advantage of thorium is that it produces less long-lived radioactive waste than does uranium. Its products are also more difficult to convert into nuclear weapons material. However, thorium still produces $^{232}$U, which was used in one nuclear bomb core produced during the **Operation Teapot** bomb tests in 1955. Thus, thorium is not free of nuclear weapons proliferation risk. In addition, thorium reactors require the same long time lag between planning and operation as uranium reactors (Section 3.3.1.1) and most likely longer because hardly any contractors or scientists have experience building or running thorium reactors. Thus, thorium reactors will produce greater emissions from the background electric grid compared with WWS technologies, which have a shorter time lag. Finally, life-cycle emissions of carbon from a thorium reactor are similar to those from a uranium reactor.

A proposed alternative to the large once-through reactor and the breeder reactor is the **small modular reactor (SMR)**. SMRs are nuclear fission reactors that are much smaller than a traditional reactor and prefabricated in a factory. The purpose of prefabricating much of the reactor is to reduce construction time, costs, and mistakes during construction. The reactor would then be moved to its final site, where construction would be completed. Many types of SMRs have been proposed, including miniature versions of current reactors as well as new designs.

One type of new design is a **fast reactor**, in which the fuel is reformulated to allow fast-moving neutrons, rather than slow-moving neutrons, to split an atom. One way to do this is to increase $^{239}$Pu, which absorbs more fast-moving neutrons than does $^{235}$U. Fast reactors can be turned into breeder reactors by surrounding the core with $^{238}$U, which absorbs a fast-moving neutron to become $^{239}$U, which decays to $^{239}$Pu.

Whereas slow reactors still produce significant radioactive waste, fast reactors produce less waste but also increase the potential for nuclear weapons proliferation by producing more $^{239}$Pu. Because slow and fast SMRs are small and modular, many countries that don't currently have nuclear energy facilities could more readily purchase them, increasing the risk of nuclear weapons proliferation. Most SMRs also have meltdown risk. They also require uranium. Slow reactors have the same resource limitation, lung cancer risk, and land despoilment risk associated with uranium mining as do non-SMRs (Section 3.3.2.4). Finally, because SMRs have not been commercialized to date, their emissions, time lag between planning and operation, and cost are still not known.

Finally, **nuclear fusion** of light atomic nuclei (e.g., protium, deuterium, or tritium) could theoretically supply power indefinitely without long-lived radioactive waste because the products are isotopes of helium. However, little prospect exists for fusion to be commercially available for at least 50 to 100 years, if ever.

Nuclear power from fission first became a source of electric power in the 1950s. The first nuclear power plant to produce electricity was an experimental reactor in Arco, Idaho. On December 20, 1951, it powered four light bulbs. On June 26, 1954, a 5-MW nuclear reactor was connected to the electric power grid for industrial use in Obninsk, Russia. Subsequently, on August 27, 1956, a 50-MW reactor was connected to the grid for commercial use in Windscale, England.

Below, the risks associated with nuclear power are discussed in detail.

### 3.3.1 Risks Affecting the Ability of Nuclear Power to Address Global Warming and Air Pollution

The first category of risk associated with nuclear power includes risks affecting nuclear power's ability to reduce global warming and air pollution. These risks include the long lag times between planning and operating and to refurbish a nuclear reactor, nuclear's high carbon equivalent emissions relative to WWS technologies, and nuclear's high cost.

### 3.3.1.1 Delays between Planning and Operation and due to Refurbishing Reactors

The longer the time lag is between the planning and operation of an energy facility, the greater the air pollution and climate-relevant emissions will be from the background electric power grid (Section 3.2.2). Similarly, the longer the time that is required to refurbish a plant for continued use at the end of its life, the greater the emissions will be from the background grid while the plant is down.

The time lag between planning and operation of a nuclear power plant includes the times to obtain a construction site, a construction permit, an operating permit, financing, and insurance; the time between construction permit approval and issue; and the construction time of the plant.

In March 2007, the United States Nuclear Regulatory Commission approved the first request for a site permit in 30 years. This process took 3.5 years. The time to review and approve a construction permit is another 2 years and the time between the construction permit approval and issue is about 0.5 years. Thus, the minimum time for preconstruction approvals (and financing) in the United States is 6 years. An estimated maximum time is 10 years. The time to construct a nuclear reactor depends significantly on regulatory requirements and costs. Although nuclear reactor **construction times** worldwide are often shorter than the 9-year median construction time in the United States since 1970 (Koomey and Hultman, 2007), they averaged 7.4 years worldwide in 2015 (Berthelemy and Rangel, 2015). As such, a reasonable estimated range for construction time is 4 to 9 years, bringing the overall time between planning and operation of a nuclear power plant worldwide to 10 to 19 years.

An examination of some recent nuclear plant developments confirms that this range is not only reasonable, but an underestimate in at least one case. The **Olkiluoto 3** reactor in Finland was proposed to the Finnish cabinet in December 2000 to be added to an existing nuclear power plant. Its latest estimated completion date is 2021, giving a **planning-to-operation (PTO)** time of 21 years. The **Hinkley Point** nuclear plant was planned starting in 2008. Construction began only on December 11, 2018. It has an estimated completion year of 2025 to 2027, giving it a PTO time of 17 to 19 years. The **Vogtle**

**3 and 4** reactors in Georgia were first proposed in August 2006 to be added to an existing site. The anticipated completion dates are November 2021 and November 2022, respectively, giving them PTO times of 15 and 16 years, respectively. Their construction times will be 8.5 and 9 years, respectively. The **Flamanville**, France, Unit 3 reactor was planned on an existing nuclear site starting in 2004. A contract was awarded in 2005. Construction started in 2007 but is not expected to be completed until 2023, for a construction time of 16 years and PTO time of 19 years. The **Haiyang 1 and 2** reactors in China were planned starting in 2005. Construction started in 2009 and 2010, respectively. Haiyang 1 began commercial operation on October 22, 2018. Haiyang 2 began operation on January 9, 2019, giving them construction times of 9 years and PTO times of 13 and 14 years, respectively. The **Taishan 1 and 2** reactors in China were bid in 2006. Construction began in 2008. Taishan 1 began commercial operation on December 13, 2018. Taishan 2 began operation on September 9, 2019, giving them construction times of 10 and 11 years and PTO times of 12 and 13 years, respectively. Planning and procurement for four reactors in **Ringhals**, Sweden, started in 1965. One took 10 years, the second took 11 years, the third took 16 years, and the fourth took 18 years to complete. In sum, PTO times for both recent and older nuclear plants have mostly been in the range of 10 to 19 years.

Some contend that France's 1974 Messmer Plan resulted in the building of its 58 reactors in 15 years. The **Messmer Plan** was a proposal, enacted without public or parliamentary debate, by then prime minister of France, Pierre Messmer, to build 80 nuclear reactors by 1985 and 170 by 2000. In fact, the plan had been in the works for years prior and was only proposed publicly following the international oil crisis of 1973 (Morris, 2015). For example, the Fessenheim nuclear reactor obtained its construction permit in 1967 and was planned before that. In addition, 10 of the reactors were completed only between 1991 and 2000. As such, the whole planning-to-operation time for the 58 reactors was at least 33 years, not 15. That of any individual reactor was 10 to 19 years.

Planning-to-operation delays are not the only cause of background emissions associated with nuclear power or any other energy technology. Nuclear reactors have an expected lifetime on the order of 40 years. To run longer, they need to be refurbished. An estimate of the time to refurbish a nuclear reactor is 2–4 years. Refurbishment of the Darlington 2, Ontario, nuclear reactor, for example, began in October 2016 and is scheduled to take 3 years and 4 months (World Nuclear News, 2018).

Equations 3.1 and 3.2 provide an estimate of the opportunity cost $CO_2e$ emissions resulting from emissions from the background due to a nuclear power plant's long PTO time and refurbishment time. Table 3.5 provides an overall estimate of this opportunity cost emissions as 64 to 102 g-$CO_2e$/kWh, which is higher than nuclear's lifecycle emissions. Opportunity cost emissions also include health-affecting air pollution emissions.

## Transition highlight

Example 3.11 illustrates how China's investment in nuclear plants, which have long planning-to-operation times, instead of wind power resulted in China's $CO_2$ emissions rising 1.3 percent from 2016 to 2017, rather than declining by an estimated average of 3 percent during that period. A similar result would be found if China invested in solar instead of nuclear.

The health impacts of such delays in China are substantial. In 2016, 1.9 million people died from air pollution particles and gases in China (Table 7.14). Assuming that air pollution emissions are proportional to $CO_2$ emissions, **82,000 (1.9 million × 4.3 percent) more people may have died in 2016 alone due to China's investment in nuclear instead of wind or solar**. Additional deaths likely occurred prior and since. Thus, opportunity cost emissions affect both climate and health.

---

## Example 3.11  Did construction of nuclear plants in China cause its emissions to rise between 2016 and 2017?

Between 2016 and 2017, the $CO_2$ emission rate in China (including Hong Kong) increased by 121 million metric tonnes (MT), or 1.3 percent, over its 2016 emission rate of 9,310 MT-$CO_2$ (British Petroleum, 2018). During that period, China had 14 GW of nuclear power under construction, with planning for all the plants starting before 2012. The capital cost of a new nuclear power plant ranges from \$6,500/kW to \$12,250/kW, whereas that of a new wind turbine ranges from \$1,150/kW to \$1,550/kW (Lazard, 2018). Assuming the capital for the nuclear plants had been invested in wind instead and the wind turbines had been installed prior to 2017 (because the planning-to-operation time of wind is 2 to 5 years versus 10 to 19 years for nuclear), estimate the 2017 $CO_2$ emissions that would have been avoided. Assume the wind turbine capacity factor ranges from 0.3 to 0.37 and that the $CO_2$ emission intensity of the grid in China is between 850 and 900 g-$CO_2$/kWh (Li et al., 2017).

### Solution:

Dividing the high (and low) capital cost of nuclear per kW by the low (and high) capital cost of wind per kW and multiplying the result by 14 GW gives a range of 58.7 to 149 GW nameplate capacity of wind that could have been installed and running prior to 2017. Multiplying by the capacity factor range of wind and 8,760 hours per year and dividing by 1,000 GW per trillion watts (TW) gives the annual energy output of the wind that could have been installed as 154 to 483 terawatt hours per year (TWh/y). Multiplying this range by the $CO_2$ emission intensity that wind would have avoided, 850 to 900 g-$CO_2$/kWh, and by $10^9$ kWh/TWh, and dividing by $10^{12}$ g/MT gives 131 to 435 MT-$CO_2$/y avoided. In other words, investing in wind instead of nuclear would have resulted in China's decreasing its $CO_2$ emissions by about 1.4 to 4.7 percent (for an average of 3.0 percent) instead of increasing it by 1.3 percent. As such, investing in nuclear has caused an opportunity cost $CO_2$ emission in China.

### 3.3.1.2 Air Pollution and Global Warming Relevant Emissions from Nuclear

Nuclear power contributes to global warming and air pollution in the following ways: (1) emissions of air pollutants and global warming agents from the background grid due to its long planning-to-operation and refurbishment times (Section 3.2.2.1); (2) **lifecycle emissions** of air pollutants and global warming agents during construction, operation, and decommissioning of a nuclear plant; (3) heat and water vapor emissions during the operation of a nuclear plant (Sections 3.2.2.2 and 3.2.2.3); (4) carbon dioxide emissions due to covering soil or clearing vegetation during the construction of a nuclear plant, uranium mine, and waste site (Section 3.2.2.5); and (5) the emissions risk of air pollutants and global warming agents due to nuclear weapons proliferation (Section 3.3.2.1).

Every one of these categories represents an actual emission or emission risk, yet most of these emissions, except for lifecycle emissions, are incorrectly ignored in virtually all studies of nuclear energy impacts on climate. Almost no study considers the impact of nuclear energy on air pollution mortality. Studies that ignore these factors distort the impacts on climate and air pollution health of nuclear power.

Table 3.5 summarizes the $CO_2e$ emissions from nuclear power from each of the five categories just described. The table indicates that the opportunity cost emissions of nuclear (64 to 102 g-$CO_2e$/kWh) are higher than the lifecycle emissions (9 to 70 g-$CO_2e$/kWh). The range of lifecycle emissions estimated in Table 3.5 for nuclear power is well within the "range of harmonized lifecycle greenhouse gas emissions reported in the literature," 4 to 110 g-$CO_2e$/kWh, from the Intergovernmental Panel on Climate Change review (Bruckner et al., 2014, p. 540). It is also conservative relative to the 68 (10 to 130) g-$CO_2e$/kWh from the review of Lenzen (2008) and relative to the 66 (1.4 to 288) g-$CO_2e$/kWh from the review of Sovacool (2008).

Emissions from the heat and water vapor fluxes from nuclear (totaling 4.4 g-$CO_2$-kWh) alone suggest that during the life of an existing nuclear power plant, **nuclear can never be a zero-carbon-equivalent technology**, even if its lifecycle emissions from mining and refining uranium were zero. On the other hand, the emissions from nuclear due to covering and clearing soil are relatively small (0.17 to 0.28 g-$CO_2e$/kWh). Finally, Table 3.5

provides a low estimate (zero) and a high estimate (1.4 g-$CO_2e$/kWh) for the 100-year risk of $CO_2e$ emissions associated with nuclear weapons proliferation due to nuclear energy. These numbers are derived in Section 3.3.2.1.

The total $CO_2e$ emissions from nuclear power in Table 3.5 are 78 to 178 g-$CO_2e$/kWh. These emissions are 9 to 37 times the $CO_2e$ emissions from onshore wind power. The ratio of health-affecting air pollutant emissions from nuclear relative to onshore wind is 7 to 25. This is determined by considering only the lifecycle, opportunity cost, and weapons proliferation emissions from nuclear and wind in Table 3.5.

Although the emissions from nuclear are lower than from coal or natural gas with carbon capture, nuclear power's high $CO_2e$ emissions coupled with its long planning-to-operation time render it an opportunity cost relative to the faster-to-operate and lower-emitting alternative WWS technologies (Jacobson, 2009).

### 3.3.1.3 Nuclear Costs

The third risk of nuclear power related to its ability to reduce global warming and air pollution is the high cost for a new nuclear reactor relative to most WWS technologies. In addition, the cost of running existing nuclear reactors has increased significantly, and the costs of new WWS technologies have dropped so much, that many existing reactors are shutting down early due to high costs. Others have requested large subsidies to stay open. In this section, nuclear costs are discussed briefly.

The levelized cost of energy (LCOE) for a new nuclear plant in 2018, based on calculations by Lazard (2018), is $151 (112 to 189)/MWh, where $100/MWh equals 10 ¢/kWh. This compares with $43 (29 to 56)/MWh for onshore wind and $41 (36 to 46)/MWh for utility-scale solar PV from the same source (Table 7.9). A good portion of the high cost of nuclear is related to its long planning-to-operation time, which in turn is partly due to construction delays.

This nuclear LCOE is an underestimate for several reasons. First, Lazard assumes a construction time for nuclear of 5.75 years. However, the Vogtle 3 and 4 reactors will take at least 8.5 to 9 years to finish construction. This additional delay alone results in an estimated LCOE for nuclear of about $172 (128 to 215)/MWh, or a cost 2.3 to 7.4 times that of an onshore wind farm (or utility PV farm).

Next, the LCOE does not include the cost of the major nuclear meltdowns in history. For example, the estimated cost to clean up the damage from three Fukushima Dai-ichi nuclear reactor core meltdowns in 2011 (Section 3.3.2.2) was $460 to $640 billion (Denyer, 2019). This is equivalent to a mean of about $1.2 billion, or 10 to 18.5 percent of the capital cost, of every nuclear reactor that exists worldwide.

In addition, the LCOE does not include the cost of storing nuclear waste for hundreds of thousands of years. In the United States alone, about $500 million is spent yearly to safeguard nuclear waste from about 100 civilian nuclear energy plants (Garthwaite, 2018). This amount will only increase as waste continues to accumulate. After the plants retire, the spending must continue for hundreds of thousands of years with no revenue stream from electricity sales to pay for the storage.

The spiraling cost of new nuclear plants in recent years has resulted in the cancelling of several nuclear reactors under construction (e.g., two reactors in South Carolina) and in requests for subsidies to keep construction projects alive (e.g., the two Vogtle reactors in Georgia). High costs have also reduced the number of new constructions to a crawl in liberalized markets of the world. However, in some countries, such as China, nuclear reactor growth continues due to large government subsidies, albeit with a 10- to 19-year time lag between planning and operation (Section 3.3.1.1) and escalating costs.

In sum, before accounting for meltdown damage and waste storage, **a new nuclear power plant costs 2.3 to 7.4 times that of an onshore wind farm (or utility PV farm), takes 5 to 17 years longer between planning and operation, and produces 9 to 37 times the emissions per unit electricity generated**. Thus, a fixed amount of money spent on a new nuclear plant means much less power generation, a much longer wait for power, and much greater emission rate than the same money spent on WWS technologies.

The Intergovernmental Panel on Climate Change similarly concluded that the economic, social, and technical feasibility of nuclear power has not improved over time,

*The political, economic, social and technical feasibility of solar energy, wind energy and electricity storage technologies has improved dramatically over the past few years, while that of nuclear energy and Carbon Dioxide Capture and Storage (CCS) in the electricity sector has not shown similar improvements.* (de Coninck et al., 2018, pp. 4–5)

Costs of existing operating nuclear plants have also escalated tremendously, forcing some plants either to shut down early or request large subsidies to stay open. Whether an existing nuclear plant should be subsidized to stay open should be evaluated on a case-by-case basis. The risk of shutting a functioning nuclear plant is that its energy may be replaced by higher-emitting fossil-fuel generation. However, the risk of subsidizing the plant is that the funds could otherwise be used immediately to replace the nuclear plant with lower-cost and lower-emitting wind or solar electricity generation. Because the nuclear plant would usually need to be replaced within a decade in any case, simply incurring the cost of new renewables now will almost always be less expensive than spending the same money on renewables in 10 years and paying nuclear a subsidy today.

For example, in 2016, three existing upstate New York nuclear plants requested and received subsidies to stay open using the argument that the plants were needed to keep emissions low. However, Cebulla and Jacobson (2018) found that subsidizing such plants may increase carbon emissions and costs relative to replacing the plants with wind or solar. For different nuclear plants and subsidy levels, the results could change, which is why each plant needs to be evaluated individually.

## 3.3.2 Risks Affecting the Ability of Nuclear Power to Address Energy and Environmental Security

The second category of risk related to nuclear power is the risk of the plant not being able to provide stable energy and environmental security. One reason for this is the risk of nuclear meltdown. Others are its risks of increasing weapons proliferation, radioactive waste exposure, and damage (cancer and land degradation) due to uranium mining. WWS technologies do not have these risks.

### 3.3.2.1 Weapons Proliferation Risk
The first risk of nuclear power related to energy and environmental security is weapons proliferation risk. The growth of nuclear energy has historically increased the ability of nations to obtain or harvest plutonium or enrich uranium to manufacture nuclear weapons. As stated by Fuhrmann (2009, pp. 12–13),

*Peaceful nuclear cooperation and nuclear weapons are related in two key respects. First, all technology and materials related to a nuclear weapons program have legitimate civilian applications. For example, uranium enrichment and plutonium reprocessing facilities are dual-use in nature because they can be used to produce fuel for power reactors or fissile material for nuclear weapons. Second, civilian nuclear cooperation builds-up a knowledge-base in nuclear matters.*

The Intergovernmental Panel on Climate Change recognizes this fact. They conclude, with "robust evidence and high agreement" that nuclear weapons proliferation concern is a barrier and risk to the increasing development of nuclear energy:

*Barriers to and risks associated with an increasing use of nuclear energy include **operational risks** and the associated safety concerns, **uranium mining risks**, financial and regulatory risks, **unresolved waste management issues, nuclear weapons proliferation concerns**, and adverse public opinion).* (Bruckner et al., 2014, Executive Summary, p. 517).

The building of a nuclear reactor for energy in a country that does not currently have a reactor increases the risk of nuclear weapons development in that country. Specifically, it allows the country to import uranium for use in the nuclear energy facility. If the country so chooses, it can secretly enrich the uranium to create weapons grade uranium as well as harvest plutonium from uranium fuel rods used in a nuclear reactor, for nuclear weapons. This does not mean any or every country will do this, but historically some have, and the risk is high, as noted by IPCC.

The next risk is whether a nuclear weapon developed in this manner is used. That risk also ranges from zero to some risk. If a weapon is used, it may kill 2 to 20 million people and burn down a megacity, releasing substantial emissions. As such, beyond the horrible risk of loss of human life, there is a risk of zero to some non-zero emission rate from nuclear weapons proliferation resulting from nuclear energy proliferation. This risk is quantified later in this section. First, the difference between weapons grade and reactor grade uranium and plutonium is described.

Uranium ore is mined in an open pit or underground and contains 0.1 to 1 percent uranium by mass. The ore is milled to concentrate the uranium in the form of a yellow power called **yellowcake**, which contains about 80 percent uranium oxide. Uranium is then processed further into uranium dioxide or uranium hexafluoride for use in nuclear reactors. However, before the uranium can be used in a reactor, it must first be enriched.

Of all uranium on Earth, 99.2745 percent is $^{238}$U, 0.72 percent is $^{235}$U, and 0.0055 percent is $^{234}$U. Thus, less than 1 percent is $^{235}$U. The $^{238}$U has a half-life of 4.5 billion years. Most commercial light water reactors use uranium consisting of 3 to 5 percent $^{235}$U. As such, the concentration of $^{235}$U in the uranium fuel rod must be increased from its ore concentration. This is done by enrichment. **Uranium enrichment** is the process of separating the isotopes of uranium to increase the percent of $^{235}$U in a batch. Enriched uranium is useful for both nuclear energy and nuclear weapons.

Enrichment is done either by gas diffusion, centrifugal diffusion, or mass separation by magnetic field. Only gas diffusion and centrifugal diffusion are commercial processes, and most enrichment today is by **centrifugal diffusion** because it consumes only 2 to 2.5 percent the energy as gas diffusion. Nevertheless, centrifugal diffusion still requires many centrifuges running for long periods, thus lots of energy. Centrifugal diffusion works by spinning a cylindrical container containing uranim. The heavier $^{238}$U atoms collect toward the outside edge of the cylinder and the lighter $^{235}$U atoms collect toward the inside.

Uranium with less than 20 percent $^{235}$U is called **low enriched uranium**. **Highly enriched uranium** contains 20 to 90 percent $^{235}$U. A nuclear weapon can be made with highly enriched uranium. However, weapons increase their destructiveness with more enrichment. Thus, 90 percent or more $^{235}$U is considered **weapons grade uranium** and is generally used with enriched plutonium in a nuclear bomb. An estimated 9,000 centrifuges can produce enough weapons grade $^{235}$U for one nuclear weapon from natural uranium in about seven months. With 5,000 centrifuges, the process takes about one year (IranWatch, 2015). Because uranium in a fuel rod used for nuclear energy has only 3 to 5 percent $^{235}$U and even less once it goes through a nuclear reactor, spent fuel rods are not considered a useful source of weapons grade uranium.

Plutonium is also used in nuclear weapons. Ten kilograms of $^{239}$Pu was used in the bomb dropped on Nagasaki. Plutonium can be obtained from a once-through nuclear reactor running on a reactor grade uranium fuel rod. When $^{235}$U decays and releases neutrons in a nuclear reactor, a neutron can bind with a $^{238}$U atom to produce $^{239}$U, which decays to produce $^{239}$P. Plutonium that is

93 percent or more $^{239}$Pu is considered weapons grade plutonium. Plutonium less than 80 percent plutonium is reactor grade. Because any plutonium can be used to make a bomb and is easier to obtain than enriching uranium (since plutonium can be harvested from a fuel rod running through a nuclear reactor), plutonium is considered the element of even greater concern than uranium with respect to nuclear weapons proliferation.

A large-scale, worldwide increase in nuclear energy facilities would exacerbate the risk of nuclear weapons proliferation. In fact, producing material for a weapon requires merely operating a civilian nuclear power plant together with a sophisticated plutonium separation facility. The historic link between nuclear energy facilities and weapons is evidenced by the development or attempted development of weapons capabilities secretly under the guise of peaceful civilian nuclear energy or nuclear energy research programs in Pakistan, India, Iraq (prior to 1981), Syria (prior to 2007), Iran, and, North Korea, among other countries.

If the world's all-purpose energy were converted to electricity and electrolytic hydrogen by 2050, the ~9 trillion watts (TW) in resulting annual average end-use electric power demand would require about 12,500 850-MW nuclear reactors (31 times the number of active reactors today), or one installed every day for 34 years. Not only is this construction time impossible given the long PTO of nuclear, but it would also result in all known reserves of uranium worldwide for once-through reactors running out in about three years. As such, there is no possibility the world will run solely on once-through nuclear energy by 2050.

Even if only 6.4 percent of the world's energy were supplied with nuclear, the number of active nuclear reactors worldwide would nearly double to around 800. Many more countries would possess nuclear reactors, increasing the risk that some of these countries would use the facilities to mask the development of nuclear weapons, as has occurred historically.

If a country were to develop a weapon as a result of its acquisition of one or more nuclear energy facilities, the risk that it would use the weapons is not zero. Here, the emissions associated with a limited nuclear exchange are quantified.

The explosion of fifty 15-kilotonne nuclear devices (a total of 1.5 megatonne, or 0.1 percent of the yield of a full-scale nuclear war) during a limited nuclear exchange in a megacity would kill 2.6 to 16.7 million people from the explosion and burn 63 to 313 teragrams (Tg) of city infrastructure, adding 1 to 5 Tg of warming and cooling aerosol particles to the atmosphere, including much of it to the stratosphere (Jacobson, 2009). The particle emissions would cause significant short- and medium-term regional temperature changes. The $CO_2$ emissions would cause long-term warming. The $CO_2$ emissions from such a conflict are projected to be 92 to 690 Tg-$CO_2$.

The annual electricity production due to nuclear energy in 2017 was 2,506 TWh/y. If that doubled to 5,000 TWh/y and if one nuclear exchange, as described above, resulted during a 100-year period, the net carbon emissions due to nuclear weapons proliferation caused by the expansion of nuclear energy worldwide would be 0.2 to 1.4 g-$CO_2$/kWh. This calculation assumes that the total energy generation is 5,000 TWh/y multiplied by 100 years. The resulting emission rate depends on the probability of a nuclear exchange over a given period and the strengths of nuclear devices used. The probability is bounded between 0 and 1 exchange over 100 years to give the range of possible emissions for one such event as 0 to 1.4 g-$CO_2$e/kWh, which is the emission rate used in Table 3.5.

### 3.3.2.2 Meltdown Risk
The second risk of nuclear power related to energy security is meltdown risk. As stated in Section 3.3.2.1, the Intergovernmental Panel on Climate Change points to **operational risks** (meltdown) as a barrier and risk associated with nuclear power.

Through 2020, about 1.5 percent of all nuclear reactors operating in history have had a partial or significant core meltdown. To date, meltdowns at nuclear power plants have been either catastrophic (Chernobyl, Russia, in 1986; three reactors at Fukushima Dai-ichi, Japan, in 2011) or damaging (Three-Mile Island, Pennsylvania, in 1979; Saint-Laurent, France, in 1980). The nuclear industry has proposed new reactor designs that they suggest are safer. However, these designs are generally untested, and there is no guarantee that the reactors will be designed, built, and operated correctly or that a natural disaster or act of terrorism, such as an airplane flown into a reactor, will not cause the reactor to fail, resulting in a major disaster.

On March 11, 2011, an earthquake measuring 9.0 on the Richter scale, and the subsequent tsunami that knocked out backup power to a cooling system, caused six nuclear reactors at the **Fukushima 1 Dai-ichi plant** in northeastern Japan to shut down. Three reactors experienced a significant meltdown of nuclear fuel rods and multiple explosions of hydrogen gas that had formed during efforts to cool the rods with seawater. Uranium fuel rods in a fourth reactor also lost their cooling. As a result, cesium-137, iodine-131, and other radioactive particles and gases were released into the air. Locally, tens of thousands of people were exposed to the radiation, and 170,000 to 200,000 people were evacuated from their homes. Approximately 1,600 to 3,700 people perished during the evacuation alone (Johnson, 2015; Denyer, 2019). At least one nuclear plant worker died of lung cancer from direct radiation exposure (BBC News, 2018).

The radiation release created a dead zone around the reactors that may not be safe to inhabit for decades to centuries. The radiation also poisoned the water and food supplies in and around Tokyo. The radiation plume from the plant spread worldwide within a week. Radioactivity spread worldwide, although levels in Japan within 100 km of the plant were extremely high, those in the rest of Japan and eastern China were lower, and those in North America and Europe were even lower (Ten Hoeve and Jacobson, 2012). It is estimated that 130 (15 to 1,100) cancer-related mortalities and 180 (24 to 1,800) cancer-related morbidities will occur worldwide, primarily in eastern Asia, over the next several decades due to the meltdown (Ten Hoeve and Jacobson, 2012). The cost of the cleanup of the Fukushima reactors and the surrounding area is estimated at $460 to $640 billion (Denyer, 2019), equivalent to about $1.2 billion for every nuclear reactor that exists worldwide.

The 1.5 percent risk of a catastrophe due to nuclear power plants is a high risk. Catastrophic risks with all WWS technologies aside from large hydropower (due to the risk of dam collapse) are zero. WWS roadmaps do not call for an increase in the number of large hydropower dams worldwide, only a more effective use of existing ones.

### 3.3.2.3 Radioactive Waste Risks

Another risk associated with nuclear power is the risk of human and animal exposure to radioactivity from fuel rods consumed by once-through nuclear reactors. Such fuel rods, once consumed, are considered **radioactive waste**. Currently, most fuel rods are stored at the same site as the reactor that consumed them. This has given rise to hundreds of radioactive waste sites in many countries that must be maintained for at least 200,000 years, far beyond the lifetimes of any nuclear power plant. Plans

in the United States, which houses about one-quarter of all nuclear reactors worldwide, to store the waste inside of Yucca Mountain in Nevada have not been approved. The more nuclear waste that accumulates, the greater the risk will be of radioactive leaks, which can damage water supply, crops, animals, and humans.

### 3.3.2.4 Uranium Mining Health Risks and Land Degradation

The final risks discussed related to nuclear power are the risk of lung cancer to miners and land degradation due to uranium mining. Such risks continue so long as nuclear power plants continue to operate, because the plants need uranium to produce electricity. WWS technologies, on the other hand, do not require the continuous mining of any material, only one-time mining to produce the WWS devices. As such, WWS technologies do not have this risk.

Uranium mining causes lung cancer in large numbers of miners because uranium mines contain natural radon gas, some of whose decay products are carcinogenic. Several studies have found a link between high radon levels and cancer (e.g., Henshaw et al., 1990; Lagarde et al., 1997). A study of 4,000 uranium miners between 1950 and 2000 (CDC, 2000) found that 405 (10 percent) died of lung cancer, a rate six times that expected based on smoking rates alone. Sixty-one others died of mining-related lung diseases, supporting the hypothesis that uranium mining is unhealthy. In fact, the combination of radon and cigarette smoking increases lung cancer risks above the normal risks associated with smoking (Hampson et al., 1998). Clean, renewable energy does not have this risk because (a) it does not require the continuous mining of any material, only one-time mining to produce the energy generators; and (b) the mining does not carry the same lung cancer risk that uranium mining does.

**Radon (Rn)** is a radioactive but chemically unreactive, colorless, tasteless, and odorless gas that forms naturally in soils. The source of radon gas is the radioactive decay of $^{238}$U. Radon formation from uranium involves a long sequence of radioactive decay processes. During radioactive decay of an element, the element spontaneously emits radiation in the form of an alpha ($\alpha$) particle, beta ($\beta$) particle, or gamma ($\gamma$) ray. An **alpha particle** is the nucleus of a helium atom, which is made of two neutrons and two protons. It is the least penetrating form of radiation and can be stopped by a thick piece of paper.

Alpha particles are not dangerous unless the emitting substance is inhaled or ingested. A **beta particle** is a high-velocity electron. Beta particles penetrate deeper than do alpha particles, but less so than do other forms of radiation, such as gamma rays. A **gamma ray** is a highly energized, deeply penetrating photon emitted from the nucleus of an atom not only during nuclear fusion (e.g., in the sun's core), but also sometimes during radioactive decay of an element.

The French physicist **Antoine Henri Becquerel** (1871–1937) discovered radioactive decay on March 1, 1896. Becquerel placed a uranium-containing mineral on top of a photographic plate wrapped by thin, black paper. After letting the experiment sit in a drawer for a few days, he developed the plate and found that it had become fogged by emissions, which he traced to the uranium in the mineral. He referred to the emissions as **metallic phosphorescence**. What he had discovered was the emission of some type of particle due to radioactive decay. He repeated the experiment by placing coins under the paper and found that their outlines were traced by the emissions. Two years later, the New Zealand–born, British physicist **Ernest Rutherford** (1871–1937) found that uranium emitted two types of particles, which he named alpha and beta particles. Rutherford later discovered the gamma ray as well.

Equation 3.11 summarizes the radioactive decay pathway of $^{238}U$ to $^{206}Pb$. Numbers shown are half-lives of each decay process.

$^{234}Pa$ decays further to uranium-234 ($^{234}U$), then to thorium-230 ($^{230}Th$), then to radium-226 ($^{226}Ra$), and then to radon-222 ($^{222}Rn$).

Whereas radon precursors are bound in minerals, $^{222}Rn$ is a gas that can be breathed in. $^{222}Rn$ has a half-life of 3.8 days. It decays to polonium-218 ($^{218}Po$), which has a half-life of 3 minutes and decays to lead-214 ($^{214}Pb$). $^{218}Po$ and $^{214}Pb$, referred to as **radon progeny**, are electrically charged and can be inhaled or attach to particles that are inhaled. In the lungs or in ambient air, $^{214}Pb$ decays to bismuth-214 ($^{214}Bi$), which decays to polonium-214 ($^{214}Po$). $^{214}Po$ decays almost immediately to lead-210 ($^{210}Pb$), which has a lifetime of 22 years and usually settles to the ground if it has not been inhaled. It decays to bismuth-210 ($^{210}Bi$), then to polonium-210 ($^{210}Po$), and then to the stable isotope, lead-206 ($^{206}Pb$), which does not decay further.

$^{222}Rn$, a gas, is not itself harmful, but its progeny, $^{218}Po$ and $^{214}Pb$, which enter the lungs directly or on the surfaces of aerosol particles, are highly carcinogenic (Polpong and Bovornkitti, 1998). Any activity, such as uranium mining, increasing the inhalation of aerosol particles (e.g., dust) enhances the risk of inhaling radon progeny. As such, exposure of uranium miners to radon is another risk associated with nuclear energy.

Like with coal, oil, and natural gas mining, uranium mining also despoils land and reduces the carbon stored in soil. In 2017, 19 countries mined uranium. Kazakhstan, Canada, Australia, Namibia, and Niger produced

$$
\begin{array}{cccccccc}
4.5\times10^9 \text{ yr} & 24 \text{ d} & 1.2 \text{ min} & 2.5\times10^5 \text{ yr} & 8\times10^4 \text{ yr} & 1620 \text{ yr} & 3.8 \text{ d} \\
^{238}U \longrightarrow & ^{234}Th \longrightarrow & ^{234}Pa \longrightarrow & ^{234}U \longrightarrow & ^{230}Th \longrightarrow & ^{226}Ra \longrightarrow & ^{222}Rn \longrightarrow & ^{218}Po
\end{array}
$$

$$
\begin{array}{cccccccc}
3 \text{ min} & 27 \text{ min} & 30 \text{ min} & 0.00016 \text{ s} & 22 \text{ yr} & 5 \text{ d} & 138 \text{ d} \\
\longrightarrow ^{214}Pb \longrightarrow & ^{214}Bi \longrightarrow & ^{214}Po \longrightarrow & ^{210}Pb \longrightarrow & ^{210}Bi \longrightarrow & ^{210}Po \longrightarrow & ^{206}Pb
\end{array}
$$

(3.11)

When it decays to produce radon, $^{238}U$ first releases an alpha particle, producing thorium-234 ($^{234}Th$), which decays to protactinium-234 ($^{234}Pa$), releasing a beta particle. $^{234}Pa$ has the same number of protons and neutrons in its nucleus as does $^{234}Th$, but $^{234}Pa$ has one less electron than does $^{234}Th$, giving $^{234}Pa$ a positive charge.

the most uranium. Mines can be open pit or underground. Open pit mines cause the most land degradation. Table 3.5 provides an estimate of the effective $CO_2e$ emissions due to clearing vegetation from land for uranium mining associated with nuclear power. The continuous mining for fuels is not needed in a 100 percent WWS world.

## 3.4 Why Not Biomass for Electricity or Heat?

**Biofuels** are solid, liquid, or gaseous fuels derived from organic matter. Most biofuels are derived from dead plants or from animal excrement. **Solid biofuels**, such as wood, grass, agricultural waste, and dung, are burned directly for home heating and cooking in developing countries and for electric power generation in most all countries. **Liquid biofuels** are generally used for transportation as a substitute for gasoline or diesel. **Gaseous biofuels**, such as methane, are used either for electricity, heat, or transportation.

Here, **biomass** (or **bioenergy**) is defined to be a biofuel that is used for electricity or heat generation. Biofuels for transportation are discussed in Section 3.5.

Biomass combustion for electricity or heat is not recommended in a 100 percent WWS world for several reasons, discussed herein. Similarly, **bioenergy with carbon capture and storage** (BECCS) also represents an opportunity cost in comparison with WWS options. Biomass combustion without and with carbon capture is discussed next.

### 3.4.1 Biomass without Carbon Capture

The main sources of biomass for energy include agricultural residues, forestry residues, energy crops, industry residues, park and garden wastes, and contaminated wastes (Kadiyala et al., 2016).

**Agriculture residues** include dry crop residue, such as straw and sugar beet leaves, and livestock waste (solid or liquid manure).

**Forestry residues** include bark, wood blocks, wood chips from treetops and branches, and logs from forest thinning.

**Energy crops** include dry wood crops (e.g., willow, poplar, eucalyptus, and short rotation coppice), dry herbaceous crops (e.g., miscanthus, switchgrass, reed, canary grass, cynara, cardu, and Indian shrub), oil energy crops (e.g., sugar beet, cane beet, sweet sorghum, Jerusalem artichoke, sugar millet), starch energy crops (e.g., wheat, potato, maize, barley, triticale, corn, and amaranth), and other energy crops (e.g., flax, hemp, tobacco stems, aquatic plants, cotton stalks, and kenaf).

**Industry residues** include wood industry residues (e.g., bark, sawdust, wood chips, slaps, and cutoffs from saw mills), food industry residues (e.g., beet root tails, used cooking oils, tallow, yellow grease, and slaughterhouse waste), and industrial products (e.g., pellets from sawdust and wood shavings, bio-oil, ethanol, and biodiesel).

**Park and garden wastes** include grass and pruned wood.

**Contaminated wastes** include demolition wood, municipal waste, sewage sludge, sewage gas, and landfill gas.

The primary reason biomass combustion is not proposed for use in a WWS world is that biomass combustion, like coal and natural gas combustion, produces air pollution. A 100 percent WWS energy infrastructure is designed to eliminate air pollution. The air pollution problems from biomass combustion are similar to those from biofuel combustion, which are discussed in Section 3.5. The problem is worse for some types of biomass, such as municipal waste, which often contains toxic chemicals. In sum, whereas biomass is partly renewable, it is not clean, and a 100 percent WWS world requires both clean and renewable energy rather than just renewable energy.

The second reason for not including biomass is its higher $CO_2e$ emissions compared with WWS technologies. Biomass grows during photosynthesis by converting $CO_2$ and $H_2O$ from the air into organic material and $O_2$, which is released back to the air. Although growing biomass takes $CO_2$ out of the air, some or all of that $CO_2$ may be returned to the air because collecting, transporting, separating, incinerating, and growing the biomass require fossil-fuel energy. In addition, biomass has emissions associated with the time lag between planning and operation of a biomass plant. It also has heat and water vapor emissions from biomass combustion and water vapor emissions from water evaporated to cool steam turbines. Finally, it has carbon emissions due to covering soil with a biomass energy facility or with a low-carbon-intensity crop instead of a high-carbon-intensity forest.

Table 3.5 shows that the overall range of $CO_2e$ emissions from biomass used for electricity is 86 to 1,788 $g\text{-}CO_2e/kWh$, or 10 to 373 times the emissions per unit energy as onshore wind. These emissions are mostly due to lifecycle emissions (43 to 1,730 $g\text{-}CO_2/kWh$). A review by Kadiyala et al. (2016) of numerous lifecycle emission studies suggests that combustion of forestry residues and industry residues may result in the least emissions (43 to 46 $g\text{-}CO_2e/kWh$) among biomass fuels. Combustion of agricultural residues and energy crops may result in higher emissions (200 to 300 $g\text{-}CO_2e/kWh$), and combustion of municipal solid waste may result in the highest emissions (mean at 1,730 g-$CO_2$e/kWh). The low emissions from forestry and industry residues are due to the fact that the feedstock does not need to be produced actively as it does with agricultural residues or energy crops. The high emissions from burning municipal solid waste are due to emissions from producing and consuming the energy required to collect, segregate, sort, transport, and incinerate the waste.

Table 3.5 also indicates that biomass energy facilities have an opportunity cost emissions of 36 to 51 g-$CO_2$/kWh, due primarily to the fact they take 4 to 9 years between planning and operation versus 2 to 5 years for onshore wind or utility PV. During the additional time, the background grid is emitting.

The main other source of emissions from biomass is the 6.6 g-$CO_2$e/kWh resulting from the combined heat and water vapor emitted from biomass combustion. Because biomass combustion is less efficient than even coal combustion (Section 3.4.2), biomass combustion releases more heat per unit electric power produced than does coal combustion (Table 3.5).

A third problem for some types of biomass, particularly energy crops, is the much greater land requirement for them than for WWS technologies. Section 3.5 discusses this issue. Given that photosynthesis is only 1 (0.25 to 3) percent efficient at converting sunlight to biomass energy, whereas solar PV panels are now over 20 percent efficient at converting sunlight to electricity, a solar panel needs only 1/20th the land to produce the same energy as a biomass crop.

An alternative to burning biomass for electricity or heat is to extract landfill gas, which contains mostly methane, and use the methane to produce hydrogen by steam reforming (Section 2.2.2.1). As discussed in Section 2.9.2, the use of methane captured from landfills and from methane digesters to produce hydrogen is one method of consuming methane that would otherwise leak to the air. The hydrogen would then be used in a fuel cell (e.g., in a vehicle) to produce electricity, thereby avoiding fossil-fuel production of the same energy. Steam reforming of methane to produce hydrogen plus the use of the hydrogen in a fuel cell to generate electricity creates little air pollution (Colella et al., 2005). As such, capturing methane from a landfill or from digesters and using it to produce hydrogen for a fuel cell is the one exception to not using bioenergy in a WWS world.

In sum, combusting forest and industry residue and other forms of biomass to provide electricity and heat results in higher $CO_2$e emissions and much more air pollution than does using WWS technologies. Some forms of biomass also require much more land than do WWS technologies. As such, using biomass for energy represents an opportunity cost. The exception is to use landfill and digester methane to produce hydrogen by steam reforming, where the hydrogen is subsequently used in a fuel cell.

### 3.4.2 Biomass with Carbon Capture

A proposed method of reducing biomass $CO_2$e emissions, and even possibly creating negative carbon emissions, is to combine it with carbon capture and storage to form **bioenergy with carbon capture and storage or use (BECCS/U)**. **Negative carbon emissions** arise if a process removes more carbon from the air than it adds to the air. BECCS can theoretically result in negative carbon emissions if, for example, forest wood residue (containing $CO_2$ from the air) is collected, little energy is used to collect, transport, and incinerate the wood, and the $CO_2$ is captured from the exhaust stream and pumped underground. If it were successful, this method would be a one-way conduit for $CO_2$ to go from the air to underground, thereby resulting in negative carbon emissions.

The problems, however, are several-fold. As with natural gas and coal with CCS/U, the carbon capture system with BECCS/U requires 25 to 55 percent more energy than without it. If that energy comes from natural gas, coal, or biomass, 25 to 55 percent more air pollution occurs than with no BECCS/U. Biomass combustion without carbon capture already results in high air pollution levels compared with WWS. Air pollution with BECCS/U is even higher than without it.

Similarly, as with CCS/U for coal and natural gas, the $CO_2$ reductions are much lower than anticipated due to the high energy requirements of BECCS/U. Leakage of $CO_2$ from underground storage is also an issue.

Second, as with gas and coal, few reliable underground storage facilities exist for BECCS. Because of the high cost of carbon capture, bioenergy with carbon capture facilities are likely to be coupled with for-profit uses of the $CO_2$, such as enhanced oil recovery. Thus, BECCU will be favored over BECCS. This will encourage more combustion of and emissions from oil products.

Third, the efficiency of biomass combustion for electricity (electricity output per unit energy in the fuel) is low (20 to 27 percent), even compared with coal combustion (33 to 40 percent). Thus, a large mass of biomass is needed to produce a small amount of electricity. As such, if BECCS/U were to provide negative emissions on a large scale, substantial land areas dedicated to bioenergy crops would be needed to maintain a continuous energy supply. Consequently, a share of agricultural land would be used for fuel instead of food, increasing the price of food. Higher food prices trigger deforestation, as high-carbon-storage forest land is turned into low-carbon-storage agricultural land.

Fourth, removing agricultural residues usually means crops need to be fertilized more since the residues contain nutrients that are no longer available once they are removed. Fertilizers contain a greenhouse gas, nitrous oxide ($N_2O$), and a major pollution, ammonia ($NH_3$), which are emitted to the air.

Finally, the cost of BECCS/U is high as of 2020, even compared with CCS for fossil fuels. To date, only six BECCS/U facilities have survived, and each has been at a high cost. One has been for capturing $CO_2$ at an ethanol refinery. The others have been for capturing $CO_2$ as municipal solid waste plants.

Thus, paying for BECCS/U instead of WWS means less energy production, a longer time lag between planning and operation, more air pollution, greater land use (for some crops), and less carbon removal (because less BECCS/U than WWS technologies can be installed for the same money).

## 3.5 Why Not Liquid Biofuels for Transportation?

Liquid biofuels are generally used for transportation as a substitute for gasoline or diesel. The most common transportation biofuels are ethanol, used in passenger cars and other light-duty vehicles, and biodiesel, used in many heavy-duty vehicles. Liquid biofuels should not be used as part of a 100 percent WWS infrastructure due to their high air pollution mortality and morbidity, climate, land, and water supply impacts. This section discusses these issues.

**Ethanol** ($C_2H_5OH$) is produced in a factory, generally from corn, sugarcane, wheat, sugar beet, or molasses. The most common among these sources are corn and sugarcane, resulting in the production of **corn ethanol** and **sugarcane ethanol**, respectively. Microorganisms and enzymes ferment sugars or starches in these crops to produce ethanol.

Fermentation of cellulose originating from switchgrass, wood waste, wheat, stalks, corn stalks, or *Miscanthus*, also produces ethanol. However, the process is more energy intensive because breakdown of cellulose (e.g., as occurs in the digestive tracts of cattle) by natural enzymes is slow. Faster breakdown of cellulose requires genetic engineering of enzymes. The ethanol resulting from these sources is referred to as **cellulosic ethanol**.

Ethanol may be used on its own, as it is frequently in Brazil, or blended with gasoline. A blend of 6 percent ethanol and 94 percent gasoline is referred to as **E6**. Other typical blends are **E10, E15, E30, E60, E70, E85,** and **E100**. In many countries, including the United States, E100 is required to contain 5 percent gasoline as a **denaturant**, which is a poisonous or foul-tasting chemical added to a fuel to prevent people from drinking it. As such, E85, for example, contains about 81 percent ethanol and 19 percent gasoline.

A proposed alternative to ethanol for transportation fuel is **butanol** ($C_4H_9OH$). It can be produced by fermenting the same crops used to produce ethanol but with a different bacterium, *Clostridium acetobutylicum*. Butanol contains more energy per unit volume of fuel than does ethanol. However, unburned butanol also reacts more quickly in the atmosphere with the OH radical than does unburned ethanol, speeding up ground-level ozone formation relative to ethanol. On average, ethanol used in internal combustion engine vehicles produces more ground-level ozone than does gasoline used in such vehicles in most regions of the United States (Jacobson, 2007; Ginnebaugh et al., 2010; Ginnebaugh and Jacobson, 2012).

**Biodiesel** is a liquid diesel-like fuel derived from vegetable oil or animal fat. Major edible vegetable oil sources of biodiesel include soybean, rapeseed, mustard, false flax, sunflower, palm, peanut, coconut, castor, corn, cottonseed, and hemp oils. Inedible vegetable oil sources include jatropha, algae, and jojoba oils. Animal fat sources include lard, tallow, yellow grease, fish oil, and chicken fat. Soybean oil accounts for about 90 percent of biodiesel production in the United States. Biodiesel derived from soybean oil is referred to as **soy biodiesel**.

Biodiesel consists primarily of long-chain methyl, propyl, or ethyl esters that are produced by the chemical reaction of a vegetable oil or animal fat (both lipids) with an alcohol. It is a standardized fuel designed to replace diesel in standard diesel engines. It can be used as pure biodiesel or blended with regular diesel. Blends range from 2 percent biodiesel and 98 percent diesel (**B2**) to 100 percent biodiesel (**B100**). Generally, only blends B20 and lower can be used in a diesel engine without engine modification. The use of vegetable oil or animal fat directly (without conversion to biodiesel) in diesel engines is also possible; however, it results in more incomplete combustion, and thus more air pollution byproducts, as well as a greater buildup of carbon residue in, and damage to, the engine than biodiesel.

A significant effort has been made to produce **algae biodiesel**, which is biodiesel from algae grown from waste material, such as sewage. However, efforts have been hampered by the fact that algae can be grown efficiently only when exposed to the sun. As such, they cannot be grown efficiently when densely layered, with one on top of the other. Instead, they require a significant surface area exposed to the sun. Each volume of oil produced from algae also requires about 100 times that volume of water. Both factors have limited the growth of the algae biodiesel industry.

Liquid biofuels (e.g., corn ethanol, cellulosic ethanol, butanol, or biodiesel) are not recommended as part of a 100 percent WWS energy infrastructure. The main reasons are that (1) nearly all biofuels are combusted to generate energy, resulting in air pollution similar to that from fossil fuels; (2) liquid biofuels do not reduce $CO_2e$ emissions nearly to the extent that WWS-powered battery-electric or hydrogen fuel cell vehicles do; (3) some liquid biofuels increase $CO_2e$ emissions relative to fossil fuels; (4) many biofuels require rapacious amounts of land; (5) many biofuels require excessive quantities of water; and (6) many biofuels are derived from food sources, increasing food shortages, food prices, and starvation (Searchinger et al., 2008; Jacobson, 2009; Delucchi, 2011). Because liquid biofuels cause greater climate, pollution, land, and water problems than do WWS technologies, biofuels represent opportunity costs.

The main issues with liquid biofuels are illustrated by comparing the impacts of using ethanol to provide energy for internal combustion engines with the impacts of using wind or solar to provide electricity for battery-electric (BE) vehicles or hydrogen fuel cell (HFC) vehicles. Figure 3.2a, for example, compares the net change in $CO_2e$ emissions and air pollution premature mortalities if all light- and heavy-duty on-road vehicles in the United States were converted from those powered by liquid fossil fuels to those powered by battery electricity, hydrogen fuel cells, or E85.

In the United States, about 30.2 percent of all $CO_2e$ emissions in 2017 were from vehicle exhaust and 10.1 percent were from the upstream production of fuel. As such, converting to BE vehicles, HFC vehicles, or E85 vehicles could reduce U.S. $CO_2e$ emissions by a maximum of 40.3 percent. Figure 3.2a shows that converting to wind-powered BE vehicles reduces U.S. $CO_2e$ emissions by 40 to 40.2 percent, which represents a 99.3 to 99.8 percent reduction compared with the maximum possible reduction.

Using wind electricity to produce hydrogen for HFC vehicles results in about three times more emission than using wind electricity to power BE vehicles directly because using hydrogen for a fuel cell vehicle requires about 3 times the number of wind turbines than using electricity from a battery for a BE vehicle. The reason is that a battery-electric vehicle converts 64 to 89 percent of the electricity from a plug outlet to motion in the car, whereas an HFC vehicle converts about 23 to 37 percent (Table 7.2).

HFC losses are due to three main factors: losses in the **electrolyzer** (used to produce hydrogen from electricity), **compressor** (used to compress hydrogen), and **fuel cell** (used to convert hydrogen to energy and water) (Table 7.2). Nevertheless, an HFC vehicle is still more efficient than is an internal combustion gasoline engine, which converts about 17 to 20 percent of the fuel in its tank to mechanical motion.

Regardless, wind-powered HFC vehicles still reduce 98 to 99 percent of the maximum possible $CO_2e$ reduction. Figure 3.2a indicates that, in comparison, corn E85 and cellulosic E85 vehicles either increase $CO_2e$ or cause much less $CO_2e$ reduction than do the WWS options.

Figure 3.2b also shows that the air pollution mortality associated with either corn or cellulosic ethanol vehicles significantly exceeds that associated with WWS-powered BE and HFC vehicles. The reason is that BE vehicles have zero tailpipe emissions of pollutants, and HFC vehicles have tailpipe emissions of water vapor only. Thus, the only pollutant emissions from BE and HFC vehicles are from the upstream production of wind turbines or solar panels. Such upstream emissions are the only cause of the mortalities for BE and HFC vehicles in Figure 3.2b. Production of the vehicles themselves also results in emissions, but such emissions are excluded for all vehicles in Figure 3.2. E85 vehicles, on the other hand, have similar tailpipe emissions as gasoline vehicles during their operation and have upstream emissions from fuel production. Whereas the tailpipe emissions of E85 vehicles differ in composition from those of gasoline or diesel vehicles, the air pollution impacts of the E85 vehicles are often greater than those of fossil-fuel vehicles (Jacobson, 2007; Ginnebaugh and Jacobson, 2012), especially at low temperature (Ginnebaugh et al., 2010).

Another problem with ethanol is water consumption. Irrigating only 13.2 percent (the U.S. average irrigation rate for corn) of the U.S. corn supply needed to power a U.S. on-road vehicle fleet with corn ethanol would require about 10 percent of the U.S. water supply (Jacobson, 2009).

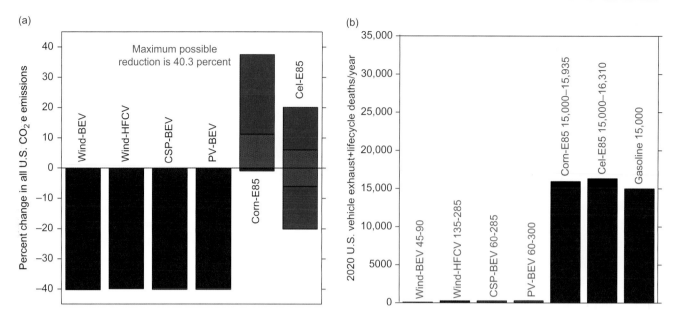

**Figure 3.2** (a) Percent changes in 2017 U.S. $CO_2$e emissions upon replacing 100 percent of on-road (light- and heavy-duty) fossil-fuel vehicles with battery-electric vehicles (BEVs) powered by wind, CSP, or solar PV electricity; with hydrogen fuel cell vehicles (HFCVs) powered by wind; or with E85 vehicles powered by corn or cellulosic ethanol. The maximum possible percent reduction in $CO_2$e due to such a conversion is 40.3 percent because 30.2 percent of U.S. $CO_2$e originates from vehicle exhaust and 10.1 percent originates from upstream fuel production. Low and high estimates are given. In all cases, except the E85 cases, blue represents the low estimate and blue plus red, the high estimate. For E85, the full bars represent the range at 100 percent penetration, and the brown bars represent the range at 30 percent penetration. For both ethanol sources, the high estimate occurs when emissions associated with price changes of fuel crops are accounted for. Such emissions occur when the price of corn increases due to corn's use as a fuel instead of food, and this triggers a conversion of forested or densely vegetated grassland to agricultural land, increasing carbon emissions (Searchinger et al., 2008). (b) Estimates of 2020 U.S. premature deaths per year due to upstream and exhaust emissions from vehicles for the scenario in (a). Low (blue) and high (blue plus red) estimates are given. In the case of corn E85 and cellulosic E85, the red bar is the additional number of deaths due to tailpipe emissions of E85 over gasoline for the United States and the black bar is the additional number of U.S. deaths per year due to upstream emissions from producing and distributing E85 fuel minus those from producing and distributing gasoline. The estimated number of deaths for gasoline vehicles in 2020 is also shown.Updated from Jacobson (2009).

Finally, because of the significant land required for either corn or cellulosic ethanol (Figure 3.3), corn or cellulosic E85 could never provide enough energy for any more than a few percent of the U.S. vehicle fleet, even ignoring the climate, air pollution, and water supply issues associated with E85. Analyses for other types of liquid biofuels give similar results.

In sum, liquid biofuels are not recommended as part of a 100 percent WWS energy infrastructure because of the climate, health, water supply, and land opportunity costs they incur.

## 3.6 Why Not Synthetic Direct Air Carbon Capture and Storage?

**Synthetic direct air carbon capture and storage (SDACCS)** is the direct removal of $CO_2$ from the air by its chemical reaction with other chemicals. Upon removal, the $CO_2$ is sequestered either underground or in a material, just as $CO_2$ from fossil fuels with carbon capture and storage (CCS) is. Alternatively, the $CO_2$ is sold for use in industry (SDACCU), just as $CO_2$ from fossil fuels with carbon capture and use (CCU) is.

SDACCS/U should not be confused with **natural direct air carbon capture and storage (NDACCS)**, which is the natural removal of carbon from the air by either planting trees or reducing permanent deforestation (by reducing open biomass burning – Section 2.9.1). Growing a tree removes $CO_2$ naturally by photosynthesis and sequesters the carbon within organic material in the tree for decades to centuries. Reducing open biomass burning similarly sequesters carbon in trees and eliminates emissions of health-affecting air pollutants and climate-affecting non-$CO_2$ global warming agents at the same time. Trees also absorb air pollutants, helping to filter them from the air.

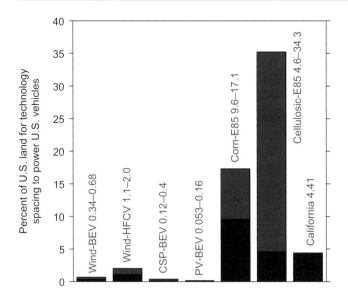

Figure 3.3 **Spacing area required for a given technology to provide energy for all U.S. vehicles in 2017 as either battery-electric vehicles (BEVs), hydrogen fuel cell vehicles (HFCVs), or E85 vehicles run on corn or cellulosic ethanol. The blue is the low estimate and the blue plus the red is the high estimate. The percentages are relative to the combined land area of all 50 U.S. states. The land area of California is shown for comparison.** Updated from Jacobson (2009).

Whereas NDAACS is recommended in a 100 percent WWS world, SDACCS is not. **SDACCS/U is basically a cost, or tax, added to the cost of fossil-fuel generation**, so it raises the cost of using fossil fuels while increasing air pollution due to its energy requirements and providing no energy security. To the contrary, it permits the fossil-fuel industry to expand its devastation of the environment and human health by allowing mining and air pollution to continue at an even higher cost to consumers than with no carbon capture.

Based on data from an existing facility powered by natural gas (Table 3.7), a **SDACCS/U plant results in 90 percent (averaged over 20 years) to 69 percent (averaged over 100 years) of the $CO_2e$ that it captures from the air being returned to the air due to the generation of energy required to run the equipment**. Even if SDACCS/U is powered by renewable electricity, it captures less $CO_2e$ than the same renewable electricity replacing a coal or natural gas plant.

Because SDACCS/U reduces little carbon, allows air pollution to continue, and incurs an equipment cost, spending on it rather than on renewables replacing fossil fuels or bioenergy always increases total social cost (equipment plus health plus climate cost). No improvement

in SDACCS/U equipment can change this conclusion, since SDACCS/U always incurs an equipment cost never incurred by renewables, and SDACCS/U never reduces, instead mostly increases, air pollution and mining.

In this section, methods of SDACCS/U and their consequences are discussed.

### 3.6.1 Discovery of Chemical Removal of $CO_2$ from the Air

In 1754, Joseph Black (1728–1799), a Scottish physician and chemist, isolated $CO_2$, which he named **fixed air**. He found that heating the odorless white powder magnesium alba (**magnesium carbonate, $MgCO_3$**) or limestone (**calcium carbonate, $CaCO_3$**) by the respective reactions,

$$MgCO_3 + heat \rightarrow CO_2 + MgO \tag{3.12}$$

$$CaCO_3 + heat \rightarrow CO_2 + CaO \tag{3.13}$$

released a gas ($CO_2$) that could not sustain life or fire. The remaining solids, magnesium usta (**magnesium oxide, MgO**) and quicklime (**calcium oxide, CaO**), respectively, weighed less than the original solids. He found further that by dissolving the gas in a solution of limewater (**calcium hydroxide, $Ca(OH)_2$**), the gas "fixed" to the CaO, reforming the calcium carbonate by

$$CO_2 + Ca(OH)_2 \rightarrow CaCO_3 + H_2O \tag{3.14}$$

The calcium carbonate precipitated as a white solid from the solution. He similarly found that adding potash (**potassium carbonate, $K_2CO_3$**) to magnesium oxide by

$$K_2CO_3 + MgO \rightarrow K_2O + MgCO_3 \tag{3.15}$$

resulted in $MgCO_3$. The mass of $MgCO_3$ exceeded that of MgO by the same mass that was lost when $MgCO_3$ was heated by Reaction 3.12 to form MgO. The difference in mass in both cases was the mass of $CO_2$. As such, Black quantified the mass of $CO_2$ for the first time.

Black soon recognized that the fixed air he had isolated was the same gas that the Belgian **John Baptist Van Helmont** (1577–1644) found by fermenting alcoholic liquor, burning charcoal, and acidifying marble and chalk. Van Helmont had called this vapor **gas silvestre** ("gas that is wild and dwells in out-of-the-way places").

Today, SDACCS/U techniques include reacting $CO_2$ from the air with (a) alkali and alkaline Earth metal oxides and hydroxides and (b) organic-inorganic sorbents consisting of amines. The $CO_2$ sequestered by these

Table 3.7 Comparison of relative $CO_2e$ emissions, electricity private costs, and electricity social costs among three scenarios related to the Carbon Engineering SDACCU plant, each over a 20-yr and 100-yr time frame. The first scenario is using an on-site natural gas combined cycle turbine to power the air capture (AC) equipment. The AC equipment does not capture the gas emissions; if it did, the results would be the same, since if the equipment captured turbine $CO_2$ emissions, it would not capture the equivalent $CO_2$ from the air. The second scenario involves using a wind turbine to power the AC equipment. The third scenario involves using the same wind turbine electricity to instead replace coal power generation without using AC equipment. All emission units (Rows a–f, i) are kg-$CO_2e$/MWh.

| | DAC with NG elec. 20 yr | DAC with NG elec. 100 yr | DAC with wind elec. 20 yr | DAC with wind elec. 100 yr | Wind replacing coal 20 yr | Wind replacing coal 100 yr |
|---|---|---|---|---|---|---|
| a) SDACCU removal from air[1] | 825 | 825 | 825 | 825 | – | – |
| b) $CO_2$ emissions combined cycle gas turbine[2] | 404 | 404 | – | – | – | – |
| c) Upstream $CO_2e$ of $CH_4$ from gas leaks[3] | 280 | 111 | – | – | – | – |
| d) Upstream $CO_2$ from gas mining, transport[4] | 54 | 54 | – | – | – | – |
| e) Emission reduction due to replacing coal with wind[5] | 0 | 0 | 0 | 0 | –1,381 | –1,168 |
| f) All emissions (b+c+d+e) | 738 | 569 | 0 | 0 | –1,381 | –1,168 |
| g) Percent $CO_2$ returned (f/a) | **89.5%** | **68.9%** | **0%** | **0%** | – | – |
| h) Percent $CO_2$ captured (100-g) | **10.5%** | **31.1%** | **100%** | **100%** | – | – |
| i) Absolute emission reduction (a–f) | 87 | 256 | 825 | 825 | 1,381 | 1,168 |
| j) Low SDACCU ($/tonne-$CO_2$-removed)[1] | 94 | 94 | 94 | 94 | – | – |
| k) High SDACCU ($/tonne-$CO_2$-removed)[1] | 232 | 232 | 232 | 232 | – | – |
| l) Low private electricity cost (aj/1000) ($/MWh)[6] | 78 | 78 | 78 | 78 | 29 | 29 |
| m) High private electricity cost (ak/1000) ($/MWh)[6] | 191 | 191 | 191 | 191 | 56 | 56 |
| n) Health cost of background grid ($/MWh)[7] | 40 | 40 | 40 | 40 | 40 | 40 |
| o) Ratio health cost of scenario to of background grid[8] | 3 | 3 | 2 | 2 | 0 | 0 |
| p) Health cost of scenario (no) ($/MWh) | 120 | 120 | 80 | 80 | 0 | 0 |
| q) Climate cost of background grid ($/MWh)[9] | 152 | 152 | 152 | 152 | 152 | 152 |
| r) Ratio climate cost of scenario to background grid[10] | 0.937 | 0.781 | 0.403 | 0.294 | 0 | 0 |

| | | | | | | |
|---|---|---|---|---|---|---|
| s) Climate cost of scenario (qr) ($/MWh) | 142 | 119 | 61.2 | 44.6 | 0 | 0 |
| t) Low social cost ($/MWh) (l+p+s) | 340 | 316 | 219 | 202 | 29 | 29 |
| u) High social cost ($/MWh) (m+p+s) | 454 | 430 | 333 | 316 | 56 | 56 |
| v) Low social cost ratio (Row t-SDACCU/u-wind) | 6.1 | 5.6 | 3.9 | 3.6 | – | – |
| w) High social cost ratio (Row u-SDACCU/t-wind) | 15.6 | 14.8 | 11.5 | 10.9 | – | – |

[1] Source: Keith et al. (2018). Assumes values for DAC with wind electricity are the same as DAC with natural gas electricity.

[2] de Gouw et al. (2014).

[3] Same methodology as in Table 3.6, Footnote 6, but using the $CO_2$ combustion emissions from Row (b) here.

[4] Source: Howarth (2014).

[5] Assumes wind that would otherwise be used to run the SDACCU equipment instead directly replaces coal electricity, its upstream $CO_2$ combustion, its upstream $CH_4$ leaks, and its stack combustion $CO_2$ emissions. The overall emission rates from coal are obtained from Table 3.6, Row d.

[6] Low and high wind electricity costs for wind-replacing coal are from Lazard (2018). Others are from the formula provided.

[7] The U.S. health cost of $40/MWh for the background grid per MWh is from Jacobsen et al. (2019).

[8] The ratio of the health cost in the scenario to that of the background grid is defined as zero for the wind-replacing-coal case, since wind produces zero emissions during its operation. In comparison, wind running SDACCU equipment allows those coal emissions, which are about twice background grid emissions, to continue, so the factor in that scenario is 2. Natural gas running SDACCU equipment not only allows those coal emissions to continue, but it also produces 50 percent more emissions, assumed equal to background grid emissions per MWh, so the factor in that scenario is 3.

[9] The U.S. climate cost of $152/MWh for the background grid is from Jacobson et al. (2017, 2019).

[10] The ratio of the climate cost of the scenario to that of the background grid is defined as zero for the wind-replacing coal case, since wind produces zero emissions during its operation. For the other cases, it is simply the absolute $CO_2$e emission reduction in the case minus that in the wind case all divided by that in the wind case, where all values are from Row i.

Source: Jacobson (2019).

methods can either be stored underground, sequestered in concrete (Section 2.4.8), or sold for use in industry. Below, methods of reacting $CO_2$ with air are discussed, followed by an examination of the issues associated with SDACCS/U.

## 3.6.2 Reaction of $CO_2$ with Alkali and Alkaline Earth Metal Oxides and Hydroxides

One way to remove $CO_2$ from the air is to react it with alkali and alkaline Earth metal oxides and hydroxides (Duan and Sorescu, 2010).

**Alkali metal oxides** include $Na_2O$ and $K_2O$.

**Alkali metal hydroxides** include NaOH and KOH.

**Alkaline Earth metal oxides** include BeO, MgO, CaO, SrO, and BaO.

**Alkaline Earth metal hydroxides** include $Be(OH)_2$, $Mg(OH)_2$, $Ca(OH)_2$, $Sr(OH)_2$, and $Ba(OH)_2$.

A classic method of removing $CO_2$ from the air while recycling the material that is removing it is by exposing the $CO_2$ to a large pool of limewater (an alkaline Earth metal hydroxide) by Equation 3.14. The resulting solid

$CaCO_3$ is heated to 700 K, releasing a concentrated stream of $CO_2$ through Equation 3.13 that can be captured and used. The CaO is then returned to limewater by

$$CaO + H_2O \rightarrow Ca(OH)_2 \qquad (3.16)$$

(Lackner et al., 1999). The problem with this process is that it needs a continuous net input of energy, which can become enormous with a large amount of $CO_2$ processed. An alternative process, which has been used in the paper industry for a long time, is

$$CO_2 + 2NaOH \rightarrow Na_2CO_3 + H_2O \qquad (3.17)$$

$$Na_2CO_3 + Ca(OH)_2 \rightarrow 2NaOH + CaCO_3 \qquad (3.18)$$

$$CaCO_3 + heat \rightarrow CaO + CO_2 \qquad (3.19)$$

$$CaO + H_2O \rightarrow Ca(OH)_2 \qquad (3.20)$$

(Sanz-Perez et al., 2016). However, this reaction sequence also requires a net input of energy that accumulates with an increasing amount of $CO_2$ processed. In general, removing $CO_2$ from the air with some hydroxides and oxides (e.g., $Na_2O$, $K_2O$, MgO, NaOH, KOH, and $Mg(OH)_2$) is more efficient than with others (Duan and Sorescu, 2010).

However, all reaction sequences result in net additions of energy that accumulate with increasing amounts of $CO_2$ processed.

### 3.6.3 Reaction of $CO_2$ with Organic-Inorganic Sorbents Consisting of Amines

Another approach to removing $CO_2$ from the air is by reacting it with an organic-inorganic sorbent containing amines. **Amines** are derived from **ammonia ($NH_3$)** by replacing one or more hydrogen atoms with an **alkyl group** ($CH_3$, $C_2H_5$, $C_3H_7$, etc.) or **aryl group** (a functional group containing an aromatic ring). In such cases, the alkyl or aryl group can be denoted simply with an R, so an organic-inorganic sorbent containing amines can take the form of $RNH_2$. Reaction of $CO_2$ with $RNH_2$ results in

$$CO_2 + 2RNH_2 \rightarrow RNH_3^+ + RNHCOO^- \qquad (3.21)$$

The advantage of this reaction is that $CO_2$ forms strong bonds with the amine group, so $CO_2$ can be absorbed effectively at low partial pressures. This method of $CO_2$ removal is used in submarines to purify air, but its application to removing $CO_2$ from the ambient atmosphere then returning the $RNH_2$ still requires a net energy input and high cost.

### 3.6.4 Opportunity Cost of SDACCS/U

By removing $CO_2$ from the air, SDACCS/U does exactly what WWS generators, such as wind turbines and solar panels, do. This is because WWS generators replace fossil generators, preventing $CO_2$ from getting into the air in the first place. **The impact on climate of removing one molecule of $CO_2$ from the air is the same as the impact of preventing one molecule from getting into the air in the first place.**

The differences between WWS generators and SDACCS/U equipment, though, are that the WWS generators also (a) eliminate non-$CO_2$ air pollutants from fossil-fuel combustion; (b) eliminate the upstream mining, transport, and refining of fossil fuels and the corresponding emissions; (c) reduce the pipeline, refinery, gas station, tanker truck, oil tanker, and coal train infrastructure of fossil fuels; (d) reduce oil spills, oil fires, gas leaks, and gas explosions; (e) substantially reduce international conflicts over energy; and (f) reduce the large-scale blackout risk associated with centralized power plants by decentralizing/distributing power.

SDACCS/U does none of that. Its sole benefit is to remove $CO_2$ from the air, but at a higher cost than using renewable energy to do the same thing. In fact, SDACCS/U is basically a cost added on to the cost of using fossil fuels.

Moreover, SDACCS/U is an opportunity cost. Because SDACCS/U removes no health-affecting air pollutants from the air, money spent on it takes funds away from the purchase of clean, renewable WWS technologies. Such technologies replace fossil-fuel power plants and vehicles while eliminating their health effects and costs and more $CO_2e$ than the SDACCS/U removes.

Second, SDACCS/U requires substantial electricity and heat to work, and this must come from the grid, a dedicated fossil-fuel source, or a dedicated WWS source. If grid electricity is used, air pollution emissions directly increase compared with no SDACCS/U and a portion of the $CO_2$ emissions reduced by SDACCS/U is re-emitted to the air due to the use of grid electricity.

Third, because SDACCS/U increases or prevents the reduction of grid electricity use, it extends the life of fossil-fuel and nuclear power plants; the upstream mining, transport, and processing of fossil fuels and uranium for those plants; and the emissions associated with the upstream mining. SDACCS/U similarly increases the energy insecurity and environmental and health consequences of the fossil-fuel and nuclear infrastructures.

Fourth, the higher cost of SDACCS/U relative to WWS electric power technologies ensures that a fixed amount of capital spent on SDACCS/U increases $CO_2$ and air pollution more than if the same money were spent on WWS technologies.

Even if the cost per unit mass of $CO_2$ removed by SDACCS/U were the same as or lower than that of WWS, SDACCS/U would still increase air pollution relative to WWS because SDACCS/U does not reduce any air pollutants, whereas all WWS technologies do. In addition, when fossil fuels are used to power SDACCS/U equipment, such fossils increase $CO_2$.

In terms of cost, one final factor is social cost, discussed in more detail in Chapter 7. The **social cost of air pollution** is the health-related cost of air pollution to society. For example, air pollution increases death and illness, both of which increase hospitalization stays, emergency room visits, lost workdays, lost school days, insurance rates, taxes, worker's compensation rates, and loss of companionship, among other costs. A worldwide mean health cost of fossil-fuel energy among all energy sectors is about $169 per MWh of energy produced but varying by country (Table 7.11). The cost of a new wind turbine is about $43 ($29–$56) per MWh of electricity produced (Lazard, 2018). This is less than the health cost that the wind turbine eliminates (a mean of $169 per MWh worldwide). Thus, a new wind turbine displacing a fossil-fuel power plant immediately reduces society's direct energy cost plus health cost. In other words, every wind turbine installed avoids a high cost to society. On the other hand, SDACCS/U does not reduce any air pollution. SDACCS/U allows air pollution and its costs to persist. So, **a wind turbine replacing a fossil plant will always provide much more benefit than the same money spent on SDACCS/U equipment.**

Table 3.7 summarizes the inefficiency of $CO_2$ removal from the air by an existing SDACCU facility. Electricity for the air capture (AC) equipment is provided by a natural gas combined cycle turbine. The table indicates that, averaged over 20 and 100 years, 89.5 percent and 69 percent, respectively, of all $CO_2$ captured by the AC equipment is returned to the air as $CO_2$e. The emissions come from mining, transporting, processing, and burning the natural gas used to power the equipment.

In comparison with taking no action, using SDACCU equipment powered by natural gas also increases air pollution due to the combustion and upstream emissions associated with natural gas. With no action, SDACCU further incurs an equipment cost. Thus, although SDACCU powered by natural gas reduces some $CO_2$e, the equipment cost and air pollution cost far outweigh that decrease, resulting in a near doubling of the total social cost per MWh of electricity use relative to the health and climate cost per MWh of coal power plant emissions (Figure 3.4).

Even when zero re-emissions occur, such as when wind powers the SDACCU equipment, the mean social cost of using SDACCU still exceeds that of doing nothing (Figure 3.4). On the other hand, using wind to replace coal electricity instead of to run the AC equipment eliminates $CO_2$e and air pollution emissions and their associated costs from the coal. The resulting social cost is ~15 percent of that from wind-powering SDACCU equipment (Table 3.7, Figure 3.4). A similar result is found when wind replaces a natural gas plant instead of a coal plant.

**In fact, there is no case where wind powering an SDACCU plant has a social cost below that of wind replacing any fossil-fuel or bioenergy power plant**

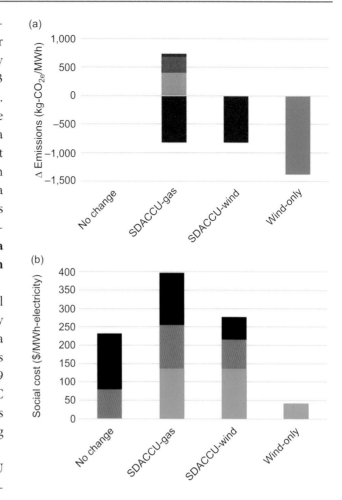

Figure 3.4 (a) Change in $CO_2$e emissions, averaged over 20 years, per unit electricity needed to run SDACCU equipment resulting from either no action (no change), using an SDACCU plant with equipment powered by natural gas (SDACCU-gas), using an SDACCU plant with equipment powered by wind (SDACCU-wind), and using the same quantity of wind required to run the SDACCU equipment but to replace coal power directly (wind-only). Blue is the removal of $CO_2$ from the air by the SDACCU equipment; orange is the natural gas turbine emissions; red is the $CO_2$e from natural gas mining and transport $CH_4$ leaks; purple is natural gas mining and transport $CO_2$e aside from $CH_4$ leaks; and green is the $CO_2$e emission reduction due to replacing coal power with wind power. (b) Mean estimate of social costs per unit electricity over 20 years for each of the four cases shown on the left. Light blue is the cost of equipment (either air capture equipment plus gas turbine, air capture equipment plus wind turbine, or wind turbine alone); brown is air pollution health cost; and black is 20-year climate cost. All data are from Table 3.7, except that the costs in the no-change case are the health and climate costs of coal power plant emissions ($80/MWh health cost and $152/MWh climate cost – Table 3.6, Footnote 13). Such emissions costs are used as the background because the wind-only case removes such emissions. From Jacobson (2019).

**directly** (Jacobson, 2019). The reasons are that wind-powering SDACCU always incurs an SDACCU equipment cost that wind alone never incurs and SDACCU always allows air pollution and mining to continue, whereas wind always eliminates air pollution and mining.

Finally, Figure 3.4 illustrates that SDACCS/U powered by wind captures less $CO_2$e than the same wind replacing a coal plant.

Example 3.12 illustrates the direct and social cost of SDACCS/U based on data from the same plant examined in Table 3.7. In the plant, a natural gas combined cycle gas turbine is used to provide the electricity needed to remove $CO_2$ from the air. In one case, combustion emissions from the gas plant are not captured. In the other, they are. In both cases, upstream emissions from mining and transporting natural gas still occur.

### Example 3.12  Costs and impacts on $CO_2$e and air pollution emissions of SDACCU

Compare the cost range of SDACCS, $94 to $232 per tonne-$CO_2$-removed (Keith et al., 2018), with the 2017 cost of onshore wind in the United States, $43 ($29–$56) per MWh of electricity produced (Lazard, 2018) under three scenarios: (a) all energy for the SDACCS plant is provided by the electric power grid, (b) all energy for the plant is provided by a dedicated natural gas-powered combined cycle gas turbine (CCGT) whose emissions are allowed to escape, and (c) all energy for the plant is provided by the same CCGT turbine, but whose combustion $CO_2$ emissions, but not upstream $CO_2$e emissions, are also captured by the plant. In each case, account for the social cost of air pollution (~$40/MWh in the United States from Table 7.11) avoided by wind but not by SDACCS and estimate the resulting difference in overall cost per MWh between the technologies.

Assume the SDACCS equipment removes 825 kg-$CO_2$ from the air per MWh of energy required to run the plant. This number is derived from Keith et al. (2018) by noting that the gas turbine used in that study emits 0.48 megatonnes-$CO_2$/y, while the plant captures 0.98 megatonnes-$CO_2$/y and that the $CO_2$ combustion emissions from a CCGT are 404 kg-$CO_2$/MWh (Table 3.1). Also assume that the average lifecycle $CO_2$e emissions (assume a 100-year time frame) on the U.S. grid in 2017 are about 557.3 kg-$CO_2$e per MWh of electricity produced and that the upstream $CO_2$e emissions (with a 100-year time frame) from the CCGT are 165 kg-$CO_2$e/MWh (Table 3.7).

### Solution:

In Case (a), the SDACCS equipment removes 825 kg-$CO_2$ from the air per MWh of electricity used to run the plant but re-emits 557.3 kg-$CO_2$e/MWh, or 67.6 percent of the $CO_2$e back to the air. Thus, it captures only 32.4 percent of what it intended to capture.

Multiplying $94 to $232 per tonne-$CO_2$-removed by 1 tonne-removed / 0.324 net-tonnes-removed = $290 to $716 per net-tonne-$CO_2$-removed. Multiplying the average U.S. grid emission rate of 557.3 kg-$CO_2$e/MWh by the cost of SDACCS per net-tonne-$CO_2$-removed gives an equivalent cost of reducing $CO_2$ from the grid with SDACCS of $162 to $399 per MWh-electricity-produced. In comparison, a wind turbine direct cost is $29 to $56 per MWh. Thus, SDACCS costs 2.9 to 14 times the direct cost of onshore wind to avoid the same $CO_2$. Adding the air pollution social cost ($40/MWh), which SDACCS continues to allow but wind does not, to the SDACCS energy cost gives the energy plus air pollution cost of a SDACCS as $202 to $439 per MWh, or 3.6 to 15.1 the cost per MWh of wind.

In Case (b), the total $CO_2$e emissions from the gas plant are 404 + 165 = 569 kg-$CO_2$e/MWh. Multiplying by 1.21 MWh per tonne-$CO_2$-removed from Case (a) gives 689 kg-$CO_2$e-emitted from the gas turbine per tonne-$CO_2$-removed. The net $CO_2$e removal from the air for every tonne captured from the air is then 1 tonne minus 0.69 tonnes = 0.31 tonnes. In other words, of every tonne of $CO_2$ removed from the air by this process, 69 percent is re-emitted due to using the gas turbine and only 31 percent is actually sequestered.

Multiplying the $94 to $232 per tonne-$CO_2$-removed by 1 tonne-$CO_2$-removed / 0.31 net-tonnes-$CO_2$-removed = $303 to $748 per net-tonne-$CO_2$-removed. Multiplying by the U.S. grid emission rate (557.3 kg-$CO_2$e/MWh) gives an equivalent cost of reducing $CO_2$ from the grid in this case with SDACCS of $169 to

$419 per MWh-electricity-produced, which is 3 to 14.4 times the direct cost of onshore wind to avoid the same $CO_2$. Adding the air pollution social cost ($40/MWh), which SDACCS continues to allow but wind does not, to the SDACCS energy cost gives the energy plus air pollution cost of a SDACCS as $209 to $459 per MWh, or 3.7 to 15.8 the cost per MWh of wind.

The result in Case (c) is the same as in Case (b). The plant emits 69 percent of what it is supposed to capture back to the air and retains only 31 percent. The reason is that the plant can remove only 825 kg-$CO_2$ from the air per MWh of electricity generated by the gas turbine. If the $CO_2$ is removed from the air (instead of from the turbine exhaust), the equivalent $CO_2$ from the turbine (404 kg-$CO_2$/MWh) will be released to the air and vice versa. In both cases, the upstream emissions from the natural gas mining and transport (165 kg-$CO_2$e/MWh) will also be released to the air. As such, while removing 825 kg-$CO_2$ from the air per MWh, the plant releases 569 kg-$CO_2$e/MWh (69 percent) back to the air. The resulting cost of SDACCS versus wind is the same as in Case (b).

Example 3.12 illustrates that using average grid electricity or a dedicated natural gas turbine to run a SDACCS/U plant results in 68 or 69 percent, respectively, of the $CO_2$ captured from the air being re-emitted back to the air due to the energy required to run the equipment, over a 100-year time frame. Table 3.7 indicates that up to 89.5 percent of the $CO_2$ captured is returned to the air over a 20-year time frame.

An argument for using SDACCS/U is that it will be needed to remove $CO_2$ from the air once all fossil fuels are replaced with 100 percent WWS. If all energy is provided by renewables at that point, SDACCS/U should reduce $CO_2$ without increasing air pollution. However, the question at that point is whether growing more trees, reducing open biomass burning, reducing agriculture and waste burning, or reducing halogen, nitrous oxide, and non-energy methane emissions (Section 2.9) is a more cost-effective method of limiting global warming. Until that time, when such an evaluation can be made, SDACCS/U will always be an opportunity cost.

In sum, as with CCS/U, SDACCS/U is not close to a zero-carbon technology. For the same energy cost, wind turbines and solar panels reduce much more $CO_2$ while also eliminating fossil air pollution, mining, and infrastructure, which SDACCS/U increases.

## 3.7 Why Not Geoengineering?

**Geoengineering** is the large-scale alteration of the natural properties of the Earth or the atmosphere in an attempt to reduce global near-surface temperatures. The two primary categories of geoengineering that have been proposed are techniques to remove carbon from the air (**carbon capture techniques**) and techniques to increase the reflectivity of the Earth or its atmosphere in order to decrease sunlight hitting the Earth's surface (**solar radiation management** techniques).

Carbon capture techniques have already been discussed. These include fossil fuels with carbon capture (Section 3.2), bioenergy with carbon capture (Section 3.4), synthetic direct air carbon capture (Section 3.6), and natural direct air carbon capture (Section 3.6). These are geoengineering techniques because they are intended to reduce the amount of $CO_2$ in the air to modulate the Earth's average temperature. Of the carbon capture techniques, only natural direct air carbon capture is recommended in a 100 percent WWS world.

The main solar radiation management techniques that have been proposed include (1) injecting reflective aerosol particles into the stratosphere to reflect sunlight directly, (2) injecting fine sea spray particles into the air just above the ocean surface to increase the number and decrease the average size of cloud drops, thereby increasing the overall cross-sectional area of cloud drops to increase their reflectivity, and (3) installing white roofs or roads.

The first problem with all these techniques is that none reduces fossil-fuel or bioenergy emissions of gases or particles that cause global warming and the 7.1 million deaths that occur annually (Table 7.13). To the contrary, with geoengineering, the public and policymakers become complacent, no longer feeling the urgency to reduce global temperatures or fossil-fuel emissions. As such, pollutant gases and particles continue to cause damage and, in fact, increase due to the complacency.

Second, geoengineering temporarily masks some warming damage, but because long-lived greenhouse gases continue to accumulate, their growth requires even more investment in geoengineering to keep up with the increase in emissions. Any interruption or stoppage of the geoengineering results in an immediate worsening of the climate problem compared with prior to starting the geoengineering because of the increased accumulation of $CO_2$e emissions during the period of geoengineering.

Third, since geoengineering does nothing to stop air pollution, air pollution mortality and morbidity continue to occur without abatement compared with no geoengineering. Such health impacts worsen if complacency allows more fossil-fuel and bioenergy combustion.

Fourth, since geoengineering does not reduce fossil-fuel or nuclear use, it does nothing to help reduce energy insecurity associated with those energy sources (Section 1.3).

Fifth, if the money spent on geoengineering were spent instead on WWS, not only would the WWS eliminate $CO_2$e emissions, but it would also eliminate air pollution emissions and resulting mortalities and morbidities, mining for fossil fuels and uranium, and energy insecurity. As such, geoengineering is an opportunity cost compared with WWS.

A sixth problem with all the proposals is the unintended consequences. For example, reducing solar radiation to the land reduces crop yields, which can result in starvation in some parts of the world. Injecting aerosol particles into the stratosphere catalyzes ozone loss in the presence of halogens currently in the stratosphere. Injecting particles into the stratosphere or into the air above the ocean results in changes in weather patterns. Injecting particles into marine air also increases the concentration of particles entering populated coastal cities, increasing morbidities and mortalities from air pollution. Particles injected into the stratosphere ultimately deposit to the ground, increasing air pollution health and acid deposition problems as well.

An example of the possible unintended consequences of a geoengineering proposal is the potential impact of white roofs and roads on global climate. Although white roofs and roads reflect radiation, cooling buildings and the ground in cities locally, they may cause large-scale global warming (Jacobson and Ten Hoeve, 2012).

The first reason is that, because white roofs cool the ground locally relative to the air, they reduce the ability of air to rise, thus clouds to form. Since clouds are reflective, reducing cloudiness increases solar radiation to the surface. This increase may be greater than the decrease resulting from white roofs and roads, especially since clouds travel and spread beyond a city, so reducing clouds over a city has the impact of increasing solar radiation reaching the ground outside of the city.

Second, black and brown carbon in the air absorb sunlight, then convert that sunlight directly to heat, which is released to the air. In the presence of white roofs or roads, black and brown carbon absorb not only downward sunlight but also sunlight reflected upward by the white surfaces.

Finally, while white roofs cool buildings, thereby reducing air conditioning energy requirements at low latitudes and during summers, the cooling increases heating energy requirements at high latitudes and during winters. In many places worldwide, heating requirements exceed cooling requirements, so adding a white roof to a building simply increases fossil-fuel use to heat buildings more.

A better solution than using a white roof is to install solar PV panels on a rooftop. The primary purpose of installing a PV panel is to generate electricity; however, panels also have several side benefits. Not only does a rooftop PV panel remove 20 percent of incoming solar radiation, converting it to electricity and cooling the underlying building, but the electricity it produces also displaces fossil-fuel use and its emissions. The reduction in solar radiation due to a solar PV panel results in a negative anthropogenic heat flux (Section 3.2.2.2), thus a negative carbon-equivalent emission ($CO_2$e) (Table 3.5). In addition, because solar panels do not reflect solar radiation upward as white roofs do, solar panels don't allow absorption of upward reflected sunlight by black and brown carbon pollution particles. Similarly, because PV panels are warmer than is a white roof, PV panels don't increase air stability, thus don't reduce cloudiness like white roofs do.

In sum, geoengineering through carbon capture is not recommended, with the exception of natural direct air capture by trees and reducing deforestation (NDACCS). Geoengineering through solar radiation management techniques is also not recommended. However, the use of rooftop PV panels, which reduce rooftop temperatures in addition to displacing fossil-fuel electricity, is recommended as part of a 100 percent WWS energy infrastructure. Similarly, wind turbines not only displace fossil-fuel emissions, but they also help to reduce globally averaged temperatures (Section 3.2.2.3, Table 3.5).

## 3.8 Summary

This chapter discussed non-WWS technologies that have been proposed to help address global warming. Almost all such technologies, though, are not recommended for two reasons. First, they do not reduce carbon emissions as much as do WWS technologies. Second, they do not simultaneously address air pollution or energy security to the same extent, if at all, as do WWS technologies. The energy technologies not recommended include natural gas for electricity or heat, any fossil fuel with carbon capture and sequestration or use (CCS/U), nuclear power for electricity, liquid biofuels for transportation, and biomass with or without carbon capture for electricity. The only exception is to capture landfill and digester methane to produce hydrogen by steam reforming. The hydrogen would then be used in a fuel cell. Additional methods that have been proposed to remediate global warming include synthetic direct air carbon capture and storage or use (SDACCS/U) and geoengineering. These technologies are also not recommended. The one exception is natural direct air carbon capture and storage. This is the natural removal of carbon from the air by either planting trees or reducing permanent deforestation. It is recommended.

## Further Reading

Bruckner T., I. A. Bashmakov, Y. Mulugetta, H. Chum, A. de la Vega Navarro, J. Edmonds, et al., 2014. Energy systems. In *Climate Change 2014: Mitigation of Climate Change. Contribution of Working Group III to the Fifth Assessment Report of the Intergovernmental Panel on Climate Change*, ed. O. Edenhofer, R. Pichs-Madruga, Y. Sokona, E. Farahani, S. Kadner, K. Seyboth, et al. Cambridge: Cambridge University Press.

Fetter, S., January 26, 2009. How long will the world's uranium supplies last?, *Scientific American*, *9*.

Fuhrmann, M., 2009. Spreading temptation: proliferation and peaceful nuclear cooperation agreements, *Int. Secur.*, *34*(1), available at SSRN: https://ssrn.com/abstract=1356091 (accessed September 9, 2019).

Howarth, R. W., 2019. Is shale gas a major driver of recent increase in global atmospheric methane?, *Biogeosciences*, *16*, 3033–3046.

IPCC (Intergovernmental Panel on Climate Change), 2005. *IPCC Special Report on Carbon Dioxide Capture and Storage*. Prepared by working group III, B. Metz, O. Davidson, H. C. de Coninck, M. Loos, and L. A. Meyer, eds. Cambridge: Cambridge University Press, 442 pp, http://arch.rivm.nl/env/int/ipcc/ (accessed June 26, 2019).

Jacobson, M. Z., 2009. Review of solutions to global warming, air pollution, and energy security, *Energy Environ. Sci.*, *2*, 148–173, doi:10.1039/b809990c.

Koomey, J., and N. E. Hultman, 2007. A reactor-level analysis of busbar costs for U.S. nuclear plants, 1970–2005, *Energy Policy*, *35*, 5630–5642.

Sgouridis, S., M. Carbajales-Dale, D. Csala, M. Chiesa, and U. Bardi, 2019. Comparative net energy analysis of renewable electricity and carbon capture and storage, *Nat. Energy*, *4*, 456–465.

Sovacool, B. K., 2008. Valuing the greenhouse gas emissions from nuclear power: a critical survey, *Energy Policy*, *36*, 2940–2953.

## 3.9 Problems and Exercises

3.1. For each proposed solution on the left, match it with one major problem on the right.

| a) Natural gas electric power | 1) Does not reduce greenhouse gas or air pollutant emissions |
| --- | --- |
| b) Nuclear power | 2) Needs more energy and mined fuel, increasing air pollution |
| c) Ethanol biofuel | 3) Needs more energy and collected fuel, increasing air pollution |
| d) Bioenergy with CCS | 4) Affects no emissions; reduces some C but not air pollution |
| e) Fossil fuels with CCS | 5) Uses lots of land/water; still emits air pollutants and GHGs |
| f) Synthetic direct air CCS | 6) Methane leaks during mining; GHG and pollutant emissions |
| g) Geoengineering by solar radiation management | 7) Weapons proliferation, meltdown, waste, and mining risks |

3.2. State three benefits of WWS electric power generators over synthetic direct air carbon capture and storage (SDACCS).

3.3. Explain what opportunity cost emissions mean. If electricity-generating technologies A, B, and C have zero emissions themselves but take 2, 4, and 6 years between planning and operation, respectively, what are the average emission rates (g-$CO_2$e/kWh) that result from building a plant for each technology over a 20-year period, inclusive of the technologies' planning-to-operation times? What are the resulting opportunity cost emissions in each case? Assume the emission rate of the background grid is 500 g-$CO_2$e/kWh and all three technologies produce equal quantities of electricity each year.

3.4. What are the opportunity cost emissions (g-$CO_2$e/kWh) over 100 years resulting from Plant B if its planning-to-operation time is 17 years, its lifetime is 40 years, and its refurbishment time is 3 years, whereas these values for Plant A are 2 years, 30 years, and 0.5 years, respectively? Assume both plants produce the same number of kWh/y once operating, and the background grid emits 550 g-$CO_2$e/kWh.

3.5. Assume the anthropogenic flux of heat (aside from heat used to evaporate water) from all coal combustion, averaged worldwide, is $A_h = 0.0015$ W/m$^2$. Estimate the $CO_2$e emissions corresponding to this coal heat flux given a worldwide energy generation of $G_{elec} = 9 \times 10^{12}$ kWh/y from coal combustion. Assume an anthropogenic $CO_2$ direct radiative forcing of $F_{CO2} = 1.82$ W/m$^2$, which corresponds to an anthropogenic mixing ratio of $CO_2$ of $\chi_{CO2} = 113$ ppmv. Also assume a $CO_2$ e-folding lifetime of $\tau_{CO2} = 50$ years.

3.6. From Example 3.3, the avoided solar flux to the surface of the earth per 1 kWh/y of electricity produced by a solar panel is $2.24 \times 10^{-16}$ W/m$^2$, averaged worldwide. How many square kilometers of the Earth would need to be covered with 390-W PV panels placed adjacent to each other for the avoided radiation by the panels to equal a direct $CO_2$ radiative forcing of 1.82 W/m$^2$? What fraction of the Earth does this area comprise, assuming the surface area of the Earth is 5.092 $\times$ 10$^{14}$ m$^2$? What percent of $CO_2$ radiative forcing would be offset if 0.2 percent of the Earth were covered with panels? Assume simplistically that the area of each panel is 1.63 m$^2$ and the mean capacity factor of solar PV is 20 percent.

3.7. Calculate the carbon-equivalent emissions from an energy technology that releases 10$^{-16}$ W/m$^2$ of latent heat in water vapor per kWh/y of electricity produced. Assume an anthropogenic $CO_2$ direct radiative forcing of $F_{CO2}$ = 1.82 W/m$^2$ corresponding to an anthropogenic mixing ratio of $CO_2$ of $\chi_{CO2}$ = 113 ppmv. Also assume a $CO_2$ e-folding lifetime of $\tau_{CO2}$ = 50 years.

3.8. Calculate the $CO_2$e emission rate of water vapor from a hydropower reservoir if the evaporation rate of water vapor is V = 1 kg-$H_2O$/kWh. Assume the latent heat of evaporation of liquid water is $L_e$ = 2.465 $\times$ 10$^6$ J/kg-$H_2O$. Also, assume an equilibrium emission rate and resulting radiative forcing of $CO_2$ from Example 3.2 of 1.809 $\times$ 10$^{16}$ g-$CO_2$/y and 1.82 W/m$^2$, respectively.

3.9. Calculate the carbon dioxide equivalent ($CO_2$e) emissions reduction per unit energy due to a wind turbine's reduction in water evaporation and in residual kinetic energy if the wind normally evaporates 0.5 kg-$H_2O$/kWh of energy in the wind and the wind turbine converts 1 kWh of energy in the wind into 1 kWh of electricity. Assume the equilibrium emission rate and resulting radiative forcing of $CO_2$ from Example 3.2 of 1.809 $\times$ 10$^{16}$ g-$CO_2$/y and 1.82 W/m$^2$, respectively. Assume the latent heat of evaporation is 2.465 $\times$ 10$^6$ J/kg and the surface area of the Earth is 5.092 $\times$ 10$^{14}$ m$^2$.

3.10. Assume a coal-fired power plant with normal emissions of 1,000 g-$CO_2$/kWh has carbon capture equipment installed that captures 85 percent of the $CO_2$ emissions, and the emissions are sent to an underground reservoir. If the e-folding lifetime against leakage is 5,000 years, calculate the average leakage rate over 100 years. Assume no $CO_2$ in the reservoir initially.

3.11. Assume a 500-MW coal facility has a 60 percent capacity factor and has a footprint on land of 6 km$^2$, including the land for the coal facility, mining, railway transport, and waste disposal. Calculate the emission rate of $CO_2$ from the soil and vegetation, averaged over 40 years, due to this facility, assuming that the facility replaces grass, 40 percent of the carbon stored in soil is lost, and all the carbon stored in the grass itself is lost. The molecular weight of $CO_2$ is 44.0095 g-$CO_2$/mol and that of carbon is 12.0107 g-C/mol.

3.12. Assume a coal-fired power plant emits 950 g-$CO_2$/kWh. Assume also that carbon capture equipment added to the plant captures 750 g-$CO_2$/kWh and sends it to an underground reservoir. The energy obtained for the carbon capture equipment is a natural gas plant that emits uncaptured 300 g-$CO_2$/kWh-coal-electricity at its stack and another 30 g-$CO_2$/kWh-coal-electricity from the upstream mining and transport of the natural gas. Finally, assume natural gas contains a 93.9 percent mole fraction of methane, the upstream leakage rate of natural gas is 2.3 percent, the molecular weight of $CO_2$ is 44.0098 g-$CO_2$/mol, the molecular weight of methane is 16.04276 g-$CH_4$/mol, and the 20- and 100-year GWPs of methane are 86 and 34, respectively. What percent of the original coal plant's $CO_2$e combustion emissions are effectively captured over a 20-year and 100-year time horizon (assuming no other emissions affect $CO_2$e aside from $CO_2$ from the coal plant and $CO_2$ and $CH_4$ from the natural gas plant)?

3.13. Compare the social cost (levelized energy plus health plus climate cost) of a new natural gas plant that has carbon capture equipment with that of a new utility-scale solar farm. Assume the levelized energy cost of the solar farm is 4 ¢/kWh and that of the natural gas plant is 6 ¢/kWh (Table 7.9). Assume the cost of the CCS equipment is 75 percent the energy cost of the natural gas plant. Assume that the CCS equipment requires 25 percent more natural gas and increases all emissions by 25 percent but captures enough $CO_2$ to reduce the overall $CO_2$ emissions by 20 percent compared with before CCS equipment is added. Assume the health and climate costs of natural gas pollution are 7 ¢/kWh and 16 ¢/kWh, respectively.

3.14. Suppose a country needs to decide in 2019 whether to invest in a 1-GW nuclear power plant in 2019 that will operate in 2034 at a capital cost of $10,000/kW-nameplate capacity versus a wind farm that will operate in 2021 at a capital cost of $1,200/kW. Assuming the total capital expenditure is the same in both cases, calculate the absolute opportunity cost emissions (in units of Tg-$CO_2$e) resulting from the nuclear plant. Assume the wind turbine capacity factor is 0.35 and the $CO_2$e emission intensity of the background grid is 850 g-$CO_2$e/kWh. Assume simplistically no opportunity cost emissions due to refurbishing.

3.15. Compare the social cost of synthetic direct air carbon capture and storage (SDACCS) with that of utility solar PV. Assume the SDACCS cost is $150 per tonne-$CO_2$-removed and the solar PV cost is $40/MWh-electricity generated. Assume also that the social cost of air pollution is $127/MWh-grid-electricity, which solar PV displaces but SDACCS does not. Further assume that the SDACCS equipment consumes 1 MWh of grid electricity for every 825 kg-$CO_2$ removed from the air. Finally, assume background grid emissions are 550 kg-$CO_2$e per MWh of grid electricity produced.

# CHAPTER

# 4 Electricity Basics

A 100 percent wind-water-solar (WWS) energy infra-structure involves electrifying or providing direct heat for all energy sectors and then providing the electricity or heat with WWS. Because electricity is such a large part of the solution, understanding how it works is important. In addition, WWS technologies convert either mechanical or solar energy into electricity. This chapter provides the basic information for understanding those conversion processes, which are elaborated on in Chapters 5 (solar energy) and 6 (wind energy). This chapter discusses the basics of electricity with a particular focus on electric power. It starts by examining different types of electricity – static electricity, lightning, and wired electricity. It then covers voltage and Kirchoff's laws of voltage and current. Next, it turns to power, resistance in series and parallel, and capacitors. This is followed by a discussion of electromagnetism, AC electricity, and inductors. Both single-phase and three-phase AC electricity, as well as generators, are then described. Finally, real and reactive power, transformers, and transmission, including high voltage AC and DC transmission, are covered.

## 4.1 Static Electricity, Lightning, and Wired Electricity

**Electricity** is the free-flowing movement of charged particles, usually either negatively charged electrons, negatively charged ions, or positively charged ions. The flow of electric charge constitutes an **electric current**. The moving charged particles in a current are called **charge carriers**. Thus, charge carriers can either be electrons, negative ions, or positive ions.

Electricity can travel in a wire or another medium, including the air. This chapter discusses primarily electricity through wires with application to electric power systems. Two other types of electricity are static electricity and lightning. These are briefly described.

### 4.1.1 Static Electricity

**Static electricity** arises by rubbing two materials, such as silk and glass, together so that one material strips electrons off the other. The result is that one material becomes negatively charged and the other, positively charged. If the charge difference is small, and if neither material is a strong conductor of electricity, such as with glass and silk, the two materials may merely stick to each other due to the attraction of the excess electrons in one material to the excess positive ions in the other. On the other hand, if one material is a strong conductor of electricity, a spark may occur. For example, a person walking on a wool rug scrapes electrons off the rug, giving his or her body a net negative charge. If the person's hand approaches a metal doorknob, which is a good conductor of electricity, the excess electrons in the person's hand induce positive charges in the metal doorknob to emanate to the knob's surface and negative charges to move in the opposite direction. Just before contact, millions of electrons fly from the person's hand through the air, smashing into the doorknob, creating a spark.

### 4.1.2 Lightning

**Lightning** similarly occurs when a charge difference between a cloud and the ground, between two parts of the same cloud, or between two clouds builds up beyond a threshold. For example, before cloud-to-ground lightning occurs, the bottom of a cloud builds up a net negative charge and the top, a net positive charge. The net negative charge is due to ice crystals bouncing off larger hail particles or due to edges of a hail stone splintering off. In both cases, hail is more positively charged at its surface because its surface is warmer than its interior, and positive charges migrate from colder areas to warmer areas, whereas negative charges migrate in the opposite direction. Hail is warmer on its surface because hail grows by its collision with supercooled liquid water drops, which freeze on contact. Freezing releases latent heat.

If a small ice crystal bounces off a larger hail particle, the hail scavenges an electron from the ice crystal. The positively charged ice crystal then rises with an updraft to the top of the cloud, while the heavier negatively charged hail settles toward the bottom of the cloud. Similarly, if a hail particle bashes into another hail particle and a small piece from the surface of either particle splinters, that piece will have a net positive charge and will rise to the top of the cloud with an updraft, and the net negatively charged large hailstone will settle to the bottom.

The net negative charge at the base of a cloud induces positive ions at high points along the ground to migrate to the ground surface and electrons in the ground to migrate in the opposite direction. Once the charge differential between the base of a cloud and the ground exceeds a threshold, a group of electrons, the **stepped leader**, streams from the base of the cloud toward the ground. As they bash into air molecules, the electrons split the air molecules into positive ions and additional electrons. The additional electrons join the downward parade of electrons, but the positively charged ions slow down the parade by attracting some of the electrons back upward. This temporary upward motion followed by the downward motion of electrons as they continue their journey toward the positively charged ground contributes to lightning's jagged shape.

As the electrons approach the ground, their attraction is so strong, a **streamer** of positive ions streams up from a high point along the ground into the air to meet them. The connection creates a channel in which electrons in the air near the ground first rush to the ground. These electrons create a flash of light near the ground as they form a hot (30,000 to 60,000 K) **plasma** of ions and electrons by bashing into air molecules. The light occurs because cooler molecules on the outside of the flash (4,000 to 8,000 K) emit photons of visible radiation. Higher-temperature molecules emit ultraviolet and even shorter wavelength radiation. The thermal expansion of the air creates a sonic boom, or **thunder**.

Electrons behind the first batch then rush to the ground, just as a line of cars released from a stoplight start moving from the front to the rear of the line. In this manner, **lightning** starts near the ground and propagates up to the cloud as each successive batch of electrons behind the previous one accelerates toward the ground. This first major flash, which is the most luminous, is the **return stroke**. Once the channel is open, additional electrons from the cloud can come through it in a downward stroke of lightning called the **dart leader**.

## 4.1.3 Wired Electricity

In a metal wire, atomic nuclei stay in a fixed position and the charge carriers are electrons, which flow freely along the wire to produce a current. The flowing electrons are sufficiently far from the nucleus of any atom in the wire that the attraction of the electrons to any nucleus is easily overcome. These conduction electrons wander from atom to atom, resulting in a current. In an electrochemical cell, positively and negatively charged ions can both move about and be charge carriers.

Electric current $i$ is quantified as the number of charges (positive or negative) dq passing by a given spot per unit time dt, thus,

$$i = dq/dt \tag{4.1}$$

The unit of current is **amperes** (A) and the unit of charge is **coulombs** (C). One coulomb (1 C) equals a charge of $6.242 \times 10^{18}$ electrons. Conversely, one electron carries a charge of $1.602176 \times 10^{-19}$ C; 1 A = 1 C/s. Thus, 1 A means that $6.242 \times 10^{18}$ electrons pass a given spot in a second.

Charge can be positive or negative. The direction of a current is defined to be the direction of positive flow (or opposite the direction of negative flow). Thus, if electrons are moving to the right, the current (positive flow) is moving to the left (Figure 4.1). In a circuit, a flow of positive charge in one direction has the same impact and the same current as a flow of negative charge in the opposite direction.

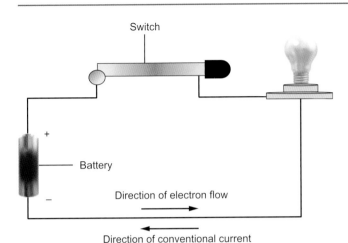

**Figure 4.1 Simple circuit with a battery, light bulb, and switch. The current flows clockwise and the electrons flow counterclockwise in this circuit. The voltage rises from zero at the negative end of the battery to the battery voltage at the positive end as the current moves through the battery. The current drops back to zero as it passes across the light bulb.**

**Direct current (DC)** is defined to be the flow of charge (e.g., electrons) at a constant rate in one direction. An example DC electricity source is a battery.

**Alternating current (AC)** is the flow of positive or negative charges back and forth, in a sinusoidal manner, over time. In **one sinusoidal cycle**, electrons first accelerate their flow to the right, reach a peak speed, decelerate to zero speed, accelerate to the left, reach a peak speed, then decelerate to zero speed. The number of complete cycles in a second is called the **AC frequency**. One cycle per second is defined as 1 **hertz** (Hz). So 3 Hz means that three complete cycles of flow and counter flow occur in 1 s. In the United States and some other countries, the AC frequency for the electric power grid is set at 60 cycles per second, or 60 Hz. **Nikola Tesla** (1856–1943) calculated

that 60 Hz was optimal for the new U.S. electric grid. In Europe and in most other countries, the AC frequency is set at 50 Hz.

In a wire conducting a direct current, electrons as a whole move relatively slowly (at their drift velocity). However, each electron carries a lot of charge. As electrons collide with each other, they transfer energy to each other, sending a stream of electricity down a wire at nearly the speed of light. As such, a wire can carry a lot of current.

Similarly, in a wire conducting alternating current at 60 Hz, electrons reverse themselves 60 times a second, so hardly move at all. Yet, they transfer energy to each other, sending a wave of electricity down a wire at nearly the speed of light.

The speed at which electrons move in a wire can be quantified with their drift velocity. **Drift velocity** is the average speed of the net flow of particles, such as electrons, in a given direction in the presence of an electric field. Electrons will move in random directions, but in the presence of an electric field, more will move in one direction than another, resulting in a net flow of electrons in one direction. Drift velocity ($v_d$) quantifies the speed of this net flow. It is calculated as

$$v_d = j/(nq) \qquad (4.2)$$

where $j$ is the current density traveling through the material ($C/m^2$-s), $n$ is the number density of the charge carrier (electrons/$m^3$-material), and $q$ is the charge on each charge carrier (C/electron). The current density is simply the current (A = C/s) divided by the cross-sectional area ($m^2$) of the wire that it is traveling through. Example 4.1 quantifies drift velocity in a wire.

## Example 4.1 Calculation of drift velocity

Find the drift velocity of electrons in a copper wire carrying a current of 20 A given the molecular weight of copper as 63.55 g/mol, the density of copper as 8,960 kg/$m^3$, the cross section of the wire as $3.31 \times 10^{-6}$ $m^2$, and assuming copper has one free electron per atom.

### Solution:

The current density flowing through the wire is the current divided by the cross-sectional area of the wire, thus $j = 20$ A / $3.31 \times 10^{-6}$ $m^2$ = $6.04 \times 10^6$ $C/m^2$-s (since 1 A = 1 C/s). The number density of charge carriers is $n = 6.023 \times 10^{23}$ atoms/mole $\times$ 1 electron/atom $\times$ 1 mole/63.55 g $\times$ $8.96 \times 10^6$ g/$m^3$ = $8.49 \times 10^{28}$ electrons/$m^3$. The charge on an electron is $q = 1.602 \times 10^{-19}$ C/electron. Substituting these values into Equation 4.2 gives $v_d = 0.00044$ m/s = 1.6 m/h. Thus, the bulk movement of electrons in a DC wire is slow.

## 4.2 Voltage and Kirchoff's Laws

Another parameter intrinsic to electricity is voltage. **Voltage** is the difference in electric field potential between two points. **Electric field potential** is the amount of work (energy) needed to move a unit of positive charge from a reference point to a specific point inside an electric field without producing acceleration of the charge. Thus, voltage is the work (energy) needed to move a single positive charge between two points in an electric field. Mathematically, voltage is defined as

$$v = dw/dq \tag{4.3}$$

where w is energy in **joules** (J). The unit of voltage is the **volt** (V), where 1 V = 1 joule (J) of energy (work) per 1 coulomb (C) of charge.

The relationship among electrical energy, voltage, and charge is analogous to the relationship among potential energy, elevation, and mass. Potential energy equals mass multiplied by gravity and by height. As a mass is lifted against gravity, it gains potential energy. Analogously, as the voltage of a charge is raised, the charge gains potential energy, referred to as electrical energy.

A simple electrical circuit can be used to illustrate voltage and current. An **electrical circuit** is a closed-loop electrical network with a return path through which a current runs. A simple circuit may consist of a battery, light bulb, wires connecting the two from either side of the battery, and a switch that connects or disconnects the circuit, such as in Figure 4.1.

Suppose the battery voltage is 12 V. This means the battery provides 12 J of energy for every 1 C of charge that it stores. Voltages are measured across components. Thus, if the voltage at one end of the battery is 0 V, the voltage at the other is 12 V, so the overall voltage across the battery is 12 V. Similarly, voltage drops 12 V when measured from one wire into the light bulb to the wire leaving the light bulb, as 12 J = 12 V × 1 C of energy from the battery used to light the bulb. Current, on the other

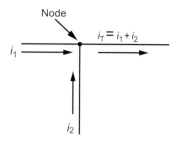

**Figure 4.2** Segment of a circuit with a node, illustrating Kirchoff's Current Law.

hand, is measured through components. So, the current through the battery and light bulb may be 10 A, for example.

A circuit may have one or more additional wires connected to a main wire, such as in Figure 4.2. The point of connection of any two wires on a circuit is a **junction** or **node**. **Gustav Kirchoff** (1824–1887) postulated that

*At every instant of time the sum of the currents flowing into any node of a circuit must equal the sum of the currents leaving the node.*

This is now known as **Kirchoff's Current Law**. Another rule of circuits, known as **Kirchoff's Voltage Law**, is

*The sum of the voltages around any loop of a circuit at any instant is zero.*

Thus, for example, in Figure 4.1, when the switch is closed, the voltage across the battery (from the negative end to the positive end) is +12 V and the voltage across the light bulb is –12 V, so the sum is zero. Kirchoff's Voltage Law can be understood in terms of the changes in elevation along a circular path. A hiker walking up a trail eventually comes back to his or her original position. The sum of elevation changes along the path always equal zero. Similarly, the sum of voltage changes around a circuit always equal zero.

---

### Example 4.2 Application of Kirchoff's Voltage Law

If a circuit has a battery and two light bulbs in series, and if the voltage across the battery is 12 V and across the first light bulb is 4 V, what is the voltage across the second light bulb?

#### Solution:

According to Kirchoff's Voltage Law, the sum of voltages around the circuit is zero, so the voltage across the second light bulb is simply 12 V – 4 V = 8 V.

## 4.3 Power and Resistance

Power (*p*) is the change in energy (dw) provided by electricity per unit time (dt). Alternatively, the differential of energy (dw) equals power (*p*) multiplied by the differential of time (dt). Since energy per time (*p* = dw/dt) equals energy per unit charge (*v* = dw/dq = voltage) multiplied by charge per time (*i* = dq/dt = current), then power (*p*) = voltage (*v*) × current (*i*):

$$p = \frac{dw}{dt} = \frac{dw}{dq}\frac{dq}{dt} = vi \tag{4.4}$$

Power has units of **watts** (W), where 1 W = 1 J/s. Thus, 1 W = 1 V multiplied by 1 A. Equation 4.4 suggests that a 12-V battery that delivers 10 A to a load supplies 120 W of power. Alternatively,

$$v = \frac{p}{i} = \frac{dw}{dt} \Big/ \frac{dq}{dt} = \frac{dw}{dq} \tag{4.5}$$

which similarly suggests that 120 W of power supplied at a current of 10 A is drawing 12 V.

Electric current flows through wires in order to power an electric load. An **electric load** (or electric demand) is a component or part of an electrical circuit that consumes electrical energy. The light bulb in Figure 4.1 is a load. Loads can also be heaters, cookers, dishwashers, car chargers, arc furnaces, etc. When loads consume energy (dw), they reduce the voltage (V = dw/dt) across the load per Equation 4.5.

Voltage is also reduced along a circuit if the current passes through an electrical insulator. An **electrical insulator** is a material that impedes the flow of a current because electric charges do not flow freely through it.

Good insulators are rubber, glass, air, plastic, paper, wood, wax, and wool. The opposite of an insulator is an **electric conductor**, through which charges flow freely. Good conductors are copper, aluminum, gold, and silver.

When a current passes through an insulator, electrical energy carried by the current turns to heat and voltage decreases, just as current does when it passes through an electrical load.

**Electrical resistance** is a parameter quantifying the opposition of flow to an electrical current. Both electrical loads and electrical insulators are forms of electrical resistance. The inverse of electrical resistance is **electrical conductance**, which quantifies how easily a current passes through a circuit. Electrical resistance (R) of an object is defined as the voltage across the object divided by the current through the object,

$$R = \frac{v}{i} \tag{4.6}$$

In other words, resistors drop the voltage across an object proportionally to the current through the object. If no voltage change occurs across a load, the resistance is zero. An alternative way to think about resistance is that the current through an ideal resistance element is directly proportional to the voltage drop across it (*i* = v/R). Thus, the higher the resistance is, the lower the current through the resistance element will be. Finally, the voltage across an object equals the current through it multiplied by the resistance (*v* = iR). Units of resistance are Ohms (Ω).

### Example 4.3 Another application of Kirchoff's Voltage Law

If a circuit has a battery and two light bulbs in series, and the voltage across the battery is 5 V, the resistance across the first light bulb is 20 Ω, and the resistance across the second light bulb is 10 Ω, what is the current through the circuit?

### Solution:

According to Kirchoff's Voltage Law, the sum of voltages around the circuit is zero, so $v_b - v_1 - v_2 = 0$, where $v_b$ is the voltage across the battery and $v_1$ and $v_2$ are voltages across the two light bulbs. Since *v* = iR, the equation can be rewritten as $v_b - iR_1 - iR_2 = 0$, or $i = v_b / (R_1 + R_2) = 5\text{ V}/(20\ \Omega + 10\ \Omega) = 0.167$ A.

**Electrical conductance** (G) is the inverse of resistance,

$$G = \frac{i}{v} \tag{4.7}$$

Units of conductance are Siemens (S). Combining Equations 4.6 and 4.4 gives the power dissipated in object of resistance as

$$p = vi = i^2R = \frac{v^2}{R} \tag{4.8}$$

## Example 4.4  Resistance of a lamp

Calculate the resistance of a filament in a lamp that produces 60 W of power if the power source is 120 V. What are the current and energy consumed over 500 hours?

### Solution:

The resistance of the filament is $R = v^2 / p = (120\text{ V})^2 / 60\text{ W} = 240\ \Omega$.
   The current through the circuit is $i = p / v = 60\text{ W} / 120\text{ V} = 0.5\text{ A}$.
   The energy consumed over 500 h is $dw = pdt = 60\text{ W} \times 500\text{ h} = 30{,}000\text{ Wh} = 30\text{ kWh}$.

## 4.4  Resistors in Series and Parallel

A circuit can have multiple resistors (loads or insulators) wired in series or in parallel or both. Figure 4.3 shows an example circuit with three **resistors wired in series**. In such a circuit, the current is constant around the whole circuit. The voltages across resistors 1, 2, and 3 are $v_1 = i$ $\times R_1$, $v_2 = i \times R_2$, and $v_3 = i \times R_3$. The total voltage drop across all three resistors is thus

$$v = v_1 + v_2 + v_3 = i(R_1 + R_2 + R_3) = iR_s \qquad (4.9)$$

where $R_s = R_1 + R_2 + R_3$. In other words, the total resistance along a circuit with resistors in series ($R_s$) is the sum of the individual resistances among all the resistors.

## Example 4.5  Calculating current through wires in a circuit with resistors in series

Calculate the current through the circuit shown in Figure 4.3.

### Solution:

The sum of resistances among the three resistors in series is $R_s = 1{,}000\ \Omega$. Given the voltage of 10 V, the current passing through the circuit is $i = v / R_s = 10\text{ V} / 1{,}000\ \Omega = 0.01\text{ A}$.

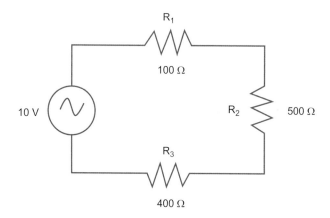

Figure 4.3 AC circuit with three resistors in series.

Figure 4.4 shows an example circuit with two resistors wired in parallel. Suppose a battery with voltage V supplies energy to these resistors, and the wire has no resistance in it. The current ($i$) flowing from the battery to the resistors splits at the junction into two separate currents, $i_1$ and $i_2$, which flow through resistors $R_1$ and $R_2$, respectively. According to Kirchoff's Current Law (Section 4.2), $i = i_1 + i_2$. After the two split wires re-join each other, the summed current $i$ continues back to the battery.

We also know that the voltage before each resistor is $v$ and after each resistor is 0. As such, according to Equation 4.6, $i_1 = v / R_1$ and $i_2 = v / R_2$. Therefore, the total current leaving and returning to the battery is

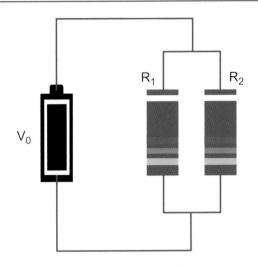
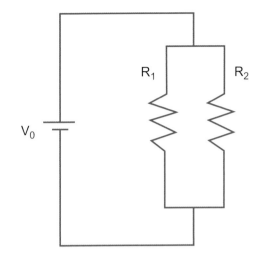

Figure 4.4 **Circuit with two resistors in parallel.**

$$i = i_1 + i_2 = \frac{v}{R_1} + \frac{v}{R_2} = v\left(\frac{1}{R_1} + \frac{1}{R_2}\right) = \frac{v}{R_p} \qquad (4.10)$$

where

$$R_p = \frac{1}{\dfrac{1}{R_1} + \dfrac{1}{R_2}} \qquad (4.11)$$

is the **effective resistance of parallel resistors**. Alternatively,

$$(4.12) \; v = i \times R_p$$

$R_p$ is always less than the resistance of either individual resistor. Since conductance ($G$) is the inverse of resistance, the **effective conductance of parallel resistors** is

$$G_p = \frac{1}{R_p} = \frac{1}{R_1} + \frac{1}{R_2} = G_1 + G_2 \qquad (4.13)$$

---

## Example 4.6 Calculating current through a circuit with resistors in parallel

Calculate the current through the circuit shown in Figure 4.4 assuming $v$ = 12 V, $R_1$ = 100 $\Omega$, and $R_2$ = 200 $\Omega$.

### Solution:

The effective resistance of the two parallel resistors is $R_p$ = 1/(1/100 $\Omega$ + 1/200 $\Omega$) = 66.67 $\Omega$. Thus, the current through the circuit is $i = v/R_p$ = 12 V / 66.67 $\Omega$ = 0.18 A. Also note that $R_p$ is less than $R_1$ or $R_2$.

---

## 4.5 Capacitors

Capacitors, like resistors, are a component of an electric circuit. **Capacitors** store electric charge when they are connected to a circuit. If they are disconnected from the circuit after accumulating charge, they can release charge like a battery. However, whereas batteries take time to charge and discharge, capacitors can charge and discharge in a fraction of a second. Capacitors are often used in electronic devices to maintain power when batteries are being charged or when voltage drops. Because voltage cannot change instantly across a capacitor,

capacitors are also used in power grids to smooth out DC voltage. They are further used to supply power to inductors in order to decrease reactive power needed on the grid (Section 4.10).

A capacitor is made of one or more pairs of parallel conducting metal plates separated by a non-conducting dielectric insulator. The simplest capacitor can consist of two sheets of aluminum foil separated by air or paper. A **dielectric insulator** is an insulator that is highly polarizable. A substance that is **polarizable** has molecules that can easily and instantly form dipoles. **Dipoles** are

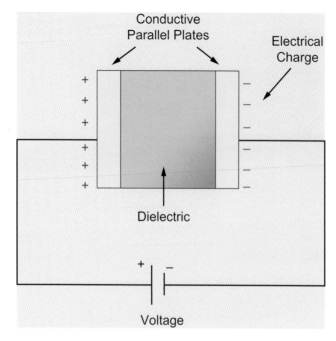

Conductive Parallel Plates

Electrical Charge

Dielectric

Voltage

Figure 4.5 **Capacitor. Two parallel conducting metal plates separated by a non-conducting dielectric.** Adapted from Electronics Tutorials (2019a).

molecules whose positive charges have become separated from their negative charges. The overall molecule is still neutrally charged, but different parts of the molecule have different concentrations of charge.

When a capacitor is connected on a circuit to a battery, electrons from the negative terminal of the battery accumulate on the plate of the capacitor attached by wire to that terminal (Figure 4.5). The plate thereby becomes negatively charged. The plate connected by wire to the positive terminal of the battery loses electrons to the positive terminal, thereby becoming positively charged. The difference in charge across the plates creates an electric field, in which electrostatic energy is stored. Once fully charged, the capacitor holds the same voltage as the battery.

If a circuit contains a battery, a light bulb, and a large capacitor, the bulb will first light up but become dimmer as the current flows to the capacitor, charging the capacitor. When the capacitor is fully charged, the bulb will darken since the voltage drop across the capacitor is now the same as that across the battery, so no more voltage is available for the light. If the battery is then disconnected from the circuit, current will flow from the capacitor to the bulb and light the bulb until the stored energy is depleted in the capacitor.

**Capacitance (C)** quantifies the ability of a capacitor to store electric charge (energy) in an electric field. It is the ratio of the electric charge held by a capacitor to its electrical potential, or voltage. If the charge on the two

plates of a capacitor are −q and +q and the voltage stored between the plates is $v$, the capacitance is

$$C = \frac{q}{v} \tag{4.14}$$

Units of capacitance are **farads** (F), where 1 farad = 1 coulomb (C) per volt (V). Thus, a capacitor with 1-F capacitance has a potential of 1 V between its plates when charged with 1 C of electric charge. A capacitor with a high capacitance holds more charge per unit voltage than a capacitor with a lower capacitance. If the distance between plates is small compared with the plate dimensions, capacitance can be estimated with

$$C = \frac{\varepsilon_0 A}{d} \tag{4.15}$$

where $\varepsilon_0 = 8.854 \times 10^{-12}$ F/m is a constant called the **permittivity of a vacuum**, A is the overlapping area of the plates (m$^2$), and d is the distance between plates (m). This equation suggests that the capacitance is linearly proportional to the area of the plates and inversely proportional to the distance between plates. Because it depends on the physical properties of the capacitor, capacitance does not vary with time.

Rearranging Equation 4.14 to q = $Cv$ and taking the derivative of q with respect to time gives the time-dependent current through a capacitor as a function of the change in voltage stored in the capacitor.

$$i = \frac{dq}{dt} = C\frac{dv}{dt} \tag{4.16}$$

Equation 4.16 suggests that, as the capacitor reaches its maximum voltage storage potential, d$v$(t)/dt goes to zero, and the current through the capacitor goes to zero. At this point, the circuit behaves like an open circuit (a circuit where the switch is disconnected). This is why a light on a circuit with a capacitor will go out once the capacitor is full.

The energy (J) stored in the electric field of a capacitor is

$$w_c = \frac{1}{2}Cv^2 \tag{4.17}$$

Taking the derivative of energy with respect to time gives the power required to initiate a change in voltage with time in a capacitor as

$$p_c = \frac{dw_c}{dt} = Cv\frac{dv}{dt} = vi \tag{4.18}$$

Equation 4.18 illustrates that capacitors resist rapid changes in voltage, which is why they are used to smoothen voltage in DC power lines. The equation shows that if d$v$/d$t$ were infinite, the power required to make d$v$/d$t$ infinite would also need to be infinite, but that is impossible. Thus, voltage cannot change instantaneously in a capacitor and capacitors smooth out variations in voltage.

Finally, when two capacitors are in parallel, the total capacitance of the circuit is $C_p = C_1 + C_2$. When two capacitors are in series, the total capacitance is $C_s = 1/(1/C_1 + 1/C_2)$.

## 4.6 Electromagnetism

Electric currents through wires can be produced not only by a battery but also by magnets. For example, a magnet moving toward or away from a coil of wire along a circuit induces an electric current in the wire. Similarly, an electric current flowing along a wire creates a circular magnetic field around the wire. The relationship between electricity and magnetism is referred to as **electromagnetism**. Understanding the basics of electromagnetism is important for understanding inductors, transformers, motors, and generators.

On April 21, 1820, the Danish physicist and chemist **Hans Christian Orsted** (1777–1851) discovered electromagnetism when he noticed that the needle of a compass was deflected in the presence of an electric current from a battery that he switched on and off. He investigated and published this phenomenon the same year, concluding that an electric current flowing through a wire produces a circular magnetic field.

In 1825, **André-Marie Ampère** (1775–1836) found that a wire carrying a current exerted an attracting force on another wire carrying a current in the same direction and a repelling force on another wire carrying a current in the opposite direction. He quantified this force with **Ampère's Force Law**, which states that the force per unit length along a wire is proportional to the product of the direct currents in the two wires and inversely proportional to the distance between the wires.

A third important advancement in electromagnetism was made by **Michael Faraday** (1791–1867). On August 29, 1831, Faraday found that by coiling two independent wires around opposite ends of a circular iron ring (a torus), then connecting one wire to a battery and the other to a **galvanometer**, which measures current, he could briefly measure an electric current through the second coiled wire (Figure 4.6). Disconnecting the battery also induced a current. Connecting the battery to the first coiled wire created a current through the wire, which created a magnetic field around the coil and iron, turning the iron into a magnet.

The pulse increase in magnetic flux ($\phi$) induced a current through and a voltage change across the second wire. Disconnecting the battery similarly changed the magnetic flux, inducing a current through and a voltage change across the second wire. Faraday called the voltage change across the second wire an **electromotive force, emf** ($e$), and postulated that it is proportional to the number of coils of wire ($N$) and the time rate of change of magnetic flux by

$$e = N\frac{\mathrm{d}\phi}{\mathrm{d}t} \qquad (4.19)$$

Faraday similarly showed that moving a magnet close to a coil of wire connected to a circuit induced a current to flow through the circuit and a voltage change across the coil. In a third experiment, he showed that moving a coil

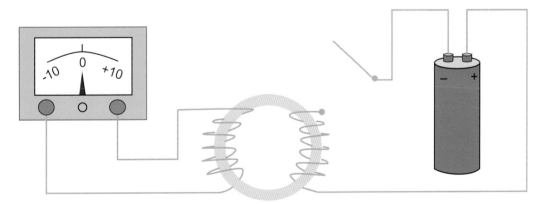

**Figure 4.6** Illustration of Faraday's August 29, 1831 experiment. Two wires were coiled independently around different ends of an iron ring. One coil was connected to a battery; the other to a galvanometer to measure current. When the battery was connected to the first coil, it created a current through the coil, which created a magnetic field around the coil and iron. The increase in magnetic field from zero induced a current through the second coil, which was measured by the galvanometer.

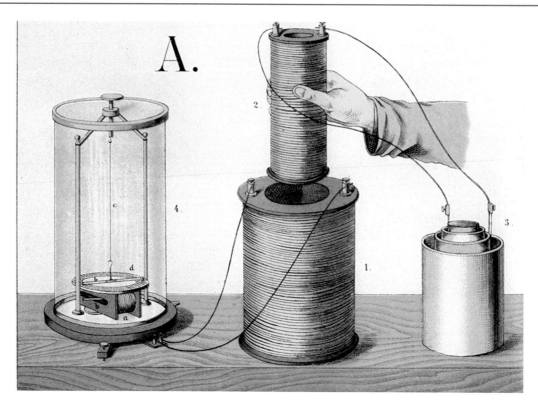

**Figure 4.7 Illustration of another Faraday electromagnetic induction experiment. The inner, smaller coil of wire is connected to a liquid battery. The outer, larger coil is connected to a galvanometer, which measures current. The battery creates an electric current that passes through the smaller coil of wire. The electric current creates a magnetic field around the smaller coil. When the smaller coil moves in and out of the larger coil, the galvanometer notes a change in current passing through the outer coil. That current is due to the moving magnetic flux inducing a current through the outer coil. The current through the outer coil stops when both coils are at rest.**
© Oxford Science Archive/Heritage Images.

of wire connected to a battery in and out of a larger coil of wire induced a current in the larger coil (Figure 4.7). The reason is that the current through the first coil created a magnetic field, and the movement of that field induced a current in the second coil. If current flowed through the smaller coil but the coil was held in place, no current through or voltage across the second coil was induced because both require the magnetic field to change with time (e.g., Equation 4.19).

## 4.7 AC Electricity and Inductors

DC electricity consists of current that flows in one direction. Both its current and voltage are independent of time (Figure 4.8). Electricity from a battery is DC electricity. AC electricity differs from DC electricity in that AC current switches direction with time (Figure 4.8). Alternating current occurs when current flows in one direction for a short period, then changes direction, then switches direction back in a repeating cycle. The alternating motion is usually in the form of a sinusoidal wave.

According to **Faraday's Law of Electromagnetic Induction**, AC voltage can be obtained by rotating a magnet around a stationary set of wire coils. In Figure 4.9, one coil is on the left and the other on the right. When the magnetic poles are vertical in the figure, no current occurs. As the magnet is rotated clockwise, the south pole approaches the right coil, the north pole approaches the left coil, and the magnetic field causes the current to accelerate in a clockwise fashion around the circuit. The peak current occurs when the poles are horizontal. As the magnet continues rotating clockwise, the poles move away from the coils, so the current decelerates. When the poles are vertical again, the current is zero. As the poles continue rotating clockwise, the south pole approaches the left coil and the north pole, the right, and the current accelerates in the opposite orientation, counterclockwise. As the magnet continues to rotate clockwise further, the current continues to oscillate direction and magnitude in a sinusoidal manner. The faster the magnet turns, the greater the frequency that the current alternates direction in the wire.

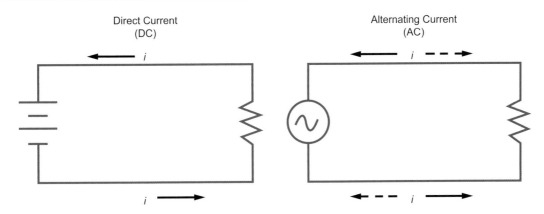

**Figure 4.8** Illustration of some differences between a DC and an AC circuit. Current (*i*) in a DC circuit flows in one direction. That in an AC circuit switches direction over time. The symbol on the left of the DC diagram is a DC battery. That on the left of the AC diagram is an AC voltage source, such as an alternator, which is a type of generator (Figure 4.9). Adapted from Kuphaldt (2019).

The time-dependent variation of the magnitude and direction of an alternating current and voltage are quantified with

$$i(t) = I_m \sin(\omega t + \phi) \tag{4.20}$$

$$v(t) = V_m \sin(\omega t) \tag{4.21}$$

respectively, where $I_m$ and $V_m$ are the magnitude of the maximum current (A) and voltage (V), respectively, $\omega$ is the angular frequency (radians per second), and $\phi$ is the **phase angle** (radians). Phase angles arise only when capacitors or inductors exist along the circuit and apply only to a current. If a phase angle is non-zero, the current and voltage are out of phase. This gives rise to reactive (imaginary) power (Section 4.10). The **angular frequency** is

$$\omega = 2\pi f \tag{4.22}$$

where $2\pi$ is the number of radians in one complete cycle and $f$ is the **frequency of oscillation** (number of cycles per second). $T = 1/f$ is the **period of oscillation** (of one complete cycle), which is the number of seconds required to complete one full wave. A cycle occurs upon a complete rotation of the magnet in Figure 4.9. The greater the number of cycles per second, the narrower the sine wave

in Figure 4.10. Figure 4.10b shows that if the phase angle is positive ($30°$ in this case), the current lags the voltage. Phase angle is discussed shortly.

Because $i(t)$ and $v(t)$ vary with time at a given location in an AC circuit (as opposed to in a DC circuit, where they are constant in time at a given location), $i(t)$ and $v(t)$ must be averaged over time to give effective values for practical use. The effective time-averaged value of current and voltage are their root-mean-square (rms) values,

$$I_{rms} = \sqrt{\frac{1}{T}\int_0^T i^2(t)dt} = \frac{I_m}{\sqrt{2}} \tag{4.23}$$

$$V_{rms} = \sqrt{\frac{1}{T}\int_0^T v^2(t)\,dt} = \frac{V_m}{\sqrt{2}} \tag{4.24}$$

With AC electricity, current and voltage are defined as root-mean-square values, so $i = I_{rms}$ and $v = V_{rms}$. Thus, 120 V AC voltage is the rms voltage. AC power, on the other hand, is a time-integrated average value, which conveniently simplifies to

$$p = P_{avg} = V_{rms}I_{rms} = vi \tag{4.25}$$

---

## Example 4.7 Calculating the resistance of and AC current through a light bulb

Calculate the resistance of and current through a 60-W light bulb powered by 120 V AC electricity.

### Solution:

$R = v^2/p = (120\ \text{V})^2 / 60\ \text{W} = 240\ \Omega$.

$i = p/v = 60\ \text{W} / 120\ \text{V} = 0.5\ \text{A}$

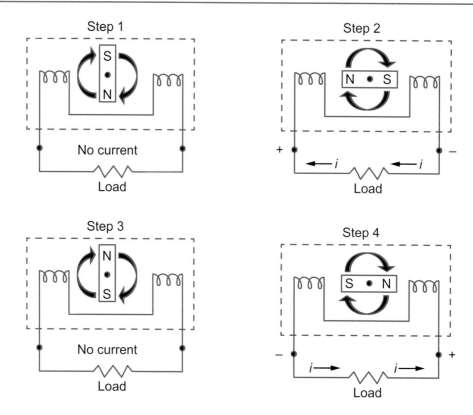

**Figure 4.9** Production of an AC current with an alternator (a type of generator) by rotating a magnet in the presence of stationary wire coils. Adapted from Kuphaldt (2019).

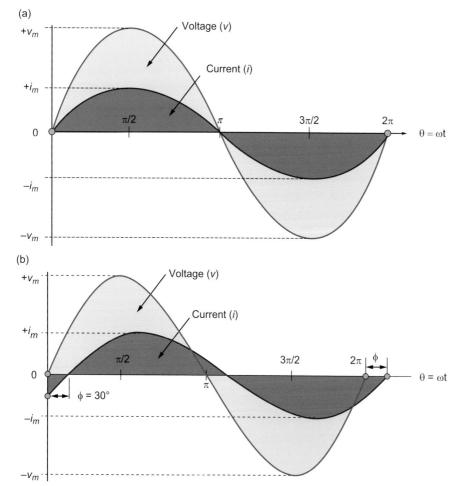

**Figure 4.10** Sine functions representing the alternating direction and magnitude of AC current and voltage. In (a), the phase angle is zero. In (b), the current is 30° out of phase with the voltage.

Equation 4.20, the sinusoidal time-dependent equation for current, includes a phase angle. A phase angle arises in the equation for current (but not voltage) if a capacitor or inductor exists along the circuit. For a capacitor along the circuit, current leads voltage (thus the phase angle is negative) because current must flow before a capacitor can show a voltage. For an inductor along a circuit, voltage leads current (the phase angle is positive, Figure 4.10) because voltage must be supplied to an inductor before current can flow through it.

Whereas capacitors store energy in an electric field, **inductors** store energy in a magnetic field. An inductor consists of insulated wire coiled around a magnetic core (e.g., iron). When a current passes through the coils of an inductor, the inductor creates a magnetic field in which voltage is stored.

Inductors oppose changes in current through them by changing the voltage proportionally to the change in current with time by

$$v(\mathrm{t}) = L \times \mathrm{d}i(\mathrm{t})/\mathrm{dt} \tag{4.26}$$

where $L$ is the inductance. **Inductance** is the ratio of voltage to the rate of change in current and has units of the **henry** (H). Inductance is thus a measure of the amount of voltage generated for a given rate of change of current.

When current that flows through an inductor increases or decreases in magnitude, the time-varying magnetic field induces a voltage change across the conductor by Faraday's Law of Electromagnetic Induction. The induced voltage opposes the change in current that created it.

For example, if the rate of change of current increases, the strength of the induced magnetic field around the inductor increases. Magnetic fields store energy, and the stronger the magnetic field, the more energy is stored within it. As the magnetic field strength increases, electrical energy is drawn from the current and converted to magnetic potential energy. The drop in electrical energy reduces the current, and the increase in magnetic potential energy increases the voltage according to Equation 4.26. Thus, energy from the increased current is absorbed in the magnetic field of the inductor, decreasing the current. Conversely, if current decreases with time, voltage stored in the magnetic field (as magnetic potential energy) is released, and electrons flow to increase the current.

Transformers are made of two or more inductors in proximity to each other. They are also used in electrical transmission systems to limit abnormal electrical currents. As in a capacitor, where voltage cannot be changed instantly, current cannot be changed instantly in an inductor. As such, inductors are useful in the electric power grid for limiting the damage due to short circuits or other abnormal currents.

## 4.8 Single-Phase and Three-Phase AC Electricity and Generators

**Generators** convert the rotational energy in a spinning shaft to electricity. Almost all rotational energy for generators worldwide is obtained from a fluid (steam, water, air, or combustion gases) passing through turbine blades connected to a rotating shaft. Generators produce either single-phase or three-phase AC electricity.

Figure 4.9 shows an example of single-phase electricity generation. The problem with single-phase AC electricity on a circuit is that the oscillating current of AC electricity causes a continuous alternating increase and decrease in current. Thus, appliances such as light bulbs continuously increase and decrease in brightness. This flickering is ameliorated by the fact that the frequency of the increases and decreases is so rapid, 50 Hz or 60 Hz, depending on country, most people cannot detect the flickering.

Three-phase AC electricity increases the apparent constancy of the current by the addition of more wire coils. Figure 4.11 illustrates two types of three-phase systems. In the system in Figure 4.11a, three equally spaced coiled wires are rotated through a magnetic field. In the system in Figure 4.11b, a magnetic field is rotated inside a circle containing three equally spaced pairs of coiled wires. Each coiled wire in Figure 4.11a or each pair of coiled wires in Figure 4.11b produces its own sine function. However, the sine functions do not overlap each other in time (Figure 4.11b). Instead, they are offset (or out of phase) with each other by 120°, or by one-third of the period $T$ of a complete cycle. Thus, instead of the maximum positive current occurring once every cycle, it occurs three times a cycle (Figure 4.11b), reducing significantly the flickering that occurs with a single-phase system. Having more than three phases increases cost with little additional benefit.

Three-phase synchronous generators produce virtually all commercial AC electricity worldwide. Three-phase generators not only smooth out current compared with single-phase generators, they also vibrate less than single-phase generators. Most three-phase AC electricity is sent along three-phase transmission lines. Three-phase transmission lines have three total wires to transmit each

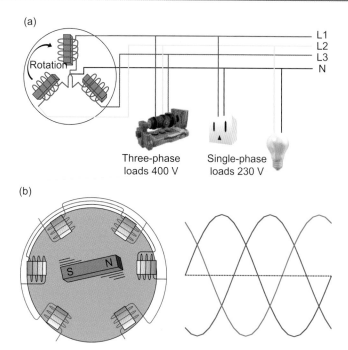

Figure 4.11 **Two methods of providing three-phase AC electricity in a generator. (a) Rotate three equally spaced coiled wires through a magnetic field. L1, L2, and L3 represent wires from each of the coils. Three-phase loads receive AC current from all three coils. Single-phase loads receive current from one coil. N is a wire completing the circuit.** Adapted from McFadyen (2012). **(b) Rotate a magnetic field through three pairs of stationary coiled wires.**

phase independently to the end load but share one wire to return current back to the generator if the return current exists. In a balanced three-phase circuit, there is no current in the neutral wire, so the neutral wire is usually not needed, except for cases where imbalances occur (Masters, 2013). If three single-phase currents are produced instead of a single three-phase current, six wires are needed: three to deliver electricity and three to return the current.

In a three-phase AC system, since each phase is transmitted along its own wire, different phases can be delivered to a household for different purposes. For example, one wire can deliver 1,400 W at 120 V; another can deliver 2,800 W at 120 V, and a third can deliver 4,800 W at 240 V. Because of the high frequency (60 Hz or 50 Hz) of the waves, flickering is not noticeable by most people with even a single phase, so using a single phase for home appliances works fine.

## 4.9 Real versus Reactive Power

**Real power** is the energy per unit time used to run a motor or a heat pump. It results from a circuit with resistive components only (no capacitors or inductors). **Reactive power** is imaginary power that does not do any useful work but simply moves back and forth within power lines. It is a byproduct of a circuit that has capacitors or inductors and arises from phase angle difference ($\phi$) between voltage and current (e.g., Figure 4.10b). It is

imaginary power that represents the product of volts and amperes that are out of phase with each other.

On a DC circuit, $p = vi$ is real power. On an AC circuit, $S = vi$ is **apparent power**, which equals the vector sum of active (real) power ($p = vi\cos\phi$) and reactive power ($Q = vi\sin\phi$) (Figure 4.12), where $\phi$ is the phase angle previously introduced. The unit of reactive power is Volt-Ampere-reactive (**VAr**). In a purely resistive AC circuit (such as power for an iron, heater, or filament bulb), $\phi = 0$, so reactive power equals zero ($Q = 0$), and apparent power equals real power ($S = p$).

Reactive power is important for three reasons. First, it smoothens voltage on the transmission grid by supplying or absorbing it. Second, it provides needed power to avoid blackouts. Third, transformers, motors, and generators need reactive power to produce a magnetic flux.

Reactive power is used to maintain voltage within a certain range (to provide **voltage control**) and overcome sudden losses along transmission lines on a timescale of

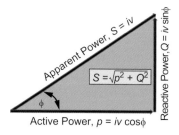

Figure 4.12 **Illustration of the relationship among apparent, active, and reactive power.** Adapted from Electronics Tutorials (2019b).

seconds. Thus, reactive power helps move power through transmission lines more efficiently. Transmission system operators modulate voltage either by installing transmission equipment (static VAr compensators, tap changers, capacitors, reactors) or by having local power generators provide reactive power (Kirby, 2004). For example, a **static VAr compensator** is a device that either provides or absorbs reactive power to stabilize voltage. Alternatively, some electric power generators are used solely to supply or absorb reactive power when voltage is too low or too high, respectively. Such generators have high heat losses so don't produce much real power. Their owners are paid for reactive power.

## 4.10 Transmission, Transformers, and the Battle of DC versus AC

Electromagnetic induction is an essential process occurring in a **transformer**, which is a device that increases or decreases voltage of AC electricity between two transmission lines. This section discusses transmission lines and transformers. It also discusses how transformers settled the battle between DC and AC electricity.

For electricity traveling down a transmission line, the power lost along the line is $p_w = i^2 R_w$ (Equation 4.8), where $R_w$ is the resistance in the wire. Since the power lost during transmission is proportional to the square of the current, the current should be kept low to avoid high line losses over a long distance. Minimizing current means increasing the voltage reaching the load to satisfy $v = p/i$ (Equation 4.5), where $p$ is the power required by the end-use load. Combining these two equations gives

$$p_w = (p/v)^2 R_w \qquad (4.27)$$

which suggests that increasing the end voltage of a system by a factor of 10 reduces power lost along a transmission line by a factor of 100. As such, transmitting electricity a long distance can be accomplished without significant line loss only if the voltage transmitted is high. Example 4.8 illustrates this fact for a simple case.

---

### Example 4.8 Calculating the power lost along a transmission line

Suppose DC electricity must travel 10 km along a copper wire from a generator to a load that needs 120 W of power. Assume the resistance along the 10-km wire is $R_w = 82.8\ \Omega$. Determine the power and voltage that the generator must produce if the load operates at 240 V versus 120 V. What are the power and voltage losses along the wire in each case?

### Solution:

If the voltage at the load is $v = 240$ V, then the current throughout the transmission wire must equal $p/v = 0.5$ A. If the voltage is 120 V, the current must equal 1 A. Since the power loss along the wire is $p_w = i^2 R_w$ (Equation 4.8), the power loss with a current of 0.5 A is 20.7 W and that with a current of 1 A is 82.8 W. The voltage loss through the wire is $v_w = i R_w$ (Equation 4.6), which equals 41.4 V for a current of 0.5 A and 82.8 V for a current of 1 A.

Thus, for the 240-V case, the generator must produce 120 W + 20.7 W = 140.7 W and 240 V + 41.4 V = 281.4 V. The power and voltage losses along the line are both 14.7 percent.

For the 120-V case, the generator must produce 120 W + 82.8 W, or 202.8 W and 120 V + 82.8 V = 202.8 V. The power and voltage losses along the line are both 40.8 percent.

As such, doubling the end-use voltage decreased line losses by a factor of 4. Thus, to minimize $i^2 R_w$ transmission loss, it is always beneficial to transmit electricity at high voltage and low current.

---

On September 4, 1882, **Thomas Edison** (1847–1931) and the Edison Illuminating Company began operating the first investor-owned electric utility in the world, the Pearl Street Power Station, in New York City. The electricity was produced from coal and provided power for 400 lamps owned by 82 customers. The power provided to customers was DC power at 110 V. Because the voltage through the transmission lines was relatively low, the current was relatively high and line losses were relatively large, limiting the distance

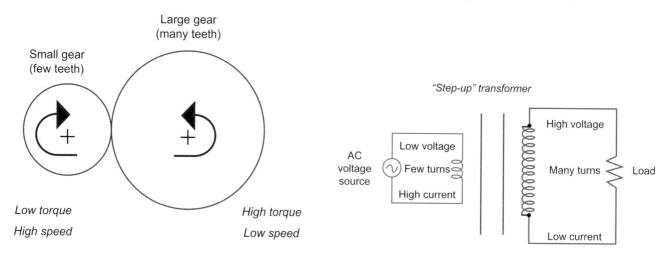

*Speed-reducing toothed gears*

Figure 4.13 **Illustration of how a step-up transformer is analogous to toothed gears.**
Adapted from Kuphaldt (2019).

between the power generation and the customers to less than a couple of kilometers.

Meanwhile, **George Westinghouse** (1846–1914), who started a DC lighting business in 1884, put his effort into marketing AC electricity in 1885. He became interested in AC electricity when he read that it had the ability to have its voltage increased (stepped up) by a **transformer** for long-distance transmission, and then decreased (stepped down) again for end use. He realized that increasing the voltage of AC electricity to produce **high voltage AC (HVAC)** electricity would decrease line loss tremendously over a long distance (e.g., Example 4.8). As such, with transformers, AC electricity could be produced far away from where it was used.

A transformer increases or decreases voltage from a powered coil to an unpowered coil. It works when an AC current winding through one coil creates a fluctuating magnetic field around a second coil not connected to the first. The fluctuating magnetic field induces a current through and voltage across the second coil. The AC voltage induced across the unpowered coil equals that across the powered coil multiplied by the ratio of the number of coil turns in the unpowered coil to those in the powered coil. The current in the unpowered coil is the opposite. It is the current through the powered coil multiplied by the ratio of the number of turns in the powered coil to that in the unpowered coil. Figure 4.13 illustrates how a step-up transformer that increases voltage and decreases current is analogous to a small toothed gear turning a larger one, thereby increasing torque but decreasing rotational speed.

A transformer does not work with DC current. Whereas DC electricity passing through one coil creates a magnetic field around the other coil, the magnetic field stays constant over time because DC current in the first coil stays constant over time. A current occurs in the second coil only if the magnetic field fluctuates with time, and this requires the current in the first coil to fluctuate with time. Because AC electricity through the first coil continuously alternates direction, the magnetic field produced by it continuously fluctuates as well, inducing a current in the second coil.

Because AC current could be sent by long-distance HVAC lines, whereas DC current could not be sent long distance without significant line losses, Westinghouse's AC electricity won the original battle of the currents. Since then, a method has been developed to transmit DC current long distances. The method is to boost AC voltage with a step-up transformer, then to convert HVAC to high voltage DC (HVDC) current. The HVDC current can then be sent a long distance along an HVDC transmission line. At the end of the line, the HVDC current is converted back to HVAC current. The AC voltage is then reduced with a step-down transformer.

## Transition highlight

George Westinghouse poured financial resources into developing an AC power system with step-up and step-down transformers. In 1891, he built the first power plant (a hydroelectric plant) to supply AC power over a long distance (5.6 km) for an industrial purpose, a gold mine. Westinghouse and Edison battled over AC versus DC powers for several years. However, AC was ultimately adopted worldwide, and the major reason was that AC power could be transmitted long distance without suffering the same line losses as DC power. AC line losses were minimized by using a transformer to step up the voltage at the start of long-distance transmission and another transformer to step it down again at the receiving end of transmission. Transformers did not work with DC current. Thus, transformers settled the battle of DC versus AC electricity.

The world's first HVDC transmission link was installed in the 1950s. From the 1950s to the 1970s, HVAC was converted to HVDC and back again with a **mercury arc valve**. Since then, **thyristors** or **transistors** have performed the conversion.

An advantage of using an HVDC transmission line over long distances (greater than 600 kilometers) is that line losses and costs are significantly less than they are for an HVAC transmission line over those distances. Over distances below ~600 km, HVDC transmission lines are more costly than are HVAC lines because expensive AC-to-DC conversion equipment is a larger share of the overall HVDC cost than it is for long-distance transmission, and the greater line losses by HVAC are less of an issue over short distances.

Today, generators produce electricity with voltages of between 12 kV and 25 kV. Step-up transformers boost the voltage to between 100 kV and 1,000 kV for HVAC transmission, and step-down transformers decrease the voltage to between 4.16 kV and 34.5 kV for distribution to the local utility. Long-distance HVDC voltages are generally between 100 kV and 1,500 kV.

Local utilities distribute AC power to neighborhood power poles. Transformers on power poles step voltage down from between 4.16 kV and 34.5 kV to 120 V or 240 V. In the United States and many countries, home wall receptors receive 60-Hz AC power at 120 V (110 to 125 V). Dryers, electric car chargers, electric water heaters, and some other appliances require 240 V. This voltage is obtained by stepping up the voltage with a transformer, reducing the current by half, and reducing wire losses by a factor of 4 (Example 4.8). In Europe and most countries, the power to homes is 50-Hz AC power at 240 V.

## 4.11 Summary

This chapter discussed the basics of electricity. It explored static electricity, lightning, and wired electricity and described voltage and Kirchoff's laws. It also discussed power, resistance, capacitors, and inducers. It further went into electromagnetism, transmission, and transformers. Single-phase and three-phase AC electricity were then covered, followed by real and reactive power. Such information is important, because a WWS infrastructure will run almost entirely on electricity. In addition, wind turbines and solar panels convert mechanical energy or solar energy, respectively, into electricity, and this chapter provided some necessary information for understanding how those conversion processes, discussed in Chapters 5 and 6, respectively, work. WWS electricity will also flow along transmission lines, including some high voltage lines. This chapter provided information on how such lines work and how transformers increase and decrease the AC voltage along a line to allow for long-distance AC transmission. It further discussed the conversion of high voltage AC to DC transmission.

## Further Reading

Chiras, D., 2019. *Solar Electricity Basics*, 2nd ed., Gabriola Island. BC: New Society Publishers, 224 pp.

Hemami, A., 2015. *Electricity and Electronics for Renewable Energy*. Boca Raton, FL: CRC Press, 824 pp.

Masters, G., 2013. *Renewable and Efficient Electric Power Systems*, 2nd ed., Hoboken, NJ: Wiley, 712 pp.

## 4.12 Problems and Exercises

4.1. What is the cross-sectional area of a copper wire that allows a drift velocity of 1.5 meters per hour when a current running through the wire is 50 A? Assume the molecular weight of copper is 63.55 g/mol, the density of copper is 8,960 kg/m$^3$, and copper has one free electron per atom.

4.2. If a circuit has a battery and three light bulbs, with resistances of 15 $\Omega$, 12 $\Omega$, and 13 $\Omega$, respectively, in series, and the current through the circuit is 0.3 A, what is the voltage across the battery?

4.3. Calculate the power emitted by a light bulb if its filament resistance is 300 $\Omega$ and its power source is 120 V. What are the current and energy consumed over 1,000 hours?

4.4. How much current flows through a curling iron if its resistance is 40 $\Omega$ and it is plugged into a 120-V power source? What is its power output and how much energy does it consume over one hour?

4.5. Calculate the current through a circuit where the power source is 10 V and three resistors in series have resistances of 20, 15, and 5 $\Omega$, respectively.

4.6. Calculate the current through a circuit with resistors in parallel, assuming a voltage source of 15 V and the resistances are 150 $\Omega$ and 50 $\Omega$, respectively.

4.7. Suppose the current through a circuit with a 100-$\Omega$ resistor in series with a 30-$\Omega$ and 70-$\Omega$ resistor in parallel is 1 A. Find the voltage giving this current.

4.8. Why was George Westinghouse's AC electricity ultimately adopted worldwide in buildings instead of Thomas Edison's DC electricity?

4.9. Suppose DC electricity must travel 100 km along a copper wire from a generator to a load that needs 300 W of power. Assume the resistance along the line is $R_w$ = 100 $\Omega$. Determine the power and voltage that the generator must produce if the load operates at 240 V. What are the power and voltage losses along the line?

4.10. Calculate the AC voltage required to light a 100-W light bulb if its resistance is 240 $\Omega$. What is the current through the circuit?

4.11. Give three reasons why reactive power is important.

4.12. What is one advantage of three-phase AC electricity over single-phase AC electricity?

# 5 Photovoltaics and Solar Radiation

Solar and wind will make up the bulk share of a 100 percent wind-water-solar (WWS) energy generation infrastructure worldwide. The main types of solar generation are solar photovoltaics (PV) on rooftops and in utility-scale power plants, concentrated solar power (CSP), and solar thermal collectors for water and air heating. The sun produces enough energy worldwide to power the world with PV for all purposes in 2050, if all energy were electrified, about 2,200 times over. Over land, PV can power all energy about 640 times over. Needless to say, the world needs only a small fraction of this. If half the world's all-purpose power were from solar PV, that would mean about 0.08 percent of the world's solar resource over land would be needed. Given the large potential of solar PV in particular for powering the world's energy needs, it is useful to understand PV panels and solar resources better. This chapter discusses both as well as how to determine the quantity of solar radiation reaching a PV panel over time and space. The chapter starts with a detailed description of solar photovoltaic cells, panels, and arrays and their efficiencies (Section 5.1). It then goes into solar resource availability and optimal tilt angles for solar panels worldwide (Section 5.2). Finally, it discusses how to calculate radiation through the atmosphere (Section 5.3).

## 5.1 Solar Photovoltaics

A solar photovoltaic (PV) is a material or device that converts photons of solar radiation to electrical voltage and direct current (DC). A photon of sunlight with a short enough wavelength causes an electron in a PV material to break free of the atom that holds it. If a nearby electric field is provided, free electrons are swept toward a metallic contact, where they become part of an electric current in wire.

PV materials are semiconductor materials that convert sunlight to electricity. The primary semiconductor materials are silicon (Si), germanium (Ge), gallium (Ga), and arsenic (As). Boron (B) and phosphorus (P) are added to silicon to make some PV cells. Gallium and arsenic are often combined to make GaAs cells.

### 5.1.1 Conduction, Forbidden, and Filled Bands

**Metals** are **conductors** because electrons can break free from atoms in a conductor material at any temperature and conduct a current. Yet, conductivity in a metal decreases with increasing temperature. **Non-metals** have low electrical conductivity, thus are **insulators**. A **semiconductor** has an electrical conductivity between that of a metal and a non-metal. Their conductivity is limited at low temperatures but much better at moderate and high temperatures. At near 0 K, for example, silicon is a perfect insulator (it has no free electrons). As the temperature rises, though, some electrons escape the nucleus and can flow in an electric current.

Pure silicon crystals consist of Si atoms bonded together with covalent bonds. Silicon has 14 total electrons, including 2 in its innermost shell, 8 in its second shell, and 4 in its third, outer (valence) shell. Thus, it has a +4 nucleus charge. **Valence electrons** in an atom are the electrons that occupy the outermost shell of the atom. If the shell is not full of electrons, valence electrons can each form a covalent bond with the valence electron of another atom. Because the third shell of an atom can

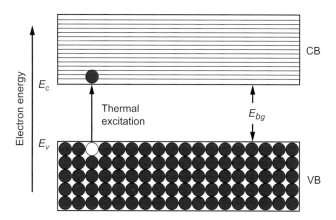

**Figure 5.1 Diagram of semiconductor energy bands. VB = valence band; CB = conduction band. In between the conduction band and valence band is the forbidden band. $E_c$ is the minimum energy needed to enter the conduction band. $E_v$ is the energy of electrons in the valence band. $E_{bg} = E_c - E_v$ is the band gap energy. If an electron in the valence band acquires energy $E_{bg}$, the electron can jump to the conduction band.**

contain up to 18 electrons and silicon's third shell contains only 4, each of silicon's 4 valence electrons can bond with one other silicon atom. Thus, 1 silicon atom covalently bonds with 4 other silicon atoms in a tetrahedral structure.

Semiconductors have three main energy bands: a valence band, a forbidden band, and a conduction band (Figure 5.1). The **valence band** of an atom is the energy band of that atom occupied by the valence electrons. If excited with enough additional energy, an electron in the valence band can jump out of the valence band to the conduction band, which is a band of energy that is otherwise vacant. Thus, the valence and conduction bands hold electrons of different energy levels, with the valence band holding electrons of a low energy level and the **conduction band** holding electrons of the highest energy level.

The **forbidden band** is an energy gap between the conduction band and the valence band. When electrons obtain a sufficient amount of energy, they jump from the valence band to the conduction band, bypassing the forbidden band. As such, no electrons ever reside in the forbidden band. In order to jump to the conduction band, an electron in the valence band must obtain enough energy, referred to as **band gap energy**. It is the energy required for a valence electron to free itself from the electrostatic force holding it to its own nucleus to jump into the conduction band.

Electrons in the conduction band contribute to current flow. For metals, the conduction band is partly filled naturally, even at low temperature, due to the fact that some electrons in metals are thermally excited enough to break free of their nucleus at low temperatures. For semiconductors, on the other hand, the conduction band is empty at 0 K. At room temperature, only 1 out of 10 billion electrons in the valence band is thermally excited enough to jump into the conduction band.

The unit of band gap energy is the electron volt (eV), which is the energy an electron acquires when voltage increases by 1 V. Because 1 V = 1 J / C and 1 C = 6.2415 $\times$ 10$^{18}$ electrons, 1 eV = 1.60218 $\times$ 10$^{-19}$ J.

The band gap energy for silicon is 1.12 eV. When an electron jumps from a silicon atom, it leaves a +4 nucleus with only 3 electrons; thus, the nucleus has a net positive charge, or a **hole**. Unless more electrons are swept away by an electric current, another electron will soon recombine with the positively charged silicon atom to fill the hole.

When electrons are freed from a semiconductor atom, other electrons in the lattice of connected semiconductor atoms may fill the hole, thus moving the hole (location of positive charge) to another atom.

In order for a photon of solar radiation to provide the band gap energy needed for a semiconductor's electron to jump from its valence band to its conduction band, the solar wavelength must be shorter than the wavelength needed to provide the band gap energy. **The band gap energy** (J) is calculated as a function of wavelength with

$$E_{bg} = h\nu = hc/\lambda_{bg} \qquad (5.1)$$

where h = 6.626 $\times$ 10$^{-34}$ J-s/cycle is Planck's constant, $c$ = 3 $\times$ 10$^6$ m/s is the speed of light, $\lambda_{bg}$ **is the wavelength (m) of a photon that gives the band gap energy**, and $\nu$ = c / $\lambda$ is frequency (Hz = cycles/s). Solving this equation for the maximum wavelength that will allow an electron to jump gives

$$\lambda_{bg} = hc/E_{bg} \qquad (5.2)$$

Thus, for silicon, with a band gap energy of $E_{bg}$ = 1.12 eV, $\lambda_{bg}$ = 1.11 μm. As such, silicon solar cells produce electricity only for solar wavelengths less than 1.11 μm. Table 5.1 gives band gap energies and maximum wavelengths that allow an electron to jump to the conduction band for several semiconductor materials.

Finally, if an electron in the conduction band recombines with certain semiconductor materials such as GaAs

Table 5.1 Band gap energies ($E_{bg}$) and maximum wavelengths ($\lambda_{bg}$) that allow an electron to jump to the conduction band for several semiconductor materials.

| Semiconductor material | $E_{bg}$ (eV) | $\lambda_{bg}$ (μm) |
|---|---|---|
| Si | 1.12 | 1.11 |
| a-Si | 1.7 | 0.73 |
| Cd-Te | 1.49 | 0.83 |
| CuInSe$_2$ | 1.04 | 1.19 |
| CuGaSe$_2$ | 1.67 | 0.74 |
| GaAs | 1.43 | 0.87 |

Source: Masters (2013).

or InAs, which are called **direct band gap materials**, the recombination releases a photon of light radiation. This is the source of light in a **light-emitting diode** (**LED**).

Silicon and germanium, on the other hand, are **indirect band gap materials**. The recombination of a conduction-band electron with a silicon atom, for example, results in the emission of a vibration rather than a photon.

## 5.1.2 Maximum Possible PV Cell Efficiency

The sun's radiation spans wavelengths ranging from <0.1 μm to 10 μm, with a peak intensity occurring at 0.5 μm, which is the wavelength that humans have evolved to see most keenly at. The visible spectrum is between 0.38 μm and 0.75 μm. Below 0.38 μm is the ultraviolet spectrum, which constitutes about 5 percent of total solar radiation energy. The solar infrared part of the spectrum is between 0.75 μm and 10 μm. As such, silicon PV cells absorb radiation in the ultraviolet, visible, and part of the solar infrared spectrum, but not across the whole solar infrared spectrum. Thus, solar cells can convert only a fraction of the solar spectrum into band gap energy necessary to push valence band electrons into the conduction band.

Further, at each wavelength below the maximum wavelength needed to create sufficient band gap energy, PV cells can use solar radiation only up to the maximum band gap energy. Thus, the useful solar energy in a PV cell is limited by both the maximum useful wavelength and the maximum useful energy at each wavelength.

For example, for silicon, only 49.6 percent of total solar radiation can theoretically be converted to useful band gap energy. This is because 20.2 percent of radiation is above its maximum wavelength, thus is unavailable, and 30.2 percent of radiation below the maximum wavelength is above the band gap energy at each wavelength ($h\nu > E_{bg}$). Of the remaining 49.6 percent, 7 percentage points are lost due to high temperatures, which result in heat radiation emissions of energy back to space, and another 9 percentage points are lost due to the recombination of electrons. The resulting maximum possible efficiency (electricity output per unit solar radiation input) of a single junction PV cell is about 33.7 percent, which is called the **Shockley–Queisser limit**.

The maximum possible efficiency of a PV cell differs for different band gap energies. With a lower band gap energy ($E_{bg}$), $\lambda_{bg}$ increases from Equation 5.2, so energy from more wavelength bands across the solar spectrum is converted. However, for each wavelength below $\lambda_{bg}$, less energy is converted since only up to the band gap energy is converted, and the band gap energy is now lower.

With a higher band gap energy, $\lambda_{bg}$ is lower, so energy from fewer wavelength bands across the solar spectrum is converted. However, for each wavelength below $\lambda_{bg}$, more energy is converted since up to the band gap energy is converted, and the band gap energy is now higher.

Based on data from many solar cells with different band gap energies, the greatest solar cell efficiency occurs with band gap energies of between $E_{bg}$ = 1.2 to 1.6 eV.

Band gap energies themselves are temperature dependent. At higher temperature, less energy is needed to send electrons into the conduction band, so band gap energy decreases. At lower temperatures, band gap energy increases.

Power output in a PV circuit equals voltage multiplied by current (Equation 4.8). When band gap energy decreases (because temperature increases), voltage drops and current rises. When band gap energy increases (because temperature decreases), voltage increases and current drops. Consequently, a curve can be drawn of power output versus $E_{bg}$. The maximum of this curve gives the band gap energy that gives the highest power output, which is around $E_{bg}$ = 1.4 eV.

## 5.1.3 Creating Electric Fields and Electricity in a PV Cell

Electric fields are built into PV cells to carry electrons in the conduction band away, preventing them from recombining with silicon atom holes. Electric fields push electrons in one direction and holes in the other. To create an

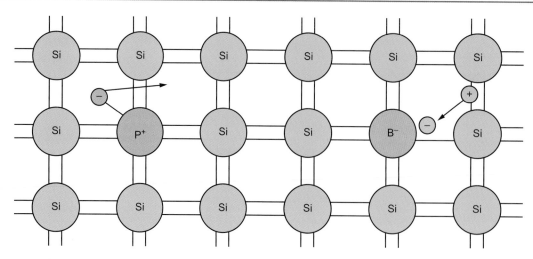

Figure 5.2 Diagram showing the substitution of a phosphorus atom (left) and boron atom (right) for a silicon atom in the lattice of a PV cell. The P atom, which attains an immobile positive charge, is an n-type material since it donates a negatively charged roaming electron. The B atom, which attains an immobile negative charge, is a p-type material since it results in a positively charged roaming hole. See text for an explanation.
Adapted from Karolkalna at the English Wikipedia.

electric field, one side of a silicon cell is contaminated with a small number (1 atom per 1,000 atoms of Si) of a trivalent element, such as boron (B), from Column III of the periodic table, and the other side, with a pentavalent element, such as phosphorus (P), from Column V.

Each phosphorus and boron atom substitute for one atom of silicon in the lattice of the PV cell (Figure 5.2). Phosphorus has 15 protons and 16 neutrons in its nucleus, 15 total electrons, and 5 electrons in its valence band. Thus, phosphorus can form five covalent bonds. Silicon can form only four covalent bonds. As such, when P is embedded in the lattice of Si, P has a free, fifth, electron in its valence band. At room temperature, this free electron escapes the atom and roams, giving the immobile phosphorus atom a positive charge. Phosphorus, however, is called an **n-type** (negative-type) material since it donates a negatively charged roaming electron.

Boron has five protons and six neutrons in its nucleus, a total of five electrons, and three electrons in its valance shell. When it is substituted into the lattice of silicon, boron forms covalent bonds with three out of its four silicon neighbors (Figure 5.2). However, the boron tries to bond covalently to the fourth neighbor, first by moving its own three valence band electrons around, then by borrowing an electron from a nearby silicon atom. This results in the immobile boron becoming net negatively charged and a positive hole forming in the silicon atom. Since a roaming electron from another silicon atom fills the hole, the hole roams to the other silicon atom and so

on. Because the boron creates a roaming hole (positive charge), the boron is called a **p-type** material.

Mobile electrons from the n-type side of the PV cell, near the border with the p-type side, drift by diffusion toward the p-type side. Similarly, mobile holes from the p-type side, near the border with the n-type side, roam toward the n-type side. In the middle, mobile electrons fill mobile holes, creating uncharged, electrically neutral atoms. This no-charge **depletion region** is about 1 μm wide and is called a **p-n junction** (Figure 5.3).

In the depletion region, the immobile positively charged P atoms on the n-type side of the p-n junction

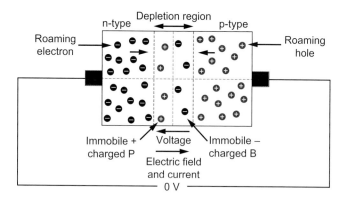

Figure 5.3 Zero-biased p-n junction diode. See text for description. The resulting electric field and current are from positive to negative (left to right) in the depletion region and the voltage is from negative to positive (right to left).
Adapted from Electronics Hub (2015).

are no longer balanced by roaming negative charges, so that region takes on a net positive charge (Figure 5.3). The immobile negatively charged B atoms on the p-type side of the p-n junction are similarly no longer balanced by roaming holes, so that side takes on a net negative charge (Figure 5.3). The immobile positive charges on the n-type side of the p-n junction in the depletion region prevent any more roaming holes (mobile positive charges) on the p-type side of the junction from moving toward them. Similarly, the immobile negative charges on the p-type side of the depletion region prevent any more roaming electrons (mobile negative charges) on the n-type side from moving toward them. As such, a charge gradient arises. The n-type side of the depletion region has immobile positive charges. The p-type side has immobile negative charges.

When an external wire is connected from a cap at the end of the p-type region to a cap at the end of the n-type region, an equilibrium charge differential arises between the n-type and p-type sides of the p-n junction in the depletion region. The charge difference due to the immobile positive charges in the n-type region and the immobile negative charges in the p-type region creates an equilibrium electric field, acting from the positive charges to the negative charges in the depletion region (from left to right in Figure 5.3) of about 0.5 V to 0.7 V for doped silicon, giving an electric field strength of about 5,000 V/cm to 7,000 V/cm.

The voltage in the depletion region acts from negative to positive (right to left in Figure 5.3). Positive current is the direction of positive ion flow, which is from left to right in the depletion region of Figure 5.3. The resulting device is called a **p-n junction diode**. It is a **zero-biased diode** because no external energy source has been provided.

In a PV cell, the n-type layer can be on top (closest to the sun) or bottom. The most efficient solar cells generally have a p-type layer on top and an n-type layer on bottom, with a p-n junction in between. In either case, photons of solar radiation energize electrons in the valence band of silicon in the layer on top, moving them into the conduction band, creating mobile electron-hole pairs. If these drift into the depletion region, the electric field will send electrons to the n-type layer and holes to the p-type layer. This creates voltage that can be used to deliver current to a load.

If a wire is attached to the top and bottom of a cell and connected to a load in between, electrons flow out of the n-type side into the wire and through the load and back to the p-type side. Wires cannot conduct holes. Electrons reaching the p-type side recombine with holes, completing the circuit.

### 5.1.4 Types of and Materials in PV Cells

The first generation of PV cells, which are still widely used, have p-n junctions 160 to 240 μm thick and are made of **single crystal silicon** (sc-Si) (also called monocrystalline silicon) or **polycrystalline silicon** (poly-Si). Silicon is the second most abundant element on Earth, comprising 20 percent of the Earth's crust. The source of silicon for PV cells is high quality silica ($SiO_2$), also called quartz, from mines or sand. Silicon processing begins with a high-temperature electric arc furnace that uses carbon to reduce $SiO_2$ to metallic grade (99 percent pure) Si. Si is purified further until it is more than 99.9999 percent pure for use in solar PV cells.

Single crystal silicon cells are uniform in structure and appear square but with clipped corners, so are really octagonal. They have a distinct pattern of small white diamonds. Polycrystalline silicon cells resemble rock-like chunks of a multifaceted metal. Because they are more uniform in structure, single crystal cells are more efficient but also more expensive to manufacture than are polycrystalline cells.

Second-generation PV cells, called **thin-film PV** cells, have 1- to 10-μm-thick junctions. They are thinner than first-generation cells to reduce the amount of active material needed. Thin-film PV cells usually consist either of a material with 2 to 4 elements, such as gallium-arsenide (**GaAs**), cadmium-telluride (**CdTe**), or copper-indium-gallium-selenium ($CuIn_xGa_{1-x}Se_2$, also called **CIGS**); or of **amorphous silicon** (a-Si). Thin-film materials are wedged between two panes of glass. Since glass is heavy and first-generation cells use only one pane of glass, second-generation cells are heavier than first-generation cells. However, the energy required to produce first-generation cells is greater.

CIGS and GaAs cells are both efficient. However, they are also expensive. Thus, despite their higher efficiency, they result in higher overall energy costs than do first-generation silicon-based cells. CdTe cells, on the other hand, are less efficient than are first-generation cells but cost less per cell and have an overall cost of energy comparable with that of first-generation cells.

Amorphous silicon is made by chemical vapor deposition of silane gas and hydrogen gas. As such, it is non-crystalline or has tiny crystals on the order of a micrometer in size (thus called microcrystalline). It has a greater band gap energy (1.7 eV) than does monocrystalline silicon (1.12 eV) (Table 5.1). Thus, amorphous silicon absorbs across a lesser portion of the solar spectrum (up to 0.73 μm versus up to 1.11 μm for monocrystalline), but it absorbs more sunlight at each wavelength.

Third-generation cells are thin-film technologies that are either emerging or still too expensive for widespread commercial use. Two of these include cells that incorporate organic materials and multijunction (tandem) cells.

**Organic cells** (cells that contain organic material) tend to be less expensive per cell than silicon-based cells but are currently about one-third the efficiency. In addition, sunlight photochemically degrades organic-based cells over time.

**Multijunction cells** (also called **tandem cells**) are either individual thin-film PV cells with multiple materials or a stack of thin-film PV cells, each with a different material, on top of each other and connected in series. In both cases, each material absorbs a different portion of the solar spectrum (thus has a different band gap energy). For example, with a two-band-gap stack of cells, the top cell may have a high band gap, thus absorb high-energy photons, but only up to a short wavelength. The bottom material, on the other hand, may have a low band gap, thus absorb only low-energy photons, but over a greater portion of the solar spectrum.

Whereas the theoretical peak efficiency of a single junction PV cell is about 33.7 percent (Section 5.1.2), that in a stack of two cells with different band gaps is around 47 percent. The theoretical peak efficiency in a stack of three cells is around 53 percent; and that in a stack of eight cells is about 62 percent (Bremner et al., 2008). To date, tandem cells have been used primarily in satellite

PV arrays and in technologies that use lenses and mirrors to focus light onto the cell to increase the intensity of sunlight hitting the cell.

Because so many alternative types of PV cells are available, it is unlikely that materials will constrain the large-scale growth of PV. For example, in multijunction cells, the limiting material is germanium (Ge); however, substituting gallium (Ga), which is more abundant, would allow terawatt expansion. In addition, the production of silicon-based PV cells is limited not by crystalline silicon (because silicon is widely abundant) but by silver, which is used as an electrode. Reducing the use of silver as an electrode would allow the virtually unlimited production of silicon-based solar cells. In sum, the development of a large global PV system will not be limited by the scarcity or cost of raw materials.

## 5.1.5 PV Panels and Arrays

One PV cell produces about 0.5 V to 0.7 V of energy potential. A **PV panel** consists of either 72, 96, or 128 pre-wired cells in series, fitted into a rectangular package. A **PV array** is a set of PV panels wired in series to increase the voltage, or in parallel to increase the current. For panels wired in series, the total voltage is the sum of voltages across all individual panels, and the total current is the current running through any one panel. For panels wired in parallel, the total current is the sum of currents through each panel, and the total voltage is the voltage across one panel.

Alternatively, an array can be wired in a combination of series and parallel to optimize power (voltage multiplied by current) output. In this case, panels are first wired in series to increase the voltage by as much as is safe, then each series string is wired in parallel to maximize power. Maximizing voltage also minimizes $i^2R$ power losses along wires. When arrays are set up in this manner, the total voltage is the sum of voltages across one string of arrays, the current through one string is the current through one panel in the string, and the total current through the array is the sum of currents through each parallel string.

## Example 5.1  Calculating current, voltage, and power through a PV array

Assume a PV array consists of 10 panels per string wired in series and 6 of these strings wired in parallel. Suppose the voltage through each panel is 36 V and the current through each panel is 6 A. Calculate the total voltage, total current, and total power output of the array as a whole.

### Solution:

The voltage through one string of the array is 10 panels multiplied by 36 V/panel = 360 V. This is also the total voltage across the whole array, since all strings are in parallel and have the same voltage. The current through one string is the same as the current through one panel in the string, 6 A. The current across all strings is 6 strings multiplied by 6 A/string = 36 A. Thus, the total power output of the array is 360 V × 36 A = 12,960 W.

### 5.1.6 PV Panel Efficiencies

Figure 5.4 shows a **current-voltage (I-V) curve** and a **power-voltage (P-V) curve** for a solar PV panel. An I-V curve is a graph of current through a solar PV panel as a function of voltage across the panel for a given level of solar insolation and at a given temperature. Multiplying the current at each voltage gives the power produced by the panel as a function of voltage. This is represented by the P-V curve.

At zero voltage, the current equals the **short circuit current** ($i_{SC}$), and the product of the two is zero power. At the **open circuit voltage** ($v_{OC}$), the current is zero, and the product of the two is also zero power. As such, the two values of zero power on the power curve in Figure 5.4 represent the extreme cases of no voltage or no current. All values in between are calculated by varying the resistive load on the circuit between that of a short circuit (zero resistance) and open circuit (infinite resistance). This results in capturing all currents between 0 and $i_{SC}$ and all voltages between 0 and $v_{OC}$

The short circuit current ($i_{SC}$) is the maximum possible current through a solar panel and occurs when zero voltage occurs across the panel. Zero voltage occurs when the positive and negative terminals are connected directly with no load (thus zero resistance) in between. Connecting the positive and negative terminals directly does not damage the panel.

The open circuit voltage ($v_{OC}$) is the maximum possible voltage across a solar panel. It occurs when the current through the panel is zero (thus terminals are not connected). The open circuit voltage is effectively the same as having an infinite resistance along the circuit.

The maximum power on the P-V curve occurs at the current $i_{MP}$ and voltage $v_{MP}$. The point on the power curve that gives this maximum power is the **maximum power point (MPP)**. The power corresponding to the maximum power point is

$$p_{MPP} = i_{MP}v_{MP} \tag{5.3}$$

It is the theoretical maximum power that a solar panel connected to a load can generate for the solar conditions and temperature represented by the power curve. Thus, a solar panel is ideally operated at its MPP. The current and voltage at the maximum power point ($i_{MP}$ and $v_{MP}$, respectively) are approximately 85 to 95 percent of the short circuit current and 80 to 90 percent of the open circuit voltage, respectively.

Two other parameters that can be derived from Figure 5.4 are the fill factor and the solar panel efficiency.

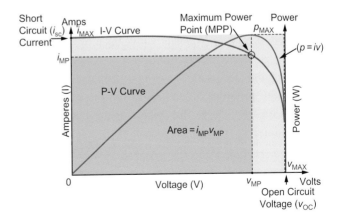

**Figure 5.4** Solar panel current-voltage (I-V) and power-voltage (P-V) curves for specific solar radiation and temperature conditions. The I-V curve gives the current as a function of voltage. The P-V curve gives the power (which equals current multiplied by voltage) as a function of voltage.

The **fill factor** (**FF**) is the ratio of the maximum power a panel can provide for the current set of conditions ($p_{MPP}$) to the product of the short circuit current and open circuit voltage. Thus, it is

$$FF = p_{MPP}/(i_{SC}v_{OC}) \tag{5.4}$$

An ideal panel has a fill factor close to 1. Typical fill factors for good solar panels are 0.7 to 0.8.

The **solar panel efficiency** ($E_{panel}$) is the actual power output obtained under standard test conditions ($p_{MPP,STC}$), divided by the maximum solar power available under those test conditions. It is calculated as

$$E_{panel} = p_{MPP,\ STC}/(F_{1000}A_{panel}) \tag{5.5}$$

where $F_{1000}$ is the solar flux under standard test conditions (1,000 W/m², defined as **1 sun**), and $A_{panel}$ is the surface area (m²) of the panel. The power output obtained under STC is also called **peak DC watts** ($W_{p,DC}$), nameplate power, or nameplate capacity.

**Standard test conditions** (**STC**) are conditions under which solar PV panels industry-wide are tested and compared with each other to evaluate their efficiency. The conditions include exposure to 1,000 W/m² of sunlight, a solar cell and air temperature of 25 °C, and a solar spectrum determined at an air mass coefficient of 1.5.

---

### Example 5.2 Calculating nameplate capacity

What is the maximum power output under standard test conditions (nameplate capacity) of a solar PV panel if the panel's efficiency is 18 percent and its area is 1.5 m²?

#### Solution:

From Equation 5.5, $p_{MPP,STC} = E_{panel}\ F_{1000}\ A_{panel} = 0.18 \times 1,000$ W/m² $\times$ 1.5 m² = 270 W. This is the nameplate capacity of the solar panel.

---

An **air mass coefficient** ($L$) is the ratio of the path length of air from the ground to the top of the atmosphere at a given solar zenith angle to the path length at zero solar zenith angle. The **solar zenith angle** ($\theta_s$) is the angle between a vertical line from the center of the Earth through a point of interest on the surface of the Earth, thus perpendicular to the surface of the Earth, and a line between the center of the Earth and the sun (Section 5.1.8.1). A solar zenith angle of 0 indicates the sun is directly overhead (perpendicular to the surface of the Earth) at the point of interest. A zenith angle of 90° indicates that a beam from the sun is along the horizon (parallel to the surface of the Earth) at the point of interest. The air mass coefficient is calculated as

$$L = 1/\cos\theta_s \tag{5.6}$$

so, an air mass coefficient of $L = 1.5$ corresponds to a solar zenith angle of 48.19°, or a solar angle of 90° − 48.19° = 41.81° above the horizon. STC conditions correspond roughly to clear-sky conditions near solar noon during the spring equinox (March 19 or 20, depending on year) and autumnal equinox (September 22 or 23) at 35 °N latitude with a panel tilted to face the sun.

In 2020, most solar panel efficiencies ranged from 15 to 24 percent. The upper limit efficiency of a single junction PV cell is about 33.7 percent, determined by the **Shockley–Queisser limit** (Section 5.1.2).

---

### Example 5.3 Calculating solar panel efficiency

Calculate the efficiency of a solar panel with a nameplate capacity of 390 W and a panel surface area of 1.63 m².

#### Solution:

From Equation 5.5, the panel efficiency is 390 W / (1,000 W/m² $\times$ 1.63 m²) = 0.239, or 23.9 percent.

---

The I-V and P-V curves in Figure 5.4 differ for each solar radiation and temperature condition. Whereas increases in solar radiation increase the maximum current ($i_{SC}$) through a panel, increases in temperature decrease the maximum voltage ($v_{OC}$) through the panel.

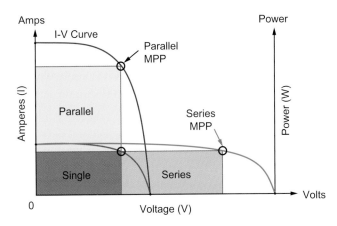

**Figure 5.5 Modification of I-V curve for two panels connected in parallel or series.**

When solar panels are connected in parallel or series or both, their I-V curves must be re-computed, as illustrated in Figure 5.5. If two panels are connected in series, then the open circuit voltage increases to the sum of the open circuit voltages of each panel, but the short circuit current stays the same, relative to one panel. Because $v_{OC}$ increases, $v_{MP}$ and $p_{MAX} = i_{MP}v_{MP}$ also increase. If two panels are connected in parallel, the open circuit voltage stays the same but the short circuit current increases to the sum of the short circuit currents of each panel. As such, $p_{MAX}$ increases in that case as well.

### 5.1.7 Correction of PV Output for Cell Temperature and Other Processes

The actual AC power output from a solar panel ($p_{AC}$) at a given time differs from the ideal output under standard test conditions. The actual output is estimated from the nameplate DC power output of the panel ($p_{MPP,STC}$) with

$$P_{AC} = p_{MPP,\ STC}C_{temp}D_FF_{cur}/F_{1000}$$
$$= F_{cur}A_{panel}E_{panel}C_{temp}D_F \tag{5.7}$$

where $C_{temp}$ is a factor correcting for cell temperature relative to the STC temperature of 25 °C, $D_F$ is the product of correction factors for several additional processes that affect solar power output, and $F_{cur}$ is the current solar flux (W/m$^2$) normal to the panel. Combining the first term in Equation 5.7 with Equation 5.5 gives the second term in Equation 5.7. The first two correction terms are discussed in this section. Section 5.1.8 discusses the correction for the current solar flux.

#### 5.1.7.1 Correction for Cell Temperature
First, above a threshold cell temperature, solar panel output decreases with increasing cell temperature. One empirically determined correction factor of solar panel output due to temperature variations from STC is

$$C_{temp} = 1 - b_{ref} \times \max \left( \min \left(T_c\text{-}T_{ref}, T_{th} \right), 0 \right) \tag{5.8}$$

where $b_{ref}$ is the temperature coefficient, which ranges from 0.0011/K to 0.0063/K for different types of panels (a typical value is 0.0025/K), $T_{ref} = 298.15$ K is the reference temperature, $T_{th} = 55$ K is the threshold difference in cell temperature over the reference temperature above which a temperature effect occurs, and

$$T_c = T_a + 0.32 \times F_{cur}/(8.91 + 2w) \tag{5.9}$$

is the cell temperature (K or °C). In Equation 5.9, $T_a$ is the air temperature (K or °C) the panel is exposed to, $w$ is the wind speed (m/s) the panel is exposed to, and $F_{cur}$ is the current solar flux (W/m$^2$) normal to the panel. The equation is empirical, so units do not equate.

---

## Example 5.4 The AC power output of a solar panel

Estimate the AC power output of a solar panel if the solar flux normal to the panel is 500 W/m$^2$, the panel efficiency is 20 percent, the panel surface area is 1.6 m$^2$, the ambient air temperature is 30 °C, the wind speed is 5 m/s, $b_{ref} = 0.0025$/K, and no other corrections aside from temperature are considered.

### Solution:
From Equation 5.9, the cell temperature is $T_c = 38.46$ °C = 311.61 K. From Equation 5.8, $C_{temp} = 1 - 0.0025/$K × max(min(311.61 − 298.15 K, 55 K), 0) = 0.966. From Equation 5.7, the AC power output = 500 W/m$^2$ × 1.6 m$^2$ × 0.2 × 0.966 × 1 = 154.6 W.

### 5.1.7.2 Corrections for Additional Processes

Table 5.2 provides correction factors for additional processes affecting PV panel output. $D_F$, used in Equation 5.7, is the product of these correction factors.

In Table 5.2, the correction for **nameplate DC rating** accounts for the fact that actual peak power output under standard test conditions of a solar panel usually differs from the nameplate capacity assigned to the panel by the manufacturer. In addition, panels degrade after their initial exposure to sunlight and before power output stabilizes. Finally, not all PV panels perform identically due to slight variations in current-voltage characteristics. A 98 percent correction factor indicates that the panel output is 2 percent less than the manufacturer's nameplate capacity rating.

**Inverters** convert DC power to AC power. The **inverter DC to AC efficiency** is the ratio of AC power output to DC power input. Inverters' efficiencies are below 1 because of power losses due to the conversion of power from DC to AC and resistive wire losses in the inverter.

The correction for **diodes and connections** accounts for losses that occur with the use of diodes or electrical connections in the system. Two main types of diodes are used in solar panel systems. One is the bypass diode and the other is the blocking diode.

Suppose two solar panels are connected in series. If one panel is shaded, it will not produce any power and have a high resistance, reducing the current flowing through it,

Table 5.2 **Correction factors (typical and range, averaged over a year) for several processes affecting PV panel output. These processes are described in the text. $D_F$ is the product of all these correction factors and is used in Equation 5.7.**

| Process affect PV output | Derating factor |
|---|---|
| Nameplate DC rating | 0.98 (0.9–1.05) |
| Inverter DC to AC efficiency | 0.98 (0.97–0.99) |
| Diodes and connections | 0.995 (0.99–0.997) |
| DC wire losses | 0.99 (0.97–0.99) |
| AC wire losses | 0.99 (0.98–0.993) |
| Panel soiling or snow cover | 0.98 (0.7–0.995) |
| System availability | 0.99 (0.7–1) |
| Degradation with age | 0.98 (0.7–1) |
| Shading | 0.97 (0.7–1) |
| **Overall derating factor ($D_F$)** | **0.864 (0.20–1)** |

Source: adapted from NREL (2018).

wiping out power produced by the unshaded panel. A **bypass diode** allows the current from the unshaded panel to bypass the high resistance of the shaded panel.

A **blocking diode** is used primarily when a solar system is connected to a battery. When the sun is shining, current flows from solar panels to the battery to charge the battery so long as the voltage produced by the panels exceeds the voltage of the battery. However, at night, when the solar production is zero, a blocking diode is needed to prevent current flowing from the battery back to the solar panels. Blocking diodes are usually built into the solar panel.

The corrections for **DC wire losses** and **AC wire losses** account for resistive wire losses between the solar panel and the inverter and between the inverter and electric meter, respectively. They do not account for transmission or distribution losses between a utility PV plant and an end-use load. Section 6.6.6 discusses such transmission and distribution losses as a fraction of load carried.

The correction for **panel soiling** or **snow cover** accounts for losses of power due to the reduction in solar radiation to the panel resulting from dust, sand, dirt, debris, or snow accumulating on the panel.

The correction for **system availability** accounts for the fact that PV systems are down part of the time for maintenance or due to power outages.

The correction for panel **degradation with age** accounts for the fact that solar panel output can decrease by 0.5 to 1 percent per year due to degradation of materials in the panel. Panels in their first year should see no degradation, so this correction factor should be 1 or near 1 for the first year.

The correction for **shading** accounts for the fact that trees, structures, or other solar panels often shade solar panels. For example, when solar panels are tilted, they cast a shadow on other panels at certain times of the day in certain seasons unless the panels are spaced sufficiently far apart. The greater the tilt angle relative to the horizontal, the further rows of panels should be separated to avoid shading, and the greater the ground or roof spacing area required per panel. An approximate equation for the **ground cover area per panel** ($A_G$) for rows of tilted east-west panels lined up one behind the other in the south-north direction is

$$A_G = (P_H \cos\beta + D_P) P_W \tag{5.10}$$

where $P_H$ is the individual panel height, $\beta$ is the panel tilt angle relative to the horizontal, $D_P$ is the horizontal distance between the end of a panel in one row and the

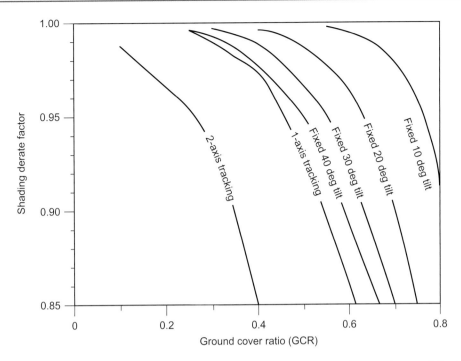

Figure 5.6 Shading derate (correction) factors for different ground cover ratios (CGRs), tracking options, and fixed tilt angles. Percent energy losses during the year = (1 − derate factor) × 100 percent. From NREL (2018).

beginning of a panel in the next row, and $P_W$ is the individual panel width.

The ratio of the **panel collector area** ($A_C = P_H P_W$) to total ground cover area per panel is called the **ground cover ratio (GCR = $A_C/A_G$)** of a solar field. A ground cover ratio of 1 occurs if the panels are flat on a surface ($\beta = 0$) and adjacent to each other ($D_P = 0$). When panels are tilted ($\beta > 0$), panels must be spaced apart more ($D_P > 0$) to avoid shading losses, so the GCR decreases. Figure 5.6 indicates that even at a GCR of 0.4, a panel with a $\beta = 40°$ fixed tilt angle still loses 2 percent of its annual energy due to shading.

---

## Example 5.5 Ground cover ratio

Find the ground cover area of a field of solar panels, where the panel width is 1.2 m, the panel height is 0.6 m, the distance between panels is 0.76 m, and the fixed tilt angle is 30°. From the ground cover area and panel tilt, estimate the loss of solar energy during the year that still occurs due to shading.

### Solution:

From Equation 5.10, the ground area per panel is $A_G$ = (0.6 m × cos(30°) + 0.76 m) × 1.2 m = 1.536 m$^2$. The panel collector area is $A_C$ = 0.6 m × 1.2 m = 0.72 m$^2$. As such, the ground cover ratio is GCR = 0.72 m$^2$ / 1.536 m$^2$ = 0.47. For a tilt angle of 30° and a ground cover ratio of 0.47, Figure 5.6 indicates that this configuration of panels results in a derate factor for shading of 0.975, thus a loss of about 2.25 percent of solar energy during the year due to shading.

---

### 5.1.8 Solar Zenith Angles and Fluxes and How They Vary with Tilted or Tracked Solar Panels

PV panel output also depends on the current solar radiation flux reaching the panel. The radiation reaching a panel at a given time of day and year depends on both the direct and diffuse solar radiation flux through the atmosphere, the tilt angle of the panel, the orientation of the panel relative to the direct solar beam, and the solar zenith angle. The radiation

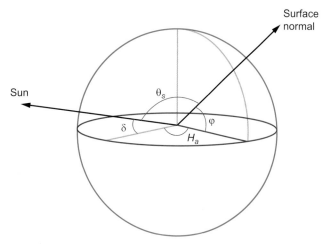

**Figure 5.7** Geometry for zenith angle calculations on a sphere. The surface normal is a line perpendicular to the surface of the Earth at the point of interest, $\varphi$ is the latitude of the point of interest, $\theta_s$ is the solar zenith angle between a direct line to the sun from the center of the Earth and the surface normal, $\delta$ is the declination angle of the sun, and $H_a$ is the hour angle of the sun.

flux itself also depends on the solar zenith angle. In this section, the solar zenith angle, direct solar radiation, and diffuse solar radiation are discussed.

### 5.1.8.1 Solar Zenith Angle

The **solar zenith angle** ($\theta_s$) was defined in Section 5.1.6 as the angle between a line from the center of the Earth to the sun and a line from the center of the Earth through a point of interest on the surface of the Earth (thus perpendicular to the surface of the Earth) (Figure 5.7). A solar zenith angle of $0°$ indicates the sun is in a line perpendicular to the surface of the Earth at the point of interest. A zenith angle of $90°$ indicates the sun is along the horizon, thus parallel to the surface of the Earth at the point of interest.

The **solar zenith angle in a vacuum** (ignoring refraction of light by air for now) relative to a solar panel laid flat (horizontally) on the surface of the Earth is

$$\cos\theta_s = \sin\varphi\,\sin\delta + \cos\varphi\,\cos\delta\,\cos H_a \qquad (5.11)$$

where $\varphi$ is the latitude of the panel, $\delta$ is the solar declination angle, and $H_a$ is the local hour angle of the sun, as

illustrated in Figure 5.7. The **declination angle** is the angle between the Earth's equator and the north or south latitude of the **subsolar point**, which is the point at which the sun is directly overhead. The **local hour angle** is the angle, measured westward, between the longitude (meridian) of the subsolar point and the longitude of a location of interest.

The solar declination angle is determined with

$$\delta = \arcsin(\,\sin\varepsilon_{ob}\sin\lambda_{ec}) \qquad (5.12)$$

where $\varepsilon_{ob}$ is the obliquity of the ecliptic and $\lambda_{ec}$ is the ecliptic longitude of the sun. The **obliquity of the ecliptic** is the angle between the plane of the Earth's equator and the plane of the Earth's orbit around the sun, the latter of which is called the **ecliptic plane**. The obliquity of the ecliptic is equal to the **Earth's tilt**, which is the angular distance between Earth's true polar north and a line through the Earth perpendicular to the sun. If the Earth had no tilt (no obliquity of the ecliptic), the sun would illuminate the North and South Poles simultaneously all year, and the Earth would have no seasons. The obliquity of the ecliptic changes slightly each day with

$$\varepsilon_{ob} = 23°.439\text{-}0°.0000004N_{JD} \qquad (5.13)$$

(NAO, 1993), where $N_{JD}$ is the number of days (including leap days) since 12 PM Greenwich Mean Time (GMT) on January 1, 2000. It is calculated as

$$N_{JD} = 364.5 + 365(Y - 2001) + D_L + D_J \qquad (5.14)$$

where $Y$ is the year of interest, $D_J$ is the day of the current year, which varies from 1 on January 1 at 00:00 GMT to 365 (for non-leap years) or 366 (for leap years) on December 31 at 00:00 GMT, and $D_L$ is the number of leap days minus 1 between January 1, 2000 and the beginning of the current year. Leap days occur once every 4 years, starting in the year 2000 for this equation. So, for example, for $Y = 2000$, $D_L = -1$; for $Y = 2001$ to 2004, $D_L = 0$; for $Y = 2005$ to 2008, $D_L = 1$, etc. With these parameters, $N_{JD} = 1.0$ on January 2, 2000, at 12:00 GMT and 366.0 on January 1, 2001, at 12:00 GMT.

## Example 5.6  Calculating the number of Julian days

Calculate the number of Julian days from 12 PM GMT on January 1, 2000, to 12 PM GMT on January 1, 2020.

### Solution:
The number of leap days minus 1 from January 1, 2000, to January 1, 2020, is 4. In addition, from 00:00 GMT January 1, 2020, to 12:00 GMT January 1, 2020, $D_J = 1.5$. Thus, from Equation 5.14, the number of Julian days through January 1, 2020, is $N_{JD} = 364.5 + 365 (2020 - 2001) + 4 + 1.5 = 7{,}305$ days.

The **ecliptic longitude of the sun** (degrees) is the angular distance between a line from the sun to the Earth when the Earth is at its current position in its orbit around the ecliptic plane, and a reference line between the sun and the Earth. The reference line is fixed in time and space and is defined based on the Earth's position during the spring equinox of the Northern Hemisphere, when the Earth passes closest to the sun, on a specific date. The ecliptic longitude is calculated empirically as

$$\lambda_{ec} = L_M + 1°.915 \sin\left(g_M\right) + 0°.020 \sin\left(2g_M\right) \quad (5.15)$$

where

$$L_M = 280°.460 + 0°.9856474\, N_{JD} \quad (5.16)$$

is the mean longitude of the sun (degrees) and

$$g_M = 357°.528 + 0°.9856003\, N_{JD} \quad (5.17)$$

is the mean anomaly of the sun (degrees). The **mean longitude of the sun** is the angular distance at which the Earth can be found at the current moment relative to the reference direction along the ecliptic plane if the

Earth's orbit were circular, rather than elliptical, and if the Earth were moving at constant speed. The **mean anomaly of the sun** is the angular distance, as seen by the sun, of the Earth from its perihelion assuming the Earth's orbit were circular, and the Earth were moving at a constant speed. The Earth's **perihelion** is the point in the Earth's real, elliptical orbit at which the Earth is closest to the sun. The mean anomaly at the perihelion is 0°. The local hour angle (in radians) is

$$H_a = 2\pi t_s/86{,}400 \quad (5.18)$$

where $t_s$ is the number of seconds past local noon, and 86,400 is the number of seconds in a day.

At noon, when the sun is highest, the local hour angle is zero, and Equation 5.11 simplifies to $\cos\theta_z = \sin\varphi \sin\delta + \cos\varphi \cos\delta$. When the sun and a solar panel are both over the equator, the declination angle and latitude are zero, and Equation 5.11 simplifies to $\cos\theta_z = \cos H_a$. Finally, the sun reaches its maximum declination ($\hat{y}23.5°$) at the summer and winter solstices and its minimum declination (0°) at the vernal and autumnal equinoxes.

## Example 5.7  Calculating solar zenith angle

Calculate the solar zenith angle at 1 PM Pacific Standard Time (PST) on February 27, 2018, at a latitude of 35 °N.

### Solution:
The number of leap days minus 1 from January 1, 2000, to January 1, 2018, is 4. The time of 1 PM PST corresponds to 21:00 GMT. From 00:00 GMT January 1, 2018, to 21:00 GMT on February 27, 2018, $D_J = 57.875$ days. Thus, from Equation 5.14, the number of Julian days from 00:00 GMT January 1, 2000, until 21:00 GMT February 27, 2018, is $N_{JD} = 364.5 + 365 (2018 - 2001) + 4 + 57.875 = 6{,}631.375$ days.

From Equation 5.16, the longitude of the sun is $L_M = 6816.65753°$; from Equation 5.17, the mean anomaly of the sun is $g_M = 6893.41319°$; from Equation 5.15, the ecliptic longitude of the sun is $\lambda_{ec} = 6818.21433°$; from Equation 5.13, the obliquity of the ecliptic is $\varepsilon_{ob} = 23.4363475°$; and from Equation 5.12, the solar declination is $\delta = -8.4885508°$. Since the local time is one hour past solar noon, the hour angle from Equation 5.18 is $H_a = 15°$. Thus, from Equation 5.11, the zenith angle is

- $\theta_s = \arccos\left[\sin(35°)\sin(-8.4885508°) + \cos 35° \cos(-8.4885508°)\cos 15°\right] = 45.7°.$

The solar zenith angle in Equation 5.11 is the zenith angle determined as if the Earth had no atmosphere. In other words, it is the **zenith angle in a vacuum**. However, light that passes from a medium of lower density (e.g., space) to higher density (air) bends (**refracts**) toward the surface normal (a line drawn vertically up from the surface of the Earth). Refraction of light is most important as the sun reaches the horizon. In fact, refraction allows light to pass through to a viewer's eye even after the sun dips below the horizon. To account for the bending of sunlight as it enters the Earth's atmosphere, the zenith angle in a vacuum from Equation 5.11 is modified according to **Snell's law of refraction**, with

$$\theta_{s,air} = \arcsin\left(\sin\theta_s / r_{air}\right) \text{ for } \theta_s \le \pi/2 \text{ and}$$
$$\theta_{s,air} = \theta_s + \theta_{crit} - \pi/2 \text{ for } \theta_s > \pi/2 \quad (5.19)$$

where $r_{air} = 1.000278$ is the real part of the index of refraction of air at a solar wavelength of 550 nm and $\theta_{crit} = \arcsin(1/r_{air}) = 88.649°$ is the critical angle. The **critical angle** is the angle of refraction through a higher density medium (air in this case) that occurs when the angle of incidence through a lower density medium (space in this case) is $90°$. At an incident angle relative to the surface normal (a line perpendicular to the surface of the Earth) of $90°$, the sun is exactly along the horizon. However, due to refraction, the light bends $1.351°$ from the horizontal toward the surface, thus the sun's light appears as if it is coming from a zenith angle in air ($\theta_{s,air}$) of $88.649°$ from the surface normal. In sum, sunlight can hit a horizontal solar panel after the sun has dipped below the horizon, but only if $\cos\theta_{s,air}$ is greater than 0.

---

## Example 5.8 Zenith angle in air

Calculate the solar zenith angle in air if the zenith angle in a vacuum is $\theta_s = 0°$, $45°$, and $91°$.

### Solution:

At $\theta_s = 0°$, Equation 5.19 gives $\theta_{s,air} = 0°$, so no refraction occurs when the sun is directly overhead. At $\theta_s = 45°$, $\theta_{s,air} = 44.98°$. At $\theta_s = 91°$, the sun is below the horizon, but $\theta_{s,air} = 89.65°$, so light still hits a panel.

---

### 5.1.8.2 Current Solar Flux to Horizontal Panels

Once the solar zenith angle in air is known, it is used to estimate the total radiation reaching a solar panel. Total radiation includes diffuse radiation ($F_{diff}$) plus a portion of direct radiation ($F_{dir}$).

The **direct solar radiation flux** ($F_{dir}$, W/m$^2$) is the flux of solar beam radiation reaching a solar panel. The solar flux at the top of the atmosphere is attenuated by absorption and scattering by gases and particles along the solar beam. **Absorption** is the process by which solar radiation enters a gas or particle, and the light is converted to heat, which is re-radiated to the surrounding air. **Scattering** is the process by which gases or particles deflect radiation in a random direction, just like a tree deflects a rock thrown at it, in a random direction. Absorption and scattering both vary with radiation wavelength and are discussed in more detail in Section 5.3.6.

The **diffuse solar radiation flux** ($F_{diff}$, W/m$^2$) is solar radiation that is scattered out of the solar beam by gases and particles but then encounters additional gases and particles, which scatter the light to the solar panel. For example, if the sun is close to the horizon, the direct solar beam may intersect the underside of a cloud, which scatters the light. Some of that scattered light may then impinge on a solar panel.

Cloud particles, pollution particles, and gases all decrease direct solar radiation but increase diffuse radiation. On average, in the absence of clouds, the diffuse radiation flux to the ground is about 10 percent of the direct flux.

In sum, diffuse radiation is due to light scattering by gases and particles in all directions, including in the direction of a solar panel. The direct beam solar flux is directed at an angle, the solar zenith angle, relative to the surface normal, so only $\cos\theta_{s,air}F_{dir}$ hits a panel lying horizontally on the surface of the Earth or on a flat rooftop. As such, the total downward solar flux onto a panel lying horizontally is

$$F_{cur} = F_{diff} + \cos\theta_{s,air}F_{dir} \quad (5.20)$$

Values of $F_{diff}$ and $F_{dir}$ vary with wavelength of light, so Equation 5.20 is repeated for each wavelength interval in the solar spectrum, and $F_{cur}$ is summed among all wavelength intervals to get the final solar radiation reaching a panel. Section 5.3 gives a method of calculating the diffuse and direct components of solar radiation.

### 5.1.8.3 Current Solar Flux to Tilted or Tracked Panels

PV panels are rarely placed horizontally. They are usually either tilted at a fixed optimal tilt angle or they are placed on a structure to track the sun. Rooftop panels are all at a fixed tilt, whereas utility-scale panels are either at a fixed tilt or track the sun. Tilted panels generally, but not always, face the south, southwest, or southeast in the Northern Hemisphere or face the north, northwest, or northeast in the Southern Hemisphere.

Panels that track the sun can generally track either on one axis vertically, one axis horizontally, or two axes. **One-axis vertical tracked PV panels** are panels that face south or north and swivel vertically around a horizontal axis. **One-axis horizontal tracked PV panels** are tilted at an optimal tilt angel and swivel horizontally between east and west around a vertical axis. **Two-axis tracked PV panels** combine one-axis vertical and one-axis horizontal tracking capabilities to follow the sun perfectly during the day.

For each of these four situations (optimal tilting, one-axis vertical tracking, one-axis horizontal tracking, and two-axis tracking), the solar zenith angle in a vacuum must be redefined to be the angle between the direct solar beam and a line normal (perpendicular) to the panel face (rather than the surface of the Earth). Under such conditions, the solar zenith angle in a vacuum can be recalculated for each tilting or tracking case as follows:

Optimal tilting
$$\cos\theta_s = \sin\varphi\ \sin(\delta+\beta) + \cos\varphi\ \cos(\delta+\beta)\ \cos H_a \tag{5.21}$$

One-axis vertical tracking
$$\cos\theta_s = \sin^2\varphi + \cos^2\varphi\ \cos H_a \tag{5.22}$$

One-axis horizontal tracking
$$\cos\theta_s = \sin\varphi\ \sin(\delta+\beta) + \cos\varphi\ \cos(\delta+\beta) \tag{5.23}$$

Two-axis tracking
$$\cos\theta_s = \sin^2\varphi + \cos^2\varphi = 1 \tag{5.24}$$

(Jacobson and Jadhav, 2018), where $\beta$ is the optimal tilt angle for fixed tilt or one-axis horizontal tracked panels. The **optimal tilt angle** is the estimated tilt angle that gives the greatest annual average incident solar radiation on a panel. It depends not only on latitude but also on cloud cover, elevation, and air pollution levels (Section 5.1.8.4).

Once the solar zenith angle in a vacuum is calculated, the solar zenith angle in air is determined from Equation 5.18, and the result is used in Equation 5.20 to find the current solar flux normal to a panel. Equation 5.20 is solved for each wavelength interval of the solar spectrum, and the result is summed over all solar wavelength intervals.

---

### Example 5.9  Zenith angle in a vacuum for tilted and tracked panels

Calculate the solar zenith angle in a vacuum for the conditions in Example 5.7, but for (a) a tilted panel with a fixed south-facing tilt of $\beta = 36°$, (b) a one-axis vertical tracking panel, (c) a one-axis horizontal tracking panel, and (d) a two-axis tracking panel.

#### Solution:

In Example 5.7, the latitude was $\varphi = 35\ °N$, the solar declination was $\delta = -8.4885508°$, and the hour angle was $H_a = 15°$. Applying these values and the tilt angle to Equations 5.21 to 5.24, respectively, gives the zenith angles in a vacuum for cases (a) to (d) respectively as

Optimal tilting
$$\theta_s = \arccos[\sin35°\sin(-8.4885508°+36°) + \cos35°\cos(-8.4885508°+36°)\ \cos15°] = 14.8°$$
One-axis vertical tracking
$$\theta_s = \arccos[(\sin35°\ )^2 + (\cos35°\ )^2\ \cos15°] = 12.3°$$
One-axis horizontal tracking
$$\theta_s = \arccos[\sin35°\sin(-8.4885508°+36°) + \cos35°\cos(-8.4885508°+36°)] = 7.49°$$
Two-axis tracking
$$\theta_s = \arccos[(\sin35°\ )^2 + (\cos35°\ )^2] = 0°$$

The smaller the zenith angle, the more direct solar radiation hits a panel. As such, two-axis tracking resulted in the most direct solar radiation hitting the panel, followed by one-axis horizontal tracking, one-axis vertical tracking, and optimal tilting, respectively.

The ideal tracking or tilting option depends not only on the incident solar radiation relative to a horizontal surface but also on the land or roof area needed to avoid shading and the cost of tracking versus tilting. For example, two-axis tracking in a utility PV plant requires more land area to avoid shading panels behind the front row of panels than do one-axis tracking or optimal tilting, and two-axis equipment is more expensive than is one-axis equipment or optimal tilting.

Shading depends not only on panel tilting, but also on the height that panels are placed relative to each other. For example, panels that track the sun placed on a south-facing hillside will likely see less shading than will panels on uniformly elevated ground. Shading further depends on the number of panels placed on each single platform that tracks the sun.

In sum, the decision about the best type of tracking or tilting option ultimately depends not only on the incident radiation received normal to each panel, but also on the land or roof area required to avoid shading and on cost (e.g., Breyer, 2012).

### 5.1.8.4 Optimal Tilt Angles

Optimal tilt angles increase with increasing latitude from equator to pole in both hemispheres. The higher the latitude, the less the vertical component of direct sunlight reaches the ground. Thus, panels need to be tilted more and more toward the low-lying sun.

Figure 5.8 provides estimates of optical tilt angles versus latitude for one or more locations in each country of the world and a polynomial fit through the data. The figure also shows results from two linear estimates of optimal tilt angle versus latitude.

For most low- and mid-latitude data values, the linear estimates and the polynomial fit in Figure 5.8 track the individual optimal tilt angle data well. However, for high latitudes in the Northern Hemisphere, the linear fits diverge substantially for most, but not all, data. The reason is that the linear estimates ignore cloud cover and pollution, but in reality, heavy cloud cover and haze exist in many high-latitude countries and regions. Greater cloud cover results in lower optimal tilt angles because clouds scatter solar radiation **isotropically** (equally in all

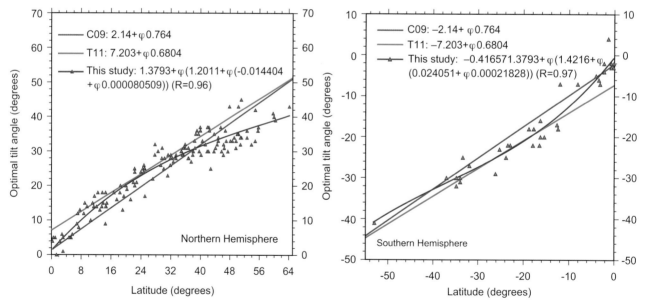

**Figure 5.8.** Estimated optimal tilt angles and 3rd-order polynomial fits through them of fixed-tilt solar collectors for one or more locations in all countries in the Northern Hemisphere and Southern Hemisphere. The values were obtained by running the PV Watts model (NREL, 2018) at each location in each country for different tilt angles until the tilt angle that gave the maximum PV power output over a year was obtained. Also shown are linear equations from Chang (2009, C09) and Talebizadeh et al. (2011, T11). To allow for rain to naturally clean panels, optimal tilt angles between −10 and +10 degrees latitude are usually limited to either −10 degrees (for negative values) or +10 degrees (for positive values).
From Jacobson and Jadhav (2018).

directions), so the closer a panel is to the horizontal under a cloudy sky, the more it receives diffuse solar radiation scattered by clouds above it. Further, clouds block the direct component of solar radiation, so the diffuse component becomes more important when clouds are present.

## Transition highlight

Figure 5.8 indicates that, at the same high latitude but different longitudes, optimal tilt angles can differ significantly from each other. For example, Calgary (51.12 $^{\circ}$N) has a higher optimal tilt angle (45$^{\circ}$) than does Beek, the Netherlands (34$^{\circ}$), which is at a similar latitude (50.92 $^{\circ}$N). The reason is that Calgary is exposed to less cloud cover so panels can take advantage of overhead sun more efficiently there. In Beek, panels are exposed to less direct sunlight due to heavier cloud cover so must take advantage of diffuse light scattered by clouds above them. The lower the optimal tilt angle, the more such panels can receive diffuse light under clouds.

### 5.1.8.5 Impacts of Tilting and Tracking versus Horizontal Panels

Figure 5.9 shows that the benefits of tilting and tracking solar PV panels relative to horizontal panels generally grow with increasing latitude, both north and south of the equator. It also shows that, in the global and annual average, two-axis tracked panels, one-axis horizontal tracked panels, one-axis vertical tracked panels, and optimally tilted panels receive about 1.39, 1.35, 1.22, and 1.19 times, respectively, the incident solar radiation that horizontal panels receive. As such, on average, two-axis tracked panels provide 39 percent more incident solar radiation than do horizontal panels and 17 percent more than do optimally tilted panels. However, these ratios vary substantially with latitude.

In addition, at virtually all latitudes, one-axis horizontal tracking panels receive within 1 to 3 percent the radiation as do two-axis tracked panels; thus, one-axis horizontal tracking panels appear more optimal than two-axis tracking panels. The main reasons are that one-axis-tracked panels require less land and cost less for the same output as two-axis panels. One-axis vertical tracking and optimal tilting are less beneficial than one-axis horizontal or two-axis tracking. One-axis horizontal tracking receives more incident solar than does one-axis vertical tracking below 65 $^{\circ}$N and S, whereas above 65 $^{\circ}$N and S, the incident solar is similar for both.

Above 75 $^{\circ}$N and 60 $^{\circ}$S, little added benefit can be found for any type of tracking relative to optimal tilting. Thus, the default recommendation for utility-scale PV is for one-axis horizontal tracking for all except the highest latitudes, where optimal tilting appears sufficient (Jacobson and Jadhav, 2018).

## Transition highlight

Where is the sunniest place on Earth from a solar panel's perspective? Figure 5.9 indicates that, over the Antarctic, horizontal panels receive relatively little radiation. However, optimally tilted and tracked panels actually receive more sunlight, averaged annually, than anywhere else on Earth. This is primarily due to the fact that the Antarctic receives sunlight 24 hours per day in the Southern Hemisphere summer. Also, much of the Antarctic is at a high altitude, thus above more air and clouds than is the Artic or other latitudes. Tilted and tracked panels over the Arctic receive more radiation than between 40 to 80 $^{\circ}$N, but less than over the Antarctic, due to the lower altitude and greater cloudiness above the surface of the Arctic.

## 5.2 Solar Resources

In this section, the spatial distribution of the world's solar resources is examined. The resource available is compared with world end-use power demand if all energy worldwide were electrified. A method of calculating radiation through the atmosphere is then provided.

Figure 5.10 shows the spatial distribution of downward solar radiation normal to the Earth's surface in the annual average. The "horizontal" panel curve in Figure 5.9a represents the result of zonally averaging the values in Figure 5.10. A **zonal average** is obtained by averaging values at each latitude over all longitudes. Both figures indicate that incident solar radiation peaks near the equator and generally decreases near the poles. However, a slight increase in surface solar radiation occurs at the poles relative to 60 $^{\circ}$N and 60 $^{\circ}$S due to lower cloud cover in polar regions. Similarly, in the United States, more solar radiation reaches the ground in the Southwest than in the Southeast due to the higher cloud cover in the southeast. The Sahara Desert also receives a substantial amount of radiation due to low cloud cover.

Worldwide, about 97,000 TW (4.57 kWh/m²/day) of solar power hits the Earth's surface in the annual average, accounting for clouds and particles in the air, but before any reflection by the ground surface is considered. Of this downward solar radiation, about 28.7 percent hits land (Table 5.3).

**Table 5.3** Annual average solar power hitting the Earth's surface, maximum annual average electricity output possible assuming no space between panels and a 20 percent panel efficiency, and practical solar potential over land and near shore close to population centers.

|  | Land | Ocean | Global |
|---|---|---|---|
| Solar power hitting ground (TW) | 27,800 | 69,200 | 97,000 |
| Maximum electricity (TW) | 5,600 | 13,840 | 19,400 |
| Practical electricity (TW) | 1,300 | 50 | 1,350 |

The surface area of the Earth is $5.092 \times 10^{14}$ m²; that of land is $1.489 \times 10^{14}$ m².

If the Earth's land and oceans were covered completely with horizontal panels with an efficiency of 20 percent (with no spacing between panels and ignoring any feedbacks of the panels to climate), the global electricity available from solar PV worldwide would be about 19,400 TW. That over land would be about 5,600 TW (Table 5.3). This is equivalent to 48.7 EWh/y of energy, where 1 EWh (exawatt-hour) = $10^6$ TWh = $10^{15}$ kWh.

If the world's all-purpose energy were electrified, the annually averaged end-use power demand in 2050 would be about 8.7 TW (Table 7.1). As such, enough solar theoretically exists over land alone to provide the world's 2050 end-use power 640 times over. However, many locations over land, such as Antarctica and the Himalaya mountains, are not easily accessible to transmission. In addition, panels require walking between rows. These factors reduce the practical solar installation potential.

However, the low solar radiation incident on horizontal panels in high northern and southern latitudes in Figure 5.10 is not a limiting factor for installing solar PV. The reason is that, as shown in Figures 5.11 and 5.9a, incident radiation on a panel is magnified by up to a factor of 2.6 if the panel tracks the sun. Figure 5.9a, in fact, shows that more solar radiation hits a tracking panel

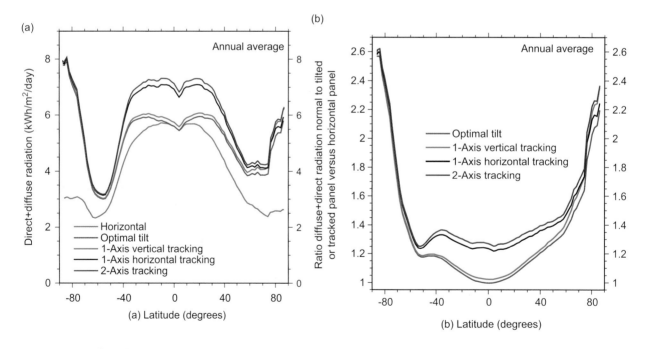

**Figure 5.9** (a) Zonally and annually averaged direct plus diffuse solar radiation received by horizontal panels, optimally tilted panels, and tracked panels; (b) zonally and annually averaged ratios of incident direct plus diffuse solar radiation normal to a tilted or tracked panel surface relative to flat (horizontal) panels. A zonally averaged value is a value averaged over all longitudes for each latitude. From Jacobson and Jadhav (2018).

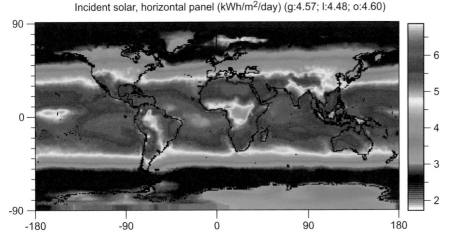

Incident solar, horizontal panel (kWh/m²/day) (g:4.57; l:4.48; o:4.60)

Figure 5.10 **Annually averaged downward solar radiation (W/m²) reaching the Earth's surface and normal to the surface. The global, land, and ocean averages correspond to 190.4, 186.7, and 191.7 W/m², respectively; 1 W/m² = 0.024 kWh/m²/ day. The letters g = global average; l = land average; o = ocean average.** From Jacobson and Jadhav (2018).

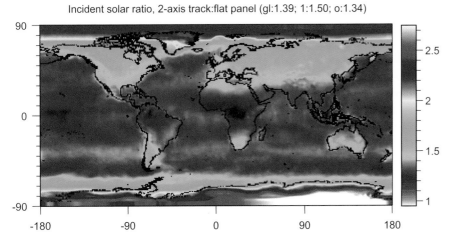

Incident solar ratio, 2-axis track:flat panel (gl:1.39; 1:1.50; o:1.34)

Figure 5.11 **Ratio of incident direct plus diffuse solar radiation normal to a two-axis tracking solar panel relative to that normal to a flat (horizontal) panel. The letters g = global average; l = land average; o = ocean average.** From Jacobson and Jadhav (2018).

over the Antarctic than anywhere else in the world in the annual average. As such, solar PV can be used to generate electric power effectively almost anywhere on land near where people live.

Solar PV can also be used effectively over inland waters (Figure 2.8), which are relatively calm. Ocean surfaces are generally more volatile, due to high waves, and more corrosive, due to salt. However, breakwaters and jetties can be constructed to separate rough from calm water, just as in a harbor. Since most people live near the coast, offshore solar may grow significantly in the future.

In sum, the solar resource is limited primarily by proximity to where people live and by the spacing required between tilted or tracked solar PV panels. The world's likely developable solar PV resource over land is about 1,300 TW (11.4 EWh/y) (Table 5.3). Thus, the ratio of accessible solar PV over land to all-purpose end-use demand is about 150. As such, solar PV over land and near shore alone could power the world many times over.

This does not include the additional power available from floating PV over coastal ocean waters.

## 5.3 Calculating Direct and Diffuse Fluxes of Solar Radiation

The direct and diffuse solar radiation fluxes in Equation 5.20 can be either measured with instruments or computed mathematically in an atmospheric computer model. This section discusses the calculation of such fluxes based on physical interactions of sunlight with the atmosphere.

### 5.3.1 Radiation Spectra

**Radiation** is the emission or propagation of energy in the form of a photon or an electromagnetic wave. A **photon** is a particle or quantum of energy that has no mass, no electric charge, and an indefinite lifetime. An **electromagnetic**

wave is a disturbance traveling through a medium, such as air or space, that transfers energy from one object to another without permanently displacing the medium itself.

Because radiative energy can be transferred even in a vacuum, it is not necessary for gas molecules to be present for radiation transfer to occur. Thus, such transfer can occur through space, where few gas molecules exist, or through the Earth's atmosphere, where many molecules exist.

All bodies in the universe that have a temperature above absolute zero (0 K) emit radiation. During emission, a body releases electromagnetic energy at different wavelengths, where a **wavelength** is the difference in distance between two adjacent peaks (or troughs) in a wave. The intensity of emission from a body varies with wavelength, temperature, and the efficiency of emission. Bodies that emit radiation with perfect efficiency are called blackbodies. A **blackbody** is a body that absorbs all radiation incident upon it. No incident radiation is reflected by a blackbody. No bodies are true blackbodies, although the Earth and the sun are close, as are black carbon, platinum black, and black gold. The term "blackbody" was coined because good absorbers of visible radiation generally appear black. However, good absorbers of infrared radiation are not necessarily black. One such absorber is white oil-based paint.

Bodies that absorb radiation incident upon them with perfect efficiency also emit radiation with perfect efficiency. The wavelength of peak emission intensity of a blackbody is inversely proportional to the absolute temperature of the body. This is **Wien's displacement law**, which was derived in 1893 by German physicist **Wilhelm Wien** (1864–1928). Wien's law states

$$\lambda_p(\mu m) \approx 2,897/T(K) \tag{5.25}$$

where $\lambda_p$ is the wavelength (in micrometers, μm) of peak blackbody emission, and $T$ is the absolute temperature (K) of the body.

---

## Example 5.10  Application of Wien's law

Calculate the peak wavelength of blackbody radiation for both the sun and the Earth assuming the effective radiating temperature of the sun is its photosphere temperature, 5,785 K, and the mean temperature of the Earth is 288 K.

### Solution:

From Equation 5.25, the peak wavelength of the sun's emissions is about 0.5 μm and that of the Earth's is about 10 μm.

---

At any wavelength, the intensity of radiative emission from an object increases with increasing temperature. Thus, hotter bodies (e.g., the sun) emit radiation more intensely than do colder bodies (e.g., the Earth). Figure 5.12 provides the radiation intensity versus wavelength for blackbodies at four temperatures. At 15 million K, a temperature at which nuclear fusion reactions occur in the sun's center, **gamma radiation** wavelengths ($10^{-8}$ to $10^{-4}$ μm) and **X radiation** wavelengths ($10^{-4}$ to 0.01 μm) are the wavelengths emitted with greatest intensity. At 6,000 K, **visible** wavelengths (0.38 to 0.75 μm) are the most intensely emitted wavelengths, although shorter **ultraviolet (UV)** wavelengths (0.01 to 0.38 μm) and longer **infrared (IR)** wavelengths (0.75 to 1,000 μm) are also emitted. At 300 K, infrared wavelengths are the wavelengths emitted with greatest intensity.

Figure 5.13 focuses on the 6,000 K and 300 K spectra in Figure 5.12. These are the radiation spectra of the sun's photosphere and of the Earth, respectively. The **solar spectrum** is divided into the UV, visible, and solar-IR spectra. The UV spectrum is divided further into the **far UV** (0.01 to 0.25 μm) and **near UV** (0.25 to 0.38 μm) spectra. The near UV spectrum is subdivided into **UV-A** (0.32 to 0.38 μm), **UV-B** (0.29 to 0.32 μm), and **UV-C** (0.25 to 0.29 μm) wavelength regions. The visible spectrum contains the colors of the rainbow. For simplicity, visible light is categorized as **blue** (0.38 to 0.5 μm), **green** (0.5 to 0.6 μm), or **red** (0.6 to 0.75 μm). Infrared wavelengths are partitioned into **solar-IR (near-IR)** (0.75 to 10 μm) and **thermal-IR (far-IR)** (4 to 1,000 μm) wavelengths. The intensity of the sun's emission is strongest in the visible spectrum, weaker in the solar-IR and UV

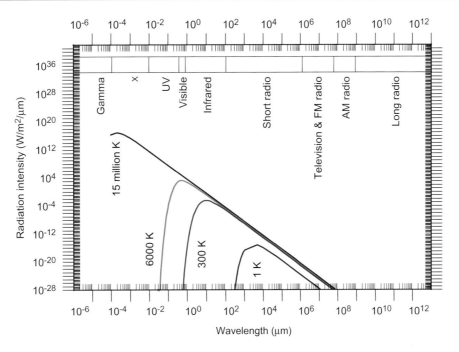

Figure 5.12 **Blackbody radiation emissions versus wavelength at four temperatures. Units are watts (joules of energy per second) per square meter of area per micrometer wavelength. The 15 million K spectrum represents emissions from the sun's center. Most of such emissions do not penetrate to the sun's exterior. The 6,000 K spectrum represents emissions from the sun's surface (photosphere) received at the top of the Earth's atmosphere (not at its surface). The 300 K spectrum represents emissions from the Earth's surface. The 1 K spectrum is almost the coldest temperature possible (0 K).**

spectra, and weakest in the thermal-IR spectrum. That of the Earth's emission is strongest in the thermal-IR spectrum.

Figures 5.12 and 5.13 provide wavelength dependencies of the intensity of radiation emissions of a body at a given temperature. Integrating intensity over all wavelengths (summing the area under any of the curves) gives the total intensity of emission of a body at a given temperature. This intensity is proportional to the fourth power of the object's kelvin temperature ($T$) and is given by the **Stefan–Boltzmann law**. This law was derived empirically in 1879 by Austrian physicist **Josef Stefan** (1835–1893) and theoretically in 1889 by Austrian physicist **Ludwig Boltzmann** (1844–1906). The law states

$$F_b = \varepsilon \sigma_B T^4 \tag{5.26}$$

where $F_b$ is the radiation intensity or flux (W/m$^2$), summed over all wavelengths, emitted by a body at absolute temperature $T$ (K), $\varepsilon$ is the emissivity of the body, and $\sigma_B = 5.67 \times 10^{-8}$ W/m$^2$/K$^4$ is the Stefan–Boltzmann constant. The **emissivity**, which ranges from

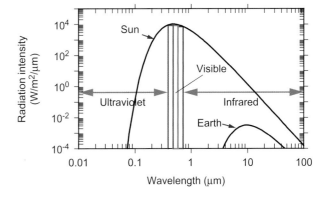

Figure 5.13 **Radiation spectrum as a function of wavelength for the sun's photosphere and the Earth when both are considered blackbodies. The sun's spectrum is received at the top of the Earth's atmosphere.**

0 to 1, is the efficiency at which a body emits radiation in comparison with the emissivity of a blackbody, which is unity. Soil has an emissivity of 0.88 to 0.99, and seawater has an emissivity of 0.95 to 0.99. All the curves in Figures 5.12 and 5.13 are emission spectra for blackbodies ($\varepsilon = 1$).

## Example 5.11 Application of the Stefan–Boltzmann law

How does doubling the kelvin temperature of a blackbody change the intensity of radiative emission of the body? What is the ratio of intensity of the sun's radiation compared with that of the Earth's?

### Solution:

From Equation 5.26, the doubling of the kelvin temperature of a body increases its intensity of radiative emission by a factor of 16. The temperature of the sun's photosphere (5,785 K) is about twenty times that of the Earth (288 K). Assuming both are blackbodies ($\varepsilon = 1$), the intensity of the sun's radiation (63.5 million W/m$^2$) is 163,000 times that of the Earth's (390 W/m$^2$).

### 5.3.2 Solar Radiation Reaching the Top of Earth's Atmosphere

The incident solar radiation at the top of the Earth's atmosphere originates from the photosphere of the sun. The **photosphere**, which is primarily gaseous, is a transition region between the sun's atmosphere and its interior. Beyond the photosphere, the sun's atmosphere consists of the chromosphere, the corona, and solar wind discharge. The photosphere is about 500 km thick. Temperatures range from about 6,400 K at its base to 4,000 K at its top, with an average of about 5,785 K. The photosphere is the source of most solar energy that reaches the planets, including the Earth. Although the sun's interior is much hotter than is its photosphere, most energy produced in its interior is confined there.

Above the photosphere lies the **chromosphere** ("color sphere"), which is a 2,500-km-thick region of hot gases. Temperatures at the base of the chromosphere are around 4,000 K. Those at the top are up to 1 million K. The name "chromosphere" arises because, at the high temperatures found in this region, hydrogen is energized and decays back to its ground state, emitting wavelengths of radiation in the visible part of the solar spectrum. For example, hydrogen decay results in radiation emission at 0.6563 μm, which is in the red part of the visible spectrum, giving the chromosphere a characteristic red coloration observed during solar eclipses.

The **corona** is the outer shell of the solar atmosphere and has an average temperature of about 1 to 2 million K. Because of the high temperature, all gases in the corona, particularly hydrogen and helium, are ionized. A low-concentration, steady stream of these ions as well as electrons escapes the corona and the sun's gravitational field and propagates through space, intercepting the planets with speeds ranging from 300 to 1,000 km/s. This stream is called the **solar wind**. The solar wind is the outer boundary of the corona and extends from the chromosphere to the outermost reaches of the solar system.

At the Earth, the solar wind temperature is about 200,000 K, and the number concentration of solar wind ions is a few to tens per cubic centimeter of space. As the solar wind approaches the Earth, the Earth's magnetic fields bend the path of the wind toward the poles. In the Earth's atmosphere above the poles, the ionized gases collide with air molecules, creating luminous bands of streaming, colored lights. In the Northern Hemisphere, these lights are called the **Northern Lights** or **Aurora Borealis** ("northern dawn" in Latin), and in the Southern Hemisphere, they are called the **Southern Lights** or **Aurora Australis** ("southern dawn"). Green or brownish-red colors are due to collisions of the solar wind with oxygen in the atmosphere. Red or blue colors are due to collisions with nitrogen.

The total power flux (W/m$^2$) from the sun's photosphere is obtained from **the Stefan–Boltzmann law**,

$$F_p = \varepsilon_p \sigma_B T_p^{\ 4} \tag{5.27}$$

where $T_p = 5,785$ K is the temperature and $\varepsilon_p = 1$ is the emissivity of the photosphere. The emissivity is unity, since stars are nearly perfect blackbodies. These values give the radiation power flux from the sun's photosphere as approximately 63.5 million W/m$^2$.

Multiplying the power flux from the photosphere by the spherical surface area of the photosphere, $4\pi R_p^{\ 2}$, where $R_p = 6.936 \times 10^8$ m (693,600 km) is the effective **radius of the sun** (the distance from the center of the sun to the top of the photosphere), gives the total power (W) emitted by the photosphere as $4\pi R_p^{\ 2} F_p$. Energy emitted

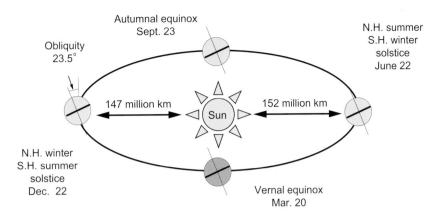

**Figure 5.14 Relationship between the sun and Earth at the times of solstices and equinoxes. The sun is positioned at one focus of the ellipse.**

from the photosphere propagates through space on the edge of an ever-expanding concentric sphere originating from the photosphere. Because conservation of energy requires that the total energy per unit time passing through a concentric sphere any distance from the photosphere equals that originally emitted by the spherical photosphere, the total energy per unit time passing through a sphere with a radius corresponding to the **Earth-sun distance** ($R_{es}$) must be

$$4\pi R_{es}^2 F_e = 4\pi R_p^2 F_p \qquad (5.28)$$

where $F_e$ is the total power flux (W/m$^2$) on a sphere with a radius corresponding to the Earth-sun distance. In other words, it is the **solar power flux at the top of the Earth's atmosphere**. Rearranging Equation 5.28 gives

$$F_e = F_p \left(\frac{R_p}{R_{es}}\right)^2 \qquad (5.29)$$

which indicates that the mean power flux from the sun decreases proportionally with the square of the distance away from the sun.

The Earth-sun distance varies daily. Figure 5.14 indicates that it is about 147.1 million km in December (Northern Hemisphere winter) and 152.1 million km in June (Northern Hemisphere summer) due to the fact that the Earth rotates around the sun in an elliptical orbit with the sun at one focus. If these distances are used in Equation 5.29, $F_e$ = 1,411 W/m$^2$ in December and 1,321 W/m$^2$ in June. Thus, a difference of 3.4 percent in Earth-sun distance between December and June corresponds to a difference of 6.9 percent in solar radiation reaching the Earth between these months. In other words, 6.9 percent

more solar radiation falls on the Earth in December than in June.

The average Earth-sun distance is $1.49598 \times 10^{11}$ m (149.598 million km), which is defined as **1 astronomical unit (AU)**. At this distance, the average solar power flux at the top of the Earth's atmosphere from Equation 5.29 is $F_s$ = 1,365 W/m$^2$, which is called the **solar constant**.

The solar constant is not really a constant, not only because it varies daily during the year but also because it varies yearly due to the sunspot cycle and due to the annual change in the Earth's orbit around the sun. A **sunspot** is a large magnetic storm that consists of a dark, cool central core, called an **umbra** and is surrounded by a ring of dark fibrils, called a **penumbra**. As a result of the magnetic activity associated with sunspots, regions near the umbra are hot, resulting in more net energy emitted by the sun when sunspots are present than when they are absent. Sunspot number and size peak every 11 years; however, because the sun's magnetic field reverses itself every 11 years, a complete **sunspot cycle** takes 22 years. The difference in solar intensity at the top of the Earth's atmosphere between times of sunspot maxima and minima is about 1.4 W/m$^2$, or only 0.1 percent of the solar constant.

The variation of Earth-sun distance can be calculated precisely each day of a year and for different years with

$$R_{es} = \left[1.0014 - 0.01671 \cos\left(g_M\right) - 0.00014 \cos\left(2g_M\right)\right] \\ \times 149.598 \text{ million km}$$

$$(5.30)$$

where $g_M$ is the mean anomaly of the sun, quantified in Equation 5.17. Equation 5.30 accounts for the change in the Earth's orbit around the sun each year.

## Example 5.12 Calculating solar flux at the top of the atmosphere

Calculate the Earth-sun distance and solar power flux at the top of Earth's atmosphere on February 27, 2018, at 1 PM Pacific Standard Time (PST).

### Solution:

From Example 5.7, the number of days since January 1, 2000, at 12:00 GMT to February 27, 2018, at 1 PM PST is $N_{JD}$ = 6,631.375 days, and the mean anomaly of the sun for this value of $N_{JD}$ is $g_M$ = 6893.41319°. Thus, from Equation 5.30, $R_{es}$ = 148.135 million km. Substituting this value into Equation 5.29 gives the solar power flux at the top of the Earth's atmosphere as $F_e$ = 1,392 W/m$^2$.

### 5.3.3 Angles on a Sphere

The solar power flux at the top of the Earth's atmosphere also varies with wavelength, as illustrated in Figure 5.13. The flux is further attenuated due to scattering and absorption by gases and aerosol particles as it passes through the atmosphere to the surface. The calculation of the solar flux reaching the surface requires the use of geometry. An important geometric parameter for this calculation is incremental solid angle.

An **incremental solid angle**, $d\Omega_a$, is the incremental surface area on a unit sphere. A unit sphere is a sphere with radius normalized to unity. Incremental solid angle has units of steradians (sr) and is analogous to an incremental arc length on a circle, which has units of radians. The equation for incremental solid angle is

$$d\Omega_a = \frac{dA_s}{r_s^2} d\Omega_a = \frac{dA_s}{r_s^2} = \sin\theta d\theta d\phi \qquad (5.31)$$

where $dA_s$ is an incremental surface area, and $r_s$ is the radius of a true sphere. The incremental surface area can be found from Figure 5.15. In the figure, a line is drawn from the center of the sphere to the center of an incremental area $dA_s$, which is a distance $r_s$ from the sphere's center. The line is directed at a zenith angle $\theta$ from the surface normal (where the surface is on the $x$-$y$ plane). The line is also located at an **azimuth angle** $\phi$, directed counterclockwise from the positive $x$-axis to a horizontal line dropped from the line. From the geometry shown, the incremental surface area is

$$dA_s = (r_s d\theta)(r_s \sin\theta d\phi) = r_s^2 \sin\theta d\theta d\phi \qquad (5.32)$$

where $d\theta$ and $d\phi$ are incremental zenith and azimuth angles, respectively. Substituting Equation 5.32 into Equation 5.31 gives the incremental solid angle as

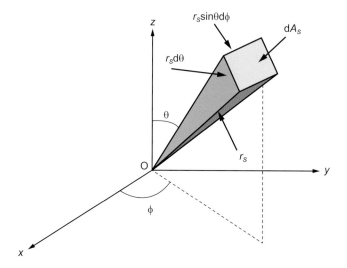

Figure 5.15 Radiance, emitted from point (O) on a horizontal plane, passes through an incremental area $dA_s$ at a distance $r_s$ from the point of emission. The angle between the $z$-axis and the angle of emission is the zenith angle ($\theta$). The horizontal angle between a reference axis ($x$-axis) and the line of emission is the azimuth angle ($\phi$). The size of the incremental surface area is exaggerated.

$$d\Omega_a = \sin\theta d\theta d\phi \qquad (5.33)$$

Integrating $d\Omega_a$ over all possible solid angles around the center of a sphere gives the solid angle around a sphere as

$$\Omega_a = \int_{\Omega_a} d\Omega_a = \int_0^{2\pi} \int_0^{\pi} \sin\theta d\theta d\phi = 4\pi \text{ steradians} \qquad (5.34)$$

### 5.3.4 Radiance and Irradiance

**Radiance (or spectral radiance)** ($I_\lambda$, W/m$^2$-μm-sr) is the energy emitted by a body per unit area per unit time (power) per unit wavelength per incremental solid angle (units of steradians). In Figure 5.15, imagine that the *x-y* plane is the surface of a blackbody object emitting radiation. If a pencil of radiation (radiance) travels through area $dA_s$, it passes through an incremental solid angle, $d\Omega_a$, which is through a cone originating at point O.

**Irradiance** (or **spectral irradiance** or **spectral power flux**) ($F_\lambda$, W/m²-μm) is the normal (perpendicular) component of spectral radiance hitting a flat plane, where the radiance originates from all directions (a hemisphere) above the plane. In terms of Figure 5.15, it is the normal component of radiance, integrated over the hemisphere above the *x-y* plane. For incoming radiance with a zenith angle of $\theta = 0°$, the irradiance impinging on the *x-y* plane is maximum. For a zenith angle of $\theta = 90°$, the irradiance impinging on the *x-y* plane is zero.

Spectral irradiance is calculated by integrating the component of radiance normal to the *x-y* plane over all solid angles of the hemisphere above the *x-y* plane. If $I_\lambda$ is the spectral radiance passing through point O in Figure 5.15, then $I_\lambda\cos\theta$ is the component of radiance normal to the *x-y* plane. Multiplying this quantity by incremental solid angle gives the **incremental spectral irradiance** (W/m²-μm) normal to the *x-y* plane as

$$dF_\lambda = I_\lambda \cos\theta d\Omega_a \tag{5.35}$$

Integrating Equation 5.35 over the hemisphere above the *x-y* plane in Figure 5.15 gives

$$F_\lambda = \int_{\Omega_a} dF_\lambda = \int_{\Omega_a} I_\lambda \cos\theta d\Omega_a$$
$$= \int_0^{2\pi}\int_0^{\pi/2} I_\lambda \cos\theta \sin\theta \, d\theta \, d\phi \tag{5.36}$$

For an isotropic emitter of radiation (a body that emits radiation with the same intensity in all directions), Equation 5.36 becomes

$$F_\lambda = I_\lambda \int_0^{2\pi}\int_0^{\pi/2} \cos\theta \sin\theta \, d\theta \, d\phi = \pi I_\lambda \tag{5.37}$$

Thus, the isotropic spectral irradiance equals $\pi$ multiplied by the isotropic spectral radiance, where units of $\pi$ are steradians. Figures 5.12 and 5.13 show spectral irradiance of blackbody radiation emissions as a function of wavelength and temperature. The fluxes in Equation 5.20 are also irradiances but summed over all wavelengths.

## 5.3.5 Optical Depth

The solar radiation reaching the Earth's surface is affected by the concentrations of gases and particles in its path and how much they scatter and/or absorb light. The extent of gas or particle attenuation of light in each wavelength interval is quantified by an extinction coefficient. An extinction coefficient is converted to an optical depth for radiation calculations.

An **extinction coefficient** measures the loss of electromagnetic radiation due to a specific process (gas scattering, gas absorption, particle scattering, or particle absorption) per unit distance though the atmosphere. Extinction coefficients, symbolized with σ, have units of inverse distance (e.g., cm⁻¹, m⁻¹, or km⁻¹) and vary with wavelength and location in the atmosphere.

An extinction coefficient as a function of wavelength λ is denoted by $\sigma_{s,g,\lambda}$ for gas scattering, $\sigma_{a,g,\lambda}$ for gas absorption, $\sigma_{s,p,\lambda}$ for particle scattering, and $\sigma_{a,p,\lambda}$ for particle absorption. Extinction coefficients are calculated from the number concentration of gas molecules or particles in the air (number of gas molecules or particles per cubic centimeter of air) multiplied by the cross-sectional area of each gas molecule or particle (cm² per gas molecule or per particle) and by an absorption or scattering efficiency (dimensionless), which varies with wavelength. The efficiencies are determined either in a laboratory or computationally from physical principles.

Since all types of gases scatter light relatively similarly, the extinction coefficient for gas scattering depends only on the number concentration of all gas molecules together, the cross-sectional area of an average molecule, and the efficiency of scattering at each wavelength.

Since different types of gases absorb differently from each other, the gas absorption extinction coefficient needs to be calculated separately for each type of gas that absorbs. The individual absorption extinction coefficients are then summed to obtain an overall gas absorption extinction coefficient. Whereas many gases absorb ultraviolet radiation, few absorb visible radiation. The main exceptions are nitrogen dioxide ($NO_2$), the nitrate radical ($NO_3$), and ozone ($O_3$). $NO_2$ absorbs primarily blue light; thus, it appears as a brown gas (because it transmits red and green, which together appear brown). $NO_3$ breaks down so quickly once the sun comes up that it can't be seen. $O_3$ weakly absorbs green light; thus, in high concentrations it appears purple (the combination of blue and red).

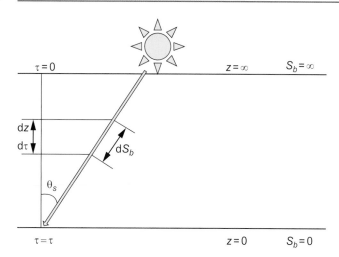

**Figure 5.16** Relationship among incremental optical depth (d$\tau$), incremental altitude (d$z$), solar zenith angle ($\theta_s$), and incremental distance along the solar beam (d$S_b$).

Since particles (both aerosol particles and cloud drop particles) have different compositions and sizes, particle scattering and absorption extinction coefficients need to be calculated separately for each particle type and each particle size. This can be done in a complex computer model that tracks particles as a function of composition and size. However, even the most complex calculation does not represent the atmosphere or particle shape exactly. The overall particle scattering and absorption extinction coefficients are obtained by summing individual extinction coefficients for each particle type and size.

Finally, the overall extinction coefficient for each wavelength and location in the atmosphere is calculated as

$$\sigma_\lambda = \sigma_{s,g,\lambda} + \sigma_{a,g,\lambda} + \sigma_{s,p,\lambda} + \sigma_{a,p,\lambda} \tag{5.38}$$

The next step in calculating solar radiative transfer through the atmosphere is to convert the overall extinction coefficient to an optical depth ($\tau$). **Optical depth** is the integral of the extinction coefficient from the top of the atmosphere down to a point of interest in the direction of the surface normal (a vertical line perpendicular to the surface of the Earth). Figure 5.16 illustrates the geometry for calculating the optical depth.

If the incremental distance along the solar beam is d$S_b$, then the vertical component of the incremental distance is

$$dz = \cos\theta_s dS_b = \mu_s dS_b \tag{5.39}$$

where $\theta_s$ is the solar zenith angle relative to the surface normal and $\mu_s = \cos\theta_s$. **Incremental optical depth** (dimensionless) is simply the product of the negative of the extinction coefficient and the incremental vertical distance,

$$d\tau_\lambda = -\sigma_\lambda dz = -\sigma_\lambda \mu_s dS_b \tag{5.40}$$

Incremental optical depth increases in the opposite direction from incremental altitude because optical depth is zero at the top of the atmosphere and maximum at the ground. Integrating Equation 5.40 from the top of the atmosphere ($z = S_b = \infty$) to any altitude $z$, which corresponds to a location $S_b$ along the beam of interest, gives the optical depth at altitude $z$ above a surface as

$$\tau_\lambda = \int_\infty^z \sigma_\lambda dz = \int_\infty^{S_b} \sigma_\lambda \mu_s dS_b \tag{5.41}$$

### 5.3.6 The Radiative Transfer Equation

The **radiative transfer equation** gives the change in spectral radiance or spectral irradiance along a beam of electromagnetic energy at a point in the atmosphere. The processes affecting radiation along a beam are scattering of radiation out of the beam, absorption of radiation along the beam, multiple scattering of indirect, diffuse radiation into the beam, and single scattering of direct solar radiation into the beam. **Single scattering** occurs when a photon of radiation is redirected into a beam after it collides with a particle or gas molecule, as shown in Figure 5.17. **Multiple scattering** occurs when a photon enters a beam after colliding sequentially with several particles or gas molecules, each of which redirects the photon. Solar radiation that has not yet been scattered is **direct radiation**. Radiation, either solar or infrared, that has been scattered is **diffuse radiation**.

The change in spectral radiance over the distance d$S_b$ along a beam is

$$dI_\lambda = -dI_{so,\lambda} - dI_{ao,\lambda} + dI_{si,\lambda} + dI_{Si,\lambda} \tag{5.42}$$

where

$$dI_{so,\lambda} = I_\lambda \sigma_{s,\lambda} dS_b \tag{5.43}$$

describes **scattering of radiation out of the beam**,

$$dI_{ao,\lambda} = I_\lambda \sigma_{a,\lambda} dS_b \tag{5.44}$$

describes **absorption of radiation along the beam**,

$$dI_{si,\lambda} = \left[ \sum_k \left( \frac{\sigma_{s,k,\lambda}}{4\pi} \int_0^{2\pi} \int_{-1}^1 I_{\lambda,\mu',\phi'} P_{s,k,\lambda,\mu,\mu',\phi,\phi'} d\mu' d\phi' \right) \right] dS_b \tag{5.45}$$

describes **multiple scattering of diffuse radiation into the beam**, and

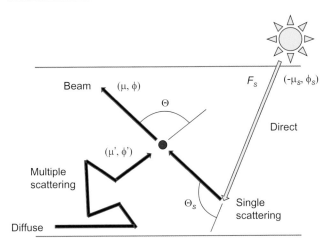

**Figure 5.17** Single scattering of direct solar radiation and multiple scattering of diffuse radiation add to the intensity along a beam of orientation $\mu$, $\phi$. The parameter $\mu = \cos\theta$ is always positive, but when a ray is directed upward, $+\mu$ is used, and when a ray is directed downward, $-\mu$ is used. Adapted from Liou (2002).

$$dI_{Si,\lambda} = \left[ \sum_k \left( \frac{\sigma_{s,k,\lambda}}{4\pi} P_{s,k,\lambda\mu,-\mu_s,\phi,\phi_s} \right) \right] F_{s,\lambda} e^{-\tau_\lambda/\mu_s} dS_b$$

(5.46)

describes **single scattering of direct solar radiation into the beam**.

In Equations 5.45 and 5.46, the summations are over all scattering processes ($k = g$ for gases and $p$ for particles), and the $P_s$ factors are scattering phase functions (Section 5.3.6). In the equations, $\sigma_{s,\lambda} = \sigma_{s,g,\lambda} + \sigma_{s,p,\lambda}$ and $\sigma_{a,\lambda} = \sigma_{a,g,\lambda} + \sigma_{a,g,\lambda}$ are extinction coefficients due to gas plus particle scattering and gas plus particle absorption, respectively.

## 5.3.7 Phase Function and Asymmetry Parameter

In Equation 5.44, the term

$$P_{s,k,\lambda,\mu,\mu',\phi,\phi'}$$

(5.47)

is the **scattering phase function for diffuse radiation**, which gives the angular frequency distribution of scattered energy as a function of direction. It relates how diffuse radiation, which has direction $\mu'$, $\phi'$, is redirected by gases or particles toward the beam of interest, which has direction $\mu$, $\phi$, as shown in Figure 5.17. In that figure, $\mu' = \cos\theta'$ and $\mu = \cos\theta$, where $\theta'$ and $\theta$ are the zenith angles of the diffuse radiation and the beam of interest, respectively. Similarly, $\phi'$ and $\phi$ are the azimuth angles of

the diffuse radiation and the beam of interest, respectively. The integral in Equation 5.45 is over all possible angles of incoming multiple-scattered radiation. Scattering phase functions vary with wavelength and differ for gases, aerosol particles, and cloud drops.

In Equation 5.46, the term

$$P_{s,k,\lambda,\mu,-\mu_s,\phi,\phi_s}$$

(5.48)

is the **scattering phase function for direct radiation**. The function relates how gases and/or particles direct solar radiation, with direction $-\mu_s$, $\phi_s$ (where $-\mu_s = -\cos\theta_{s,air}$), to direction $\mu$, $\phi$, as shown in Figure 5.17. This phase function is not integrated over all solid angles, since single-scattered radiation originates from only one angle.

The scattering phase function is defined such that

$$\frac{1}{4\pi} \int_{4\pi} P_{s,k,\lambda}(\Theta) d\Omega_a = 1$$

(5.49)

In this equation, $\Theta$ is the angle between directions $\mu'$, $\phi'$ and $\mu$, $\phi$, as shown in Figure 5.17. Thus,

$$P_{s,k,\lambda}(\Theta) = P_{s,k,\lambda,\mu,\mu',\phi,\phi'}$$

(5.50)

In Figure 5.17, $\Theta_s$ is the angle between the solar beam $-\mu_s$, $\phi_s$ and the beam of interest $\mu$, $\phi$. In that case,

$$P_{s,k,\lambda}(\Theta_s) = P_{s,k,\lambda,\mu,-\mu_s,\phi,\phi_s}$$

(5.51)

Substituting $d\Omega_a = \sin\theta d\theta d\phi$ from Equation 5.33 into Equation 5.49 gives

$$\frac{1}{4\pi} \int_0^{2\pi} \int_0^{\pi} P_{s,k,\lambda}(\Theta) \sin\Theta d\Theta d\phi = 1$$

(5.52)

The integral limits are defined so that the integration is over a full sphere. For **isotropic scattering** (equal scattering in all directions), the phase function is

$$P_{s,k,\lambda}(\Theta) = 1$$

(5.53)

which satisfies Equation 5.49. For **Rayleigh (gas) scattering**, the phase function is

$$P_{s,k,\lambda}(\Theta) = \frac{3}{4}\left(1 + \cos^2\Theta\right)$$

(5.54)

which also satisfies Equation 5.49. Figure 5.18 shows the scattering phase functions for isotropic and Rayleigh scattering. The phase function for isotropic scattering projects equally in all directions. Thus, light hitting an isotropic scatterer is scattered equally in all directions. The phase function for Rayleigh scattering is symmetric,

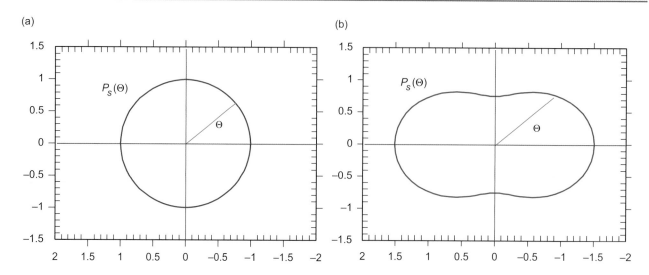

**Figure 5.18** Scattering phase functions distribution in polar coordinates for (a) isotropic and (b) Rayleigh scattering. The diagrams were generated from Equations 5.53 and 5.54, respectively. The phase functions give the frequency distribution of light scattering to different angles after it hits an isotropic scatterer or a gas molecule (for Rayleigh scattering).

but projects mostly in the forward and backward directions. Thus, light hitting a gas molecule is scattered symmetrically.

The **asymmetry parameter**, or first moment of the phase function, is a parameter derived from the phase function that gives the relative direction of light scattered by particles or gases. Its analytical form is

$$g_{a,k,\lambda} = \frac{1}{4\pi} \int_0^{2\pi} \int_0^{\pi} P_{s,k,\lambda}(\Theta) \cos\Theta \sin\Theta \, d\Theta \, d\phi \qquad (5.55)$$

The asymmetry parameter approaches +1 for scattering strongly peaked in the forward direction and –1 for scattering strongly peaked in the backward direction. If the asymmetry parameter is zero, such as with isotropic or Rayleigh scattering, scattering is equal in the forward and backward directions. Particle scattering (called Mie scattering) is strongly peaked in the forward direction. In sum,

$$g_{a,k,\lambda} \begin{cases} > 0 & \text{forward (Mie) scattering} \\ = 0 & \text{isotropic or Rayleigh scattering} \\ < 0 & \text{backward scattering} \end{cases}$$

$$(5.56)$$

In an atmospheric model, the asymmetry parameter for aerosol particles and cloud particles is obtained as a function of particle size and composition from a separate Mie-scattering algorithm (e.g., Toon and Ackerman, 1981). The calculation accounts for the ability of particles to scatter and absorb radiation. The primary **particle scattering processes** are reflection, refraction, and diffraction (Figure 5.19). Absorption also affects scattering.

**Reflection** is the process by which light bounces off a particle with an angle of incidence equal to the angle of reflection (Figure 5.19).

**Refraction** is the process by which light enters a particle from air and bends toward the surface normal due to the fact that the density of a particle exceeds that of air. Shorter wavelengths of light bend more than do longer wavelengths. As a result, white light that enters a liquid drop separates into the colors of the spectrum, with red bending the least, green bending a moderate amount, and blue bending the most. The separation of white light into colors due to refraction is called **dispersion**. Thus, refraction plus dispersion is referred to as **dispersive refraction**. Once inside a particle, light can either internally reflect one or more times, be absorbed, or exit the particle. When light exits a particle, it refracts away from the surface normal (Figure 5.19).

**Absorption** is the process by which electrons in absorbing components of a particle convert light radiation into internal energy, thereby raising the temperature of the particle. At higher temperatures, particles give off more thermal-infrared (heat) radiation to the air around them. As such, absorption converts light to heat, reducing the amount of light available in the particle. Only a few particle components absorb light radiation much. These include black carbon, brown carbon, iron, and aluminum.

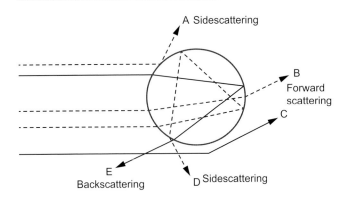

**Figure 5.19 Example of light scattering by a sphere. Ray A is reflected off the particle; Ray B is refracted into the particle and refracted out of the particle; Ray C is diffracted around the particle; Ray D is refracted into the particle, internally reflected twice, and then refracted out of the particle; and Ray E is refracted into the particle, reflected once internally, and then refracted out of the particle. Rays A through D scatter in the forward or sideward direction, whereas Ray E scatters in the backward direction.**

**Diffraction** is the bending of light as it passes by the edge of a particle. Some of the light that bends can bend into the particle. When it does so, the light refracts into the particle, then either internally reflects, is absorbed, or refracts out of the particle in a different direction than it came in. Other light bends around the particle (Figure 5.19). Since diffraction of a straight beam of light occurs on both sides of a particle, diffraction causes light to spread out on the forward side of a particle, contributing to the fact that particles scatter light primarily in the forward direction.

About 60 to 85 percent of light directed toward an aerosol particle scatters in the forward direction. The rest scatters to the side or the backward direction. For clouds, about 70 to 85 percent of light is scattered in the forward direction (Gerber et al., 2000).

A Mie-scattering algorithm calculates single-particle total scattering efficiencies ($Q_{s,i,\lambda}$), forward scattering efficiencies ($Q_{f,i,\lambda}$), and absorption efficiencies ($Q_{a,i,\lambda}$), where the subscript $i$ indicates particles of different size. Particles of different size each have a different composition.

A **single-particle scattering efficiency** is the ratio of the effective scattering cross section of a particle to its actual cross section. The scattering efficiency can exceed unity because a portion of the light diffracting around a particle can be intercepted and scattered by the particle.

Scattering efficiencies above one account for the additional scattering. The **single-particle forward scattering efficiency** is the efficiency with which a particle scatters light in the forward direction. It is always less than the single-particle scattering efficiency. A **single-particle absorption efficiency** is the ratio of the effective absorption cross section of a particle to its actual cross section. The absorption efficiency can exceed unity because a portion of the light diffracting around a particle can be intercepted and absorbed by the particle. Absorption efficiencies above one account for the additional absorption.

For a single particle of size $i$, the asymmetry parameter is simply the ratio of the forward scattering efficiency to the total scattering efficiency: $g_{a,i,\lambda} = Q_{f,i,\lambda} / Q_{s,i,\lambda}$. For the combination of all particles in a layer of air in the atmosphere, the asymmetry parameter $g_{a,p,\lambda}$ is simply the weighted sum among all particle sizes of forward scattering efficiencies divided by the weighted sum among all particle sizes of total scattering efficiencies. The weighting in both cases is the number concentration of particles of size $i$ (particles per cubic centimeter of air) multiplied by the cross-sectional area of each particle of size $i$ (square centimeters per particle).

### 5.3.8 Solutions to the Radiative Transfer Equation

Equations 5.42 to 5.46 can be combined to give the radiative transfer equation as

$$\frac{dI_{\lambda,\mu,\phi}}{dS_b} = -I_{\lambda,\mu,\phi}(\sigma_{s,\lambda} + \sigma_{a,\lambda})$$
$$+ \sum_k \left( \frac{\sigma_{s,k,\lambda}}{4\pi} \int_0^{2\pi} \int_{-1}^{1} I_{\lambda,\mu',\phi'} P_{s,k,\lambda,\mu,\mu',\phi,\phi'} d\mu' d\phi' \right)$$
$$+ F_{s,\lambda} e^{-\tau_\lambda/\mu_s} \sum_k \left( \frac{\sigma_{s,k,\lambda}}{4\pi} P_{s,k,\lambda,\mu,-\mu_s,\phi,\phi_s} \right)$$

(5.57)

The fraction of total extinction that is due to scattering is the **single-scattering albedo**,

$$\frac{\sigma_{s,\lambda}}{\sigma_\lambda} = \frac{\sigma_{s,g,\lambda} + \sigma_{s,p,\lambda}}{\sigma_{s,g,\lambda} + \sigma_{a,g,\lambda} + \sigma_{s,p,\lambda} + \sigma_{a,p,\lambda}} = \omega_{s,\lambda} \qquad (5.58)$$

Substituting Equation 5.58, $\sigma_\lambda = \sigma_{s,\lambda} + \sigma_{a,\lambda}$, and $dS_b = -d\tau_\lambda / (\sigma_\lambda \mu)$ from Equation 5.40 (but for a generic beam rather than the solar beam) into Equation 5.57 gives the radiative transfer equation as

$$\mu \frac{dI_{\lambda,\mu,\phi}}{d\tau_\lambda} = I_{\lambda,\mu,\phi} - J^{diffuse}_{\lambda,\mu,\phi} - J^{direct}_{\lambda,\mu,\phi} \tag{5.59}$$

where

$$J^{diffuse}_{\lambda,\mu,\phi} = \frac{1}{4\pi} \sum_k \left( \frac{\sigma_{s,k,\lambda}}{\sigma_\lambda} \int_0^{2\pi} \int_{-1}^1 I_{\lambda,\mu',\phi'} P_{s,k,\lambda,\mu,\mu',\phi,\phi'} \, d\mu' \, d\phi' \right) \tag{5.60}$$

$$J^{direct}_{\lambda,\mu,\phi} = \frac{1}{4\pi} F_{s,\lambda} e^{-\tau_\lambda/\mu_s} \sum_k \left( \frac{\sigma_{s,k,\lambda}}{\sigma_\lambda} P_{s,k,\lambda,\mu,-\mu_s,\phi,\phi_s} \right) \tag{5.61}$$

The radiative transfer equation must be solved numerically, except for idealized cases. For example, an analytical solution can be derived when absorption is considered but scattering is neglected. In this case, Equation 5.59 simplifies for downward directions to

$$-\mu \frac{dI_{\lambda,-\mu,\phi}}{d\tau_{a,\lambda}} = I_{\lambda,-\mu,\phi} \tag{5.62}$$

where $-\mu$ is used for downward radiation, and $\tau_{a,\lambda}$ is the optical depth due to absorption only. Integrating the downward equation from an optical depth of zero (at the top of the atmosphere) to the optical depth of interest, $\tau_{a,\lambda}$, gives the spectral radiance at the optical depth of interest relative to that at the top of the atmosphere as

$$I_{\lambda,-\mu,\phi}(\tau_{a,\lambda}) = I_{\lambda,-\mu,\phi}(0) e^{-\tau_{a,\lambda}/\mu} \tag{5.63}$$

which is **Beer's law**. Beer's law states that the absorption of radiation increases exponentially with increasing absorption optical depth.

When particle scattering is included in the radiative transfer equation, analytical solutions are not available and numerical solutions are needed. One numerical solution is found with the **two-stream method**. With this method, radiance is divided into an upward ($\uparrow$) and downward ($\downarrow$) component, each of which is approximated with a forward and a backward scattering term in the diffuse phase function integral of Equation 5.60. One approximation to the integral is

$$\frac{1}{4\pi} \int_0^{2\pi} \int_{-1}^1 I_{\lambda,\mu',\phi'} P_{s,k,\lambda,\mu,\mu',\phi,\phi'} \, d\mu' \, d\phi'$$

$$\approx \begin{cases} \dfrac{(1+g_{a,k,\lambda})}{2} I \uparrow + \dfrac{(1-g_{a,k,\lambda})}{2} I \downarrow & \text{upward} \\[2mm] \dfrac{(1+g_{a,k,\lambda})}{2} I \downarrow + \dfrac{(1-g_{a,k,\lambda})}{2} I \uparrow & \text{downward} \end{cases} \tag{5.64}$$

where $I\downarrow$ is the **downward radiance**, $I\uparrow$ is the **upward radiance**, and $(1+g_{a,k,\lambda})/2$ and $(1-g_{a,k,\lambda})/2$ are integrated fractions of the forward- and backward-scattered energy, respectively. Wavelength subscripts have been omitted on $I\uparrow$ and $I\downarrow$. Equation 5.64 is the **two-point quadrature** approximation to the phase function integral (Meador and Weaver, 1980; Liou, 2002). Substituting Equation 5.64 into Equation 5.60 gives

$$\frac{1}{4\pi} \sum_k \left( \frac{\sigma_{s,k,\lambda}}{\sigma_\lambda} \int_0^{2\pi} \int_{-1}^1 I_{\lambda,\mu',\phi'} P_{s,k,\lambda,\mu,\mu',\phi,\phi'} \, d\mu' \, d\phi' \right)$$

$$\approx \begin{cases} \omega_{s,\lambda}(1-b_\lambda)I\uparrow + \omega_{s,\lambda}b_\lambda I\downarrow \\ \omega_{s,\lambda}(1-b_\lambda)I\downarrow + \omega_{s,\lambda}b_\lambda I\uparrow \end{cases} \tag{5.65}$$

where $\omega_{s,\lambda}$ is the single-scattering albedo from Equation 5.58,

$$1 - b_\lambda = \frac{1+g_{a,\lambda}}{2} \qquad b_\lambda = \frac{1-g_{a,\lambda}}{2} \tag{5.66}$$

are effective integrated fractions of forward- and backward-scattered energy, and

$$g_{a,\lambda} = \frac{\sigma_{s,a,\lambda} g_{a,p,\lambda}}{\sigma_{s,g,\lambda} + \sigma_{s,p,\lambda}} \tag{5.67}$$

is an **effective asymmetry parameter** which is the extinction-coefficient-weighted sum of asymmetry parameters for gases and particles. Asymmetry parameters for gases are zero.

The **phase function for the single scattering of solar radiation**, used in Equation 5.61, can be estimated with the Eddington approximation of the solar phase function (Eddington, 1916),

$$P_{s,k,\lambda}(\Theta_s) \approx 1 \pm 3g_{a,\lambda}\mu_1\mu_s \tag{5.68}$$

where the diffusivity factor, $\mu_1$, $= 1/\sqrt{3}$ when the two-point quadrature approximation for diffuse radiation is used.

With the parameters above, the **spectral radiance** from Equation 5.59 can be written for solar wavelengths (where the emission term is neglected) in terms of an upward and a downward component as

$$\mu_1 \frac{dI\uparrow}{d\tau} = I\uparrow - \omega_s(1-b)I\uparrow - \omega_s b I\downarrow$$
$$- \frac{\omega_s}{4\pi}\left(1 - 3g_a\mu_1\mu_s\right)F_s e^{-\tau/\mu_s} \tag{5.69}$$

$$-\mu_1 \frac{dI\downarrow}{d\tau} = I\downarrow - \omega_s(1-b)I\downarrow - \omega_s b I\uparrow$$
$$- \frac{\omega_s}{4\pi}\left(1 + 3g_a\mu_1\mu_s\right)F_s e^{-\tau/\mu_s} \tag{5.70}$$

respectively (Liou, 2002), where all wavelength subscripts have been omitted. The term $-\mu_1$ is used for direct solar radiation on the right side of Equation 5.69 because solar radiation is downward relative to $/\uparrow$. The term $+\mu_1$ is used on the right side of Equation 5.70 for direct solar radiation since solar radiation is in the same direction as $/\downarrow$.

Equations 5.69 and 5.70 can be written in terms of **spectral irradiance** with the conversions

$$F\uparrow = 2\pi\mu_1 I\uparrow \quad F\downarrow = 2\pi\mu_1 I\downarrow \tag{5.71}$$

and generalized for the two-point quadrature, which is a **two-stream approximation** (one downward stream and one upward stream of radiation). The resulting equations are

$$\frac{dF\uparrow}{d\tau} = \gamma_1 F\uparrow - \gamma_2 F\downarrow - \gamma_3 \omega_s F_s e^{-\tau/\mu_s} \tag{5.72}$$

$$\frac{dF\downarrow}{d\tau} = -\gamma_1 F\downarrow + \gamma_2 F\uparrow + (1-\gamma_3)\omega_s F_s e^{-\tau/\mu_s} \tag{5.73}$$

(Meador and Weaver, 1980; Toon et al., 1989; Liou, 2002), where

$$\gamma_1 = \frac{1 - \omega_s(1 + g_a)/2}{\mu_1}$$

$$\gamma_2 = \frac{\omega_s(1 - g_a)}{2\mu_1} \quad \gamma_3 = \frac{1 - 3g_a\mu_1\mu_s}{2} \tag{5.74}$$

Equations 5.72 and 5.73 may be solved in a model after the equations are discretized, a vertical grid with multiple discrete layers is defined, and boundary conditions are defined. The derivatives can be discretized over such a grid with a finite-difference expansion of any order. A discretization requires boundary conditions at the ground and top of the atmosphere. At the ground, boundary conditions for irradiance at a given wavelength are

$$F\uparrow_B = A_e F\downarrow_B + A_e \mu_s F_s e^{-\tau_{N_L + 1/2}/\mu_s} \tag{5.75}$$

where $A_e$ is the surface albedo and the subscript $B$ indicates the bottom of the model. Equation 5.75 states that the upward irradiance at the surface equals the reflected downward diffuse irradiance plus the reflected downward direct solar irradiance. At the top (subscript $T$) of the model atmosphere, the boundary condition is $F\downarrow_T = \mu_s F_s$; thus, the source function is the solar flux multiplied by the cosine of the solar zenith angle.

Terms from the discretization of Equations 5.72 and 5.73 among all atmospheric layers are placed in a matrix. If a second-order central difference discretization is used, the matrix is tridiagonal and is solved without iteration to obtain the upward and downward fluxes at the boundary of each layer (e.g., Toon et al., 1989). The downward flux, summed over all solar wavelengths, has the form of Equation 5.20, $F_{cur} = F_{diff} + \cos\theta_{s,air} F_{dir}$.

When the atmosphere absorbs significantly, the two-stream approximations underestimate forward scattering because the expansion of the phase function is too simple to obtain a strong peak in the scattering efficiency. As a partial remedy, the effective asymmetry parameter, single-scattering albedo, and optical depth can be adjusted with

$$g_a' = \frac{g_a}{1 + g_a} \quad v(z) = v_{10}\frac{\ln\left(\frac{z}{z_0}\right)}{\ln\left(\frac{z_{10}}{z_0}\right)} \quad \tau' = \left(1 - \omega_s g_a^2\right)\tau$$

$$\tag{5.76}$$

respectively (e.g., Liou, 2002), which replace terms in Equations 5.59 to 5.61 and Equation 5.74.

## 5.4 Summary

This chapter discussed solar photovoltaics and solar radiation in depth. It described how PV cells work and how to determine their maximum efficiencies. It also described how an electric field is created and used to generate electricity in a PV cell. The chapter then touched upon materials used in PV cells. It next described PV panels and arrays, including the determination of panel efficiency. Panel output and its correction for a variety of factors were subsequently discussed. The next major subject was solar radiation. This coverage included a discussion of how to calculate solar zenith angle and the zenith angles for tilted and tracked PV panels. Solar resources worldwide were then quantified. Finally, a method of calculating direct and diffuse radiation with the radiative transfer equation was described. Such radiation is used to determine the energy available for solar PV panels.

## Further Reading

Boxwell, M., 2019. *Solar Electricity Handbook*. London: Greenstream Publishing, 186 pp.

Fthenakis, V., and M. Raugei, 2017. Environmental life-cycle assessment of photovoltaic systems. In *The Performance of Photovoltaic (PV) Systems: Modelling, Measurement, and Assessment*, ed. N. Pearsall. Duxford: Woodhead, pp. 209–232.

Masters, G., 2013. *Renewable and Efficient Electric Power Systems*, 2nd ed., Hoboken, NJ: Wiley, 712 pp.

Toon O. B., C. P. McKay, and T. P. Ackerman, 1989. Rapid calculation of radiative heating rates and photodissociation rates in inhomogeneous multiple scattering atmospheres, *J. Geophys. Res.*, **94**,16, 287–301.

## 5.5 Problems and Exercises

5.1. Explain why it is possible for one semiconductor material to have a lower band gap energy than another yet have a higher PV cell efficiency.

5.2. Assume a PV array consists of 3 strings wired in parallel, where each string consists of 12 panels wired in series. Suppose the voltage through each panel is 40 V and the current through each panel is 5 A. Calculate the total voltage, total current, and total power output of the array as a whole.

5.3. If the efficiency of a solar panel is 22 percent and the panel's surface area is 1.5 m$^2$, what is the nameplate capacity of the panel?

5.4. What is a solar panel's efficiency if its AC power output is 200 W, the solar flux normal to the panel is 700 W/m$^2$, the panel surface area is 1.6 m$^2$, the ambient air temperature is 32 °C, the wind speed is 1 m/s, $b_{ref}$ = 0.0025/K, and no other corrections aside from temperature are considered?

5.5. What is the horizontal distance between the end of a solar PV panel in one row and the beginning of a panel in a second row of a solar farm that gives a solar panel ground cover area per panel of 1.6 m$^2$? Assume the panel width is 1.3 m, the panel height is 0.7 m, and the tilt angle is fixed at 25°. From the ground cover area and panel tilt, estimate the loss of solar energy during the year that still occurs due to shading.

5.6. Calculate the number of Julian days from 12 PM GMT on January 1, 2000, to 12 PM GMT on March 20, 2002.

5.7. Calculate the solar zenith angle in a vacuum at 4 AM Pacific Standard Time (PST) on March 20, 2002, at a latitude of 65 °S.

5.8. Calculate the solar zenith angle in air if the zenith angle in a vacuum is $\theta_s$ = 82°.

5.9. Calculate the solar zenith angle in a vacuum for the conditions in Problem 5.6, but for (a) a tilted panel with a fixed north-facing tilt of $\beta$ = –60°, (b) a one-axis vertical tracking panel, (c) a one-axis horizontal tracking panel, and (d) a two-axis tracking panel.

5.10. Calculate the peak wavelength of emissions and the radiation flux from a blackbody whose temperature is 310 K. Assume the emissivity is 1. What is the percent change in radiation flux if the temperature increases to 320 K?

5.11. If the peak wavelength of radiation from an object is 8 μm, what is the radiation flux emitted by the object if the emissivity of the object is 0.98?

5.12. Calculate the Earth-sun distance and solar power flux at the top of the Earth's atmosphere on March 20, 2002, at 12 PM GMT.

5.13. Calculate the number of solar PV panels and the fractional area of the United States required to power a 100 percent battery-electric (BE) vehicle fleet with utility-scale PV power plants. Assume that the end-use energy required to run a BE vehicle fleet after the plug-to-wheel efficiency of each vehicle is accounted for is $E_v = 1.15 \times 10^{12}$ kWh/y (2017). Also make the following assumptions: the plug-to-wheel efficiency of an electric vehicle is $\eta_e = 0.85$, a solar panel's rated power is $P_s = 327$ W, the footprint and spacing areas of a solar panel are both $A_s = 4.0$ m$^2$, the panel's capacity factor is $CF_s = 0.21$, the transmission plus distribution efficiency of energy from a PV power plant is $\eta_s = 0.95$, and the area of the 50 U.S. states is $9.162 \times 10^6$ km$^2$. *Hint*: The single-panel annual energy output (kWh/y) is $E_s = P_s \times CF_s \times H \times \eta_s$, where $H$ is the number of hours in a year.

5.14. What are the three main components of light scattering by particles? What light attenuation process affects scattering?

5.15. What would cause light sent toward an airborne particle to backscatter?

5.16. What are the four main components of light extinction in the atmosphere?

5.17. Summarize the main radiative processes that the radiative transfer equation solves.

5.18. What type of scattering is an asymmetry parameter of 0 associated with? What about an asymmetry parameter of 0.7?

# 6 Onshore and Offshore Wind Energy

After solar, onshore and offshore wind have the potential to supply the greatest portion of the world's all-purpose energy demand. Not only are wind resources abundant in almost every country of the world, but the cost of onshore wind energy has also declined so much in recent years, that it is, in 2020, the least expensive form of new electric power in many countries of the world. The low cost has resulted in massive installations of wind to replace fossil-fuel power plants and to provide new energy demand.

This chapter discusses wind turbine history (Section 6.1), types (Section 6.2), parts (Section 6.3), mechanics (Section 6.4), generators (Section 6.5), and output (Section 6.6). It also discusses wind farm footprint and spacing areas (Section 6.7), wind physics and resources (Section 6.8), and the impacts of wind turbines on wind speed, temperatures, and hurricanes (Section 6.9). The discussion of output includes a section on calculating the efficiency of a wind turbine. The discussion of wind resources includes a section on the forces acting on air to form winds and a section on the land area needed for wind turbines. The discussion of impacts includes a section on why wind turbines reduce temperatures in the global average and a section on the relative impacts, versus other hazards, of wind turbines on birds and bats.

## 6.1 Brief History of Windmills and Wind Turbines

A **windmill** extracts kinetic energy from the wind and turns it into mechanical energy. A **wind turbine** converts the wind's kinetic energy into electricity.

The first known use of a circular wheel driven by the wind to provide mechanical energy was in the first century AD. The engineer **Heron of Alexandria** (*c.* AD 10 to *c.* 70), in Roman Egypt, invented such a windmill to provide mechanical power to a musical instrument, an organ. Windmills were also developed in the fourth century AD in India, Tibet, and China to rotate prayer wheels. Between the seventh and ninth century AD, the Persians, near the border of present-day Iran and Afghanistan, developed a windmill with 6 to 12 rectangular sails rotating horizontally around a **vertical axis** (an axis that is normal to the ground) to pump water and to grind grain. The use of windmills for these purposes spread across the Middle East, Central Asia, India, and China. By AD 1000, the Chinese and Sicilians were using windmills to pump seawater to make salt. Windmills rotating around a **horizontal axis** (parallel to the ground) were developed to grind flour in northwestern Europe beginning around AD 1180.

In the United States, a windmill was built on Cape Cod to pump seawater for making salt during the American Revolution. On August 29, 1854, the inventor **Daniel Halladay** (1826–1916) patented the first commercially viable windmill worldwide that he built in his Connecticut machine shop. The differences between his and previous windmills were that he added a tail fin (**wind vane**) to allow his windmill automatically to change direction to face the wind, and his turbine maintained a constant speed by changing the **pitch** (steepness of slope) of its sails without human oversight. Halladay's windmills were sold by the thousands and were used to pump water and grind grain on farms and ranches and to help open up the western United States to rail transport. Trains required water for their steam engines, and windmills were used to pump water along rail routes for this purpose. Between

1854 and 1970, over 6 million windmills were installed in the United States.

In July 1887, **Professor James Blyth** (1839–1906), an electrical engineer at Anderson's College, Glasgow, built and operated the world's first wind turbine. The turbine used cloth sails to turn a rotor. It was used to charge an energy storage device, and the stored power provided electricity for lights in his cottage.

Near the same time, between 1887 and 1888, **Charles F. Brush** (1849–1929) of Cleveland, Ohio, built a larger wind turbine with a unique design. It had 144 rotor blades, each 17 m in diameter, made of cedar wood. The blades were mounted on an 18-m tower and generated a peak of 12 kW of electric power that he used to charge 12 batteries and to provide electricity for 100 light bulbs, 3 arc lamps, and several motors in his mansion. The turbine and battery bank provided power for the mansion for 20 years.

In 1891, **Paul la Cour** of Denmark invented a wind turbine that he used to electrolyze water to produce hydrogen gas that he used to power hydrogen gas lamps.

Subsequently, between 1900 and 1940, hundreds of thousands of small wind turbines, with peak power outputs of 5 to 25 kW, were installed to produce electricity in rural areas of the United States. These areas lacked access to the electric grid.

Large-scale commercial wind turbine installations began in California. Between 1981 and 1986, the state installed 15,000 wind turbines in three locations – **Altamont Pass, Tehachapi Pass**, and **San Gorgonio Pass**. The turbines had a combined nameplate capacity exceeding 1 GW. From 1990 to 2000, the center of wind farm development, though, moved to Europe. During that period, 10 GW were installed in Europe, compared with 2.2 GW in the United States.

In 1991, the first offshore wind turbine, 225 kW in nameplate capacity, was installed in 25 m of water offshore of Sweden. Denmark installed the first offshore wind farm in 1992. It consisted of eleven 450-kW turbines in 2 to 4 m of water 3 km from shore.

Since 2000, onshore and offshore wind turbine development has taken off worldwide.

## 6.2 Types of Wind Turbines

Wind turbines that spin vertically around a horizontal axis are called **horizontal axis wind turbines (HAWTs)**. They are by far the most common type and can be mounted on tall towers, some with a hub height of up to 150 m. The **hub height** is the height where the horizontal axis, or **rotor**, of a turbine is located. Some HAWTs have low hub heights for use in backyards and in locations where wind turbine heights or nameplate capacities are restricted. Almost all HAWTs today are **3-blade turbines**, and all today face the wind (thus are **upwind turbines**).

In the past, some HAWTs faced away from the wind (**downwind turbines**). The advantage of a downwind turbine is that it naturally lets the wind rotate it horizontally to the opposite direction of the wind. The horizontal rotation of a horizontal axis wind turbine around its vertical axis is called **yaw control**. Yaw control on an upwind HAWT (to help it face the wind whenever the wind direction changes) requires a much more complex mechanical system. Wind tails, such as those used on the Halladay windmill, can only help with yaw control for small turbines.

The main problem with downwind turbines, though, is that they do not receive power when each blade passes by the tower, since the tower itself blocks the wind. Also, because the wind speed is first high, then zero, then high again as each blade passes the tower, the blades are subject to enormous **wind shear** (variation of wind speed with distance), which increases wear and tear on and failure of the blades. Because of these two problems (power loss and wear), no turbines built today are downwind turbines.

Another type of HAWT that was previously used but is no longer manufactured is the **2-blade turbine**. The advantages of the 2-blade over the 3-blade turbine are that the fewer blades a turbine has, the less expensive the turbine is to build and the less each blade's turbulence causes wear, tear, and efficiency loss on the next blade. The reduced turbulence in a 2-blade system also causes the blades to spin faster, decreasing generator costs. For these reasons, a 2-blade turbine would seem to be more efficient and less costly than a 3-blade turbine. However, 3-blade turbines absorb horizontal and vertical wind shear more evenly, thus require fewer repairs and less downtime than do 2-blade turbines. Finally, because of their lower spin rate and more even distribution of wind shear, 3-blade turbines are quieter. For those reasons, only 3-blade turbines are manufactured on a large scale today.

Wind turbines that spin horizontally around a vertical axis are **vertical axis wind turbines** (VAWTs). These turbines are usually short; thus, they can be used in urban regions or regions with height restrictions. An early VAWT was the **Darrieus turbine**, which had two or more curved blades attached at each end to the top and bottom of a vertical rotating shaft. This turbine looked like an eggbeater and was named after the French aeronautical engineer **Georges Jean Marie Darrieus** (1888–1979), who was granted a patent on it in 1927. Its main problem was that its blades fatigued easily. Darrieus' patent covered many other types of VAWTs, including an **H-rotor**, which has straight vertical blades connected horizontally (like the letter *H*). The center of the horizontal bar is connected to a vertical rotating shaft. Today, many variations of VAWT designs exist.

Advantages of VAWTs over HAWTs are that VAWTs do not need to rotate to track the wind, they operate better under turbulent conditions, they are easier to maintain because their gearbox is closer to the ground, they cost less to install because their tower is shorter, and they can be grouped more closely to each other in a wind farm. In addition, they can be deployed in places with legal height restrictions, such as near airports and in some urban areas, or where tall turbines are a disadvantage, such as military bases in combat zones.

The main disadvantages of VAWTs (and short HAWTs) are that, because their heights are limited, they are exposed to slower winds than are tall HAWTs. Since instantaneous wind power is proportional to the cube of the instantaneous wind speed, this means a lot less power for VAWTs, increasing their costs per unit energy significantly. A second issue is that VAWT blades can become fatigued easily due to the variation of forces acting on them during each rotation and because winds near the ground are more turbulent than are wind aloft. This can cause greater wear and tear and repair requirements for VAWTs than for tall HAWTs.

## 6.3 Wind Turbine Parts

A horizontal axis, upwind wind turbine consists of a tower, nacelle, rotor with blades, low-speed shaft, high-speed shaft, gearbox, generator, heat exchanger, controller, brake, pitch system, yaw drive, yaw motor, wind vane, and anemometer (Figure 6.1). The **nacelle** houses the components of the wind turbine, aside from the tower, rotor, blades, wind vane, and anemometer.

The kinetic energy in the wind turns the **blades**, which are connected to the **rotor**. The spinning rotor turns a low-speed shaft about 3 to 20 revolutions per minute (rpms) for large turbines and up to 400 rpms for residential turbines. A gearbox connects the low-speed shaft to a **high-speed shaft**, which rotates 750 to 3,600 rotations per minute, sufficient for a **generator** to convert rotational mechanical energy into electricity. **Direct-drive generators** can operate at low rotational speed, so they do not need a gearbox. Turbines with direct-drive generators are called **gearless wind turbines**. In both cases, the generator produces 50- or 60-Hz AC electricity (depending on country) although some generators produce DC electricity. A **heat exchanger** keeps the generator cool.

The **anemometer** measures wind speed and communicates the information to the controller. When the wind speed measured by the anemometer first increases above a low threshold, the **controller** starts the rotor up from rest. The controller also shuts the rotor off when the wind speed increases beyond a high threshold. The **pitch system** changes the steepness of the slope of (**pitches**) the blades to minimize their direct contact with the wind in order to control the rotor speed or to stop the rotor from spinning when the wind speed exceeds the high threshold or falls below the low threshold wind speed of the turbine. The **brake** stops the rotor in an emergency as a backup to the pitch system or as a parking brake during maintenance. The **wind vane** measures wind direction and communicates the information to the yaw drive. The **yaw drive** rotates the turbine horizontally around a vertical axis to face the wind. The **yaw motor** provides power to the yaw drive.

The primary materials needed for wind turbines include pre-stressed **concrete** (for towers), **steel** (for towers, nacelles, and rotors), **copper** (for generator coils and electricity conduction), **aluminum** (for nacelles), wood epoxy (for rotor blades), glass-fiber-reinforced plastic (GRP; for rotor blades), carbon-filament-reinforced plastic (CFRP; for rotor blades), and **neodymium** (for permanent magnets in generators).

The large-scale growth of wind power will not be constrained by limits in the quantities of these materials. The major components of concrete – gravel, sand, and limestone – are abundant, and concrete can be recycled and reused. The world does have somewhat limited reserves of economically recoverable iron ore (on the order of 100 to 200 years at current production rates), but the steel used to make towers, nacelles, and rotors for wind turbines is 100 percent recyclable. The production

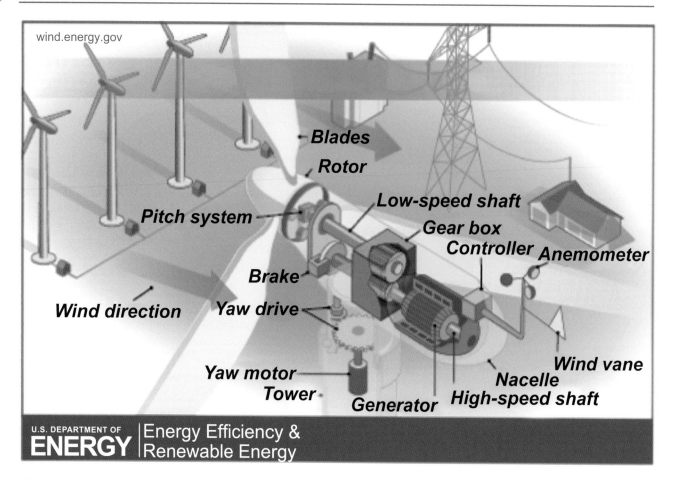

Figure 6.1 **Components of a wind turbine. The function of each component is described in the text.** From DOE (2019).

of millions of wind turbines would consume less than 10 percent of the world's low-cost copper reserves. Other conductors could also be used instead of copper. Section 6.5 quantifies neodymium worldwide availability.

on the top and higher pressure on the bottom of the airfoil creates an upward **lift force** perpendicular to the airflow.

## 6.4 Wind Turbine Mechanics

A wind turbine blade rotates when the wind flows around it. A blade is like an airfoil that is round (convex) on the top and flatter or more concave on the bottom (Figure 6.2). The main requirement of a wind turbine airfoil blade is that the distance from the front tip to the back tip must be greater on the top than on the bottom. Wind hitting the airfoil splits. Some of it flows over the top and the rest, under the bottom. Because the top distance is greater, air flowing over the top must travel faster than air flowing under the bottom in order for the two air parcels to meet at the back tip. Because air flowing over the top must travel further and faster, it spreads out more, causing air pressure above the foil to drop relative to air pressure under the foil. The lower pressure

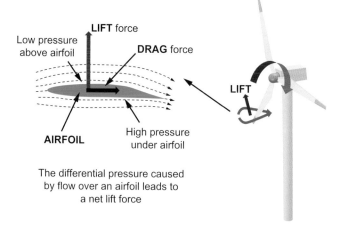

Figure 6.2 **Lift and drag forces acting on the cross section of a wind turbine blade when the wind flows around it. Also shown is the direction of lift in a wind turbine that rotates in the clockwise direction.**

The lift force causes the blade to move in the direction of the lift force, just as the lift force acts upward over an airplane wing to keep the airplane afloat against the force of gravity. Because a wind turbine is constrained to rotate in a circle, and the lift force always acts perpendicular to the direction of the air flow on the more rounded side of the turbine blade, the lift force accelerates the spin of the turbine in the clockwise direction so long as the more rounded top of the blade faces the clockwise direction (e.g., Figure 6.2). There is no advantage or disadvantage to turbines flowing clockwise versus counterclockwise, but all commercial turbines since 1978 have been built to spin clockwise. Previous to that, wind turbines and windmills mostly spun counterclockwise.

The lift force (N = kg-m/s$^2$) is quantified with

$$F_L = 0.5C_L \rho A_F v^2 \qquad (6.1)$$

where $C_L$ is the dimensionless lift coefficient, $\rho$ is air density (kg/m$^3$), $A_F$ is the one-sided surface area of the airfoil (m$^2$), and $v$ is the wind speed (m/s). The one-sided surface area is roughly the product of the airfoil (blade) width (distance between the front tip and back tip of the airfoil in Figure 6.2) and length. Thus, lift is proportional to the surface area of the airfoil, air density, and the square of the wind speed. Near the surface of the Earth, the air density averages around 1.225 kg/m$^3$ at 15 $^\circ$C.

For one particular wind turbine airfoil, the lift coefficient ranges from 0.3 at a 0$^\circ$ angle of attack to 1.4 at a 10$^\circ$ angle of attack (Pires et al., 2016). The **angle of attack** is the angle between the airflow and the airfoil. If the front tip and back tip of the airfoil are parallel to the wind, the angle of attack is 0$^\circ$. If the front tip is then lifted so that the wind slightly impinges on the underside of the airfoil, the angle of attack is positive. Angles of attack of wind turbines are usually less than 10$^\circ$.

In order to impart a lift force on the airfoil, the wind must give up some kinetic energy, converting it to mechanical energy. The removal of kinetic energy slows down the wind past the turbine.

The other force the wind imparts on a wind turbine is the **drag force**, which consists of two components. First, viscous friction on the surface of the airfoil slows down the wind, removing more kinetic energy from it. That energy from the wind is converted into an equal and opposite drag force of the wind pushing the airfoil slightly in the direction of the wind. Second, because the wind speed downstream of the turbine is slower than the wind speed upstream, the air pressure downstream is lower

than that upstream. This results in a pressure force acting from the higher pressure upstream to the lower pressure downstream part of the airfoil. This force is also a drag force that is added to the first drag force in the direction of the wind.

The overall drag force acts parallel to the direction of airflow, thus perpendicular to the lift force (Figure 6.2). The drag force (N) is

$$F_D = 0.5C_D \rho A_F v^2 \qquad (6.2)$$

where $C_D$ is the dimensionless coefficient of drag. For one particular wind turbine airfoil, the drag coefficient ranges from about 0.006 at zero angle of attack to 0.014 at a 10$^\circ$ angle of attack (Pires et al., 2016). Equation 6.2 indicates that the drag force is proportional to the one-sided surface area of the airfoil that the wind contacts, air density, and the square of the wind speed. Thus, doubling the wind speed quadruples the drag force. Also, drag increases on colder days because air is denser on colder days than on warmer days. Lift also increases with denser air.

Wind turbines operate best with high lift-to-drag-force ratios. Thus, they are built to maximize lift and minimize drag. This can be accomplished by increasing the angle of attack up to a point. For example, increasing the angle of attack from 0$^\circ$ to 5$^\circ$ increases the lift-to-drag ratio ($F_L$:$F_D$ = $C_L$:$C_D$) from about 50 to 123 for one particular turbine. However, increasing the angle of attack further to 10$^\circ$ decreases the ratio down to about 98 (Pires et al., 2016). Higher angles of attack decrease the ratio further.

The lift-to-drag ratio decreases at high angles of attack because turbulence builds up on the top of the airfoil under such angles, causing the air flowing over the top to no longer stick to the airfoil. This results in the air pressure above the top of the airfoil increasing, reducing lift, and slowing the spinning motion of the blade. This slowdown is referred to as **stall**. A benefit of stall is that, when wind speeds are above a threshold, the wind turbine controller and pitch system (Figure 6.1) can increase the angle of attack to increase stall until the blades slow down sufficiently or stop spinning. This method of reducing or stopping power output by a turbine is called **active stall control**.

Related to active stall control is pitch control. **Pitch control** is the computer-controlled change in the angle of attack of wind turbine blades to maximize power output for all wind speeds. With pitch control, if wind power output exceeds the maximum nameplate capacity of the generator, but the wind speed is still below the maximum allowable wind speed set by the manufacturer of the turbine, the controller and pitch system can reduce the angle of attack. This reduces the lift-to-drag ratio, reducing output without stopping the turbine from spinning or producing power. Reducing the angle of attack in this manner is called **feathering**.

Finally, a method of limiting the power output of a wind turbine to the maximum that the generator can handle without involving moving parts or electronics is passive stall control. With **passive stall control**, blades are designed so that they twist with increasing distance from the rotor so as to induce stall automatically when the wind speed exceeds the speed that generates the maximum allowable power. The turbine will continue to spin but not any faster than the speed that gives the maximum allowable power. The main problem with passive stall control is that it is difficult to design a turbine that ensures stall occurs under all conditions, so a safety margin is needed. As a result, passive stall control turbines do not operate under optimal conditions, and they are limited primarily to smaller turbines (Schubel and Crossley, 2012).

---

## Example 6.1  Calculating lift and drag forces on an airfoil

Calculate the lift force, drag force, and lift-to-drag force ratio on an airfoil if the wind speed is 10 m/s, the air density is 1.23 kg/m$^3$, the blade length is 60 m, the blade width is an average of 1 m, the drag coefficient is 0.01, and the lift coefficient is 1.0.

### Solution:

From Equation 6.1, $F_L = 0.5 \times 1 \times 1.23$ kg/m$^3$ $\times 60$ m $\times 1$ m $\times (10$ m/s$)^2 = 3{,}690$ N.
From Equation 6.2, $F_D = 0.5 \times 0.01 \times 1.23$ kg/m$^3$ $\times 60$ m $\times 1$ m $\times (10$ m/s$)^2 = 36.9$ N.
Thus, the ratio $F_L{:}F_D = 100$. The lift force is enough to suspend a weight of 376 kg against gravity (9.8 m/s$^2$).

---

## 6.5  Wind Turbine Generators

Wind turbine generators convert the rotational energy of a wind turbine's high-speed shaft into electricity. The more the wind rotates the blade, the more the generator can produce electricity, up to the nameplate capacity of the generator.

The three main categories of generators used to generate electricity in general are DC generators (called **dynamos**), AC synchronous generators, and AC asynchronous generators. Dynamos are used primarily in small wind turbines. **AC synchronous generators** are used in wind turbines where the high-speed shaft operates at a fixed speed. **AC asynchronous generators** are used in wind turbines where the high-speed shaft operates at a variable speed. Most generators in the world are synchronous generators used to produce electricity in coal, gas, nuclear, and other power plants that produce energy constantly. Most wind turbines use AC asynchronous generators due to the variable nature of wind energy production.

All generators for wind turbines in these three categories operate based on electromagnetic induction, thus are **induction generators**. Induction generators can also act as **motors** by operating in reverse, turning electricity into rotating motion. In fact, an induction generator acts as a motor to start up a wind turbine. Once the wind begins rotating the turbine blades, the generator returns to producing electric power as a generator.

The three main types of induction generators are wound field generators, permanent magnet generators, and squirrel-cage induction generators.

A 2-pole **wound field generator** consists of two sets of coils mounted on opposite sides of a rotating structure called a **rotor** (Figure 6.3), not to be confused with a wind turbine's rotor. A wind turbine's rotor is connected to a low-speed shaft, which is connected to a gearbox, which is connected to a high-speed shaft, which is connected to the generator's rotor (Figure 6.1). A 4-pole generator is similar, except that it has four sets of coils mounted 90° apart on the rotor. In both cases, the high-speed rotating shaft, operating at 750 to 3,600 rpms, turns the rotor when the wind is blowing. The rotor is surrounded by a stationary structure, called a **stator,** that has three sets of coils spaced 120° apart. The rotor first creates a magnetic field when a DC current passes through its two sets of

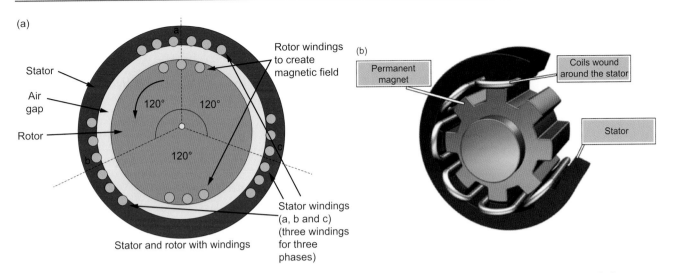

**Figure 6.3** (a) A 2-pole wound field generator. An external DC current passing through two sets of coils, which spin with the rotor, creates a magnetic field. Three sets of equally spaced coils on the stator see the magnetic field as fluctuating, thus the magnetic field induces an AC current through the stator coils. The current is sent to the grid. Adapted from Electrical Systems (2019). (b) Permanent magnet generator, as described in the text. Adapted from HSMag (2016).

wound coils. This current must be created by an outside power source. When the rotor starts spinning and the coils on the rotor pass the coils on the stator, the stationary stator coils see a changing magnetic field from the spinning rotor coils. The change in magnetic flux induces a 3-phase AC current in the stator coils. The faster the rotor spins, the greater will be the voltage generated across the coils. The AC current is then sent by wire to its destination.

A **permanent magnet generator** is similar. But in this case, magnets, instead of coils, are mounted around the rotor (Figure 6.3b). The magnets are separated so that, as the rotor spins at high speed, the stator coils see a changing magnetic field. The changing magnetic field induces an alternating current in the stator coils (Section 4.8). The AC current is sent to its destination. The advantage of a permanent magnet generator is that, unlike a wound field generator, it does not need an external DC power source to initiate a magnetic field. This is useful particularly if the wind farm is located far from an electric power grid. Reducing the external DC power requirement also reduces the need for batteries and capacitors. Further, magnets in a permanent magnet generator weigh less than the copper windings in a wound field generator that they replace.

On the flip side, controlling voltage is difficult with permanent magnets. Permanent magnets also don't operate well under high temperature and require greater cooling. Finally, most permanent magnets used today

contain neodymium, which is subject to price fluctuations and shortages due to the fact that it is mined in only a few places worldwide. Some concern also exists as to whether enough neodymium exists for permanent magnet generators to be used on a large scale in wind turbines. Aside from the fact that other types of asynchronous generators, including wound field generators, can be used instead of permanent magnet generators, the answer is yes.

Most permanent magnets are made of a neodymium-iron-boron alloy in the form of $Nd_2Fe_{14}B$. This magnet was developed first in 1982 and is the strongest permanent magnet available commercially. Although **neodymium** is abundant in the Earth's crust, it is not an isolated element in nature. Instead, it appears in ores, such as **monazite** and **bastnasite**, which are mineral group names. These ores contain small amounts of all rare-earth metals. **Rare-earth metals** are a group of 17 elements in the periodic table (15 lanthanides plus scandium and yttrium). The rare-earth metals tend to be found together. Although they are abundant in nature, they are concentrated enough for mining in only a few geographically dispersed locations and are rarely found in economically viable ore deposits.

Wind turbines with permanent magnet generators require about 650 kg of magnet per MW of wind nameplate capacity (Pavel et al., 2017). Since neodymium comprises 26.7 percent of $Nd_2Fe_{14}B$ by mass, about 173.4 kg-Nd (or 202.3 kg-$Nd_2O_3$) is required per MW

| Country or group of countries | Reserve base (Tg) |
| --- | --- |
| China | 16 |
| Commonwealth of Independent States | 3.8 |
| United States | 2.1 |
| Australia | 1 |
| India | 0.2 |
| Others | 4.1 |
| World | 27.3 |

Source: Jacobson and Delucchi (2011).

Tg = teragram.

of wind nameplate capacity. With an estimated 13 TW of installed wind power needed to provide 37.1 percent of the world's all-purpose end-use power demand in 2050 with wind (Jacobson et al., 2017), the amount of Nd$_2$O$_3$ needed among all wind turbines in 2050 is about 2.6 Tg. This represents only about 9.5 percent of the world's known reserve base, 27.3 Tg (Table 6.1).

A third type of induction generator is a **squirrel cage induction generator**. This consists of a fixed cylindrical cage-shaped rotor with several conducting copper or aluminum bars electrically connected together at their ends by aluminum end rings. Centered inside the rotor is the high-speed shaft, which spins the rotor. The rotor spins inside of a stationary stator that contains coiled wires powered by an external AC electricity source. The AC source sent through the coiled wires produces a magnetic field in the stator. This magnetic field appears to move relative to the rotating rotor. The changing magnetic field from the rotor's point of view induces a strong current in the cage bars, which, in turn, creates a magnetic field within the rotor. The rotor's magnetic field is dragged by the magnetic field of the stator, so it begins to rotate itself. Above a certain rpm speed, the rotor moves faster than the rotating magnetic field of the stator. This causes the stator to induce a strong current in the rotor. This AC current is sent to the grid.

All wind turbine generators produce 50-Hz (Europe and most countries) or 60-Hz (United States and other countries) 3-phase AC power. The power is sent to a step-up transformer so that the voltage can be increased from 10,000 V to 30,000 V for AC transmission. The AC

voltage may then be converted to DC voltage for extra long distance, as described in Section 4.7.

## 6.6 Power in the Wind and Wind Turbine Power Output

Like solar PV panels, individual wind turbines generate electricity **intermittently** during a year. In other words, their energy output fluctuates up and down over short timescales (sometimes seconds) as well as over longer timescales (hours, days, weeks, and months). Chapter 8 discusses how wind power intermittency is reduced and how wind is used to help balance power on the electric grid. Namely, Chapter 8 discusses aggregating geographically dispersed wind energy over large regions to reduce intermittency and combining wind with other renewables, storage, transmission, and demand response to match power demand with supply continuously.

In this section, wind turbine manufacturer parameters are first defined. Wind speed frequency distributions, the power in the wind, the maximum possible power output of a wind turbine (Betz limit), wind turbine capacity factors, and wind turbine energy and power outputs are then discussed.

### 6.6.1 Wind Turbine Power Curve

Figure 6.4 shows the **power curve** of a wind turbine with a 5-MW nameplate capacity. The nameplate capacity, or **rated power**, of a wind turbine is the maximum instantaneous power that the turbine can produce. The turbine's nameplate capacity is limited by its generator's nameplate capacity. Wind turbines produce power as a function of wind speed. Between zero wind speed and the **cut-in wind speed** (generally 2 to 3.5 m/s), a turbine produces no power because the power generated would be so low that it is uneconomical to produce. Above the cut-in wind speed, the instantaneous power output increases roughly proportionally to the cube of the wind speed.

A wind turbine's **rated wind speed** is the wind speed at which the turbine reaches its maximum power output, also called the **rated power**. The power output stays at the rated power for all wind speeds above the rated wind speed due to pitch control or passive stall control. However, wind turbines operating under realistic conditions have difficulty maintaining exactly constant power with increasing wind speed because of the time delay associated with pitch control and the

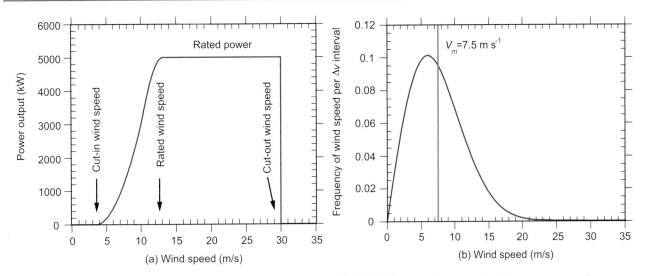

Figure 6.4 (a) Power curve for a Senvion wind turbine rated at 5,000 kW (5 MW) nameplate capacity. The curve shows the power output of the turbine as a function of instantaneous wind speed. The turbine rotor diameter is 126 m, and the hub height is 100 m above the surface. The cut-in wind speed is 3.5 m/s, the rated wind speed is 13 m/s, and the cut-out wind speed is 30 m/s. From Archer and Jacobson (2007). (b) Frequency distribution of wind speed per wind speed interval, f(v), for a Rayleigh distribution of wind speeds that have a mean of 7.5 m/s.

approximation of turbine blade design for passive stall control (Section 6.4).

With increasing wind speed above the rated wind speed, the power output remains roughly constant until the **cut-out wind speed** is reached, at which point the power output is reduced to zero. The shutoff is accomplished through pitch control or active stall control and is needed to prevent damage to the turbine. Most turbines, even when shut off, are designed to survive wind speeds up to a **destruction wind speed** of 50 m/s for up to 10 minutes.

### Transition highlight

A new class of turbines is being designed to withstand sustained 10-minute wind speeds of up to 57.5 m/s (Sirnivas et al., 2014). These **typhoon-class** wind turbines will be placed primarily offshore in locations prone to hurricanes or typhoons. For comparison, sustained 1-minute wind speeds in a Category 4 hurricane are 58.6 to 69.3 m/s.

Figure 6.4a can be used to illustrate how a wind turbine may be optimized for use at a slow or fast wind site. The figure shows that, if the mean wind speed at a given location is low, an ideal wind turbine would reach its rated power at a lower wind speed. This can be accomplished by increasing the blade diameter at constant generator size (constant rated power). The reason is that

the instantaneous power output of a wind turbine is proportional to the swept area of the turbine blade so increases with increasing blade diameter. Thus, increasing blade diameter at constant generator power allows the maximum power to be generated at a lower wind speed.

The figure also shows that increasing generator size (rated power) but keeping the blade diameter constant allows more of the power in fast winds to be captured, increasing overall power output at fast wind sites.

### 6.6.2 Rayleigh and Weibull Frequency Distributions

Determining a wind turbine's annual average power output requires characterizing how often the wind moves at different speeds. Figure 6.5 shows the frequency distribution of wind speed at two locations from data and a curve fit to the data. A **frequency distribution** (or probability distribution) gives the percentage or fraction of all wind speeds measured over a period of time (e.g., a year) per unit wind speed interval. If $v$ (m/s) is the average instantaneous wind speed in the interval $\Delta v = v_2 - v_1$, where $v_1$ is a low instantaneous wind speed and $v_2 = v_1 + \Delta v$ is a high instantaneous wind speed, and if $f(v)$ is the fractional occurrence (probability) of wind speed $v$ (m/s) during a time period (e.g., a year) per unit wind speed interval $\Delta v$, then

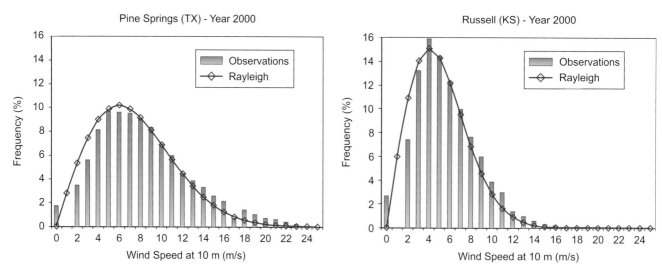

Figure 6.5 Frequency distribution of wind speed data in the annual average at two locations and a Rayleigh frequency distribution curve fit to the data. The wind speed measurements were taken at 10 m above ground level (AGL). From Archer and Jacobson (2003).

$$\sum_{v=0}^{\infty} f(v)\Delta v = 1 \qquad (6.3)$$

In other words, $f(v)\Delta v$ is the probability of wind speed $v$ occurring during a time period, and the sum among all wind speeds of frequencies at which a wind speed occurs must equal one. The curve fit shown in Figure 6.5 is that of a Rayleigh frequency distribution.

A **Rayleigh frequency distribution** of wind speed is a probability distribution of wind speed that looks similar to a bell curve but skewed toward higher wind speed, as in Figures 6.4b and 6.5. The Rayleigh frequency distribution is a specialized case of the **Weibull frequency distribution** of wind speed,

$$f(v) = \frac{k}{c}\left(\frac{v}{c}\right)^{k-1}\exp\left[-\left(\frac{v}{c}\right)^{k}\right] \qquad (6.4)$$

where $k$ is the shape parameter (an integer) and $c$ is a scale parameter that includes the term $V_m$, which is the mean wind speed (m/s) among all wind speeds in the probability distribution. The mean wind speed is determined by summing, among all wind speeds, the product of the wind speed and the fractional occurrence of the wind speed. Thus,

$$V_m = \sum_{v=0}^{\infty} vf(v)\Delta v \qquad (6.5)$$

For a Rayleigh distribution, $k = 2$, giving

$$f(v) = \frac{2v}{c^2}\exp\left[-\left(\frac{v}{c}\right)^{2}\right] \qquad (6.6)$$

Substituting Equation 6.6 into Equation 6.5 and solving for $c$ in the limit of tiny increments of $v$ gives $c = 2V_m/\sqrt{\pi}$. **Wind speed distributions are often Rayleigh in nature**, so measurements of a wind speed frequency distribution over a year often result in a Rayleigh distribution, as illustrated in Figures 6.4b and 6.5. Figure 6.4b shows the location of the mean wind speed in a Rayleigh frequency distribution. However, wind speed probability distributions can also be described as Weibull distributions with values of $k$ ranging from 1.1 to 3.4 (Hu and Cheng, 2007).

---

### Example 6.2 Mean wind speed

Calculate the mean wind speed if 15 percent of wind speeds are 3 m/s, 60 percent are 6 m/s and 25 percent are 9 m/s.

### Solution:

From Equation 6.5, the mean wind speed is $V_m = 3$ m/s $\times$ 0.15 + 6 m/s $\times$ 0.6 + 9 m/s $\times$ 0.25 = 6.3 m/s.

## 6.6.3 Power in the Wind

The wind at a given speed passing through the swept area of a wind turbine at a given instant in time contains a certain amount of power. Wind turbines can extract only a portion of this power. In this section, the instantaneous and average power in the wind are quantified.

The swept area (m²) of a wind turbine is simply

$$A_t = \frac{\pi D_t^2}{4} \tag{6.7}$$

where $D_t$ (m) is the **blade diameter** of the turbine (2 times the distance from the center of the rotor to the end of a single blade for 3-blade turbines). If a horizontal wind, with a speed $v$ (m/s), passes through the swept area $A_t$ (in the absence of a wind turbine) for a given time $t$ (s), then the distance (m) it travels is $x = vt$. Further, the mass (kg) of air that moves through that distance $x$ is

$$M_a = \rho A_t x \tag{6.8}$$

where $\rho$ is air density (kg/m³). Further, the **kinetic energy (J) in the wind** passing through that distance $x$ is

$$KE = \frac{1}{2} M_a v^2 = \frac{1}{2} \rho A_t x v^2 = \frac{1}{2} \rho A_t v^3 t \tag{6.9}$$

Because 1 W = 1 J/s, kinetic energy (J) can also be expressed in watt-seconds (Ws), or **watt-hours (Wh)** (by dividing Ws by 3600 s/h) or **kilowatt-hours (kWh)** (by dividing Wh by 1,000 W/kW).

Because power is energy per unit time, taking the time derivative of Equation 6.9 gives the **instantaneous power (W) in the wind** as

$$P_W = \frac{dKE}{dt} = \frac{1}{2} \rho A_t \frac{dx}{dt} v^2 = \frac{1}{2} \rho A_t v^3 \tag{6.10}$$

Equation 6.10 indicates that the power in the wind is linearly proportional to air density and proportional to the cube of instantaneous wind speed. Thus, doubling wind speed increases the instantaneous power in the wind by a factor of eight if all else is the same.

If the probability distribution of wind speeds over a period of time is Rayleigh in nature, then the mean power in the wind (W) over that period of time is derived as

$$P_m = \frac{1}{2} \rho A_t \sum_{v=0}^{\infty} f(v) v^3 = \frac{6}{\pi} \frac{1}{2} \rho A_t V_m^3 \tag{6.11}$$

Thus, the mean power in the wind that exhibits a Rayleigh distribution of wind speed is proportional to the cube of the mean wind speed of the distribution rather than of the instantaneous wind speed.

---

### Example 6.3 Energy and power in the wind

Calculate the kinetic energy passing through a 10-m² area for 10 hours if the wind speed is (a) 4 m/s and (b) 8 m/s, assuming the air density is 1.23 kg/m³. Determine the power in the wind in both cases.

### Solution:
From Equation 6.9, KE = 0.5 × 1.23 kg/m³ × 10 m² × (4 m/s × 10 h) × (4 m/s)² = 3,936 Wh in Case (a) and KE = 0.5 × 1.23 kg/m³ × 10 m² × (8 m/s × 10 h) × (8 m/s)² = 31,488 Wh in Case (b). The instantaneous power in the wind in both cases is $P_W$ = KE / 10 h = 393.6 W in Case (a) and 3,148 W in Case (b). Thus, doubling the wind speed increased the power in the wind by a factor of 8.

---

### 6.6.3.1 Impacts of the Variation of Day and Night Wind Speed with Altitude on Power in the Wind

Wind speeds are always zero at the Earth's surface or at the height of the lowest roughness element (e.g., water, soil, rocks, trees, buildings) above the surface. Wind speeds increase with increasing height above the surface logarithmically, on average (Figure 6.6). Wind turbine developers have an incentive to build taller and taller turbines to extract the additional kinetic energy of the wind at higher altitudes. Two standard equations for estimating the increase in wind speed with increasing altitude in the bottom few hundred meters of the atmosphere are the power law profile and the log law profile.

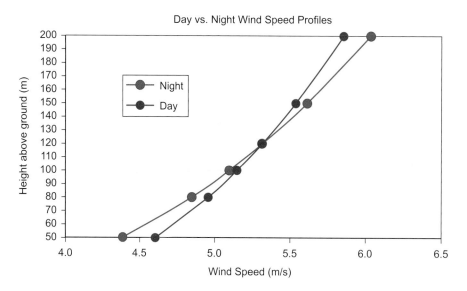

**Figure 6.6 Daytime versus nighttime variation of wind speed between 50 m and 200 m altitude above ground level, based on data from 446 sounding stations worldwide in the year 2000.** From Archer and Jacobson (2005).

The **power law profile** is

$$v(z) = v_{10} \left( \frac{z}{z_{10}} \right)^{\alpha} \tag{6.12}$$

where $v(z)$ is the wind speed (m/s) at altitude $z$ (m) above ground level (AGL), $v_{10}$ is the wind speed at $z_{10} = 10$ m altitude, and $\alpha$ is the **power law friction coefficient**. The power law friction coefficient varies with location, but a default value for open terrain is 1/7. As such, Equation 6.12 is often called the **1/7th power law profile**. Equation 6.12 gives a zero wind speed at $z = 0$, a wind speed of $v = v_{10}$ at $z = z_{10}$, and a wind speed larger than $v_{10}$ at higher altitudes. It is by no means an exact equation. It is merely an approximation to the vertical profile of wind speed in the lowest part of the atmosphere.

The **log law profile** is

$$v(z) = v_{10} \frac{\ln \left( \dfrac{z}{z_0} \right)}{\ln \left( \dfrac{z_{10}}{z_0} \right)} \tag{6.13}$$

where $z_0$ (m) is the **surface roughness length for momentum**. This is the height above the Earth's surface at which the log law profile of wind speed versus altitude extrapolates to zero wind speed. It also gives a measure of mechanical turbulence that occurs when a horizontal wind flows over a rough surface. Mechanical turbulence arises when objects in the path of a wind produce swirling motions of air, or **eddies**. The same swirling motion can be found by rolling a pencil between your hands. The lower hand represents the zero wind speed at the top of a stationary object. The hand rolling the pencil represents the non-zero wind speed in the air above the object. The rolling pencil represents the eddy. **Turbulence** is the combined effect of a group of eddies of different sizes. Turbulence mixes energy, pollutants, and winds primarily vertically. Strong winds produce strong eddies and turbulence. Turbulence from wind-generated eddies is called **mechanical turbulence**.

The larger the roughness length for momentum is, the greater the mechanical turbulence will be when wind passes over a roughness element. For a perfectly smooth surface, the roughness length is zero, and turbulence is zero. For smooth water (which occurs at low wind speed), the roughness length is about 0.0002 m. For cropland, it is about 0.1 m. For dense urban areas or forests, it can be about 1.6 m. In general, it can be approximated as 1/30th the height of the average roughness element protruding from the surface. Equation 6.13 gives a zero wind speed at $z = z_0$, a wind speed of $v = v_{10}$ at $z = z_{10}$, and a wind speed larger than $v_{10}$ at higher altitudes.

## Example 6.4  Power law and log law wind speed profiles

Calculate the wind speed at a 100-m hub height above ground level from the 1/7th power law profile and from the log law profile assuming a 10-m wind speed of 5 m/s and a roughness length of momentum of both 0.0002 m (smooth water) and 0.2 m (a rough surface).

## Solution:

From Equation 6.12, the power law profile gives $v(100 \text{ m}) = 5 \text{ m/s} \times (100 \text{ m} / 10 \text{ m})^{1/7} = 6.95$ m/s. From Equation 6.13, the log law profile for the smooth surface gives $v(100 \text{ m}) = 5 \text{ m/s} \times \ln(100 \text{ m} / 0.0002 \text{ m}) / \ln(10 \text{ m} / 0.0002 \text{ m}) = 6.06$ m/s. That for the rough surface gives $v(100 \text{ m}) = 7.94$ m/s. As such, the power law profile gives a 100-m wind speed between that of a smooth surface and rough surface from the log law profile.

Most modern-day wind turbine hub heights are between 80 m and 150 m altitude **above ground level (AGL)**. Figure 6.6 illustrates that, on average, if a wind turbine hub height is below about 120 m AGL, then the turbine will be exposed to faster daytime winds than nighttime winds. Because Figure 6.6 shows average values, it obscures the fact that the crossover point varies at each location worldwide, sometimes substantially.

The reason daytime wind speeds are faster near the surface than are nighttime wind speeds and the reverse occurs aloft is as follows: Wind speeds increase, on average, logarithmically from zero near the surface to some finite value aloft. During the day, the faster winds aloft mix down toward the surface, speeding up near-surface winds and slowing winds aloft. Vertical mixing of horizontal winds occurs during the day due to greater air instability caused by strong thermal turbulence during the day than at night. During the night, thermal turbulence shuts off, stabilizing the air, preventing the vertical mixing of fast horizontal winds aloft down to the surface.

A **stable atmosphere** means that air does not rise or sink easily. The air is strongly stable when a temperature inversion occurs. A **temperature inversion** occurs when warm air lies on top of cold air (air temperature increases with increasing altitude). At night, the ground cools radiatively, cooling the bottom layers of air, but layers higher up do not cool so much, so a temperature inversion forms. This type of inversion is called a **radiation inversion** or **nocturnal inversion**. This temperature inversion stabilizes the air (shutting off vertical turbulence), preventing air and its properties (including horizontal winds) from rising or sinking. The reason is that

cold air is denser than warm air. Thus, if warm, less dense air from aloft were pushed down into cold, more dense air below, the less-dense warm air would buoyantly rise back up to where it started, just as a plastic air-filled ball pushed down into water jumps back up to the surface of the water.

During the day, on the other hand, the sun warms the ground, which in turn warms the bottom layers of air. This creates an **unstable atmosphere** of hot air underneath colder air. When hot air sits under cold air, the hot air rises and the cold air sinks. The circular motion of this rising and sinking air is an eddy. Groups of eddies are turbulence. Turbulence due to ground heating is **thermal turbulence**. This thermal turbulence mixes not only pollutants but also horizontal wind vertically. Thus, during the day, thermal turbulence mixes fast horizontal winds from aloft down to the surface and slow winds at the surface, aloft. The net result is faster near-surface winds and slower winds aloft during the day than at night, as illustrated in Figure 6.6.

The results in Figure 6.6 are highly relevant to wind turbine power production. The price paid to energy producers for daytime energy often exceeds that for nighttime energy, particularly in locations with high summer air conditioning demand. When prices paid are higher during the day, it may be beneficial to have a turbine hub height below the crossover point in Figure 6.6. In other locations, such as those with high nighttime heating demands in high latitudes, the cost of nighttime electricity may exceed that of daytime electricity, and a taller turbine may be more cost effective. In both cases, the extra cost of a higher tower must be weighed against the extra benefit of higher wind speeds at higher altitudes and

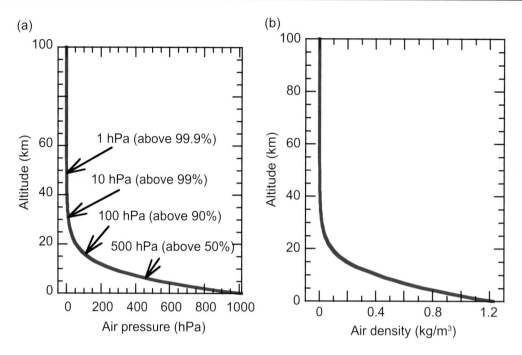

Figure 6.7 Average variation of (a) air pressure and (b) air density with increasing altitude above sea level. Under standard conditions, the sea level air pressure is 1013.25 hPa and the sea level air density is 1.23 kg/m³.

the difference in price paid for electricity during the night versus during the day.

### 6.6.3.2 Impacts of the Variation in Air Density and Pressure with Altitude on Power in the Wind

**Air pressure** is the summed weight of all gas molecules between a horizontal plane at a given altitude and the top of the atmosphere, divided by the area of the plane. Because weight is a force, air pressure is the force exerted by air molecules above a given altitude divided by the area over which the air lies. The more molecules that are present above a plane, the greater the air pressure. Because the weight of air per unit area above a given altitude is always larger than that above any higher altitude, air pressure decreases with increasing altitude. In fact, pressure decreases exponentially with increasing altitude (Figure 6.7). **Standard sea level pressure is** 1013.25 hPa (hectaPascal, where 1 hPa = 100 Pascal (Pa) = 1 millibar (mb) = 100 N/m² = 100 kg/m-s²). The sea level pressure at a given location and time typically differs by +10 to –20 hPa from standard sea level pressure. In a strong low-pressure system, such as at the center of a hurricane, the actual sea level pressure may be 50 to 100 hPa lower than standard sea level pressure.

Air density is the mass of air per unit volume of air. Thus, unlike air pressure, air density depends only on the local number of molecules and the volume they reside in. Like pressure, though, air density also decreases exponentially with increasing altitude (Figure 6.7). As such, the increase in the power in the wind with increasing altitude due to the increase in wind speed with increasing altitude is slightly offset by the decrease in power in the wind due to the decreasing air density with increasing altitude. Air density ($\rho$, kg/m³) is related to air pressure ($p_a$, hPa) and absolute temperature ($T$, K) by the **equation of state,**

$$\rho = \frac{p_a}{R_a T} \tag{6.14}$$

where

$$R_a = R'\,(1 + 0.608q_v) \tag{6.15}$$

is the gas constant for moist air (dry air plus water vapor). In this equation, $R'$ is the **gas constant for dry air** (2.8704 m³-hPa/kg-K), and $q_v$ is the **water vapor specific humidity** (kg-water vapor per kg of moist air) (Jacobson, 2005b). Dry air includes all gases in the atmosphere aside from water vapor. Figure 6.8 shows the variation in the annual and zonal (over all longitudes) average near-surface specific humidity (in units of g-$H_2O$/kg-moist

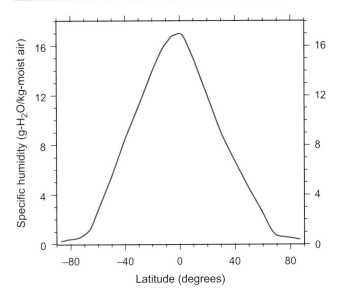

**Figure 6.8 Plot of the variation in the average near-surface specific humidity with latitude. Each value is an average over all longitudes and over the year.**

air) with latitude worldwide. Specific humidity is small near polar regions, but it reaches 0.017 kg/kg-moist air near the equator. As such, it has a small but non-zero impact on air density.

From Equation 6.14, doubling the absolute air temperature reduces air density by half. Similarly, a decrease in air pressure decreases air density. The equation also indicates that, at average atmospheric temperature and pressure ($T = 15\ °C = 288.15$ K and $p_a = 1013.25$ hPa), $\rho = 1.23$ kg/m$^3$.

Figure 6.7 illustrates that pressure itself decreases exponentially with increasing altitude. Over short altitude distances, such as between the surface and 300 m above the surface, the decrease in pressure with increasing altitude can be approximated as a linear decrease. A simple rule of thumb is that, for every 10 m of altitude increase, air pressure declines by about 1 hPa.

## Example 6.5  Air density and pressure variation with altitude

Estimate the air density under three conditions: (a) air temperature is 15 °C and air pressure is 1013.25 hPa, (b) air temperature is 30 °C and air pressure is 1013.25 hPa, and (c) the same as (a) but at an altitude 100 m higher than the altitude in (a). Assume the specific humidity is zero and temperature is isothermal (does not change) with increasing altitude.

### Solution:

Absolute temperature (K) equals Celsius temperature + 273.15. Thus, for (a), $T = 288.15$ K and air density $\rho = 1013.25$ hPa / (2.8704 m$^3$-hPa/kg-K × 288.15 K) = 1.225 kg/m$^3$ from Equation 6.14. For (b), $\rho = 1013.25$ hPa / (2.8704 m$^3$-hPa/kg-K × 303.15 K) = 1.164 kg/m$^3$. For (c), air pressure decreases 10 hPa with an altitude increase of 100 m; thus, air pressure is now 1003.25 hPa and air density is $\rho = 1003.25$ hPa / (2.8704 m$^3$-hPa/kg-K × 288.15 K) = 1.213 kg/m$^3$. Thus, the increase in air temperature of 15 °C had a greater impact on air density than the decrease in altitude of 100 m.

Temperatures in the lowest 300 m of the atmosphere vary substantially with altitude between day and night. As discussed in Section 6.6.3.1, temperature generally increases with increasing altitude at night and decreases with increasing altitude during the day near the ground. As such, there is no reliable rule about how a surface temperature can be extrapolated to a temperature at turbine hub height. However, because the average temperature change in the free troposphere above the boundary layer decreases 6.5 K or °C per km increase in altitude, this temperature change with altitude can be applied all the

way down to the surface from around noon to a few hours after sunset for a rough approximation. Thus, for example, if the temperature at the surface is 300 K, the temperature 100 m above the surface would be 299.35 K.

Example 6.6 illustrates that the change of wind speed and air density (which depends on pressure and temperature) with altitude results in a difference in the power in the wind at the highest versus lowest extent of a wind turbine's blades. This difference in power in the wind with height across a turbine can cause wear and tear on the turbine's blades over time.

## Example 6.6 Difference in power in the wind at the highest versus lowest extent of a wind turbine's blades

Estimate the power in the wind at the highest and lowest extent of a wind turbine blade if the turbine hub height is 100 m, the blade diameter is 126 m, the wind speed at 10 m AGL is 6 m/s, the surface pressure (at 0 m AGL) is 1,000 hPa, the surface temperature is 288 K, and the air temperature decreases 6.5 K per km increase in altitude. Assume a 1/7th power law wind speed profile, air pressure decreases 1 hPa for each 10-m altitude increase, and no water vapor in the air. Also, assume the power in the wind at the highest extent applies to the top half of the turbine swept area and that at the lowest extent applies to the bottom half.

## Solution:

The turbine swept area is $\pi \times (126 \text{ m} / 2)^2 = 12{,}469 \text{ m}^2$. Half (6,234 m$^2$) is applied to each upper and lower power in the wind.

The highest extent of the turbine blades is 100 m + 0.5 × 126 m AGL = 163 m; the lowest is 37 m AGL.
Air pressure at 163 m is 1,000 hPa – 163 m × 1 hPa / 10 m = 983.7 hPa; that at 37 m is 996.3 hPa.
The air temperature at 163 m is 288 K – 6.5 K/km × 0.163 km = 286.94 K; that at 37 m is 287.76 K.
From Equation 6.14, the density at 163 m is 1.194 kg/m$^3$; that at 37 m is 1.206 kg/m$^3$.
From Equation 6.12, the wind speed at 163 m is 8.94 m/s; that at 37 m is 7.23 m/s.
From Equation 6.10 the power in the wind at 163 m = 0.5 × 1.194 kg/m$^3$ × 6234 m$^2$ × (8.94 m/s)$^3$ = 2.66 MW; that at 37 m = 1.42 MW.

Thus, the power at 163 m is 87 percent higher than at 37 m due to the faster winds at 163 m, despite the lower density at 163 m.

### 6.6.4 Betz Limit

The maximum percent of power in the wind that can be extracted by a wind turbine, regardless of its design, is 59.3 percent. This is **Betz's law**. The German physicist **Albert Betz** (1885–1968) derived this law in 1919 by combining the principles of conservation of mass and momentum through a theoretical disk that extracts energy from the air.

The basic idea is that, because a wind turbine extracts kinetic energy from it, the wind on the downwind side of a wind turbine has a lower speed and air pressure than on the upstream side of the turbine. The pressure difference causes air passing through the turbine to spread out, expanding in volume. If the turbine extracted all the kinetic energy, air would stop behind the turbine, preventing any more wind from passing through the blades. If the wind speed downwind of the turbine equaled the wind speed upwind of the turbine, then no kinetic energy would be extracted. Betz's law states that the wind speed downwind of a turbine can be no less than one-third that upstream of the turbine, and the resulting power extraction can be no more than 59.3 percent of the power in the wind.

The Betz limit is derived assuming that the mass flow rate of air ($dM_a/dt$) upstream, downstream, and at the blade of a wind turbine must remain constant. Assume $v$ is the upstream wind speed, $v_d$ is the downstream wind speed, $v_b = (v + v_d)/2$ is the wind speed at the blade, $P_u$ is the power in the wind upstream, $P_d$ is the power in the wind downstream, and $P_b = P_u - P_b$ is the power extracted by the wind turbine blades. Since the mass flow rate is constant everywhere along the flow, it can be determined at the blade as

$$\frac{dM_a}{dt} = \rho A_t v_b = \rho A_t \frac{(v + v_d)}{2} \qquad (6.16)$$

From Equations 6.9 and 6.10, the power in the wind can be determined from the time derivative of kinetic energy as

$$P_w = \frac{dKE}{dt} = \frac{1}{2} \frac{dM_a}{dt} v^2 \qquad (6.17)$$

Substituting Equation 6.16, which is constant throughout the flow, into Equation 6.17 for both upstream and downstream wind speeds, gives the power in the wind upstream and downstream, respectively, as

$$P_u = \frac{1}{2}\rho A_t \frac{(v + v_d)}{2} v^2 \qquad (6.18)$$

$$P_d = \frac{1}{2}\rho A_t \frac{(v + v_d)}{2} v_d^2 \qquad (6.19)$$

Substituting Equations 6.18 and 6.19 into $P_b = P_u - P_d$ gives the power extracted by the turbine as

$$P_b = \frac{1}{2}\rho A_t \frac{(v + v_d)}{2} \left(v^2 - v_d^2\right) \qquad (6.20)$$

Betz noticed that Equation 6.20 could be rewritten as

$$P_b = \frac{1}{2}\rho A_t v^3 \left[\frac{1}{2}(1 + \lambda)(1 - \lambda^2)\right] = \frac{1}{2}\rho A_t v^3 C_p \qquad (6.21)$$

where $\lambda = v_d/v$ and

$$C_p = \left[\frac{1}{2}(1 + \lambda)(1 - \lambda^2)\right] \qquad (6.22)$$

is the **rotor efficiency**. It is the fraction of upstream power in the wind that is extracted by the turbine rotor and blades. The maximum rotor efficiency occurs when $dC_p/d\lambda = 0$. The solution occurs when $\lambda = v_d/v = 1/3$. Thus, the rotor efficiency is maximized when the downstream wind speed is one-third the upstream wind speed. Substituting $\lambda = 1/3$ into Equation 6.22 gives the maximum rotor efficiency as 0.5926, or about 59.3 percent. This is the Betz limit. It means that no wind turbine can extract more than 59.3 percent of the power in the wind that it is exposed to.

The best wind turbines today have a rotor efficiency of 45 to 47 percent, thus 76 to 79 percent of the Betz limit. A turbine's rotor efficiency depends largely on its revolutions per minute (rpms). Blades that spin too slowly allow too much wind to pass. Blades that spin too quickly create turbulence that affects other blades, thus reducing lift and rotor efficiency. By plotting rotor efficiency versus tip speed ratio of a turbine, the ideal number of rpms can be selected for a turbine. The **tip speed ratio (TSR)** equals the rotor tip speed divided by the upwind wind speed, thus

$$TSR = \frac{\pi D \times \text{rpm}}{v \times 60s/\text{min}} \qquad (6.23)$$

Based on such plots, two-bladed turbines are most efficient with a TSR between 5 and 7. Their maximum rotor

efficiency, which is in the middle of this range, is about $C_p$ = 45 percent. Three-bladed turbines are most efficient with a TSR between 4 and 5. Their maximum rotor efficiency in the middle of this range is about $C_p$ = 47 percent (Masters, 2013).

### 6.6.5 Wind Turbine Energy Output and Capacity Factor

Now that we know how to determine the power in the wind, the maximum power a wind turbine can extract, and the efficiency of a turbine when it encounters the wind, the next question to ask is, how much energy and power do wind turbines actually extract during the year?

The **gross annual energy output** (kWh/y) of a single wind turbine, before transmission and distribution losses, downtime losses due to repairs, or losses due to competition among multiple turbines for the same kinetic energy are accounted for, is

$$E_t = P_r \times CF \times H \qquad (6.24)$$

where $P_r$ is the rated power (nameplate capacity) (kW) of the wind turbine, $CF$ is the capacity factor of the turbine (expressed as a fraction), and $H$ is the number of hours in a year (8,760 for non-leap years and 8,784 for leap years). Equation 6.24 can be written in terms of **annual average power output** (kW) with

$$P_t = \frac{E_t}{H} = P_r \times CF \qquad (6.25)$$

Rearranging Equation 6.25 gives the capacity factor as

$$CF = \frac{P_t}{P_r} = \frac{E_t}{P_r H} \qquad (6.26)$$

The **capacity factor** is the annually averaged power produced by a wind turbine divided by the rated power (nameplate capacity) of the turbine. Alternatively, it is the annual energy produced by the turbine divided by the maximum possible energy produced ($P_r H$).

If a wind turbine could theoretically run for a full year at its rated power, its annual average power output would be $P_t = P_r$, and its capacity factor would be 1. However, the capacity factor of a turbine is always less than 1 because annually averaged wind speeds are always less than the rated wind speed of the turbine. Real capacity factors range from 15 percent for old turbines at locations with low wind speeds to 56 percent in the annual average for modern turbines at high-wind-speed offshore locations. For example, the 2018 capacity factor of the Hywind floating wind farm (five 6-MW turbines 29 km offshore of Peterhead, Scotland) was 56 percent. The annual average capacity factor of wind turbines in the United States increased from 23.8 percent for those installed between 1998 and 2001 to 41.9 percent for those installed between 2014 and 2017 (Wiser et al., 2019).

The capacity factor of a wind turbine at a given location can be estimated with a spreadsheet if the wind turbine power curve (e.g., Figure 6.4a) and the mean annual wind speed at the location are known. The annual energy output of the turbine at that location is simply

$$E_t = \sum_{v=0}^{\infty} P_b(v)Hf(v)\Delta v \qquad (6.27)$$

where $P_b(v)$ is the instantaneous power output of the turbine as a function of instantaneous wind speed $v$ from the turbine power curve. Also, $f(v)\Delta v$ is the probability of wind speed $v$ occurring during a year at the location, determined from Equation 6.6 for a Rayleigh distribution of wind speeds. Once $E_t$ is known, it is substituted into Equation 6.26 to obtain the turbine's capacity factor at that location. Table 6.2 shows an example calculation of the capacity factor determined in this way.

A less tedious method of determining the capacity factor of almost any turbine is with the clever equation of Masters (2013). This equation gives the capacity factor as a function of three parameters, the mean Rayleigh-distributed wind speed ($V_m$), the rated power of a turbine ($P_r$), and the turbine blade diameter ($D$). It works for nearly every geared wind turbine worldwide. The equation was applied to estimate the number of wind turbines needed to replace 60 percent of U.S. coal production (Jacobson and Masters, 2001).

The equation was derived in the following manner: The capacity factor of one wind turbine was calculated at one mean Rayleigh distributed wind speed ($V_m$), as done in Table 6.2. Next, the calculation was repeated for a range of mean wind speeds. All the capacity factors for that turbine were then plotted as a function of mean wind speed, as in Figure 6.9a. A careful evaluation of the curve indicates that, between a mean wind speed of 4 and 9 m/s, the slope of the curve is linear and equal to 0.087. When the slope is extrapolated to zero mean wind speed, the intercept is a negative value whose magnitude turns out

**Table 6.2** Example calculation of capacity factor using Equations 6.27 and 6.26. The probability distribution of wind speed is from Equation 6.6 for a Rayleigh distribution with a mean wind speed of 7.5 m/s. The turbine is a $P_r$ = 5,000 kW Senvion turbine with a blade diameter of 126 m and the power curve shown in Figure 6.4a. The sum of all probabilities does not exactly equal 1.0 and the sum of hours per year does not exactly equal 8,760 because of the coarse resolution of the wind speed intervals. The annual average power production $P_t$ = 1,682 kW is the total energy output divided by 8,760 hours per year. The capacity factor (Equation 6.26) is then 1,682 kW / 5,000 kW = 33.6 percent.

| (a) Wind speed (v, m/s) | (b) Probability of wind speed [f(v)Δv, fraction] | (c) Hours per year at wind speed = 8,760 × (b) | (d) Turbine power output at wind speed (kW) | (e) Annual energy output = (c) × (d) (kWh/y) |
|---|---|---|---|---|
| 1 | 0.041 | 361.9 | 0 | 0 |
| 3 | 0.111 | 970.8 | 0 | 0 |
| 4 | 0.134 | 1,173.9 | 30.77 | 36,121 |
| 6 | 0.203 | 1,775.8 | 500.1 | 888,001 |
| 8 | 0.183 | 1,601.5 | 1,500 | 2,402,679 |
| 10 | 0.138 | 1,211.0 | 3,000 | 3,632,982 |
| 12 | 0.090 | 786.2 | 4,720 | 3,710,541 |
| 14 | 0.051 | 443.7 | 5,000 | 2,218,738 |
| 16 | 0.025 | 219.4 | 5,000 | 1,097,146 |
| 18 | 0.011 | 95.5 | 5,000 | 477,608 |
| 20 | 0.004 | 36.7 | 5,000 | 183,643 |
| 22 | 0.001 | 12.5 | 5,000 | 62,517 |
| 24 | 0.000 | 3.8 | 5,000 | 18,876 |
| 26 | 0.000 | 1.0 | 5,000 | 5,062 |
| 28 | 0.000 | 0.2 | 5,000 | 1,207 |
| 30 | 0.000 | 0.1 | 5,000 | 256 |
| 32 | 0.000 | 0.01 | 0 | 0 |
| **Total** | **0.992** | **8,694** | | **14,735,376** |

to equal $P_r/D^2$. When the experiment is repeated for different turbines, the same slope is found and the magnitude of the intercept similarly equals $P_r/D^2$ for those turbines.

In sum, the simple, generalized equation for the capacity factor of a wind turbine from Masters (2013) is

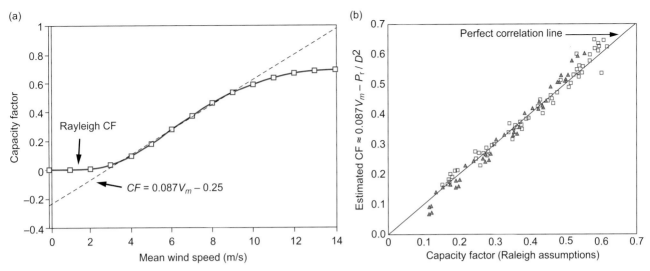

**Figure 6.9** (a) Plot of capacity factor versus mean (average) wind speed of a Rayleigh distribution ($V_m$) for a specific wind turbine. Also shown is a linear fit through the curve that is valid for mean wind speeds of 4 to 9 m/s. The intercept of the linear fit (0.25) equals $P_r/D^2$ for the turbine. (b) Comparison of capacity factors calculated using Equation 6.28 with capacity factors calculated as in Table 6.2 for 17 turbines ranging in nameplate capacity from 250 kW to 7 MW. The triangles indicate year 2000 turbines (NEG 1.5 MW, 64-m blade diameter; Nordex 1.3 MW, 60-m; NEG 1 MW, 60-m; NEG 1 MW, 54-m; Vestas 600 kW, 42-m; Bonus 300 kW, 33.4-m; Windworld 250 kW, 29.2-m). The squares indicate year 2012 turbines (Vestas 7 MW, 164-m blade diameter; Vestas 3 MW, 112-m; Siemens 3 MW, 101-m; GE 2.5 MW, 103-m; Siemens 2.3 MW, 101-m; Suzlon 2.1 MW, 97-m; Suzlon 2.1 MW, 88-m; GE 1.6 MW, 100-m; GE 1.5 MW, 77-m; Vergnet 275 kW, 32-m). From Masters (2013).

$$CF = 0.087 \times V_m - \frac{P_r}{D^2} \qquad (6.28)$$

where $V_m$ is the mean annual wind speed in units of m/s, $P_r$ is the rated power in kW, and $D$ is the turbine rotor diameter in m. The units must be those specified here and do not equate because the equation is empirical. Figure 6.9b shows that Equation 6.28 gives capacity factors extremely close to those calculated tediously with the method used in Table 6.2. In fact, Equation 6.28 is accurate to within 1 to 3 percent for most turbines and accurate to within 10 percent for all turbines tested.

## Example 6.7 Capacity factor

Calculate the capacity factors of a 5-MW turbine with a 126-m blade diameter operating in 7, 7.5, and 8.5 m/s annually averaged wind speeds. What is the percent difference in capacity factor at 7.5 m/s when compared with the result from Table 6.2?

### Solution:

Substituting $P_r$ = 5,000 kW, $D$ = 126 m, and $V_m$ = 7, 7.5 m/s, and 8.5 m/s into Equation 6.28 gives capacity factors of $CF$ = 0.294, 0.338, and 0.425, respectively. Thus, during the year, this turbine produces 29.4, 33.8, or 42.5 percent of its maximum possible energy output at those three mean wind speeds. The capacity factor from Table 6.2 was 33.6 percent, which is only 0.56 percent different from the value calculated from Equation 6.28, 33.8 percent, at a wind speed of 7.5 m/s.

Finally, the **turbine efficiency** is the annual average power output by a wind turbine divided by the mean power in the wind over the year. Thus, it is

$$TE = \frac{P_t}{P_m} \qquad (6.29)$$

For Rayleigh distributed winds, Equation 6.11 gives the mean power in the wind. The turbine efficiency differs from the capacity factor, which is the annual average power output divided by the rated power of the turbine (Equation 6.26). The turbine efficiency is the same as the rotor efficiency, $C_p$, which is the maximum instantaneous power extracted from the wind relative to the power in the wind (Section 6.6.4), except that the turbine efficiency is an annually averaged value of rotor efficiency.

## Example 6.8  Turbine efficiency

Calculate the turbine efficiency for the conditions in Table 6.2. Compare it with the capacity factor. Assume air density is 1.23 kg/m³.

### Solution:

The mean wind speed of the Rayleigh distribution and the rotor diameter from Table 6.2 are $V_m$ = 7.5 m/s and $D$ = 126 m. Substituting this into Equation 6.11 gives the mean power in the wind as $P_m$ = 6,179 kW. The turbine efficiency from Equation 6.29 is then 1,682 kW / 6,179 kW = 27.2 percent. This compares with the capacity factor from Table 6.2 of 33.2 percent.

### 6.6.6 Factors Reducing Wind Turbine Gross Annual Energy Output

Some of a wind turbine's gross annual energy output, determined from Equation 6.24, is lost due to four factors: transmission and distribution losses, downtime losses, curtailment losses, and array losses.

#### 6.6.6.1 Transmission and Distribution Losses

Figure 6.10 illustrates the AC **transmission and distribution (T&D) system**. AC electricity flows along a transmission line from an electric power–generating facility to a step-up transformer station. The station boosts the voltage to produce high voltage AC (HVAC) electricity in order to reduce long-distance AC transmission losses (Example 4.8). Along the HVAC line, AC electricity may or may not be converted to DC electricity for extra-long-distance HVDC transmission (not shown in Figure 6.10). At the end of the HVDC line, the DC electricity is converted back to HVAC electricity. The HVAC electricity is then transmitted to a step-down transformer station in a neighborhood, where the voltage is decreased, and the electricity is sent to local distribution lines. Electricity then goes to a transformer near buildings, where the AC voltage is dropped further for use in the buildings.

Losses along transmission and distribution (T&D) lines arise due to five factors. First, resistance along transmission and distribution lines converts some electricity to heat. Second, losses arise from step-up and step-down transformers to convert low voltage to high voltage AC (HVAC) electricity and back again. Losses similarly arise in local transformers, which reduce voltage further for electricity use in buildings, at the end of distribution lines. Third, losses occur in equipment converting HVAC

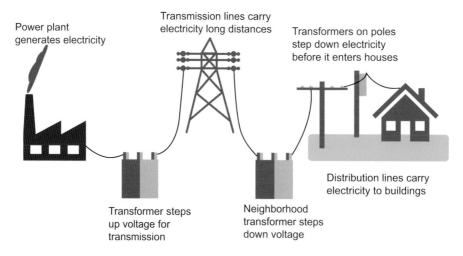

**Figure 6.10** Diagram of transmission and distribution system. See text for a description.

electricity to HVDC electricity and back again. Fourth, losses arise from downed power lines. Fifth, in countries with transmission and distribution loss rates above 15 percent (e.g., Table 6.3), electricity theft from power lines is a major source of loss (Sadovskaia et al., 2019).

Of all transmission and distribution losses in the current energy system in countries without theft of power, about 16 to 33 percent are short- and long-distance transmission losses from the electricity generator station to the step-down transformer substation, 32 to 40 percent are distribution losses between the step-down substation and the end user, and 27 to 52 percent are transformer losses (IEC, 2007).

When HVAC electricity is converted to HVDC electricity for extra-long-distance transmission (beyond 600 km), transmission losses are reduced compared with HVAC transmission (Section 4.7). For example, the overall loss of electricity along an 800-kV HVDC line ranges from 2.5 to 4 percent per 1,000 km compared with twice that for an HVAC line (ABB, 2004, 2005). However, a portion of the benefit of long-distance HVDC transmission is offset by losses arising from the HVAC-to-HVDC-to-HVAC conversion process. Such converter station losses are about 0.6 percent of energy transmitted (ABB, 2004).

Table 6.3 summarizes average transmission and distribution losses by country of the world in 2014. The losses range from 2 percent in Singapore to 72.5 percent in Togo. Fifty-four percent of the countries have T&D losses of 10 percent or higher. Losses in some large countries and regions are as follows: China (4 percent), the United States (5.9), the European Union (6.4), the Russian Federation (10), Brazil (15.8), and India (19.4). The world

average is 8.3 percent (World Bank, 2018). An independent analysis suggests that total transmission and distribution losses in the United States between 2012 and 2016 were similarly about 5 percent of electricity generation (EIA, 2018e).

Table 6.3 indicates that a lot of room exists for reducing electricity losses in many countries of the world, particularly countries with T&D losses exceeding 10 percent, including Brazil and India. Reducing losses will result in substantial low-cost benefits. Equation 6.30, derived here mathematically, quantifies some of the benefits of reducing T&D losses. It gives the percent reduction in electric power generation ($\Delta G_{elec}$) needed, compared with the original generation, if T&D losses, originally $L_{TD}$ percent, are reduced by $\Delta L_{TD}$ percentage points:

$$\Delta G_{elec} = \frac{100 \times \Delta L_{TD}}{100 - L_{TD} + \Delta L_{TD}} \qquad (6.30)$$

Equation 6.30 indicates that, if a country's current T&D loss rate is 10 percent, reducing the loss rate by 1 percentage point to 9 percent will reduce electricity generation requirements in the country by 1.1 percent. If the current loss rate is 20 percent, a 1 percentage point reduction will reduce generation requirements by 1.23 percent. If the loss rate is 72.5 percent, as in the case of Togo, simply reducing the loss rate to 71.5 percent will reduce the generation requirement by 3.5 percent (Example 6.9). Figure 6.11 shows a plot of Equation 6.30 for T&D loss rates that cover the range of all countries in Table 6.3. It illustrates that benefits of reducing T&D losses occur all the way down to a current loss rate of 1 percent.

---

### Example 6.9 Reducing electricity generation requirements by reducing transmission and distribution losses

Calculate the reduction in electricity generation that is possible, without reducing end-use electricity availability in Togo if T&D losses are reduced by $\Delta L_{TD} = 1$ percentage point from their 2014 value of $L_{TD} = 72.5$ percent.

### Solution:
From Equation 6.30, $\Delta G_{elec} = 100 \times 1 / (100 - 72.5 + 1) = 3.5$ percent.

---

Reducing the loss rate by 1 percentage point allows even more fossil-fuel generation to retire than is indicated by Equation 6.30 and Figure 6.11. The reason is that fossil fuels usually do not supply 100 percent of electricity in a

country. As such, if all electricity reductions that result from reducing T&D losses are obtained by shutting fossil-fuel plants, a higher percent of fossil-fuel plants will be shut than Equation 6.30 indicates.

**Table 6.3 Percent of electricity production lost due transmission and distribution losses by country in 2014.**

| Country | Loss | Country | Loss | Country | Loss | Country | Loss |
|---|---|---|---|---|---|---|---|
| Singapore | 2.0 | Paraguay | 6.6 | Peru | 11.0 | Nigeria | 16.1 |
| Trinidad/Tobago | 2.3 | Switzerland | 6.7 | Egypt | 11.2 | Zimbabwe | 16.4 |
| Slovak Republic | 2.5 | Kazakhstan | 6.7 | Angola | 11.3 | Tajikistan | 17.0 |
| Iceland | 2.7 | Estonia | 6.8 | El Salvador | 11.3 | Algeria | 17.1 |
| Israel | 2.9 | Saudi Arabia | 6.8 | Bangladesh | 11.4 | Pakistan | 17.4 |
| Gibraltar | 3.0 | Italy | 7.0 | Sri Lanka | 11.4 | Montenegro | 17.5 |
| South Korea | 3.3 | U.A. Emirates | 7.2 | Dominican Rep. | 11.6 | Kenya | 17.6 |
| Germany | 3.9 | Ireland | 7.9 | Kuwait | 11.7 | Tanzania | 17.7 |
| Bahrain | 3.9 | Bosnia & Herzegovina | 8.2 | Armenia | 12.0 | Ethiopia | 18.5 |
| Cyprus | 4.0 | Greece | 8.2 | Hungary | 12.4 | India | 19.4 |
| Finland | 4.1 | United Kingdom | 8.3 | Turkmenistan | 12.5 | Myanmar | 20.5 |
| Japan | 4.4 | South Africa | 8.4 | Hong Kong | 12.5 | Nicaragua | 20.8 |
| Czech Republic | 4.5 | Bulgaria | 8.6 | Iran | 12.6 | Congo, DR | 21.5 |
| Malta | 4.7 | Suriname | 8.7 | Senegal | 12.8 | Moldova | 21.5 |
| Netherlands | 4.8 | Uzbekistan | 8.8 | Ecuador | 12.9 | Lithuania | 22.0 |
| Sweden | 4.8 | Canada | 8.9 | Croatia | 13.1 | Ghana | 22.6 |
| Australia | 4.8 | Latvia | 9.0 | Eritrea | 13.5 | Cambodia | 23.4 |
| Slovenia | 4.8 | Belarus | 9.2 | Azerbaijan | 13.6 | Gabon | 23.4 |
| Austria | 5.3 | Bolivia | 9.2 | Mexico | 13.7 | Albania | 23.7 |
| Belgium | 5.4 | Vietnam | 9.2 | Sudan | 14.3 | Kyrgyz Rep. | 23.7 |
| China | 5.5 | Indonesia | 9.4 | Panama | 14.3 | Yemen | 25.8 |
| South Sudan | 5.7 | Philippines | 9.4 | Ivory Coast | 14.3 | Jamaica | 26.7 |
| Georgia | 5.8 | Guatemala | 9.5 | Argentina | 14.3 | Niger | 26.8 |
| Malaysia | 5.8 | Spain | 9.6 | Morocco | 14.7 | Nepal | 32.2 |
| United States | 5.9 | Uruguay | 9.6 | Mozambique | 14.7 | Honduras | 34.9 |
| Qatar | 6.1 | Cameroon | 9.8 | Mongolia | 14.8 | Venezuela | 36.0 |
| Norway | 6.1 | Portugal | 10.0 | Turkey | 14.8 | Namibia | 36.2 |
| Thailand | 6.1 | Russian Fed. | 10.0 | Tunisia | 14.9 | Congo, Rep. | 44.5 |
| Denmark | 6.1 | Lebanon | 10.5 | Zambia | 15.0 | Iraq | 50.6 |
| Mauritius | 6.2 | Colombia | 10.7 | Kosovo | 15.0 | Haiti | 60.1 |
| Luxembourg | 6.3 | Jordan | 10.7 | Cuba | 15.3 | Libya | 69.7 |
| France | 6.4 | Ukraine | 10.8 | Syria | 15.4 | Togo | 72.5 |
| Brunei Darussalam | 6.4 | Botswana | 10.8 | Serbia | 15.4 | | |
| Poland | 6.5 | Costa Rica | 10.8 | Brazil | 15.8 | | |
| New Zealand | 6.5 | Oman | 10.9 | North Korea | 15.8 | European Union | 6.4 |
| Chile | 6.5 | Romania | 10.9 | Curacao | 16.0 | **World** | **8.3** |

Source: World Bank (2018).

## Example 6.10 Reducing fossil-fuel generation requirements when reducing T&D losses

Calculate the reduction in fossil-fuel electricity generation that is possible upon a 1 percentage point reduction in T&D losses for a country that currently has 20 percent T&D losses and provides 60 percent of its electricity from fossil fuels.

### Solution:

From Equation 6.30, $\Delta G_{elec} = 100 \times 1 / (100 - 20 + 1) = 1.23$ percent, which is the percent reduction in electricity generation that can be obtained in the country with a 1 percentage point reduction in T&D losses. Since fossil fuels produce 60 percent of the electricity, this means that 1.23 percent / 0.6 = 2.05 percent of fossil-fuel generation can be reduced. In other words, for every 1 percentage point of T&D loss reductions, the country can eliminate over 2 percent of its fossil-fuel electricity generation fleet.

In 2020, about 65 percent of all electricity worldwide was produced from fossil fuels. If the world average T&D loss were reduced from 8.3 to 7.3 percent, world electricity generation could be reduced by 1.08 percent and fossil-fuel generation could be reduced by 1.66 percent. The range in fossil-fuel reductions possible for individual countries under the same assumption is 1.55 to 5.4 percent.

### Transition highlight

Example 6.10 shows that if a country with 60 percent of its electricity from fossil fuels and a 20 percent T&D loss reduces its T&D loss by 1 percentage point, the country will reduce fossil-fuel electricity needed by 2.05 percent and overall electricity needed by 1.23 percent.

Replacing T&D equipment costs only a fraction of the cost savings of reducing electricity generation. Coal and gas plants spend money on mining, transporting, and processing fossil fuels and on the rest of their operations. They charge consumers for the resulting electricity. If the electricity is wasted as heat or theft in the T&D system, consumers are paying for electricity that no one uses. Only a portion of wasted energy costs is needed to upgrade the T&D system, and the upgrades will last up to 70 years for some of the T&D components.

By eliminating extraneous coal and gas plants, a country also reduces health and climate costs due to unnecessary pollution that is emitted to produce electricity that is not used because it is wasted along the T&D system. Eliminating extraneous fossil-fuel plants further reduces

the devastation to land and water due to unnecessary mining for fossil fuels used in the plants.

Finally, in a 100 percent WWS world, a portion of new electricity generation, including from offshore wind, tidal and ocean current, wave, and floating solar power, will be offshore. Offshore renewables often require short transmission distances to load centers because most people in the world live along the coasts, and most offshore resources will be sited 0 to 200 km offshore. As such, the growth of offshore renewables should increase the efficiency of the T&D system.

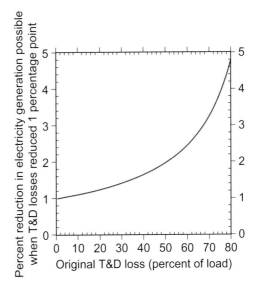

Figure 6.11 Percent reduction in electricity generation from power plants that is possible if transmission and distribution losses are reduced by 1 percentage point below the original T&D loss. Thus, for example, if the original T&D loss is 50 percent and reduced by 1 percentage point to 49 percent, 1.96 percent less electricity generation is needed from power plants. The numbers in the figure are obtained from Equation 6.30.

### 6.6.6.2 Downtime Losses

A wind turbine's gross annual energy output is also reduced by downtime losses. Downtime occurs due to regularly scheduled maintenance, equipment failure, or refurbishment of the turbine. Section 3.2.2.1 discussed refurbishment times for onshore and offshore wind turbines.

An analysis of repair data from 1,500 onshore wind turbines over about 10 years found the average downtime for repairs was 1.6 percent (6 days) of the year. Minor failures (such as with the electrical system, electrical controls, the hydraulic system for pitch control, and yaw system), which represent 75 percent of problems, caused only about 5 percent of the downtime. Major failures (such as with the rotor blades, rotor hub, drive train, gearbox, and generator), which represent 25 percent of problems, caused about 95 percent of the downtime (Faulstich et al., 2011).

Offshore wind turbines generally require more repair-related downtime than do onshore turbines because harsh weather conditions, including high waves and strong winds, prevent access to offshore turbines for several days per month during stormy months. In addition, merely scheduling a boat ride to an offshore wind turbine can take additional time compared with driving to an onshore turbine. Third, the harsher weather conditions and faster wind speeds offshore increase wear and tear compared with onshore wind turbines. On the other hand, because of the lack of terrain and cooler ocean surface than land, offshore turbines are exposed to less turbulence; thus, they have less wear and tear due to turbulence than onshore turbines. Offshore turbines are down for scheduled maintenance 4 to 6 days per year (Zhang et al., 2012) and unscheduled maintenance (for repairs) 4 to 8 days for a total of 8 to 14 days (2.2 to 3.8 percent of the year).

In comparison, the average coal plant and combined cycle gas plant in the Eastern United States (PJM regional transmission region) were down 10.6 percent and 3.0 percent, respectively, of the time between January and June 2015 for unscheduled maintenance (Monitoring Analytics, 2015). Coal plants are down another 6 percent of the year for scheduled maintenance (Jacobson, 2012a).

A difference, though, between outages of **centralized power plants** (coal, nuclear, natural gas) and outages of **distributed power plants** (wind, solar, wave) is that when individual solar panels or wind turbines are down, only a small fraction of electrical production is affected. When a centralized plant is down, a large fraction of the grid is affected (Section 1.3.2). When more than one large, centralized plant is offline at the same time, an entire grid can be affected.

### 6.6.6.3 Curtailment Losses

A third factor that reduces a wind turbine's gross annual energy output is curtailment. **Curtailment**, or **shedding**, is the reduction in output of an electricity generator below what it could otherwise produce. Grid operators may order curtailment under contract with a large electricity consumer when electricity supply exceeds demand. Curtailment is not a T&D loss because the electricity never reaches the grid. Curtailment is avoided if excess electricity is stored or used on site to produce heat, cold, or hydrogen, which are either stored or used immediately (Chapter 8).

### 6.6.6.4 Array Losses due to Competition among Wind Turbines for Available Kinetic Energy

Finally, competition among wind turbines for available kinetic energy in the wind reduces the annual wind turbine energy output compared with the gross energy output calculated from Equation 6.24. Losses due to competition among turbines are commonly called **array losses**. Equation 6.24 uses the wind speed upstream of a wind farm, usually taken from measurement or model simulation. However, in reality, wind turbines past the first row of turbines in a farm (downstream turbines) receive wind speeds slower than wind speeds received by the first row because the first row extracts kinetic energy from the wind. As such, the annual energy output of a wind farm is always less than that calculated, assuming the wind speed that each turbine sees is the upstream wind speed of the farm.

Section 6.8.6 discusses the maximum possible wind power extraction worldwide in the presence of competition among wind turbines for available kinetic energy. Such competition gives rise to array losses.

Figure 6.12 provides an estimate of the loss in wind power output due to array losses, averaged over a year, for a case with a lower penetration of wind than in Section 6.8.6. This case explores the scenario where 37.1 percent of the world's all-purpose end-use energy is generated by wind in 2050. Competition among wind turbines for limited kinetic energy reduces end-use

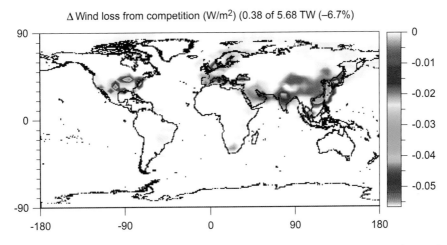

**Figure 6.12** Computer-modeled difference in power extracted per square meter of ground worldwide by wind turbines when competition among wind turbines for available kinetic energy in the wind is accounted for versus not accounted for. The model simulations include 1.58 million onshore and 935,000 offshore 5-MW wind turbines needed to power 37.1 percent of the world's all-purpose end-use energy in 2050 after all energy has been electrified. The turbines are placed among 139 countries. The 6.7 percent reduction in annually averaged wind turbine power output (0.38 TW out of 5.68 TW produced on average without competition) represents the worldwide power loss (array loss) due to competition among wind turbines for available kinetic energy. From Jacobson et al. (2018a).

annually averaged wind power output by about 6.7 percent. Reductions in densely packed farms (or multiple farms close to each other) are greater than in less densely packed farms (or farms spread apart from each other) (Figure 6.25).

The result in Figure 6.12 suggests that, for purposes of providing sufficient power for the world from wind, more turbines will be needed compared with no array losses having occurred. However, the additional number of turbines and costs are less than those arising due to the 8.3 percent transmission and distribution losses that wind turbines face (Section 6.6.6.1).

### 6.6.6.5 Overall Loss of Wind Energy Output

The overall percent loss in wind energy output due to the four processes just discussed compared with the gross energy production from Equation 6.24 is

$$L_T = 100\left[1 - \left(1 - \frac{L_{TD}}{100}\right)\left(1 - \frac{L_D}{100}\right)\left(1 - \frac{L_C}{100}\right)\left(1 - \frac{L_A}{100}\right)\right]$$
(6.31)

where $L_{TD}$ is the percent of annual energy lost due to T&D losses, $L_D$ is the percent of energy lost due to downtime, $L_C$ is the percent of energy lost due to curtailment, and $L_A$ is the percent of energy lost due to array losses. T&D losses are expected to decline from a world average of 8.3 percent in 2014 to about 6.2 (2 to 10) percent in 2050 (Sadovskaia et al., 2019). Maintenance downtime for onshore and offshore wind turbines should decline slightly from today's losses to about 1.5 (1 to 2) percent in 2050. Curtailment of wind, which is high in some countries today because excess wind is not used for non-electric-grid purposes, should go to zero in a 100 percent WWS world in which excess wind is used to provide heat, cold, or hydrogen for storage or direct use. Finally, array losses in a 100 percent WWS world in 2050 are expected to increase to about 6.7 percent (Figure 6.12). Example 6.11 suggests that the overall loss in 2050 under these conditions is estimated to be $L_T$ = 13.8 (9.5 to 18.6) percent. Thus, a worldwide mean of 13.8 percent of all energy produced may be lost due to T&D, maintenance, curtailment, and array losses in 2050.

## Example 6.11  Overall reduction in annual wind energy output

Calculate the overall reduction in wind energy output compared with the gross annual energy output before losses for onshore and offshore wind turbines in 2050 for (a) a low case of 2 percent T&D losses, 1 percent downtime, and 6.7 percent array losses, (b) a middle case of 6.2 percent T&D losses (Sadovskaia et al., 2019), 1.5 percent downtime, and 6.7 percent array losses, and (c) a high case of 10 percent T&D losses, 2 percent downtime, and 6.7 percent array losses. Assume curtailment losses in all three cases are 0 percent.

### Solution:

From Equation 6.31, $L_T = 100[1 - (1-2/100)(1-1/100)(1-0/100)(1-6.7/100)] = 9.5$ percent in the low case, 13.8 percent in the middle case, and 18.6 percent in the high case.

## 6.7  Wind Turbine Footprint and Spacing Areas

Two types of land or water areas associated with wind farms are their footprint area and spacing area.

The **footprint area** of a wind farm is primarily the topsoil or water area touched by all the wind turbine towers and bases in the wind farm. It does not include the areas of the bases under the ground or underwater because the purpose of knowing a wind farm's footprint area is knowing how much land above the ground can be used for other purposes. Thus, footprint area includes areas of any permanent roads (e.g., covered with pavement), installed due to the wind farm, but it does not include areas of unpaved roads, which can readily be reverted back to their natural conditions. Transmission lines between the wind farm and grid that are underground also do not count toward footprint. However, the land areas of aboveground transmission line pads touching the ground due solely to the wind farm do count toward footprint. However, the areas of transmission tower pads are trivially small, as indicated by the fact that vegetation can grow under transmission tower lattices.

## Example 6.12  Wind farm footprint area

Calculate the range in footprint area of a wind farm with 200 turbines that have circular cement bases 4 to 5 m in diameter securing the turbines' tubular towers. The farm also has unpaved access roads and underground transmission.

### Solution:

The footprint area among all turbines in the low case is 200 turbines $\pi \times (4 \text{ m} / 2)^2 = 2{,}513 \text{ m}^2 = 0.0025 \text{ km}^2$. In the high case, it is $3{,}927 \text{ m}^2 = 0.0039 \text{ km}^2$. The unpaved access roads and underground transmission lines do not count toward footprint.

### 6.7.1  Defining Wind Farm Spacing Area

The **spacing area** of a wind farm is the land or water area between wind turbines and beyond the edge of the farm required to (1) prevent one turbine's blades from touching another's, (2) prevent turbines from falling onto one another or onto nearby structures, (3) minimize wear and tear on a downstream turbine resulting from the turbulent wake of an upstream turbine, and (4) minimize array losses (competition among all turbines for limited available kinetic energy in the wind). Spacing area between onshore turbines can be used for multiple purposes, including for agriculture, cattle grazing, ranching, solar arrays, forests, and open space. Water between offshore is similarly open ocean water. Because footprint areas in a wind farm are small (Example 6.12), wind farms leave 98 to 99.5 percent of land or water undisturbed or usable for another purpose (e.g., Figure 6.13).

**Figure 6.13** Illustration of the small footprint area of wind turbines and the spacing between wind turbines that can be used for multiple purposes. © David Collection/AdobeStock.

Spacing area is an important parameter because it affects a wind farm's installed power density and output power density. The **installed power density** ($\rho_I$, MW-nameplate/km$^2$-ground) of a wind farm is the nameplate capacity summed among all wind turbines in the farm ($P_{F,N}$, MW) divided by the spacing area of the farm ($A_s$, km$^2$):

$$\rho_I = \frac{P_{F,N}}{A_s} \qquad (6.32)$$

Greater installed power densities imply more potential installations of wind over smaller land or ocean areas. The **output power density** of a wind farm ($\rho_O$, W-output/m$^2$-ground) is the annual average power output of the farm ($P_{F,O}$, W) divided by the spacing area of the farm (in m$^2$):

$$\rho_O = \frac{P_{F,O}}{A_s} \qquad (6.33)$$

The **capacity factor of a wind farm** ($CF_F$, dimensionless) is simply the output power density divided by the installed power density of the farm; thus, it is independent of spacing area:

$$CF_F = \frac{\rho_O}{\rho_I} = \frac{P_{F,O}}{P_{F,N}} \qquad (6.34)$$

No unique way exists to define spacing area. At one extreme, it could be defined as the surface area of the

Earth. This would give trivially small installed and output power densities but the same capacity factor as a small spacing area. At the other extreme, it could be defined as the circular area on the ground around a wind turbine tower, with the circle's radius equal to the turbine tip height, multiplied by the number of wind turbines in the wind farm. The **tip height** of a turbine equals its hub height plus one blade radius. The tip height would be used to define the minimum spacing area of a single turbine because that is the distance a turbine will extend to if it falls to the ground in any direction. Laws in many locations require structures to lie beyond the tip height of a turbine.

In practice, many approximations for spacing have been made, most of them applicable to some but not other wind farms. The most common type of approximation has been to assign each turbine a rectangular spacing area of $yD \times zD$, where $D$ is rotor diameter, $y$ and $z$ are integers, and one turbine lies in the center of each rectangle. This approximation assumes that turbines are in a rectangular grid. One such approximation is $4D \times 7D$, where the longer distance is the distance between rows of turbines, such that the wind approaches the first row perpendicular to the turbine blades. The shorter distance is the distance between columns of turbines (parallel to the blades). In this configuration, the second row is shifted so that the turbines are not directly behind those in the first row. Third row turbines are directly behind

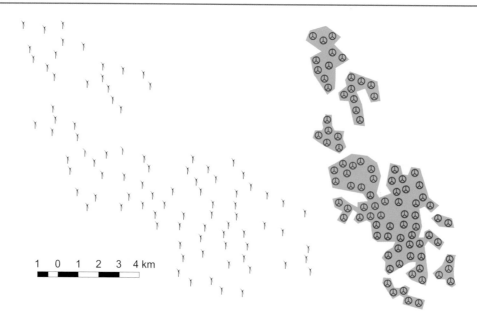

Figure 6.14 Left: Location of turbines in one particular wind farm. Right: Spacing area of wind turbines determined by accounting for and combining three areas. These are (1) the circular area around each turbine, where the radius of the circle is the turbine's tip height, (2) the area of each cluster, where the outer edges of the circular areas define the outer boundaries of each cluster, and (3) the summed area among all clusters, which defines the spacing area. Each of the seven noncontiguous blue areas is a cluster. From Enevoldsen and Jacobson (2020).

first row turbines, and fourth row turbines are staggered again, and so on.

Although there is no correct method to determine spacing area, methods that assign specific areas (e.g., rectangles, circles, or polygons) to each turbine erroneously include areas that are not part of a wind farm, so cannot be correct. For example, they erroneously count, as part of a wind farm, space outside a wind farm's boundary, space between clusters of turbines in a farm, and overlapping space that results when each turbine is assigned a large spacing area.

A way to define spacing that overcomes these three issues is as follows: The methodology involves defining and combining three areas (Enevoldsen and Jacobson, 2020). The areas assume that a wind farm consists of one or more clusters of individual wind turbines. The first area is simply a circular area around each turbine (the red circles in Figure 6.14), where the radius of each circle is the tip height of the turbine. This distance is chosen because it is the maximum physical horizontal distance a turbine can extend to if it falls to the ground. This distance also ensures that, if a wind turbine is on the edge of a wind farm, no space beyond the tip height will be counted as part of the wind farm spacing. Thus, the edge of a farm extends horizontally to no more than one tip height from each tower at the edge of the farm.

The second area is the spacing area within each cluster of turbines in the wind farm. The cluster area is obtained by tracing a line around the edges of the circles around each wind turbine on the outside of the cluster. This cluster area includes the circular areas of all wind turbines within the cluster. Figure 6.14 shows a wind farm with seven clusters separated by small or large distance. Each noncontiguous blue area is a cluster.

A **cluster** of turbines within a wind farm is separated from another cluster when the distance from a turbine tower in the first cluster to the closest neighboring turbine tower exceeds three tip heights (Enevoldsen and Jacobson, 2020). The area in between clusters is not included as part of the spacing area.

One argument for including space between clusters as part of a wind turbine spacing area is that a developer paid for the land. However, the counter argument is that the land is open space that can be used for non-wind purposes so should not automatically be included as part of a wind farm's spacing if it is merely separating clusters of turbines.

The third and final area is the total spacing area of the wind farm. This is simply the sum of the spacing areas of all the clusters within the farm, as illustrated in Figure 6.14.

The method just described avoids counting areas outside of cluster boundaries as part of a wind farm, avoids counting areas between clusters as part of a wind farm, and avoids counting overlapping areas, previously assigned to individual turbines, as part of a wind farm. It also accounts for actual distances between wind turbines.

The main conceptual difference between the method just discussed and previous methods is that a spacing area calculated with the present method reflects near the minimum spacing area needed for a wind farm. Thus, the addition of a turbine to the farm always increases the spacing area required by the farm, as it should in reality.

With other methods, so much extraneous space exists between wind turbines in a farm and outside of the wind farm boundaries that many additional turbines can be added to the farm without increasing the spacing area. This artificially deflates both the installed and output power densities of wind farms in which turbines are spread apart or separated into clusters. In other words, a definition of spacing area that allows turbines to be added to a farm without increasing the spacing area of the farm incorrectly implies that a wind farm *needs* the entire spacing area assigned to it. Further, the fact that an additional wind turbine can be added to a wind farm without increasing the overall spacing area proves that the area occupied by the new turbine was not a necessary part of the spacing area of the original farm. With the definition proposed here, each wind farm needs most of the spacing area assigned to it, and each new turbine added to a wind farm increases the spacing area of the farm.

### 6.7.2 Estimates of Wind Farm Spacing Areas

Table 6.4 summarizes results from a study that used the methodology described in Section 6.7.1. It considered 16 onshore and 7 offshore wind farms among 13 countries across 5 continents and found that installed power densities ranged from 7.2 MW/km$^2$ for offshore European farms to 19.8 MW/km$^2$ for onshore European farms to 20.5 MW/km$^2$ for onshore farms across four other continents.

Figure 6.15 illustrates visually that the mean European value (19.8 MW/km$^2$) gives a more realistic estimate of wind turbine spacing than do two other estimates. One of the other estimates of spacing, 7.2 MW/km$^2$, is the mean from most previous studies of wind farm spacing area. That estimate erroneously accounts for space between clusters of turbines not associated with the wind farm turbines. The other estimate of spacing, which gives an installed power

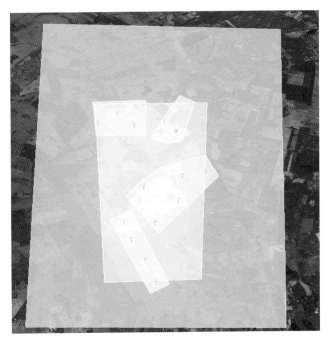

Figure 6.15 Spacing areas determined with three different methods for the Gawlovice wind farm in Poland. The largest area represents an average installed power density of 1.5 MW/km$^2$ from one specific study. The second area represents an average installed power density of 7.2 MW/km$^2$ from multiple previous studies that have estimated wind installed power densities, and the innermost shading represents 19.8 MW/km$^2$, which is the mean onshore installed power density for European wind farms from Table 6.4. From Enevoldsen and Jacobson (2020).

density of 1.5 MW/km$^2$, is the average estimate for U.S. wind farms from one particular study. It erroneously includes not only areas between clusters of turbines, but also areas far beyond the outer edge of the wind farm.

In sum, spacing area of a wind farm is important for determining wind farm installed and output power density. Spacing area must be determined carefully to avoid erroneously counting areas outside of wind farm boundaries and between clusters of turbines and to avoid double counting areas assigned to individual turbines in a farm.

### 6.7.3 Application of Spacing Area

The wind turbine installed and output power density can be used to estimate the number of wind turbines needed and their spacing requirements for specific scenarios. Table 6.5 provides one such scenario, which is an estimate of the number of wind turbines required to power all U.S. vehicles if they are converted to

Table 6.4 **Mean installed and output power densities and capacity factors of 12 onshore wind farms among 9 countries in Europe, 7 offshore wind farms in Europe, and 4 wind farms in 4 countries, each on a different continent outside of Europe. The method for calculating the spacing areas is described in Section 6.7.1.**

| Location | Mean installed power density (MW/km$^2$) | Mean output power density (W/m$^2$) | Mean capacity factor (%) |
|---|---|---|---|
| Onshore Europe | 19.8 | 6.64 | 33.5 |
| Offshore Europe | 7.20 | 2.94 | 40.8 |
| Onshore outside Europe | 20.5 | 6.84 | 33.4 |

Source: Enevoldsen and Jacobson (2020).

battery-electric vehicles. The table indicates that between 65,000 and 148,000 large (5-MW) wind turbines could power the entire U.S. on-road vehicle fleet, and such turbines would require 0.18 to 1.1 percent of U.S. land area. The factor of five difference in low versus high estimate of the land area required reflects the possible differences in mean wind speed available to the turbines and the range in the installed power density of turbines from Table 6.4.

## 6.8 Wind Physics and Resources

This section discusses the forces acting on the air to form winds, how horizontal winds form, the locations of high onshore and offshore wind speeds worldwide, and the

Table 6.5 **Low and high estimates of the number of 5-MW, 126-m blade diameter turbines needed and the resulting land or water spacing area required to provide electricity if all on-road vehicles in the United States were converted to battery-electric vehicles.**

| Parameter | Low case | High case |
|---|---|---|
| a) Mean annual wind speed (m/s) | 8.5 | 7 |
| b) Turbine capacity factor (Equation 6.28) | 0.425 | 0.294 |
| c) T&D, downtime, array losses (Example 6.11) | 9.5 | 13.8 |
| d) One turbine gross energy output (kWh)/y (Equation 6.24) | 18.6 million | 12.9 million |
| e) One turbine end-use energy output (kWh/y) = d × (1-c) | 16.8 million | 11.1 million |
| f) 2016 U.S. on-road vehicle miles travelled/y | 3.2 trillion | 3.2 trillion |
| g) 2016 fleet-averaged gasoline mileage (mi/gal) | 20 | 18 |
| h) Gallons of fuel consumed/y = f/g | 160 billion | 178 billion |
| i) Lower heating value of gasoline (MJ/gal) | 124.9 | 124.9 |
| j) Energy in fuel in all vehicles (MJ/y) = h × i | 2 trillion | 2.2 trillion |
| k) Tank-to-wheel efficiency of gasoline (–) | 0.17 | 0.20 |
| l) Energy actually used by vehicles (MJ/y) = j × k | 3.4 trillion | 4.4 trillion |
| m) Energy actually used by vehicles (kWh/y) = l /3.6 | 944 billion | 1.2 trillion |
| n) Battery-to-wheel efficiency of electric vehicle (–) | 0.91 | 0.8 |
| o) Efficiency of car charger | 0.95 | 0.85 |
| p) Energy to power all electric vehicles (kWh/y) = m/(n × o) | 1.1 trillion | 1.76 trillion |
| q) Number of wind turbines to power all vehicles = p/e | 65,000 | 159,000 |
| r) Nameplate capacity of turbines (MW) = q × 5 MW/turbine | 325,000 | 795,000 |
| s) Installed power density of turbines (MW/km$^2$) | 20 | 7.2 |
| t) Spacing area needed for turbines (km$^2$) = r/s | 16,250 | 110,000 |
| u) 50-state U.S. land area (km$^2$) | 9.16 million | 9.16 million |
| v) Percent of U.S. land area for all vehicles = 100 × t/u | 0.18 | 1.2 |

maximum power that can be extracted from the wind at a given time.

## 6.8.1 Forces Acting on the Air

Winds form because forces act on the air above a rotating Earth. This section describes the four major forces that act on the air to produce horizontal winds. These are the pressure gradient force, apparent Coriolis force, friction force, and apparent centrifugal force.

### 6.8.1.1 Pressure Gradient Force

When air pressure is high in one location and low nearby, air moves horizontally from high to low pressure. The force causing this motion is the **pressure gradient force** (PGF), which is proportional to the difference in pressure divided by the distance between the two locations and always acts from high to low pressure. Air pressure differences arise due to winds converging or diverging and due to differential heating by the sun.

When winds from opposite directions move toward each other (**converge**), air piles up in a column, increasing surface air pressure if all else is the same. When winds move in opposite directions (**diverge**), then air is removed from a column, decreasing surface air pressure if all else is the same. However, to determine whether surface convergence or divergence increases or decreases surface pressure, it is also necessary to know the extent of divergence or convergence aloft. For example, if air converges horizontally at the surface but diverges horizontally more aloft, air pressure at the surface will decrease because more air is removed from a column than added.

Shadows or clouds that reduce sunlight in one location versus another cause differential heating. If the ground is warmer in one location than another, more air will rise in the warmer location. That air will pile up aloft then be carried away by fast winds aloft, decreasing surface air pressure in the warm region. Conversely, air in the cold air region will sink, and fast winds aloft will replace the sinking air, increasing surface air pressure in the cold region.

Thus, warm regions are often associated with surface low pressure and cold regions, with surface high pressure. In this way, differential heating can cause a horizontal pressure gradient, which leads to winds.

### 6.8.1.2 Apparent Coriolis Force

When air is in motion over a rotating Earth, it appears to an observer fixed in space to accelerate to the right in the Northern Hemisphere and to the left in the Southern Hemisphere by the **apparent Coriolis force** (ACoF; Figure 6.16). The ACoF is not a real force; rather, it is an acceleration that arises when the Earth rotates under a body (air in this case) in motion. The ACoF is zero at the equator, maximum at the poles, zero for bodies at rest, proportional to the speed of the air, and always acts 90 degrees to the right (left) of a moving body in the Northern (Southern) Hemisphere.

Figure 6.16 illustrates the ACoF. If the Earth did not rotate, an object thrown from point A directly north would be received at point B, along the same longitude as point A. Because the Earth rotates, objects thrown to the north have a west-to-east velocity equal to that of the Earth's rotation rate at the latitude from which they originate. The Earth's rotation rate near the equator (low latitude) is greater than that near the poles (high latitudes); thus, objects thrown northward from a low latitude have a greater west-to-east velocity than does the Earth below them when they reach a high latitude. For example, an object thrown from point A to point B in the north will end up at point C, instead of at point B' by the time the person at point A reaches point A'. The object will appear as if it has been deflected to the right (from point B' to point C). Similarly, an object thrown southward from point B to point A will end up at point

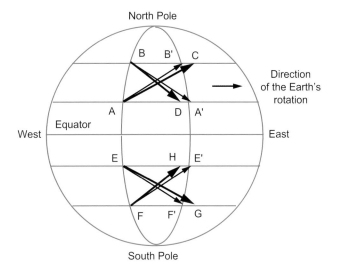

**Figure 6.16 Example of the apparent Coriolis force (ACoF), described in the text. Thin arrows represent intended paths, and thick arrows are actual paths.**

D, instead of at point A' by the time the person at point B reaches point B'. The Coriolis effect, therefore, appears to deflect moving bodies to the right in the Northern Hemisphere and to the left in the Southern Hemisphere.

### 6.8.1.3 Friction Force

A third force that acts on moving air is the **friction force** (FF). This force is important near the surface of the Earth only. The FF slows the wind. Its magnitude is proportional to the wind speed, and it acts in exactly the opposite direction from the wind. The rougher the surface is, the greater the FF will be. The FF over water and flat desert surfaces is small, whereas the FF over forests and buildings is large. The opposite of the friction force is the drag force, which the wind imparts on an object, such as a wind turbine blade (Section 6.4).

### 6.8.1.4 Apparent Centrifugal Force

A fourth force, which also acts on moving air, is the **apparent centrifugal force** (ACfF). This force is another fictitious force; it arises when an object rotates around an axis. The apparent force is directed outward, away from the axis of rotation. When a passenger in a car rounds a curve, for example, a viewer traveling with the passenger sees the passenger being pulled outward, away from the axis of rotation, by this force. In contrast, a viewer fixed in space sees the passenger accelerating inward due to a **centripetal acceleration**, which is equal in magnitude to, but opposite in direction from, the apparent centrifugal force.

### 6.8.2 How Winds Form

The major forces acting on the air in the horizontal direction are thus the PGF, ACoF, FF, and ACfF. In the vertical, the major forces are the upward-directed vertical pressure gradient force and the downward-directed force of gravity. These forces drive winds. Examples of horizontal winds arising from force balances are discussed next.

### 6.8.2.1 Geostrophic Wind

The type of wind involving the least number of forces is the **geostrophic** ("Earth-turning") **wind**. Only the pressure gradient force and the apparent Coriolis force are needed to drive this wind. The wind occurs above the boundary layer, where surface friction is negligible, and

along straight isobars, where curvature, thus the apparent centrifugal force, is negligible.

An **isobar** is a line of constant pressure drawn on a constant altitude map. Suppose a horizontal pressure gradient, represented by two parallel isobars, exists, such as in the top part of Figure 6.17. The PGF causes still air to move from high pressure to low pressure. As the air moves, the ACoF deflects the air to the right in the Northern Hemisphere. The ACoF continues deflecting the air until the ACoF exactly balances the magnitude and is in the opposite direction from the PGF. This condition is referred to as **geostrophic balance**. Figure 6.17 shows that the resulting wind (the geostrophic wind) flows parallel to the isobars. The closer the isobars are together, the faster the geostrophic wind. In reality, geostrophic balance occurs following a process called **geostrophic adjustment**, during which the wind overshoots and then undershoots its ultimate path in an oscillatory fashion.

In the Southern Hemisphere, the ACoF deflects moving air to the left, so the geostrophic wind flows in the opposite direction from that shown at the top of Figure 6.17.

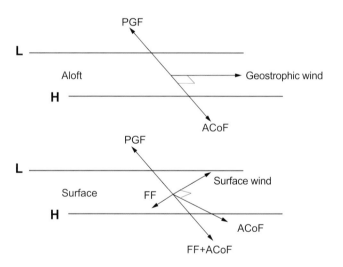

**Figure 6.17 The figure shows forces acting to produce winds aloft and at the surface along straight isobars in the Northern Hemisphere. The isobar denoted with an L is a line of low pressure; that denoted with an H is a line of high pressure. Two forces drive the geostrophic wind. Three forces drive the surface wind. The pressure gradient force (PGF) is perpendicular to the isobars, acting from high to low pressure. The apparent Coriolis force (ACoF) acts 90° to the right of the wind. The friction force (FF) acts in the opposite direction from the wind.**

The magnitude of the geostrophic wind (m/s) is determined from a balance between the pressure gradient force and the Coriolis force. The result is

$$v_g = \frac{C}{\rho_a 2\Omega \sin\varphi} \frac{\Delta p}{\Delta x} \qquad (6.35)$$

where $\rho_a$ is air density (kg/m$^3$), $\Omega$ **is the rotation rate of the Earth** (= $2\pi$ radians per day / 86,164 seconds per day = $7.292 \times 10^{-5}$ radians/s), $\varphi$ is latitude (degrees), $\Delta p/\Delta x$ is the pressure gradient, or change in pressure per unit distance (hPa/m), and $C$ = 100 kg/m-s$^2$-hPa converts units. Thus, the geostrophic wind speed increases with an increasing pressure gradient, a decreasing air density, and a decreasing latitude. It does not exist at the equator because the Coriolis force is zero at the equator.

---

## Example 6.13  Calculating the geostrophic wind speed

Calculate the geostrophic wind speed for a pressure gradient of 4 hPa over 150 km at a latitude of 30 °N at a pressure altitude of 570 hPa, where the air density is 0.73 kg/m$^3$.

### Solution:
From Equation 6.35, the geostrophic wind speed is 100 kg/m-s$^2$-hPa × (4 hPa / 150,000 m) / (0.73 kg/m$^3$ × 2 × 0.0000729 rad/s × sin30°)] = 50.1 m/s.

---

### 6.8.2.2  Surface Winds along Straight Isobars

When isobars are straight near the surface of the Earth, the friction force also affects the equilibrium wind speed and direction. The bottom of Figure 6.17 shows the wind direction that results from a balance among the PGF, ACoF, and FF at the surface in the Northern Hemisphere. Friction, which acts in the opposite direction from the wind, slows the wind. Because the magnitude of the ACoF is proportional to the wind speed, a reduction in wind speed reduces the magnitude of the ACoF. Because the sum of the FF and the ACoF must balance the magnitude of and be in the opposite direction from the PGF, the equilibrium wind direction shifts toward low pressure. On average, surface friction turns winds 15° to 45 ° toward low pressure, with lower values corresponding to smooth surfaces and higher values corresponding to rough surfaces. As such, if the wind is at your back in the Northern Hemisphere, turn 15° to 45° clockwise, and low pressure will be on your left.

In the Southern Hemisphere, surface winds are also angled toward low pressure, but symmetrically in the opposite direction from the wind shown in the bottom of Figure 6.17. Thus, if the wind is at your back in the Southern Hemisphere, turn 15° to 45° counterclockwise, and low pressure will be on your right.

### 6.8.2.3  Gradient Wind

When centers of low and high pressure (relative to nearby pressures at the same altitude) appear above the boundary layer, three forces, the PGF, ACoF, and ACfF, control the wind speed. The resulting wind is called the **gradient wind**, which is a circular wind around centers of low and high pressure aloft.

Gradient winds aloft flow counterclockwise around a center of low pressure in the Northern Hemisphere. An easy rule for determining the direction of flow in the Northern Hemisphere is the **left-hand rule**. Point your left thumb down (for low pressure) and follow your fingers to determine the direction of the gradient wind around a low-pressure center aloft. For a high-pressure center, point your left thumb up (for high pressure) and follow your fingers. They will indicate that the gradient wind flows clockwise around high-pressure centers aloft in the Northern Hemisphere.

Figure 6.18a illustrates the forces acting on the air and the resulting gradient wind around a low-pressure center in the Northern Hemisphere. The PGF must point toward the center of the low, and the ACfF must point away from it. However, the ACfF cannot, on its own, balance the magnitude of the PGF, because any moving wind has an ACoF associated with it. The PGF must be balanced in magnitude and opposite in direction from the

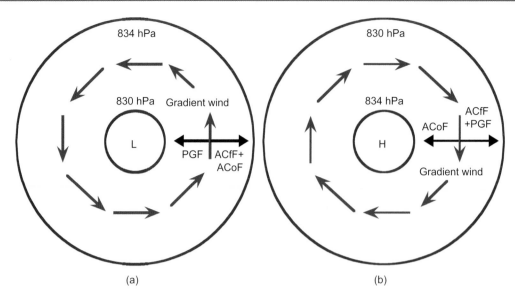

**Figure 6.18** Gradient winds (in green) around a center of (a) low and (b) high pressure in the Northern Hemisphere and the forces (in blue) affecting them. The pressures given represent those along the circular isobars (in red).

sum of the ACfF and the ACoF. Because the wind must point 90 degrees to the left of the ACoF (or the ACoF must point 90 degrees to the right of the wind), the resulting wind must flow counterclockwise around the center of low pressure.

Around a high-pressure center aloft in the Northern Hemisphere, the PGF and ACfF point opposite from the center of high. To balance the sum of these two forces, the ACoF must point toward the center of high. Because the wind is always pointed 90 degrees to the left of the ACoF in the Northern Hemisphere, the resulting gradient wind must flow clockwise around the center of high in the Northern Hemisphere, as illustrated in Figure 6.18b.

In the Southern Hemisphere, the **right-hand rule** is used. The gradient winds flows clockwise around a low-pressure center (in the direction of your fingers when your thumb is pointed down for low) and counterclockwise around a high-pressure center.

### 6.8.2.4 Surface Winds along Curved Isobars

The large-scale circulation of the wind around a surface low-pressure center is called a **cyclone**, and that around a surface high-pressure center is called an **anticyclone**. **Cyclonic flow** is flow around a low-pressure center (either at the surface or aloft), and **anticyclonic flow** is flow around a high-pressure center (either at the surface or aloft).

Near the Earth's surface, the friction force slows and turns the wind toward the center of low pressure in a cyclone and away from the center of high pressure

in an anticyclone. The flow of air into the center of low pressure from all directions around the center is a type of **convergence**, whereas the flow of air away from the center of high pressure is **divergence**. Air converges into centers of surface low pressure and diverges from centers of surface high pressure in both hemispheres.

In the Northern Hemisphere, surface winds converge while flowing counterclockwise around the center of low pressure and diverge while flowing clockwise around the center of high pressure (left-hand rule). Figure 6.19 shows the force balances and resulting winds in the presence of a surface (a) low-pressure system and (b) high-pressure system in the Northern Hemisphere.

In the low-pressure case near the surface, the PGF is balanced by the sum of the ACoF, ACfF, and FF. The resulting surface wind converges counterclockwise into the center of the low. The converging air rises, and the rising air expands and cools. If sufficient cooling occurs, clouds form. As such, surface cyclones are frequently associated with stormy weather. Near-surface wind speeds are fast because pressure gradients are strong in surface cyclones.

In the high-pressure case, the sum of the PGF and ACfF is balanced by the sum of the FF and ACoF. The resulting wind diverges clockwise out of the center of the high. The diverging air pulls more air downward into the center of the high. The descending air compresses and warms, potentially evaporating clouds. As such, surface high-pressure centers, or anticyclones, are often associated with sunny skies and warm weather. Near-surface

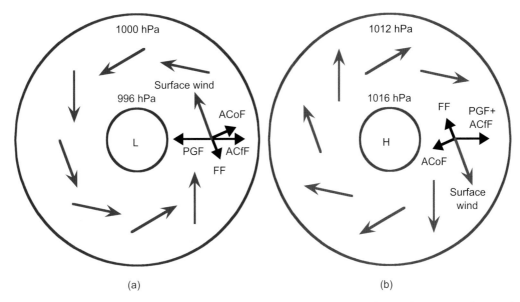

**Figure 6.19** Surface winds (in green) around centers of (a) low and (b) high pressure in the Northern Hemisphere and the forces (in blue) affecting them. The pressures given represent those along the circular isobars (in red).

wind speeds are usually slow because pressure gradients are weak in surface anticyclones.

In the Southern Hemisphere, surface winds converge while flowing clockwise around the center of low pressure and diverge while flowing counterclockwise around the center of high pressure (right-hand rule).

### 6.8.3 Global Circulation of the Atmosphere

The global circulation of the atmosphere drives large-scale pressure systems, which in turn, drive global wind patterns. Figure 6.20 shows features of the global circulation, including the major circulation cells, the belts of low and high pressure, and the predominant wind directions.

Winds have a west-east (**zonal**), south-north (**meridional**), and vertical component. The three circulation cells in each hemisphere shown in Figure 6.20 represent the meridional and vertical components of the Earth's winds, averaged zonally (over all longitudes) and over a long time period. The cells are symmetric about the equator and extend up to the tropopause, which is near 18 km altitude over the equator and near 8 km altitude over the poles.

Two cells, called **Hadley cells**, extend from 0 °N to 30 °N and 0 °S to 30 °S latitude, respectively. These cells were named after **George Hadley** (1685–1768), an English physicist and meteorologist, who, in 1735, first proposed the cells. Hadley's original cells, however, extended between the equator and the poles.

In 1855, **William Ferrel** (1817–1891), an American schoolteacher, meteorologist, and oceanographer, published an article pointing out that Hadley's one-cell model did not fit observations as well as did the three-cell model shown in Figure 6.20. Today, the middle cell in the three-cell model is called the **Ferrel cell**. A Ferrel cell extends from 30 °N to 60 °N and from 30 °S to 60 °S, whereas a **Polar cell** extends from 60 to 90 degrees in each hemisphere as well.

#### 6.8.3.1 Equatorial Low-Pressure Belt

Circulation in the three cells is controlled by heating at the equator, cooling at the poles, and the rotation of the Earth. In the Hadley cells, air rises over the equator because the sun heats this region intensely. Much of the heating occurs over water, some of which evaporates. As air containing water vapor rises, the air expands and cools, and the water vapor re-condenses to form clouds of great vertical extent. Condensation of water vapor releases latent heat, providing the air with more buoyancy. Over the equator, the air can rise up to about 18 km before it is decelerated by the stratospheric inversion. Once the air reaches the tropopause, it cannot rise much farther, so it diverges to the north and south. At the surface near the equator, air is drawn in horizontally to replace the rising air. So long as divergence aloft exceeds

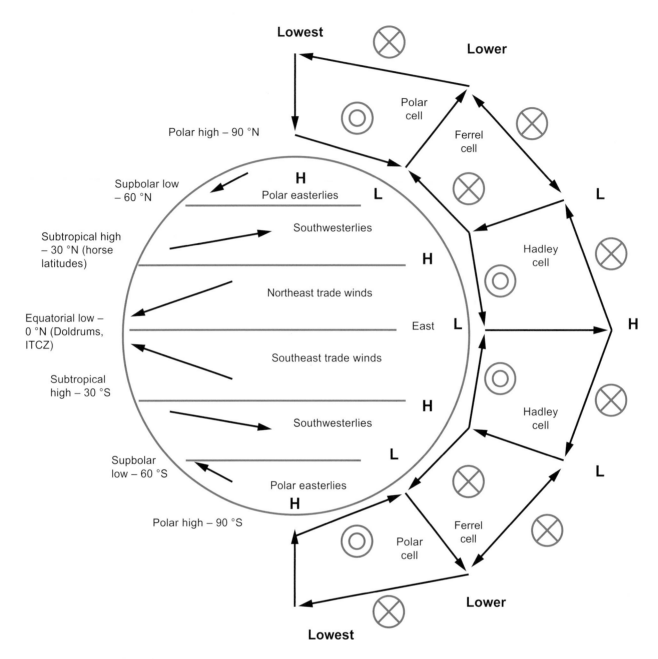

**Figure 6.20** Diagram of the three major circulation cells, the predominant surface pressure systems, and the predominant surface wind systems on the Earth; H at the surface indicates high pressure. H aloft is high altitude at a given pressure above the surface, which is similar to saying it is high pressure at a given altitude above the surface; L at the surface indicates low pressure. L aloft is low altitude at a given pressure above the surface (or low pressure at a given altitude). "Lower" indicates lower altitude at the same pressure as L aloft (or lower pressure at the same altitude). "Lowest" indicates lower altitude than "Lower" at the same pressure as L aloft (or lower pressure at the same altitude). Thus, elevated high and low altitudes are relative to other altitudes at the same pressure. This is similar to saying that elevated high and low pressures are relative to other pressures at the same altitude. The circles with crosshairs (×) denote winds going into the page (west to east). The concentric circles denote winds coming out of the page (east to west).

convergence at the surface, surface air pressure decreases. The surface low-pressure belt at the equator is called the **equatorial low-pressure belt**. Because pressure gradients are weak near the equator, winds are light, and the weather is often rainy. Thus, this region is also called the **doldrums**.

### 6.8.3.2 Winds Aloft in the Hadley Cells
As air diverges toward the north in the elevated part of the Northern Hemisphere Hadley cell, the ACoF force deflects much of it to the right (to the east), giving rise to **westerly winds** aloft (winds are generally named after the direction that they originate from). Westerly winds aloft in the Northern Hemisphere Hadley cell increase in magnitude with increasing distance from the equator until they meet air moving toward the equator from the Ferrel cell at 30 °N. The intersection of these two masses of air is called the **subtropical front**. The front is a region of sharp temperature contrast. The winds at the front are strongest at the tropopause, where they are called the **subtropical jet stream** winds. Winds aloft in the Southern Hemisphere Hadley cell are also westerly and culminate in a tropopause subtropical jet stream at 30 °S.

### 6.8.3.3 Subtropical High-Pressure Belts
As air converges aloft at the subtropical fronts at 30 °N and 30 °S, much of it descends. Air is then drawn in horizontally aloft to replace the descending air. As long as inflow aloft exceeds outflow at the surface, surface air pressure builds up. The surface high-pressure belts at 30 °N and 30 °S are called **subtropical high-pressure belts**. Because descending air compresses and warms, evaporating clouds, and because pressure gradients are relatively weak around high-pressure centers, surface high-pressure systems are characterized by sunny skies and light winds. Sunny skies and the lack of rainfall at 30 °N and 30 °S are two reasons why many deserts of the world are located at these latitudes. The light winds forced some ships sailing at 30 °N to lighten their cargo, the heaviest and most dispensable component of which was often horses. Thus, the 30 °N latitude band was also known as the **horse latitudes**.

### 6.8.3.4 The Trade Winds
At the surface near 30 °N and 30 °S, descending air diverges, some toward the equator and the rest toward a pole. Most of the air moving toward the equator is deflected by the ACoF to the right (toward the west) in the Northern Hemisphere and to the left (toward the west) in the Southern Hemisphere, except that friction reduces the extent of ACoF turning. The resulting winds in the Northern Hemisphere are called the **northeast trade winds** because they originate from the northeast (Figure 6.20). Those in the Southern Hemisphere are called the **southeast trade winds** because they originate from the southeast. Sailors from Europe have used the northeast trades to speed their voyages westward since the fifteenth century. The trade winds are consistent winds. The northeast and southeast trade winds converge at the **Intertropical Convergence Zone** (ITCZ; Figure 6.20), which moves north of the equator in the Northern Hemisphere summer and south of the equator in the Southern Hemisphere summer, generally following the direction of the sun. At the ITCZ, air convergence and surface heating lead to the rising arm of the Hadley cells.

### 6.8.3.5 Subpolar Low-Pressure Belts
As surface air moves poleward in the Ferrel cells, the ACoF turns it toward the right (east) in the Northern Hemisphere and left (east) in the Southern Hemisphere. However, surface friction reduces the extent of turning, so that near-surface winds at **midlatitudes** (30 °N to 60 °N and 30 °S to 60 °S) are generally westerly to southwesterly (from the west or southwest) in the Northern Hemisphere and westerly to northwesterly in the Southern Hemisphere.

In both hemispheres, near-surface air moving toward the pole in the Ferrel cell meets air from the Polar cell moving toward the equator at the **polar front**, which is a region of sharp temperature contrast between these two cells. Converging air at the surface front rises and diverges aloft, reducing surface air pressure and increasing air pressure aloft relative to pressures at other latitudes but at the same altitude. The surface low-pressure regions at 60 °N and 60 °S are called **subpolar low-pressure belts**. Regions of rising air and surface low pressure are associated with storms. Thus, the intersection of the Ferrell and Polar cells is associated with stormy weather. Unlike at the equator, surface pressure gradients and winds at the polar front are relatively strong. West-east wind speeds also increase with increasing height at the polar fronts. At the tropopause in each hemisphere, they culminate in the **polar front jet streams**. Although the subtropical jet streams do not meander to the north or south over great distances, the polar front jet streams do. Their predominant direction is still from west to east.

### 6.8.3.6 Westerly Winds Aloft at Midlatitudes

One might expect air in the elevated portion of the Ferrel cells to move toward the equator and for the ACoF to deflect such air to the west, creating easterly winds (from east to west) aloft at midlatitudes in both hemispheres. In fact, winds aloft in the Ferrel cell are generally westerly (from west to east), although they meander between south and north. Part of the reason for the westerly winds aloft in the Ferrel cells is that heights of constant pressure (or pressures at a constant height) decrease between the equator and the poles in the upper troposphere, as shown in Figure 6.20. Winds tend to start flowing down the height (or pressure) gradient, toward the poles in the Ferrel cell aloft. The Coriolis force then acts on this moving air, turning it toward the east in both hemispheres, creating westerly winds aloft in the Ferrel cell in both hemispheres.

The reason heights of constant pressure (or pressures at a constant height) decline from equator to poles is that temperatures transition from warm to cold between the equator and poles. Warm air rises and cold air sinks; thus, near the equator, the rising air pushes up the height of a constant pressure level (or increases the pressure at a constant height), and near the poles, sinking air pushes down the height of a constant pressure level (or decreases the pressure at a constant height).

The second reason for westerly winds aloft in the Ferrel cell relates to the presence of centers of low and high pressure. Descending air at 30 °N and 30 °S creates centers of surface high pressure, and rising air at 60 °N and 60 °S creates bands or centers of surface low pressure. Surface winds moving around a Northern Hemisphere surface high-pressure center accelerate clockwise (diverging away from the center of the high), and surface winds moving around a surface low-pressure center accelerate counterclockwise (converging into the center of the low) (Figure 6.19).

In the Northern Hemisphere, low-pressure centers at 60 °N are staggered between high-pressure centers at 30 °N (e.g., Figure 6.21). Three to five such low- and high-pressure centers circle the Earth. Since winds travel clockwise around the highs and counterclockwise around the lows, the winds between 30 °N and 60 °N will always meander sinusoidally from west to east around the globe, both near the surface and aloft. The same holds true in the Southern Hemisphere. The flow created by these highs and lows is consistent with the expectation that near-surface and elevated winds in the Ferrel cell are predominantly westerly (west to east).

### 6.8.3.7 Polar Easterlies

Air moving poleward aloft in the Polar cells is turned toward the east in both hemispheres by the ACoF, causing elevated winds in the Polar cells to be westerly. At the poles, cold air aloft descends, increasing surface air pressure. The surface high-pressure regions are called **Polar highs**. Air at the polar surface diverges toward the equator. The ACoF turns this air toward the west. Friction is weak over the Arctic because polar surfaces are either sea ice, snow on top of sea ice, or water, and all three have relatively smooth surfaces. Due to the lack of friction, surface winds are relatively easterly. The Antarctic is a continent with high mountains and rough surfaces, so the surface winds are turned more toward low pressure by friction, thus are more southeasterly (coming from the southeast). Nevertheless, in both hemispheres, the resulting surface winds in the Polar cells are called **polar easterlies** (Figure 6.20).

## 6.8.4 Local Winds

Two important types of local winds relevant to wind power are sea/land breezes and gap winds. These are discussed briefly.

### 6.8.4.1 Sea/Land Breezes

**Sea breezes** are daytime flows of air from the ocean to the land. They also occur between a lake and land and between a bay and land. In those cases, they are called **lake breezes** and **bay breezes**, respectively. At night, the reverse flow, a **land breeze**, occurs in all cases. However, nighttime land breezes are much weaker than are daytime sea, lake, or bay breezes.

Figure 6.22 illustrates a basic sea breeze circulation. During the day, land heats up relative to water because soil has a lower specific heat than does water. Rising air over land forces air aloft to diverge horizontally, decreasing surface air pressures over land. As a result of the pressure gradient between land and water, air moves from the water, where the pressure is now relatively high, toward the land. In the case of ocean water meeting land, the movement of near-surface air is the sea breeze. Although the ACoF acts to turn the sea breeze air slightly to the

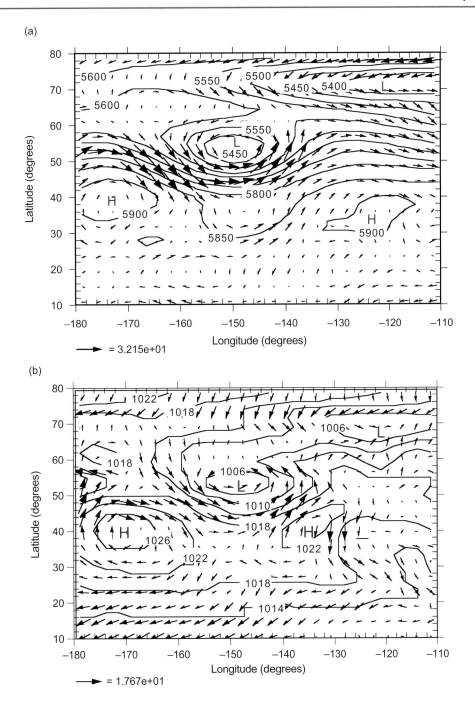

(a)

(b)

**Figure 6.21** Maps of (a) 500-hPa height contours (m) and wind vectors (m/s) and (b) sea level pressure contours (hPa) and near-surface wind vectors (m/s) for August 3, 1990, at 12 GMT for the northern Pacific Ocean. Height contours on the constant pressure map (a) are analogous to isobars on a constant height map; thus, high (low) heights in map (a) correspond to high (low) pressures on a constant height map. The surface low-pressure system at −148 °W, 53 °N in (b) is the Aleutian low. The surface high-pressure system at −134 °W, 42 °N in (b) is the Pacific high. The arrow below each map gives the scale of the wind speed arrow in m/s.

right, the distance traveled by the sea breeze is too short (a few tens of kilometers) for the air to be turned noticeably.

Meanwhile, some of the diverging air aloft over land returns toward the water. The convergence of air aloft over water increases surface air pressure over water, prompting a stronger flow of surface air from the water to the land, completing the basic sea breeze circulation cell. At night, land cools to a greater extent than does water, and all the pressures and flow directions in Figure 6.22 reverse themselves, creating a

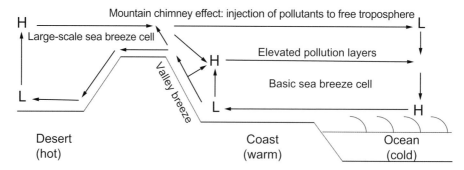

**Figure 6.22** Illustration of a basic sea breeze circulation cell embedded in a large-scale sea breeze circulation cell. Also shown are a valley breeze, the mountain chimney effect, and the formation of elevated pollution layers. Pressures shown (L and H) are relative to other pressures at the same altitude.

land breeze, a near-surface flow of air from land to water.

Figure 6.22 illustrates that a basic sea breeze circulation cell can be embedded in a large-scale sea breeze cell. The Los Angeles Basin, for example, is bordered on its southwestern side by the Pacific Ocean and on its eastern side by the San Bernardino Mountains. The Mojave Desert lies to the east of the mountains. The desert heats up more than does land near the coast during the day, creating a thermal low-pressure region over the desert, drawing air in from the coast, and creating the circulation pattern shown.

Sea breeze wind speeds peak in the afternoon, when land-ocean temperature differences peak. Similarly, sea breeze winds peak during summer and are at a minimum during winter. Land breezes are weaker than sea breezes because the land-ocean temperature difference is small at night.

The regularity of sea breeze winds is one reason that offshore wind is so attractive. In many locations, near-coast offshore wind speeds peak late in the afternoon due to the peaking of sea breeze winds. As such, offshore winds can be peak coincident with electricity demand, increasing their economic attractiveness (Dvorak et al., 2013).

### 6.8.4.2 Gap Winds, Valley Breezes, and Mountain Breezes

When air flows through a mountain pass or a narrow valley, it speeds up to conserve mass. For example, air mass (kg) equals $\rho_a A v$, where $\rho_a$ is air density (kg/m$^3$), and $A$ is the cross-sectional area (m$^2$) through which the air is moving at wind speed $v$ (m/s). When air moves from an open plain to a narrow pass, its density remains constant, but the cross-sectional area through which it

flows decreases. As such, wind speed $v$ must increase to conserve mass. This increase in wind speed is called the **tunneling effect**, and the fast winds are referred to as **gap winds**. The first three commercial wind farms in the world, at Altamont Pass, Tehachapi Pass, and San Gorgonio Pass, were all located in mountain passes with fast gap winds.

Gap winds should not be confused with valley or mountain breezes. A **valley breeze** is a wind that blows from a valley up a mountain slope and results from the heating of the mountain slope during the day. Slope heating causes air on the mountain slope to heat and rise, drawing air up from the valley to replace the rising air. Figure 6.22 illustrates that in the case of a mountain near the coast, a valley breeze can become integrated into a large-scale sea breeze cell. The opposite of a valley breeze is a **mountain breeze**, which originates from a mountain slope and travels downward. Mountain breezes typically occur at night, when mountain faces cool rapidly. As a mountain face cools, air above the face also cools and drains downslope.

### 6.8.5 Global and Regional Wind Resources

Figure 6.23 shows calculated world wind resources at 100 m above ground level, which is in the range of hub heights of modern wind turbines. Large regions of fast onshore winds worldwide include the Great Plains of the United States and Canada; northern Europe; parts of Russia; the Gobi and Sahara deserts; much of the Australian desert areas; New Zealand; parts of South Africa; parts of Peru; and southern South America. Windy offshore regions include the North Sea, the east and west coasts of North America, offshore Australia, and the east coast of Asia, among other locations.

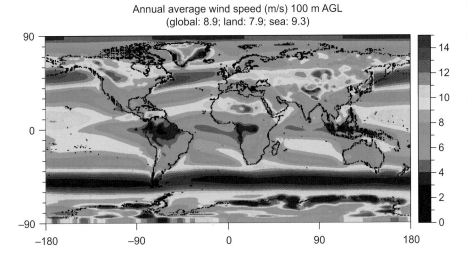

Annual average wind speed (m/s) 100 m AGL
(global: 8.9; land: 7.9; sea: 9.3)

Figure 6.23 **Annual average wind speed (m/s) at 100 m above ground level (AGL), modeled by computer at 2.5 degrees west-east × 2.0 degrees south-north horizontal resolution. The numbers in parentheses are globally, land-, and ocean-averaged values.** From Jacobson et al. (2017).

Wind turbine costs have dropped significantly since 2000, so the mean annual wind speed at which wind energy is cost competitive has dropped as well. Wind speeds 7 m/s or faster in the annual average are still ideal for inexpensive wind. Table 6.6, which is based on data, summarizes the percent land or coastal area, in different regions of the world, where the mean wind speed at 80 m above the ground level is 7 m/s or faster. The table indicates that 15 percent of the land area and 17 percent of the land plus coastal offshore areas in the United States have such wind speeds. In Europe, 14.2 percent of the land has such wind speeds. In South America, Africa, and Asia, though, the areas with fast winds are smaller. Globally, 13 percent of stations are above the threshold.

Figure 6.24 shows modeled wind resources at 80 m and 90 m above the ocean surface offshore of California and the U.S. East Coast, respectively, for water depths down to 200 m. The figure indicates that, along both coasts, offshore resources increase from south to north. This reflects the fact that northern latitudes are closer to major semi-permanent low-pressure systems near 60 °N – namely, the **Aleutian low** in the Pacific Ocean and the **Icelandic low** in the Atlantic Ocean. These low-pressure systems are the source of strong winds and storms, particularly during winter. The slower offshore winds toward the southern United States reflect the proximity of these locations to semi-permanent high-pressure systems, namely, the **Pacific high** in the Pacific Ocean and the **Bermuda high** in the Atlantic Ocean. These high-pressure systems are sources of weak winds and sunny skies. During hurricane season (June 1 to November 30), however, the Southeast and Gulf coasts of the United States are subject to increased wind speeds due to tropical storms and hurricanes forming in the tropical Atlantic Ocean.

Table 6.7 summarizes key results from Figure 6.24. The table indicates that, along the United States East Coast, the ratio of annually averaged deliverable power in shallow water (0 to 50 m depth) to that in shallow plus deeper water (0 to 200 m depth) is about 1:3. Along the West Coast, it is about 1:8. This occurs because much less shallow-water wind resource exists along the West Coast, since water along the West coast becomes much deeper much faster than does water along the East Coast. In fact,

Table 6.6 **Percent of land area and near-shore area with a mean annual wind speed exceeding 6.9 m/s at 80 m above ground level, as determined from sounding wind speed data combined with near-surface wind speed data. Because the data are limited and sparser over some continents than others, the results are only approximate.**

| Location | Percent |
|---|---|
| Europe | 14.2 |
| North America | 19.0 |
| United States over land | 15.0 |
| United States over land and near shore | 17.0 |
| South America | 9.7 |
| Oceania | 21.2 |
| Africa | 4.6 |
| Asia | 2.7 |
| Antarctica | 60 |
| Global over land | 13 |

Source: Archer and Jacobson (2005).

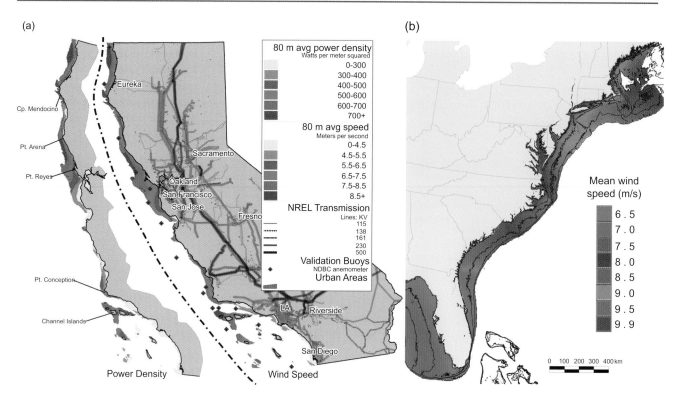

**Figure 6.24** (a) Modeled 2005 to 2007 wind power density and wind speed offshore the California coast at 80 m altitude above sea level. Values shown are in water 0 to 200 m depth. The transmission infrastructure is also shown. From Dvorak et al. (2010). (b) Modeled 2006 to 2010 annual average wind speeds offshore of the East Coast of the United States at 90 m altitude above sea level. Values shown are in water 0 to 200 m depth. From Dvorak et al. (2013).

during the last ice age, much of the water area currently offshore the East Coast was land. Due to the shallow nature of offshore East Coast water, it is easier to install wind turbines into the sea floor with bottom-fixed foundations there than along the West Coast. With the advancement of floating wind turbines, it is now possible to obtain much more annually averaged deliverable offshore wind power along both coasts than shown in Table 6.7.

A 100 percent WWS roadmap calls for 12.8 GW of delivered offshore wind power from California to provide 10 percent of California's end-use electric power demand after all energy sectors have been electrified (Jacobson et al., 2015a). Table 6.7 indicates that five times the resource needed is available from 0 to 200 m depth. Thus, meeting this goal can be accomplished with a combination of bottom-fixed foundation and floating offshore wind turbines.

A U.S. roadmap calls for 154.3 GW of delivered offshore wind power to provide 16.4 percent of all U.S. end-use electricity demand after all energy sectors have been

**Table 6.7 Summary of modeled California and U.S. East Coast offshore annually averaged deliverable power (GW) as a function of water depth. The California results are for locations with mean annual wind speeds ≥7.0 m/s. The East Coast results are for locations with a wind turbine capacity factor ≥40 percent (Dvorak et al., 2013). Both studies assume 33 percent of water area is excluded from development due to conflicting water uses.**

| Water depth (m) | Offshore California deliverable GW | Offshore East Coast deliverable GW |
|---|---|---|
| 0 to 30 | 2.3 | 18 |
| 0 to 50 | 8.3 | 59 |
| 0 to 200 | 64.9 | 157 |

Source: Dvorak et al. (2010).

electrified (Jacobson et al., 2019). Table 6.7 indicates that East Coast offshore wind resources 0 to 200 m depth alone can supply all of this demand. However, this demand will, in reality, be satisfied not only with offshore East Coast wind but also with wind offshore the West, Gulf, Alaskan, and Hawaiian coasts.

## 6.8.6 World Saturation Wind Power Potential

Wind turbines compete with each other for the same kinetic energy in the wind. When one wind turbine converts kinetic energy to mechanical energy to spin a turbine's blades, and its generator converts the mechanical energy to electricity, less kinetic energy is available for other wind turbines in the wind farm and in the world. As more turbines become operational, each turbine is able to extract less and less energy. At some point, the addition of one more turbine worldwide results in no additional power generation (kinetic energy extraction). At that point, the annual average power extracted by the existing turbines is called the **saturation wind power potential** (**SWPP**) (Jacobson and Archer, 2012). The SWPP is important because it gives the upper limit to how much power is available from wind worldwide (land plus ocean) or over land alone for turbines at a given hub height.

The reduction in wind speed due to wind turbines can be described mathematically as follows. The kinetic energy (J) in the wind at a given time $t$ and location is

$$E(t) = \frac{1}{2} M_a v(t)^2 \tag{6.36}$$

where $M_a$ is the mass of air (kg) and $v(t)$ is the instantaneous wind speed (m/s) at time $t$. Suppose, during a

time increment $\Delta t$ (s), a wind turbine extracts an amount of energy (J) from the wind equal to

$$\Delta E(t) = P_b[v(t)]\Delta t \tag{6.37}$$

where $P_b$ is the instantaneous power (W) generated by the turbine's blades as a function of wind speed $v(t)$. $P_b$ is determined from the turbine's power curve. The remaining kinetic energy in the wind at time $t + \Delta t$ is thus

$$E(t + \Delta t) = E(t) - P_b[v(t)]\Delta t = \frac{1}{2} M_a v(t + \Delta t)^2 \tag{6.38}$$

Solving for the resulting wind speed at the new time gives

$$v(t + \Delta t) = \sqrt{\frac{2[E(t) - P_b[v(t)]\Delta t]}{M_a}} \tag{6.39}$$

As such, the extraction of kinetic energy by a wind turbine reduces the wind speed seen by other turbines.

Depending on the purpose of the calculation, the air mass used in the above equations can be either the mass of the air flowing through the turbine during a specific time interval (e.g., $M_a = \rho_a A_t v(t)\Delta t$, where $\rho_a$ is air density [kg/m³] and $A_t$ is turbine swept area [m²]), or it can be the mass of all air in a large volume that has mean wind

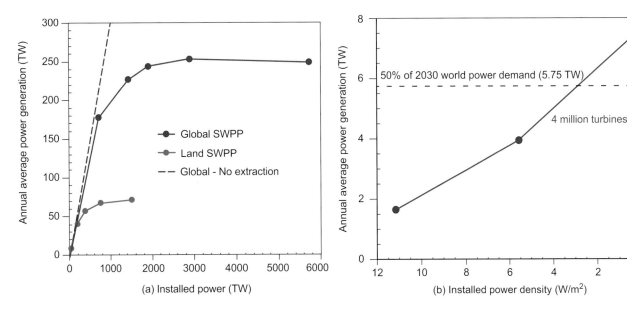

**Figure 6.25** (a) Annual average electric power generation by wind turbines as a function of their installed (nameplate) capacity worldwide (Global SWPP curve) and over all land outside Antarctica (Land SWPP). Also shown is the power output if no competition among turbines were allowed (Global – No extraction). (b) Annual average power generation at three installed power densities, each consisting of 4 million 5-MW wind turbines (20 TW total nameplate capacity). Also shown is a line indicating the power output needed to provide 5.75 TW of power from wind worldwide. From Jacobson and Archer (2012).

speed $v(t)$. The former would be used to estimate the wind speed immediately downstream of one turbine. The latter would be used to estimate the change in mean wind speed over a large volume of air encompassing one or more turbines. In the former case, Equation 6.39 is independent of the time increment $\Delta t$ because that cancels out all of terms on the right side of the equation. In the latter case, final wind speed varies with the time increment. Example 6.14 illustrates the results in the two cases.

## Example 6.14  Extracting kinetic energy from the wind

Estimate the wind speed in two cases: (a) downstream of a single turbine and (b) averaged over a large 5 km × 5 km horizontal area × 126 m vertical thickness region in which one turbine resides. In both cases, assume the upstream wind speed is 10 m/s, the turbine extracts 3,000 kW at that wind speed, the turbine blade diameter is 126 m, and the air density is 1.23 kg/m$^3$. For each case, find the downstream wind speed after 1 minute and 10 minutes.

### Solution:

The wind turbine swept area is $\pi \times (126 \text{ m} / 2)^2 = 12{,}469 \text{ m}^2$. In Case (a), the mass of air passing through the turbine blades over 1 minute is $M_a = 1.23 \text{ kg/m}^3 \times 12{,}469 \text{ m}^2 \times 10 \text{ m/s} \times 1 \text{ min} \times 60 \text{ s/min} = 9.20 \times 10^6 \text{ kg}$. The initial kinetic energy in the wind, from Equation 6.36, is $E = 0.5 \times 9.20 \times 10^6 \text{ kg} \times (10 \text{ m/s})^2 = 4.6 \times 10^8 \text{ J}$. The energy extracted by the wind turbine is $3 \times 10^6 \text{ J/s} \times 60 \text{ s} = 1.8 \times 10^8 \text{ J}$. From Equation 6.39, the downstream wind speed is 7.8 m/s, so the turbine reduces the wind speed by 22.2 percent. Over 10 minutes, the downstream wind speed is also 7.8 m/s because all terms in Equation 6.39, $M_a$, $E(t)$, and $P_b \Delta t$ have a $\Delta t$ term that cancels out, so the result is independent of time.

In Case (b), the mass of air is the mass of all air in the region, not just the air that goes through the turbine blade. As such, $M_a = 1.23 \text{ kg/m}^3 \times (5{,}000 \text{ m})^2 \times 126 \text{ m} = 3.97 \times 10^9 \text{ kg}$. The initial kinetic energy in the wind, from Equation 6.36, is $E = 0.5 \times 3.97 \times 10^9 \text{ kg} \times (10 \text{ m/s})^2 = 1.9 \times 10^{11} \text{ J}$. From Equation 6.39, the volume averaged wind speed is 9.995 m/s, so the turbine reduced the volume averaged wind speed by 0.05 percent. Over 10 minutes, the volume averaged wind speed is 9.95 m/s, so the turbine reduced the overall wind speed by 0.47 percent.

In sum, whereas the wind speed downwind of an individual turbine stays constant with time if the upstream wind speed stays constant, the mean wind speed averaged over a volume of air decreases if the extraction of kinetic energy is allowed to affect the overall kinetic energy in the volume of air.

The extraction of kinetic energy and reduction in wind speed by wind turbines must reach a limit. Figure 6.25a provides an estimate of this limit worldwide and over all world land. The figure was obtained by running global, three-dimensional computer model simulations for several model years. Each simulation contained a different number of wind turbines with a hub height of 100 m over land or over land plus ocean.

In the model, wind turbines extracted kinetic energy to produce electric power, and wind speeds were adjusted accordingly for each time increment in a manner similar to Equations 6.35 to 6.38. Because the model was three dimensional, it accounted for the increased vertical transport, due to turbulence, of faster horizontal winds aloft down to the hub height, where winds were depleted behind the turbines. As shown in Figure 6.26, this resulted in wind speeds above hub height also decreasing due to energy extraction by wind turbines at hub height.

Figure 6.25a indicates that, as the installed (nameplate) capacity of wind turbines increases over land plus ocean worldwide, the extractable power among all wind turbines increases, but with diminishing returns. In fact, above 3,000 TW of nameplate capacity, no additional power from the wind at 100 m can be extracted. Table 6.8 indicates that the worldwide limit to extractable power is about 253 TW. Over land outside of Antarctica, the limit is about 72 TW.

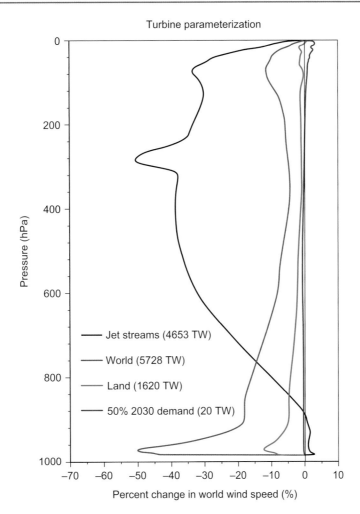

Turbine parameterization

Figure 6.26 Percent wind speed reduction averaged globally as a function of altitude (air pressure) due to using wind turbines to extract kinetic energy to produce electricity. The jet stream case is with 930.6 million 5-MW wind turbines at 10 km from 10 °S to 70 °S and 10 °N to 70 °N. The world case is with 1.146 billion 5-MW turbines at 100 m over the world's land and oceans. The land case is with 324.5 million 5-MW turbines over the world's land, including Antarctica. The 50 percent 2030 demand case is with 4 million 5-MW turbines over the world's land from 15 °S to 60 °S and from 15 °N to the Arctic Circle. In all cases, the turbines have 126-m blade diameters. From Jacobson and Archer (2012).

## Transition highlight

The world's near-surface winds contain far more power than is needed for all humanity. For example, the world needs only about 3.9 TW of annual average power output from wind in 2050 for wind to provide 45 percent of the world's end-use power demand after all energy sectors have been electrified (Jacobson et al., 2019). As such, **18.5 times more extractable wind power over land is available than is needed, and 65 times more extractable power over land plus ocean worldwide is available than is needed to meet this demand**. Thus, there is no wind resource barrier to obtaining even 100 percent of the world's 2050 all-purpose electric power from wind.

Figure 6.25a also shows the power output from a case (dashed line) where wind turbines covering the world extract kinetic energy to produce electricity at the modeled wind speeds, but the wind speeds are not reduced. In this case, wind turbines have no impact on the wind speeds and turbines extract 1,750 TW of power (Table 6.8). This compares with 253 TW when wind turbines do reduce wind speeds (Table 6.8). Thus, not accounting for wind turbine reduction in wind speed can result in a factor of 7 error in the overall wind power output in the limit of complete coverage of the world with wind turbines.

Finally, Table 6.8 indicates that, in the jet streams, about 378 TW of annual average power is available. However, as of 2020, it is not cost effective to extract

**Table 6.8** Saturation wind power potential (SWPP) at 100 m above ground level (AGL) globally, at 100 m AGL over land outside Antarctica, and at 10 km in the jet streams (from 10 °S to 70 °S and 10 °N to 70 °N). Also shown is the annual average power available worldwide at 100 m AGL if kinetic energy extraction by wind turbines were not accounted for.

| Region | Annual power output (TW) |
|---|---|
| Global (No extraction) | 1,750 |
| Global-SWPP | 253 |
| Land-SWPP | 72.0 |
| Jet streams | 378 |
| Wind needed in 2050 | 3.9 |

Source: Jacobson and Archer (2012). Also shown ("Wind needed in 2050") is the wind power supply needed over land plus ocean in 2050 for wind to provide 45.0 percent of the world's end-use power demand for all purposes after all energy has been electrified. That supply would be obtained with a projected nameplate capacity of 12.1 TW (Jacobson et al., 2019).

wind power from the jet streams commercially, although some companies have tried.

Figure 6.25b examines the impact on total power output with different installed wind power densities but with the same overall nameplate capacity. It examines a situation with 4 million 5-MW turbines (20 TW total nameplate capacity) in three configurations. In the first configuration, the 20 TW are compressed into three wind farms globally. In the second, they are expanded slightly to 8 wind farms globally. In the third, the 20 TW are spread out over land from 15 °S to 60 °S and from 15 °N to 60.56 °N (Arctic Circle). The figure indicates that, **when wind farms are separated from each other, their output power can increase by a factor of up to 4.6 for the same nameplate capacity.** This is due to the reduced competition for available kinetic energy among wind turbines when wind farms are separated.

Finally, Figure 6.26 shows the vertical profile of the percent change in world wind speed for each kinetic energy extraction case in Table 6.8. It shows that saturating the world with wind turbines at 100 m or at 10 km reduces the global average wind speeds at those altitudes by 50 percent. Saturating land reduces the global average 100-m wind speed by about 12.2 percent. Using 4 million 5-MW turbines reduces the global average 100-m wind speed by only about 0.36 percent.

## 6.9 Wind Turbine Impacts on Climate, Hurricanes, and Birds

This section examines the impacts of wind turbines on global and local climate, hurricanes, and birds. Briefly, wind turbines cause a net cooling of global climate due to their reduction in water vapor, a greenhouse gas. Lots of offshore wind turbines can help to dissipate a hurricane from the outside in, thereby reducing hurricane wind speeds and storm surge. Whereas wind turbines kill birds, the numbers are small compared with other sources of bird death, including fossil-fuel electricity generation, buildings, and cats.

### 6.9.1 Wind Turbine Impacts on Climate

Whereas wind turbines can cause a local warming of the ground downstream of a wind farm, they cause a net cooling of climate in the global average. The reasons are as follows.

Wind turbines extract kinetic energy from the wind, reducing wind speeds downstream of a wind farm. A reduction in wind speed reduces evaporation of liquid water from soil or a water body (Section 3.2.2.3). Since evaporation is a cooling process, a reduction in evaporation warms the ground or the ocean or a lake surface. Some studies have found an increase in ground temperature near a wind farm (e.g., Zhou et al., 2012).

Normally, water vapor in the air would condense to form clouds. However, because wind turbines reduce evaporation, less condensation and cloud formation now occur. Because condensation releases heat to the air, less condensation cools the air.

In addition, because a wind turbine slows the wind in its wake, it drops the air pressure in its wake (Section 6.4). Lower pressure decreases temperature, as evidenced by the increase in fog thickness in the wakes of turbines at the Horns Rev offshore farm (Hasager et al., 2013). The increase in fog thickness results from a slight increase in the relative humidity in a turbine's wake upon a slight drop in temperature, which is due to the drop in pressure.

The ground warming (due to reduced evaporation) and air cooling (due to reduced condensation and to the temperature drop in the turbine wake) effectively cancel each other out in the global average.

However, water vapor is a greenhouse gas that traps thermal-infrared (TIR) radiation emitted by the surface

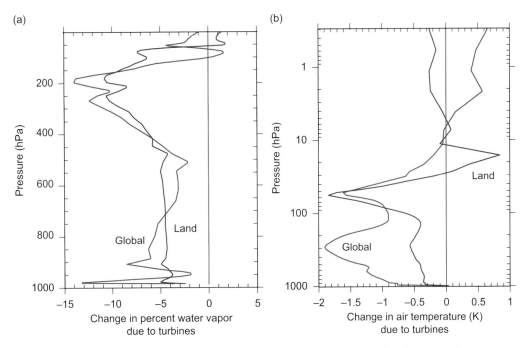

Figure 6.27 (a) Percent reduction in water vapor mixing ratio (kg-H$_2$O/kg-dry-air) averaged globally as a function of altitude (air pressure) due to kinetic energy extraction from the wind by 1.146 billion 5-MW wind turbines (5,730 GW total) at 100 m over the world's land and oceans (blue, Global) and by 299 million 5-MW turbines (1,495 GW total) over the world's land alone outside of Antarctica (red, Land). (b) Change in the vertical profile of globally averaged air temperature for the same conditions as in (a). From Jacobson and Archer (2012).

of the Earth, in the lower atmosphere. Thus, less water vapor means that TIR radiation can now escape to the upper atmosphere or outer space. The net removal of energy from the lower atmosphere by wind turbines through this mechanism causes a net global cooling. However, such cooling is measurable only with a large number of turbines.

Wind turbines cause other impacts that affect temperatures, but the water vapor effect dominates. For example, wind turbines convert kinetic energy to electricity. The electricity dissipates to heat, which increases the internal energy of the air (Section 3.2.2.3), raising air temperature. However, some of that higher air temperature is used to lift air, converting internal energy to gravitational potential energy (GPE) (Section 3.2.2.3). Differences in GPE over a horizontal distance reproduce some kinetic energy, thus winds.

Thus, if water vapor were not a dominant factor (e.g., if no TIR radiation leaked to space due to wind turbine reduction of water vapor), electricity production from wind turbines (and other electricity sources) might increase temperatures slightly. However, in the absence of wind turbines, the kinetic energy in the wind would also dissipate to heat. This is because winds bash into rocks, trees, houses, waves, and soil. The resulting friction converts kinetic energy to

heat. This dissipation occurs over a longer time than does the dissipation caused by wind turbine extraction of energy followed by electricity use.

Figures 6.27 and 6.28 show additional results from two of the global simulations that were used to calculate the maximum extractable power in the atmosphere by wind turbines at a given height (Section 6.8.6). Each figure shows the difference in water vapor and temperature when wind turbines covering either all world ocean and land (Figure 6.27) or all world land outside of Antarctica (Figures 6.27 and 6.28) are present versus absent. The results represent the impact of wind turbines on either water vapor or temperature.

In the simulations with turbines, the turbines extract kinetic energy from the wind and convert it to electricity. The electricity dissipates to heat. The heat causes air to rise. Thus, the heat is partly converted to gravitational potential energy. Differences in GPE recreate some kinetic energy, but the net result of wind turbines is to reduce kinetic energy and wind speed. Reductions in wind speed reduce water vapor, allowing more TIR radiation to escape to space, cooling the Earth's surface.

Figure 6.27 indicates that wind turbines reduced near-surface water vapor by 4 to 13 percent in the global case

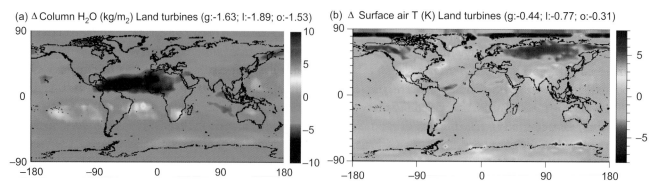

(a) Δ Column H$_2$O (kg/m$_2$) Land turbines (g:-1.63; l:-1.89; o:-1.53)

(b) Δ Surface air T (K) Land turbines (g:-0.44; l:-0.77; o:-0.31)

**Figure 6.28 (a) Computer-modeled difference in annual average column water vapor due to wind turbines covering land and their extraction of available kinetic energy versus when no turbines are present. The model simulation with turbines included 299 million 5-MW turbines (1,495 GW total) over the world's land alone outside of Antarctica. (b) Change in annual average near-surface air temperature from the same simulation pair. The letters g, l, and o are global, land, and ocean averaged values, respectively. Figure 6.27 shows the globally averaged vertical profiles of the change in water vapor and temperature for this simulation pair.** From Jacobson and Archer (2012).

and 2 to 5 percent in the land case. Figure 6.28a indicates that the column reduction in water vapor in the land case was 4.7 percent. The water vapor reduction led to a globally averaged near-surface air temperature reduction of 0.44 K in the land case (Figure 6.28b).

### Transition highlight

Whereas the global and land turbine simulations just discussed are extreme cases that will never be realized on Earth, Figure 6.29 shows results from a more realistic pair of simulations. In this pair, 2.5 million 5-MW turbines are distributed over 139 countries to provide a projected 37.1 percent of the world's all-purpose end-use energy in 2050 after all energy sectors have been electrified. This is the same pair of simulations that resulted in the extracted wind power shown in Figure 6.12. Figure 6.29 indicates that, in this case, wind turbines reduced column water vapor by 0.29 kg/m$^2$ and globally averaged near-surface air temperatures by about 0.03 K. This temperature change represents about 3 percent of the 1 K global warming through 2020. **Thus, wind turbines powering 37.1 percent of all the world's 2050 end-use energy may directly reduce on the order of 3 percent of net global warming** (Jacobson et al., 2018a).

In sum, with the installed wind energy needed for humankind in 2050, wind turbines may serve an additional benefit beyond replacing the electricity from polluting fossil fuels. They may also directly cool the surface,

thereby reducing global warming beyond that caused by their reduction in greenhouse gas emissions.

## 6.9.2 Wind Turbine Impacts on Hurricanes

Because wind turbines extract kinetic energy from regular winds, wind turbines can extract kinetic energy from hurricane winds as well. However, if hurricane winds are too strong, they will topple a wind turbine. Also, a hurricane contains so much kinetic energy, many more than a few turbines are needed to have any measurable impact on the hurricane.

Wind turbines in the past have been designed to withstand only 50-m/s wind speeds. Some of those now are built to withstand 57.5 m/s (Sirnivas et al., 2014). Both of these wind speeds correspond to Category 3 hurricane wind speeds on the Saffir–Simpson scale. Yet, hurricanes can reach Category 4 (58.6 to 69.3 m/s) and Category 5 (>69.3 m/s) status. Thus, an important question to address is whether many turbines collectively can keep hurricane wind speeds below the destruction wind speed of a turbine while extracting sufficient power from the hurricane to dissipate it.

Because hurricane paths are not predictable, offshore wind farms will always be built primarily to generate electric power year-round in order to pay for themselves. However, if enough offshore farms are built, some farms can be placed strategically in front of a city to maximize protection of the city as well, thus serving a dual benefit.

It is impossible to know how a real wind farm might diminish a hurricane, because once a hurricane has

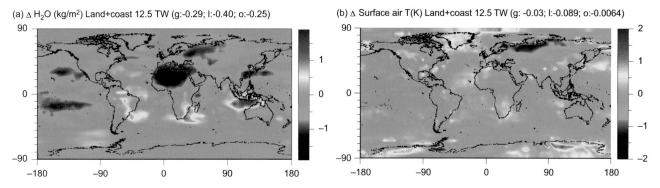

**Figure 6.29** (a) Computer-modeled difference in annual average column water vapor due to wind turbines and their extraction of kinetic energy in the wind versus when no turbines are present. The power extracted by the wind turbines is given in Figure 6.12. The model simulations with turbines include 1.58 million onshore and 935,000 offshore 5-MW wind turbines proposed to power 37.1 percent of the world's all-purpose end-use energy in 2050 after all energy has been electrified. The turbines are placed among 139 countries. (b) Change in annual average near-surface air temperature from the same simulations as (a). The letters g, l, and o are global, land, and ocean averaged values, respectively.
From Jacobson et al. (2018a).

passed the farm, it is not possible to replicate the hurricane in the absence of the farm. However, such an experiment can be carried out with a computer model that predicts hurricane formation and accounts for extraction of kinetic energy by wind turbines. In such a case, two hurricane simulations are needed – one with turbines and one without. Table 6.9 summarizes results from a computer model study of three hurricanes – Katrina, Sandy, and Isaac.

**Hurricane Katrina** occurred from August 23 to 31, 2005. It significantly damaged New Orleans, Louisiana, and caused 1,836 deaths. Peak wind speeds offshore were 78.2 m/s, making it a Category 5 hurricane offshore. Its lowest pressure was 902 hPa. When it hit land, it was a Category 4 hurricane. About half of hurricane damage is usually due to high winds and the other half, due to storm surge.

**Storm surge** is the rise in water levels generated by a storm above and beyond water levels caused by normal tides under normal conditions. Storm surge in a hurricane is enhanced by three factors. One is that warm ocean temperatures due to global warming cause water to expand, increasing the background sea level height. Second, the low surface air pressure in a hurricane raises the ocean sea level further compared with the sea level height under normal atmospheric pressure. Third, the strong **cyclonic** (counterclockwise in the Northern Hemisphere; clockwise in the Southern Hemisphere) flow of winds around a hurricane's core over long distances creates tall waves that lift water to great heights when the waves hit land.

Storm surge caused flooding in New Orleans, Louisiana, which is a city below sea level. The flooding was exacerbated due to the breach of many levies, which were built to prevent flooding.

**Hurricane Sandy** formed October 22, 2012, and dissipated November 2, 2012. It hit the U.S. East Coast between Washington, DC, and New York. Its peak measured wind speed was 35.8 m/s and lowest pressure was 966 hPa. It was one of the widest hurricanes to hit the United States and one of the few to hit the Northeast coast. Its damage was due primarily to flooding of the low elevation coast caused by storm surge.

**Hurricane Isaac** formed August 21, 2012, and dissipated September 3, 2012. It hit Louisiana east of New Orleans. Its peak measured wind speed was 51.4 m/s, lowest pressure was 940 hPa, and peak storm surge was 3.4 m. It caused 41 deaths. Table 6.9 indicates that large arrays of offshore wind turbines within 100 km of the coast could have reduced peak wind speeds and storm surge significantly in all three hurricanes. In the cases of Katrina and Isaac, large wind farms (totaling 78,000 7.58-MW wind turbines) to the southeast of New Orleans could have avoided the damage. In the case of Sandy, large wind farms (totaling 112,000 7.58-MW turbines) between Washington, DC, and New York City could have ameliorated much damage.

For Katrina, the farms could have reduced peak wind speeds by about 36 percent and storm surge by 6 to 71 percent, depending on location. For Sandy, wind farms could have reduced peak wind speeds 36 percent and storm surge 12 to 21 percent. For Isaac, wind speed

Table 6.9 **Characteristics of wind turbines used in three hurricane computer model simulation pairs (where one simulation included wind turbines and the other did not) and summary of results from each simulation pair.**

| | Katrina | Sandy | Isaac |
|---|---|---|---|
| [a]Turbine rated power (MW) | 7.58 | 7.58 | 7.58 |
| [b]Cutout wind speed (m/s) | 50 | 50 | 50 |
| [c]Spacing area (A) (m$^2$) | $28D^2$ | $28D^2$ | $28D^2$ |
| Installed power density (W/m$^2$) | 16.78 | 16.78 | 16.78 |
| Number of turbines | 78,286 | 112,014 | 78,286 |
| Nameplate capacity (TW) | 0.593 | 0.849 | 0.593 |
| [d]Normal delivered power (TW) | 0.221 | 0.316 | 0.221 |
| [e]Peak extracted hurricane power (TW) | 0.450 | 0.767 | 0.417 |
| [f]Peak 15-m wind decrease (m/s) | –36.1 | –36.0 | –25.5 |
| [f]Storm surge reduction (%) | 6–71 | 12–21 | 18–60 |

[a] The 7.58-MW turbine is the Enercon E-126 ($D$ = 127-m-diameter blade).
[b] These simulations assumed the cutout wind speed, normally 34 m/s, was increased to the destruction wind speed of the turbine, 50 m/s.
[c] All turbines were placed in a rectangular grid within 100 km of the coast. For Katrina and Isaac, all turbines were placed 87.5 °N to 89.5 °W (southeast of New Orleans). For Sandy, they were placed from Washington, DC, to New York City (38.8 °N to 41 °N).
[d] Normal delivered power was estimated assuming a mean annual Rayleigh-distributed offshore wind speeds of 8.5 m/s at hub height and the turbine power curve, without considering reduced wind speeds due to power extraction by turbines.
[e] This is the peak power extracted by wind turbines at any time during the simulation, accounting for reduced wind speeds due to extraction.
[f] From model simulation results, accounting for reduced wind speeds, reduced wave heights, and increased central pressure due to power extraction by turbines.
Source: Jacobson et al. (2014a).

and storm surge reductions could have been 26 percent and 18 to 60 percent, respectively.

Figure 6.30 shows the wind speeds 100 m above the ocean surface during Hurricane Katrina without and with wind farms present, at two different times. The figure indicates that the wind farms first reduce the outer rotational winds of the hurricane, then subsequently increase central pressure and decrease the peak wind speeds in the core of the hurricane. As such, **wind farms dissipate a hurricane from the outside in**. The reason is as follows.

Wind turbines are exposed first to the outer hurricane's rotational winds, which are slower than eye-wall winds (near the center of the hurricane). The reduction in outer rotational wind speeds decreases wave heights there, because wave heights are proportional to wind speed, a reduction in wind speed reduces wave height. Since waves are a source of friction, decreasing wave height reduces surface friction.

The angle at which cyclonic (counterclockwise in the Northern Hemisphere) surface winds converge to the eye wall in a hurricane is governed by the balance among the pressure-gradient force, Coriolis force, apparent centrifugal force, and friction forces (Figure 6.19). The decrease in the friction force decreases convergence (winds become more circular, cyclonically) toward the eye wall.

The decrease in convergence of winds toward the eye wall decreases the forced spiraling and lifting of air around the eye wall and divergence aloft above the eye wall. Because fast divergence aloft relative to slower convergence at the surface drops a hurricane's surface pressure, strengthening the hurricane, a decrease in divergence aloft conversely increases surface air pressure, weakening the hurricane. In the case of Hurricane Katrina, the modeled addition of wind farms increased central pressure by more than 40 hPa compared with no wind farms.

The increase in surface air pressure reduces the horizontal pressure gradient between the center of the hurricane and the outside of the hurricane. The reduction in pressure gradient slows down hurricane winds. Slower winds, in turn, reduce wave heights further, reducing convergence to the center, reducing forced lifting of air and divergence aloft, increasing central pressure further in a positive feedback. In this way, slowing the outer rotational winds of a hurricane with wind turbines dissipates the hurricane from the outside in.

Because wind turbine arrays dissipate a hurricane from the outside in, peak hurricane wind speeds reaching the turbines never reach the destruction wind speed of a wind turbine (Figure 6.30). In other words, the risk of a hurricane destroying a wind farm is reduced if a sufficient number of wind turbines is available to battle the hurricane. When the outer rotational winds of a hurricane first encounter a wind farm, the turbines in the farm reduce wave heights and thus convergence of outer rotational winds moving toward the eye wall and thus eye-wall wind speeds themselves. As such, if many wind turbines are present, the wind turbines never experience wind speeds exceeding the destruction wind speed of a turbine.

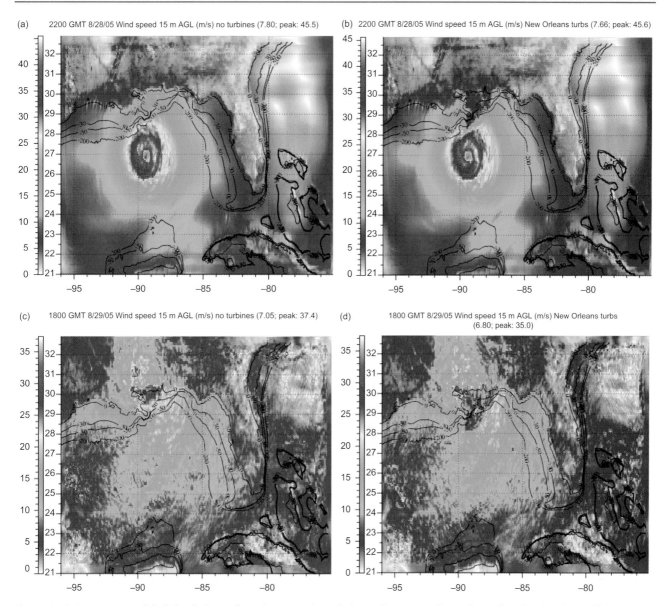

**Figure 6.30** Computer-modeled simulations of Hurricane Katrina wind speeds at 15 m above the surface level without offshore wind turbines (a, c) and with offshore wind turbines (b, d) at two times: August 28, 22:000 GMT (a, b) and August 28, 18:00 GMT, 2005 (c, d). The wind farms are located to the southeast of New Orleans (Table 6.9). From Jacobson et al. (2014a).

Because wind turbines reduce hurricane wind speeds, they also reduce storm surge (Table 6.9), which is proportional to the square of the wind speed. Storm surge is also proportional to the **fetch**, which is the length of water over which the wind blows.

Whereas the simulations summarized in Table 6.9 included an enormous number of wind turbines, any smaller number of turbines has a proportionally smaller benefit and thus can still help. The United States may need about 88,000 offshore 5-MW turbines to provide 16.4 percent of its 2050 all-purpose end-use energy from offshore wind (Jacobson et al., 2019). Dividing this number into wind farms of 100 to 300 turbines, placing those farms in separate clusters (where each farm is separated by distance to minimize competition for available kinetic energy), and placing each cluster of wind farms strategically in front of major cities at risk of a hurricane may reduce future risk of hurricane damage.

Because offshore wind farms will be built to generate year-round electric power but may also reduce potential damage due to hurricanes, they are cost effective in comparison with other techniques of reducing hurricane damage, such as building sea walls. Whereas turbines pay for themselves from the sale of electricity they

**Table 6.10 Estimated number of United States bird fatalities due to infrastructure, electric power plants, and cats in 2017.**

| Source of bird mortality | All bird deaths/y | All bird deaths/GWh | Raptor deaths/y | Raptor deaths/GWh | Bat deaths/y | Bat deaths/GWh |
|---|---|---|---|---|---|---|
| [a]Wind turbines | 76,000 | 0.3 | | | | |
| [b]Wind turbines | 178,000 | 0.7 (0.07–4.9) | 8,900 | 0.035 (−0.3) | 267,000 | 1.05 (0–14) |
| [c]Wind turbines | 990,000 | 3.9 | 142,000 | 0.56 | 1.5 million | 6.0 |
| [d]Communication towers | 7 million | | | | | |
| [d]Transmission/distribution lines | 8–57 million | | | | | |
| [d]Windows/buildings | 1 billion | | | | | |
| [d]Cats | 2.4 billion | | | | | |
| [a]Coal+natural gas electricity | 13.2 million | 5.2 | | | | |
| [a]Nuclear for electricity | 102,000 | 0.4 | | | | |

[a] GWh/y from Sovacool (2009) for wind, coal+gas, and nuclear. The deaths/GWh for wind were multiplied by 254,303 GWh produced from wind in 2017 to obtain estimated bird deaths. The deaths from coal+gas and nuclear were obtained by multiplying the deaths/GWh by 2,502,000 GWh produced by coal+gas and 805,000 GWh produced by nuclear, respectively, in 2017.

[b] Median and range of all bird, raptor, and bat deaths per GWh were from NWCC (2010), who provide deaths per MW-nameplate per year. These were converted to GWh per year assuming a capacity factor of 32.6 percent (the 2017 average for U.S. wind). The median all-bird death rate is among 45 wind farms for which data were provided. The median number of deaths/GWh were multiplied by 254,303 GWh produced in 2017 to obtain the total deaths in 2017.

[c] All bird, raptor, and bat deaths from U.S. wind in 2012 from Smallwood (2013) were converted to deaths/GWh and multiplied by 254,303 GWh from wind in 2017.

[d] Source: ABC (2019).

produce, sea walls have no other function than to reduce storm surge. Sea walls do not even reduce damaging hurricane wind speeds, as wind turbines do.

### 6.9.3 Wind Turbine Impacts on Birds and Bats

Birds and bats can collide with the spinning blades of a wind turbine, often resulting in fatalities. Some bats also die from the drop in pressure behind the wind turbine, which causes their organs and blood vessels to expand to equalize the pressure. Of particular concern is the collision of **raptors** (birds of prey) with wind turbines. Raptors include eagles, ospreys, kites, hawks, buzzards, harriers, vultures, falcons, caracaras, and owls.

Table 6.10 summarizes statistics from a few analyses of bird and bat mortalities in the United States due to wind turbines. The range in mortalities, based on 2017 U.S. wind electricity production, differs from 76,000 to 990,000, depending on the study. These seem large until other sources of avian deaths are compared.

In the United States, cats are the number one threat to birds, killing an estimated 2.4 billion per year (ABC,

2019). Buildings kill about 1 billion birds per year. Transmission and distribution lines and communication towers also kill many more birds than wind turbines do. **The number of birds that died from wind turbines in the United States in 2017 ranged from 0.002 percent to 0.028 percent of all human and cat causes of bird death.**

Table 6.10 also indicates that coal and natural gas electric power plants kill more birds per unit energy and in total than do wind turbines. As such, **transitioning from coal or gas to wind reduces bird kills**. Coal and natural gas used for electric power kills birds in three major ways. One is through the destruction in bird habitat due to invasive mining, such as mountaintop removal in the case of coal and conventional drilling and fracking in the case of gas. The second is through air and water pollution, acid deposition, and climate change caused by fossil-fuel power plants. The same air pollution from fossil plants that kills birds contributes to the 4 to 9 million premature human deaths worldwide each year (Section 1.1). The third way coal and natural gas kill birds is through power plant equipment, which birds collide with or are electrocuted by (Sovacool, 2009).

Even nuclear power kills a similar number of birds per kWh as wind turbines, according to one of the studies in Table 6.10. Nuclear power causes bird fatalities in two major ways. One is from contaminated ponds associated with uranium mining and milling. Abandoned open-pit uranium mines also form hazardous lakes. Second, birds collide with nuclear power plant buildings and power plant equipment.

Raptor deaths comprise between 5 and 15 percent of all bird deaths (Table 6.10). Bat deaths per unit energy are about 50 percent higher than are all bird deaths. Many bat deaths occur near wind turbines in the Northeast and upper Midwestern United States. A non-wind-turbine major source of bat death since 2006 has been white nose syndrome, which has decimated populations of bats in North America.

Wind turbines today are safer for birds and bats than are those built in the 1980s and 1990s. Early wind turbine towers often had lattices for birds to perch on, increasing their proximity to wind turbine blades. Also, spin rates were not previously controlled, resulting in higher tip speeds at high wind speeds. The removal of lattices and the controlling of the spin rate at high wind speeds have contributed to safer (but not completely safe) turbines. Because many birds fly low, taller turbines have also allowed birds to fly under turbine blades. In addition, radar systems are now being used to alert wind farm operators of approaching flocks of birds, including raptors. Some endangered birds, such as condors, are fitted with **global positioning satellite (GPS)** transmitter. Wind farm operators can slow down or stop wind turbines if they are alerted that a condor is approaching. Devices that emit high-frequency, ultrasonic sounds are now being tested to deter bats from entering a wind farm. One way of reducing bird mortalities is to site wind turbines out of migratory paths of birds.

## 6.10 Summary

Wind will play a major role in a 100 percent WWS future. This chapter discussed the technical details of wind turbines and wind energy resources. It started with a brief history of windmills and wind turbines. It then described types of wind turbines, the components of a wind turbine, how wind turbines work, and wind turbine generators. The next topic was on determining wind turbine output. This involves examining both wind speed and wind turbine characteristics. A wind turbine power curve was described, followed by the definitions of Rayleigh and Weibull frequency distributions. An equation for the power in the wind was then given. The Betz limit, or the maximum theoretical fraction of the wind's power that a turbine can capture, was derived. Wind turbine energy output and capacity factor were then defined. Factors affecting turbine output were then discussed. Wind turbine footprint and spacing areas were defined and quantified. The next major section was on wind energy resources. This section discussed the forces acting on air, how winds form, the global circulation of the atmosphere, local winds, and global wind resources. The maximum possible extractable wind (saturation wind potential) was quantified next. The last section of the chapter described the beneficial impacts of wind turbines on climate and hurricanes, and the impacts of wind turbines on birds and bats.

## Further Reading

Ahrens, C. D., and R. Henson, 2019. *Meteorology Today: An Introduction to Weather, Climate, and the Environment*, 12th ed., Boston: Cengage Learning, 587 pp.

Archer, C. L., and M. Z. Jacobson, 2005. Evaluation of global wind power, *J. Geophys. Res.*, **110**, D12110, doi:10.1029/2004JD005462.

Dvorak, M. J., B. A. Corcoran, J. E. Ten Hoeve, N. G. McIntyre, and M. Z. Jacobson, 2013. U.S. East Coast offshore wind energy resources and their relationship to peak-time electricity demand, *Wind Energy*, **16**, 977–997, doi:10.1002/we.1524.

Gipe, P., 2016. *Wind Energy for the Rest of Us: A Comprehensive Guide to Wind Power and How to Use It*. Bakersfield, CA: wind-works.org, 560 pp.

Masters, G., 2013 *Renewable and Efficient Electric Power Systems*, 2nd ed., Hoboken, NJ: Wiley, 712 pp.

## 6.11 Problems and Exercises

6.1.   What is the difference between a windmill and a wind turbine?

6.2.   Under which, if either, condition is the lift-to-drag ratio of a wind turbine airfoil higher? At 10 m height where the air density is 1.23 kg/m$^3$ and the wind speed is 5 m/s, or at 100 m height, where the air density is 1.21 kg/m$^3$ and the wind speed is 9 m/s? Also, under which condition is the lift greater? Assume the blade length is 60 m, the blade width is an average of 1 m, the drag coefficient is 0.01, and the lift coefficient is 1.0.

6.3.   Calculate the mean wind speed if 10 percent of wind speeds are 3 m/s, 25 percent are 5 m/s, 30 percent are 7 m/s, 25 percent are 9 m/s, and 10 percent are 11 m/s.

6.4.   If a wind turbine's blade diameter is 100 m, calculate the kinetic energy passing through the blade's swept area over 1 hour if the wind speed is 10 m/s. Assume the air density is 1.21 kg/m$^3$. Also determine the power in the wind.

6.5.   If the wind speed at 100 m height above the ground is 8 m/s and that at 10 m height is 5 m/s, calculate the power law coefficient for this profile. Using the coefficient, estimate the wind speed at 200 m above the ground.

6.6.   Estimate the air density when the air temperature is 30 °C, the air pressure is 950 hPa, and the specific humidity is 10 g-H$_2$O/kg-moist air.

6.7.   Assume a wind turbine has a hub height of 110 m above the ground and a blade diameter of 130 m. Calculate the instantaneous power in the wind in two cases. In the first case, the power is calculated based on the wind speed at hub height applied to the whole turbine. In the second case, the power is calculated based on the wind speed at the highest height of the turbine blade applied to half the blade's swept area and the wind speed at the lowest height of the turbine blade applied to the other half of the swept area. Assume wind speed increases logarithmically with height with a 1/7th power law and the wind speed at 10 m AGL is 5 m/s. Assume also that the surface air pressure (at 0 m AGL) is 1,010 hPa and decreases 1 hPa per 10-m altitude increase. Further, assume that air temperature at the surface is 270 K and increases 2 K per 1 km increase in altitude. Finally, assume no water vapor in the air.

6.8.   If the power in the wind passing through a turbine swept area is 10 MW, what is the maximum extractable power by the turbine?

6.9.   Using the empirical capacity factor equation, estimate the capacity factors of a 5-MW turbine with a 126-m blade diameter and of a 1.5-MW turbine with a 77-m blade diameter when the mean wind speed (assuming a Rayleigh distribution) is 8 m/s in both cases. What is the annual energy output in both cases (assuming a non-leap year)?

6.10.  For the same turbines and mean wind speed in Problem 6.9, calculate the turbine efficiencies and compare them with the turbine capacity factors. Assume the air density is 1.23 kg/m$^3$.

6.11.  Calculate the reduction in electricity generation that is possible in Haiti, without reducing end-use electricity availability, if the transmission and distribution loss is reduced by 5 percentage points from its 2014 value.

6.12. Calculate the reduction in fossil-fuel electricity generation that is possible upon a 2-percentage-point reduction in transmission and distribution losses for a country that currently has a 15 percent T&D loss and obtains 50 percent of its electricity from fossil fuels.

6.13. Calculate the annual average wind power output after losses are accounted for if the gross annual average power output of a wind turbine before loss is 5 MW, and the turbine suffers 5 percent T&D loss, 1 percent downtime loss, 0 percent curtailment loss, and 5 percent array loss.

6.14. Explain the difference between footprint and spacing of an energy technology. What are some of the uses of the spacing?

6.15. Calculate the footprint area of a wind farm with 100 turbines, each of which has a circular cement base of 5 m in diameter. The farm also has unpaved access roads and underground transmission.

6.16. How many 5-MW onshore wind turbines with a rotor diameter of 126 m operating in a mean annual wind speed of 7.5 m/s are needed to power the U.S. on-road vehicle fleet consisting of battery-electric (BE) vehicles if the end-use energy required to run such a fleet is $E_v = 1.15 \times 10^{12}$ kWh/y (2017) and the plug-to-wheel efficiency of an electric vehicle is $\eta_e = 0.85$? Assume the system efficiency of each wind turbine is $\eta_t = 0.9$. *Hint*: First determine the total electrical energy required to run the fleet by dividing the end-use energy required to run vehicles by the plug-to-wheel efficiency.

6.17. If all the wind farms used to provide electric power for the U.S. vehicle fleet in Problem 6.16 have an installed power density of 10 MW/km², calculate the wind turbine spacing area required for all turbines needed to power the fleet. What percent of the 50-state U.S. land area ($9.162 \times 10^6$ km²) does this area represent? Assuming each turbine has a 5-m-diameter circular base, also calculate the footprint area of all the turbines.

6.18. Using the information obtained from Problem 6.16 and the following equations, calculate the number of wind turbines required to power a 100 percent U.S. hydrogen fuel cell (HFC) vehicle fleet, where the hydrogen is produced by wind electrolysis. Account for energy requirements of the electrolyzer, compressor, and fuel cell. Assume the efficiency of the fuel cell converting hydrogen to energy and water is $\eta_f = 0.6$, the lower heating value of hydrogen is $L_h = 33.3$ kWh/kg-$H_2$, the hydrogen leakage rate is $l_h = 0.01$ (fraction), the electrolyzer energy required is $L_z = 53.4$ kWh/kg-$H_2$, and the compressor energy required is $L_c = 5.64$ kWh/kg-$H_2$.

   Energy required to move all U.S. vehicles as HFC vehicles: $E_f$ (kWh/y) $= E_v/\eta_f$
   $H_2$ mass to power all vehicles, accounting for leaks: $M_h$ (kg-$H_2$/y) $= E_f/[L_h(1-l_h)]$
   Energy needed to power a U.S. HFC vehicle fleet: $E_h$ (kWh/y) $= (L_z+L_c) \times M_h$
   Number of wind turbines needed to power a U.S. HFC vehicle fleet $= E_h/E_t$

6.19. Estimate the wind speed downstream of a wind turbine after 1 minute, assuming the upstream wind speed is 9 m/s and the turbine extracts 2,500 kW at that wind speed. Also assume that the turbine blade diameter is 115 m and the air density is 1.23 kg/m³.

# CHAPTER

# 7

# Steps in Developing 100 Percent All-Sector WWS and Storage Roadmaps

So far, this book has examined several components of a 100 percent clean, renewable energy and storage system. This chapter focuses on integrating the components together in countries, states, cities, and towns to provide end-point roadmaps for a transition. Such roadmaps provide scenarios for meeting all-purpose, annually averaged power demand with 100 percent WWS in 2050. Chapter 8 discusses methods of matching time-dependent power demand with supply and storage. The subjects discussed in this chapter are projecting annually averaged power demand in all energy sectors to 2050 (Section 7.1), quantifying the transition of all business-as-usual (BAU) energy in all sectors to electricity, electrolytic hydrogen, and some heat, all sourced by WWS (Section 7.2), reducing end-use power demand due to such a transition (Section 7.3), performing a renewable energy resource analysis (Section 7.4), selecting a WWS energy mix in each location to meet end-use demand in the annual average while also meeting resource constraints (Section 7.5), calculating changes in energy costs due to such a transition (Section 7.6.1), calculating changes in air pollution mortality and morbidity and their associated costs due to such a transition (Section 7.6.2), and estimating the climate-relevant emissions and their associated costs due to such a transition (Section 7.6.3). The methods in this chapter are applicable to roadmaps for towns, cities, states, provinces, and countries but are derived here for countries as an example.

## 7.1 Projecting End-Use Energy Demand

The purposes of transitioning towns, cities, states, countries, and the world to 100 percent clean, renewable

WWS energy and storage for everything and to eliminate non-energy sources of emissions are to eliminate air pollution mortality and morbidity, reduce and then eliminate global warming, and provide energy security.

In a 100 percent WWS world, all energy sectors are electrified or powered with direct heat, where the electricity and heat are provided by WWS. Some electricity is used for hydrogen, primarily for transportation, but also to make steel. Some electricity and heat are stored in electricity, heat, cold, and hydrogen storage. Some heat is provided by district heating. Energy efficiency and energy reduction measures are put in place to reduce demand. Grid operators also shift the times of peak demand with demand response to help keep the grid stable.

The first step in developing a roadmap to transition the energy infrastructure of a country, province, state, city, or town is to project annually averaged, end-use energy (or power) demand across all energy sectors forward from current demand to a future year in a business-as-usual (BAU) case. In general, a **BAU case** is one in which the energy infrastructure does not change drastically in the future, but demand does change due to increases in population and natural improvements in energy efficiency over time. A BAU case also includes some transition from fossil fuels to renewables, but only at recent historical rates, which are relatively slow.

**End-use energy** is energy directly used by a consumer. It is the energy embodied in electricity, natural gas, gasoline, diesel, kerosene, and jet fuel that people use directly. It equals primary energy minus the energy lost in converting primary energy to end-use energy, including energy lost during transmission and distribution.

**Primary energy** is the energy naturally embodied in chemical bonds in raw fuels, such as coal, oil, natural gas, biomass, uranium, or renewable (e.g., hydroelectric, solar, wind) electricity, before the fuel has been subjected to any conversion process. The conversion of primary energy to end-use energy differs for different energy sectors and types of fuels.

In the **electricity sector**, for example, end-use energy is primary energy minus the energy lost during the generation, transmission, and distribution of electricity. For instance, when coal is burned to produce electricity, only about one-third of the energy embodied in the coal is converted to electricity. The rest is waste heat. Further, some of the electricity produced is lost during transmission and distribution. The end-use electricity in this case is the electricity that consumers use in the end, not the primary energy that was contained in the coal nor the energy in the electricity at the power plant itself.

In another example, solar electricity that is produced by a PV panel is primary energy. Some of that electricity is lost during transmission and distribution. Thus, in this case, the solar electricity that actually reaches a consumer after transmission and distribution losses is end-use energy. Similarly, hydroelectric power produced at a hydropower plant is primary energy. The hydroelectricity remaining after transmission and distribution losses is end-use energy.

In the **transportation sector**, the energy embodied in crude oil is primary energy. Converting crude oil to end-use products, including gasoline, diesel, kerosene, refinery gas, and jet fuel involves almost no loss of the primary energy in crude oil, so the end-use energy available in all products is close to, but not exactly the same as, the primary energy in crude oil. Oil and its products have different chemical structures from each other.

**Natural gas** used for heating and cooking is similar to the gas when it is recovered from a well, so primary energy and end-use energy are about the same. On the other hand, when natural gas is burned for electricity production (in the electricity sector), only a portion of the natural gas is converted to electricity and some of that electricity is lost during transmission and distribution, so the end-use energy in the electricity, as with coal, differs substantially from the primary energy in the natural gas creating the electricity.

Table 7.1 shows an example projection of 2016 BAU all-purpose annual-average, end-use power demand (load) by sector to 2050, summed among 143 countries.

**Annual average power demand** is just annual energy consumption divided by the number of hours per year. This projection was performed starting with end-use power consumption data by sector and fuel type for 2016 from the International Energy Agency (IEA, 2019). The consumption for each fuel type in each sector and in each country was then projected forward to 2040 based on *reference* scenario projections from the U.S. Energy Information Administration (EIA, 2016a) for each fuel type in each sector in 16 world regions.

EIA's reference scenario is one of moderate economic growth. It accounts for policies in different countries, population growth, economic and energy growth, the use of some renewable energy, modest energy efficiency measures, and reduced energy use between 2016 and 2040. EIA refers to this reference scenario as their BAU scenario. These projections were extrapolated from 2040 to 2050 using a 10-year moving linear extrapolation for each fuel type in each sector in each world region. The 2050 BAU end-use energy for each fuel type in each energy sector in each country then equals the 2016 end-use energy from IEA multiplied by the EIA 2050-to-2016 energy consumption ratio, available after the extrapolation, for the fuel type, energy sector, and region containing the country. Table 7.1 shows the resulting 2050 end-use demand, summed over all fuel types, sectors, and countries.

In 2016, the 143-country, annually averaged, end-use power demand for energy among all energy sectors was about 12.6 TW. Of this, 2.6 TW (20.7 percent) was electricity demand. The projection just described suggests that, in 2050, all-purpose, end-use demand in the BAU case may grow to 20.3 TW (by 61 percent) if no large-scale transition to WWS occurs. The growth is due to a population increase and an increase in energy demand per person due to the lifting of many people out of poverty, partly mitigated by reduced energy use resulting from some modest shifts from coal to gas, biofuels, bioenergy, some WWS, and some energy efficiency.

## 7.2 Transitioning Future Energy to WWS Technologies

The second step is to transition 2050 BAU end-use energy for each fuel type in each energy sector to electricity, electrolytic hydrogen, or heat, where the electricity and heat are provided by WWS. In addition, energy

Table 7.1 Reduced end-use demand upon a transition to 100 percent WWS. The table shows 2016 BAU, 2050 BAU, and 2050 100 percent WWS annually averaged end-use power demand (GW) by sector, summed among 143 countries. The last column shows the total percent reduction in 2050 BAU end-use power demand due to switching from BAU to WWS, including the effects of reduced energy use caused by (a) the higher work output to energy input ratio of electricity over combustion, (b) eliminating energy used to mine, transport, and/or refine coal, oil, natural gas, biofuels, bioenergy, and uranium, and (c) assumed policy-driven increases in end-use energy efficiency beyond those in the BAU case, which is a case of moderate economic growth and moderate energy efficiency improvements. Four 2050 WWS cases are included. Case WWS-A eliminates the energy used to mine, transport, and refine fossil fuels and uranium and includes energy efficiency measures, but it does not change the work-output-to-energy-input ratio relative to fossil fuels. Case WWS-B is the same as case WWS-A, except that case WWS-B includes the higher work-output-to-energy-input ratio of electric vehicles and hydrogen fuel cell vehicles powered by WWS over internal combustion vehicles. Case WWS-C is the same as case WWS-B, except that case WWS-C accounts for the higher work-output-to-energy-input ratio of high-temperature industrial processes with WWS. Case WWS-D is the same as case WWS-C, except that case WWS-D accounts for the higher work-output-to-energy-input ratio of heat pumps over internal combustion heating for low-temperature heat. The result indicates that, of the 38.3 percent demand reduction due the higher work-output-to-energy-input ratio of electricity over combustion, 21.7, 3.4, and 13.2 percentage points are due to the efficiency of WWS transportation, the efficiency of WWS electricity for industrial heat, and the efficiency of heat pumps, respectively.

| Scenario | Total end-use demand (GW) | Residential percent of total | Commercial percent of total | Industrial percent of total | Transport percent of total | Ag-forestry/ fishing percent of total | Military/ other percent of total | (a) 2050 change in demand (percent) due to higher work: energy ratio of WWS | (b) 2050 change in demand (percent) due to eliminating upstream w/WWS | (c) 2050 change in demand (percent) due to efficiency beyond BAU w/WWS | Total 2050 change in demand (percent) w/WWS |
|---|---|---|---|---|---|---|---|---|---|---|---|
| BAU 2016 | 12,628 | 21.1 | 8.13 | 38.4 | 28.7 | 2.1 | 1.5 | | | | |
| BAU 2050 | 20,255 | 19.1 | 7.80 | 37.4 | 32.3 | 1.9 | 1.5 | | | | |
| WWS-A 2050[a] | 15,932 | 20.2 | 8.50 | 34.9 | 32.6 | 2.2 | 1.6 | 0 | -13.7 | -7.6 | -21.3 |
| WWS-B 2050[b] | 11,968 | 27.0 | 11.3 | 46.4 | 11.8 | 1.6 | 1.9 | -21.7 | -12.4 | -6.8 | -40.9 |
| WWS-C 2050[c] | 11,294 | 28.6 | 12.0 | 43.2 | 12.5 | 1.7 | 2.0 | -25.1 | -12.3 | -6.8 | -44.2 |
| WWS-D 2050[d] | 8,693 | 17.7 | 10.5 | 52.0 | 16.2 | 1.7 | 1.8 | -38.3 | -12.1 | -6.6 | -57.1 |

[a] Scenario involves eliminating energy used in the mining, transporting, and processing of fossil fuels, increasing energy efficiency beyond BAU (changing all values for *extra efficiency* in Table 7.3 to current values from unity), but assuming the efficiency of electrification is the same as that of fossil fuels (leave the *electricity-to-fuel ratio* = 1 for all fuels in all sectors in Table 7.3).

[b] Scenario involves electrifying transport (including the use of hydrogen fuel cell vehicles) (reducing the *electricity-to-fuel ratios* from 1 to their current values for oil, natural gas, and biofuels/waste in the transportation sector and for oil in the agriculture/forestry/fishing and military/other sectors in Table 7.3) in addition to making the changes in Footnote a.

[c] Scenario involves electrifying industrial high-temperature heat (reducing the *electricity-to-fuel ratios* from 1 to their current values for oil, natural gas, coal, and biofuels/waste in the industrial sector in Table 7.3) in addition to making the changes in Footnote b.

[d] Scenario involves using electric heat pumps for all building air and water heating (reducing the *electricity-to-fuel ratios* from 1 to their current values for all remaining values in Table 7.3 – oil, natural gas, coal, and biofuels/waste in the residential and commercial sectors; heat for sale in all sectors; natural gas, coal, and biofuels/waste in the agriculture/forestry/fishing and military/other sectors) in addition to making the changes in Footnote c.

Source: Jacobson et al. (2019).

Source: EIA (2016a).

efficiency measures beyond those in the BAU case are invoked.

The sectors to transition are electricity, transportation, building heating and cooling, industry, agriculture/forestry/fishing, and the military. Sections 2.1 to 2.4 discuss WWS technologies proposed for electric power, transportation, heating and cooling, and industry, respectively, in a WWS world. Section 2.5 discusses additional electric appliances and machines that would replace fossil-fuel ones. Section 2.6 discusses energy efficiency measures. The technologies discussed are almost all available today.

The main WWS electricity-generating technologies proposed include onshore and offshore wind, concentrated solar power (CSP), geothermal power, solar PV on rooftops and in power plants, tidal and ocean current power, wave power, and hydropower.

Proposed vehicles for transportation include battery-electric (BE) vehicles, hydrogen fuel cell (HFC) vehicles, and HFC hybrid vehicles, where the hydrogen is produced by electrolysis. BE vehicles will dominate 2- and 3-wheel transportation; short- and long-distance light-duty transportation; and most truck, construction machine, and agricultural equipment transportation. Batteries will also power most short- and moderate-distance trains, short-distance boats and ships (e.g., ferries, speedboats), short-distance military equipment, and aircraft traveling less than 1,500 km. Some short-distance trains will run on overhead-wire electricity. Of all commercial aircraft flight distances traveled worldwide, about 53.9 percent are short-haul flights (less than 3 hours in duration, with a mean distance of 783 km) (Wilkerson et al., 2010). As such, approximately half the aircraft flights worldwide may be electrified with batteries.

Hydrogen fuel cell or battery-electric-HFC hybrid vehicles will likely dominate transportation by medium- and heavy-duty trucks, long-distance trains, long-distance ships, long-haul aircraft, and long-distance military equipment.

Air heating and cooling will be performed with ground-, air-, or water-source electric heat pumps. Hot water will be generated with heat pumps. In some cases, an electric resistance element will be added for low temperatures. In other cases, rooftop solar hot water heaters will preheat the water for the heat pump. In some tropical countries, only rooftop solar hot water heaters are needed. Cook stoves will be electric induction. Clothes dryers will all be electric heat pump or resistance.

Electric arc furnaces, induction furnaces, dielectric heaters, and resistance furnaces will provide high temperatures for industrial processes.

## 7.3 Calculating End-Use Energy Reductions due to a Transition

The third step is to calculate reductions in annually averaged, end-use power demand that arise from transitioning to 100 percent WWS electricity and heat for all purposes. The reductions will occur for five main reasons:

(1) The efficiency of electricity and electrolytic hydrogen over combustion for transportation;
(2) The efficiency of electricity over combustion for high-temperature industrial heat;
(3) The efficiency of moving low-temperature building air and water heat with heat pumps instead of creating heat with combustion;
(4) Eliminating the energy needed to mine, transport, and process fossil fuels, biofuels, bioenergy, and uranium; and
(5) Improving end-use energy efficiency and reducing energy use beyond what will occur under BAU.

These reductions are discussed in turn.

### 7.3.1 Efficiency of Electricity and Electrolytic Hydrogen over Combustion for Transportation

First, replacing end-use energy from fossil fuels (natural gas, gasoline, diesel, kerosene, jet fuel), biofuels, and bioenergy for transportation with WWS electricity eliminates waste heat of combustion. This is because electricity and electrolytic hydrogen have higher energy-to-work conversion efficiencies than do fossil fuels, biofuels, and bioenergy. This factor is embodied in the **electricity-to-fuel ratio**, which is the energy required for an electric or hydrogen fuel cell machine to perform the same work as a BAU machine running on fossil fuels, biofuels, or bioenergy. This ratio is calculated below for battery-electric vehicles and hydrogen fuel cell vehicles versus **internal combustion engine (ICE)** vehicles.

#### 7.3.1.1 Efficiency of Battery-Electric Vehicles over Fossil-Fuel Vehicles

An example of the greater efficiency of electricity over combustion arises with battery-electric (BE) vehicles in comparison with ICE vehicles. Only 17 to 20 percent of

**Table 7.2** Efficiencies of individual factors affecting the overall plug-to-wheel efficiency of a battery-electric passenger vehicle and a hydrogen fuel cell vehicle. Low and high estimates of efficiency are given. See text for a discussion of the numbers.

| Type of efficiency | Battery-Electric (BE) Vehicle | | Hydrogen Fuel Cell (HFC) Vehicle | |
| --- | --- | --- | --- | --- |
| | Low efficiency | High efficiency | Low efficiency | High efficiency |
| Battery charging | 0.80 | 0.96 | | |
| Battery discharging (to DC) | 0.98 | 0.99 | | |
| Electrolyzer | | | 0.738 | 0.738 |
| Compressor | | | 0.904 | 0.904 |
| Leakage | | | 0.99 | 0.997 |
| Fuel cell producing DC electricity | | | 0.5 | 0.7 |
| Fuel cell latent heat loss | | | 0.846 | 0.846 |
| DC to AC inverter/wiring/power electronics | 0.97 | 0.98 | 0.97 | 0.98 |
| Electric AC motor | 0.84 | 0.96 | 0.84 | 0.96 |
| **Overall plug-to-wheel** | **0.64** | **0.89** | **0.23** | **0.37** |

the end-use energy embodied in a fossil fuel or liquid biofuel is used to move an ICE passenger vehicle. This is the **tank-to-wheel efficiency** of the vehicle. The rest of the energy (80 to 83 percent) is waste heat. As such, gasoline and diesel, for example, contain potential energy that is 5 to 5.9 times the end-use energy (work) that is actually used to move the vehicle.

A BE vehicle, on the other hand converts 64 to 89 percent of electricity at the plug (before charging the car) into motion (**plug-to-wheel efficiency**), and the rest is waste heat. This efficiency is determined as follows and summarized in Table 7.2.

First, a **permanent magnet electric motor** has an efficiency of 89 to 96 percent. An **induction electric motor** has an efficiency of 84 to 94 percent (Bistak and Kim, 2017). Thus, the range of efficiencies of electric car motors is 84 to 96 percent. Whereas the Tesla Model S and Model X use an induction motor, the Tesla Model 3, for example, uses a permanent magnet motor.

In addition, efficiency losses occur due to converting electricity from the grid to chemical energy in a battery. These vehicle-charging losses can range from 4 to 20 percent of the electricity going into the battery, depending on the current and voltage used to charge the vehicle. Another 1 to 2 percent of energy is lost during conversion of the battery's chemical energy to DC electricity. In addition, 2 to 3 percent is lost converting DC to AC electricity in an inverter, adjusting the voltage for use in the motor, and using power electronic controls in the

vehicle. Inverter losses are only 1 percent of energy (thus an efficiency of up to 99 percent) when the inverter uses silicon carbide as the semiconductor material (Rahman et al., 2016).

Accounting for all losses gives the overall **plug-to-wheel efficiency of a battery-electric passenger vehicle** as 64 to 89 percent, with an average of 77 percent (Table 7.2).

The tank-to-wheel efficiency of a fossil-fuel passenger vehicle divided by the plug-to-wheel efficiency of an electric vehicle is the fraction of a fossil-fuel vehicle's end-use energy needed for electricity to move an electric vehicle the same distance. This electricity-to-fuel ratio ranges from 0.19 (= 0.17/0.89) to 0.31 (= 0.2/0.64), or an average of 0.25. In other words, **a fossil-fuel vehicle requires 4 (3.2 to 5.3) times the energy in gasoline than an electric vehicle needs in electricity at the plug (before charging) to drive the same distance**. By 2050, the average electricity-to-fuel ratio is expected to be closer to 0.19 due to improvements in vehicle charging efficiencies and battery efficiency, and the greater use of permanent magnet motors. This is the value used in Table 7.3 (the electricity-to-fuel ratio for WWS battery-electric vehicles replacing oil for transportation).

### 7.3.1.2 Efficiency of Hydrogen Fuel Cell Vehicles over Fossil-Fuel Vehicles

Whereas the efficiency of an electric passenger vehicle over an ICE vehicle results in a significant reduction in end-use power requirements in the transportation sector,

Table 7.3 Factors to multiply BAU end-use energy consumption by in different IEA (2019) energy categories to obtain equivalent WWS end-use energy consumption. The factors are the ratio of BAU work-output/energy-input to WWS work-output/energy-input, by fuel and sector.

| Fuel | Residential | | Commercial/Government | | Industrial | | Transportation | | Agriculture/Forestry/Fishing | | Military/Other | |
|---|---|---|---|---|---|---|---|---|---|---|---|---|
| | Elec:fuel ratio | Extra efficiency | Elec:fuel ratio | Extra efficiency | Elec:fuel ratio | Extra efficiency | Elec:fuel ratio | Extra efficiency | Elec:fuel ratio | Extra efficiency | Elec:fuel ratio | Extra efficiency |
| Oil | 0.2[a] | 0.84 | 0.2[a] | 0.95 | 0.82[e] | 0.98 | 0.19/0.46[f] | 0.96 | 0.19/0.46[f] | 0.96 | 0.19/0.46[f] | 0.96 |
| Natural gas | 0.2[a] | 0.81 | 0.2[a] | 1 | 0.82[e] | 0.98 | 0.19/0.46[g] | 0.88 | 0.2[a] | 0.91 | 0.2[a] | 0.91 |
| Coal | 0.2[a] | 1 | 0.2[a] | 1 | 0.82[e] | 0.97 | 0.19[h] | 0.96 | 0.2[a] | 1 | 0.2[a] | 1 |
| Electricity | 1[b] | 0.77 | 1[b] | 0.78 | 1[b] | 0.92 | 1[b] | 1 | 1[b] | 0.78 | 1[b] | 0.78 |
| Heat for sale | 0.25[c] | 1.0 | 0.25[c] | 1 | 0.25[c] | 1 | – | – | 0.25[c] | 1 | 0.25[c] | 1 |
| WWS heat | 1[d] | 1 | 1[d] | 1 | 1.0[d] | 1 | – | – | 1[d] | 1 | 1[d] | 1 |
| Biofuels/waste | 0.2[a] | 0.87 | 0.2[a] | 1 | 0.82[e] | 1 | 0.19[h] | 0.96 | 0.2[a] | 0.93 | 0.2[a] | 0.93 |

*Residential* loads include electricity and heat consumed by households, excluding transportation.

*Commercial/Government* loads include electricity and heat consumed by commercial and public buildings, excluding transportation.

*Industrial* loads include energy consumed by all industries, including iron, steel, and cement; chemicals and petrochemicals; non-ferrous metals; non-metallic minerals; transport equipment; machinery; mining (excluding fuels, which are treated under Transportation); food and tobacco; paper, pulp, and print; wood and wood products; construction; and textile and leather.

*Transportation* loads include energy consumed during any type of transport by road, rail, domestic and international aviation and navigation, or by pipeline, and by agricultural and industrial use of highways. For pipelines, the energy required is for the support and operation of the pipelines. The Transportation category excludes fuel used for agricultural machines, fuel for fishing vessels, and fuel delivered to international ships, since those are included under the Agriculture/Forestry/Fishing category.

*Agriculture/Forestry/Fishing* loads include energy consumed by users classified as agriculture, hunting, forestry, or fishing. For agriculture and forestry, it includes consumption of energy for traction (excluding agricultural highway use), electricity, or heating in those industries. For fishing, it includes energy for inland, coastal, and deep-sea fishing, including fuels delivered to ships of all flags that have refueled in the country (including international fishing) and energy used by the fishing industry.

*Military/Other* loads include fuel used by the military for all mobile consumption (ships, aircraft, tanks, on-road and non-road transport) and stationary consumption (forward operating bases, home bases), regardless of whether the fuel is used by the country or another country.

*Elec:fuel ratio* (electricity-to-fuel ratio) is the ratio of the energy input of end-use WWS electricity to energy input of BAU fuel needed for the same work output. For example, a value of 0.5 means that the WWS device consumed half the end-use energy as did the BAU device to perform the same work.

*Extra efficiency* is the effect of the additional efficiency and energy reduction measures assumed in the WWS system beyond those used in the BAU system, which is based on a moderate economic growth assumption by EIA (2016a). For example, in the case of natural gas, oil, and biofuels for residential air and water heating, it is additional efficiency due to better insulation of pipes and weatherizing homes. For residential electricity, it is due to more efficient light bulbs and appliances. In the industrial sector, it is due to faster implementation of more energy-efficient technologies than in the BAU case. The improvements are calculated as the product of (a) the ratio of energy use, by fuel and energy sector, of the EIA's *high efficiency all scenarios* (HEAS) case and their *reference* (BAU) case and (b) additional estimates of slight efficiency improvements beyond those in the HEAS case (Jacobson et al., 2019).

*Oil* includes end-use energy embodied in oil products, including refinery gas, ethane, liquefied petroleum gas, motor gasoline (excluding biofuels), aviation gasoline, gasoline-type jet fuel, kerosene-type jet fuel, other kerosene, gas oil, diesel oil, fuel oil, naphtha, white spirit, lubricants, bitumen, paraffin waxes, petroleum coke, and other oil products. Does not include oil used to generate electricity.

*Natural gas* includes end-use energy embodied in natural gas. Does not include natural gas used to generate electricity.

*Coal* includes end-use energy embodied in hard coal, brown coal, anthracite, coking coal, other bituminous coal, sub-bituminous coal, lignite, patent fuel, coke oven coke, gas coke, coal tar, brown coal briquettes, gas works gas, coke oven gas, blast furnace gas, other recovered gases, peat, and peat products. Does not include coal used to generate electricity.

*Electricity* includes end-use energy embodied in electricity produced by any source.

*Heat for sale* is end-use energy embodied in any heat produced for sale. This includes mostly waste heat from the combustion of fossil fuels, but it also includes some heat produced by electric heat pumps and boilers.

*WWS heat* is end-use energy in the heat produced from geothermal heat reservoirs and solar hot water heaters.

*Biofuels and waste* include end-use energy for heat and transportation from solid biomass, liquid biofuels, biogas, biogasoline, biodiesel, bio jet kerosene, charcoal, industrial waste, and municipal waste.

[a] The ratio 0.2 assumes electric heat pumps (coefficient of performance, COP = 3.2 to 5.2) replace oil, gas, coal, biofuel, and waste combustion heaters (COP = 0.80) for low-temperature air and water heating in buildings. The ratio is calculated by dividing the COP of BAU heaters by that of heat pumps to give 0.15 to 0.25, where the low number is closer to ground-source heat pumps and the high number is closer to air-source heat pumps (Section 7.3.3). The final ratio is just an average of the two.

[b] Since *electricity* is already end-use energy, there is no reduction in end-use energy (only in primary energy) from using WWS technologies to produce electricity.

[c] Since *heat for sale* is low-temperature heat, it will be replaced by heat from electric heat pumps (COP = 3.2 to 5.2) giving an electricity-to-fuel ratio of 0.19 (=1/5.2) to 0.31 (=1/3.2), or an average of 0.25. Heat for sale is also low-temperature heat in the industrial sector, so it is replaced in that sector with heat pumps as well.

[d] Since *WWS heat* is already from WWS resources, there is no reduction in end-use or primary energy upon a transition to 100 percent WWS for this source.

[e] The ratio 0.82 for industrial heat processes assumes electric resistance heaters, arc furnaces, induction furnaces, and dielectric heaters replace oil, gas, coal, biofuel, and waste combustion heaters. The industrial sector electricity-to-fuel ratio and extra efficiency measure factors are applied only after industrial sector BAU energy used for mining and processing fossil fuels, biofuels, bioenergy, and uranium (industry "own use") has been removed from each fuel sector. The amount of industry own use is determined in IEA (2019) for each country and fuel sector.

[f] The electricity-to-fuel ratio for a battery-electric (BE) vehicle is 0.19; that for a hydrogen fuel cell (HFC) vehicle is 0.46. Two percent of BAU energy in the form of *oil* in the *transportation* sector is used to transport fossil fuels, biofuels, bioenergy, and uranium. That BAU energy is eliminated in a 100 percent WWS world (no elimination is needed in the *Agr./For./Fishing* or *Military/Other* sectors). Of the remaining transportation, 76 percent is electrified. Thus, 76 percent is multiplied by the electricity-to-fuel ratio for BE vehicles to determine the WWS electricity used for BE transportation replacing oil and 24 percent is multiplied by the electricity-to-fuel ratio for HFC transportation replacing oil.

[g] About 80 percent of *natural gas* energy in the transportation sector is used to transport fossil fuels, biofuels, bioenergy, and uranium (e.g., through pipelines or other means). That BAU energy is eliminated in a 100 percent WWS world. Of the remainder, 95 percent is electrified with BE vehicles and 5 percent is electrified with HFC vehicles.

[h] *Coal* is still used in the transportation sector in some industrial locomotives in China (where over 90 percent of all coal transportation occurs) and in a few other countries. About 50 percent of coal in the transportation sector is used to transport fossil fuels, biofuels, bioenergy, and uranium. That BAU energy is eliminated in a 100 percent WWS world. Of the remainder, which is used to transport other industrial goods, 100 percent will be electrified. Similarly, 100 percent of the *biofuels and waste* currently used in transportation will be electrified in 2050.

Source: Jacobson et al. (2019).

that reduction is limited by the fact that a portion of future transportation will use hydrogen fuel cells (HFCs), which are less efficient than electric vehicles (but more efficient than ICEs).

HFCs convert hydrogen to electricity, which is then used in a motor to produce rotation to move wheels or a propeller. Pure HFC or HFC-battery-electric (BE) hybrid vehicles will be used primarily for long-distance, heavy transportation.

The overall **plug-to-wheel efficiency of an HFC vehicle** ranges from 23 to 37 percent, or an average of 30 percent. In other words, only about 30 percent of the electricity used to produce and use $H_2$ actually moves the car. The reason is that electricity is needed to produce the hydrogen by electrolysis and compress it or liquefy it for storage in a holding tank and vehicle fuel tank. In addition, some of the hydrogen leaks between production and the fuel tank. Some energy in hydrogen's chemical bonds is then lost as heat in the fuel cell. In addition, water vapor produced by the hydrogen reaction in the fuel cell carries away latent heat. Finally, some electricity produced by the fuel cell is lost in wires between the fuel cell and electric motor, and more electricity is lost as heat in the motor. Table 7.2 summarizes efficiencies of different processes affecting the overall plug-to-wheel efficiency of an HFC vehicle.

The **electrolyzer efficiency** is the efficiency at which an electrolyzer converts electricity into energy stored in hydrogen. It is calculated as the higher heating value of hydrogen (141.8 MJ/kg-$H_2$, which equals 39.39 kWh/kg-$H_2$) divided by the energy per unit mass of hydrogen required to produce hydrogen by electrolysis (53.37 kWh/kg-$H_2$-produced) (Jacobson et al., 2005). The result is an electrolyzer efficiency of 73.8 percent.

Electricity is needed to compress hydrogen for temporary storage in an immobile tank and eventual storage in an HFC vehicle tank. The electricity required for compression is about 5.64 kWh/kg-$H_2$ (Jacobson et al., 2005). If this is added to the electrolyzer energy needed, the total is 59.0 kWh/kg-$H_2$. Dividing the higher heating value of hydrogen by this number gives an overall electrolyzer plus compressor efficiency of 66.8 percent. Dividing the overall efficiency by the electrolyzer efficiency of 73.8 percent gives a **compressor efficiency** of 90.4 percent. In other words, of the total energy needed to produce and compress a kilogram of $H_2$, 26.2 percent is lost due to electrolysis and 9.6 percent is lost due to compression.

The remainder is stored in the bonds of a hydrogen molecule.

Once hydrogen is produced, some of it may leak. Hydrogen is a tiny molecule, much smaller than methane, so it can leak easily from pipes unless the pipes are well sealed. In a 100 percent WWS system, hydrogen will be produced in a controlled environment and mostly locally. Instead of hydrogen being produced far away and sent by pipeline, most will be produced locally after electricity is transmitted long distances. This will minimize hydrogen pipeline loss. Leaks, however, will still occur. An estimate of **hydrogen leakage** provided in Table 7.2 is 0.3 to 1 percent.

**Fuel cell efficiencies** (electrical output of a fuel cell divided by the lower heating value of hydrogen) were discussed in Section 2.2.2 to be between 50 and 70 percent.

The **latent heat loss efficiency** is simply the lower heating value of hydrogen (119.96 MJ/kg-$H_2$) divided by the higher heating value (141.8 MJ/kg-$H_2$), which equals 84.6 percent. Of the energy in hydrogen, 15.4 percent is lost evaporating water produced in the fuel cell. That energy is stored as latent heat in the water vapor produced by the hydrogen reaction in the fuel cell and is released back to the air when the water vapor ultimately condenses to liquid cloud water in the atmosphere.

**Inverter/wiring/power electronic losses** within an HFC vehicle are estimated to be slightly less than for a BE vehicle, because such losses in a BE vehicle include converting energy in the battery to DC electricity. The analogous loss in a fuel cell is already accounted for in the fuel cell efficiency. Thus, inverter, wiring, and power electronics losses are estimated to be 1 to 3 percent.

The **motor efficiency** of an HFC vehicle is the same as for a BE vehicle.

Table 7.2 suggests that the efficiency of the fuel cell system inside an HFC vehicle is 34 to 56 percent. However, accounting for electrolyzer, compressor, and leakage losses decreases the overall plug-to-wheel efficiency of an HFC vehicle to about 23 to 37 percent.

Dividing the average tank-to-wheel efficiency of a fossil-fuel vehicle by the plug-to-wheel efficiency of an HFC vehicle gives the **electricity-to-fuel ratio of an HFC vehicle** replacing oil as 0.46 (= 0.17/0.37) to 0.87 (= 0.20/0.23). The lower number (0.46) is used in Table 7.3 in the transportation sector because this is closer to what is expected in 2050 due to efficiency improvements along the whole process of producing and using hydrogen.

A disadvantage of a BE vehicle is that the greater its range and the heavier the vehicle, the heavier the battery pack that needs to be carried around permanently. For example, most of the energy stored in batteries in a long-distance aircraft would be used just to carry the batteries. An HFC vehicle ameliorates this problem somewhat, since the fuel itself (hydrogen) is extremely light, and as travel commences, the fuel weight burden decreases since $H_2$ is converted to oxygen and water, which are emitted from the aircraft. However, an HFC vehicle needs a large volume to store $H_2$, even if the $H_2$ is liquefied. Further, the greater the range of an HFC or HFC-BE hybrid vehicle, the greater the fuel cell stack size needed, which also increases weight and volume.

All in all, though, a vehicle with HFCs is more practical than a pure BE vehicle for very-long-distance aircraft, ships, trains, and some trucks. For these modes of transportation, recharging a BE vehicle with electricity during travel is either not an option or not an attractive option. For short- and moderate-distance transport, on the other hand, the plug-to-wheel efficiency of an HFC vehicle is only 26 to 60 percent that of a BE vehicle. In other words, **the efficiency of a BE vehicle is 1.7 to 3.8 that of an HFC vehicle**, so an HFC vehicle needs 70 to 280 percent more wind turbines or solar panels to run than does a BE vehicle to go the same distance. Thus, most short- and moderate-distance transportation in a 100 percent WWS world will be BE transportation. Most long-distance, heavy transportation will be with HFC vehicles.

## 7.3.2 Efficiency of Electricity over Combustion for High-Temperature Heat

Replacing fossil fuels and bioenergy for high-temperature industrial heat with WWS electricity also eliminates waste heat of combustion. For example, a natural gas furnace used for producing high-temperature heat is about 80 percent efficient. In other words, about 80 percent of the energy in the chemical bonds of the natural gas is converted to useful heat. The rest of the energy is lost either as waste heat or incompletely combusted natural gas that escapes as exhaust. An electric resistance furnace for producing high temperatures, on the other hand, is about 97 percent efficient. Thus, about 97 percent of the electricity going into an electric resistance furnace

gets converted to useful heat for industry. The rest is waste heat that is lost due to conduction out of the furnace without the heat being used.

As such, the WWS-to-BAU energy input ratio for obtaining the same work output in both cases is 0.82 (= 0.80/0.97). In other words, an electric resistance furnace needs 82 percent of the raw energy input as does a gas furnace to obtain the same work output. This electricity-to-fuel ratio appears in Table 7.3 for industrial sector WWS technologies replacing oil, natural gas, coal, and biofuels/waste.

## 7.3.3 Reducing Energy Use by Moving Heat with Electric Heat Pumps Instead of Creating Heat

Air-source heat pumps move low-temperature heat from one place to another rather than create new heat. As a result, they are much more efficient than are fuel combustion or electric resistance heaters, both of which create low-temperature heat. Air-source heat pumps have a coefficient of performance (COP) of 3.2 to 4.5, whereas ground-source heat pumps have a COP of 4.2 to 5.2 (Fischer and Madani, 2017). This compares with electric resistance heaters, which have a COP of 0.97 and natural gas-powered boilers, which have a typical COP of 0.8. Since only 1 joule (J) of electricity is needed to move 3.2 to 5.2 J of hot or cold air with a heat pump, a heat pump reduces power demand compared with an electric resistance heater or a natural gas boiler substantially.

For example, in Table 7.3, the electricity-to-fuel ratio for WWS electric heat pumps replacing natural gas in the residential sector is 0.2, which is an average of 0.15 and 0.25. The range is calculated by dividing the coefficient of performance of a gas heater (0.8) by the coefficient of performance of a heat pump (3.2 to 5.2). Thus, electric heat pumps reduce end-use energy demand by 75 to 85 percent compared with natural gas heaters and by 70 to 81 percent compared with electric resistance heaters.

The electricity-to-fuel ratio of 0.15 to 0.25 applies to electric heat pumps replacing natural gas, oil, coal, renewables, and biofuels, which are all used for low-temperature heating in the residential and commercial sectors.

The use of electric heat pumps for cooling does not change the electricity-to-fuel ratio for the portion of *electricity* in Table 7.3 used for air conditioning because an electric air conditioner already acts like a heat pump – it removes heat from a room and transfers it to the outside air. As such, an air conditioner has a similar coefficient of performance as does a heat pump. It just doesn't run in reverse to produce heat. Refrigerators, like air conditioners, also act like heat pumps, but only for cooling, not for heating.

### 7.3.4 Eliminating Energy to Mine, Transport, and Process Fossil Fuels, Biofuels, Bioenergy, and Uranium

Producing all energy with WWS eliminates the need to mine, transport, and process fossil fuels, biofuels, bioenergy, and uranium. Worldwide, about 12 to 14 percent of all energy is used for this purpose (Table 7.1). A WWS energy economy eliminates the need for this energy. Instead, wind comes right to the wind turbine and sunlight comes right to the solar panel, so no energy is needed to mine wind or sunlight.

Eliminating energy to mine, transport, and process fuels requires going into the sector-by-sector energy inventory for each country from IEA (2019) and eliminating energy used for industrial and transportation sector "own-use." In the industrial sector, this includes energy for mining operations and petroleum refining. In the transportation sector, this includes energy for natural gas pipelines, oil tankers, coal trains, and gasoline trucks. About 2 percent of oil, 50 percent of coal, and 80 percent of natural gas in the transportation sector is used just to transport fossil fuels, biofuels, and uranium (Jacobson et al., 2017). The need for this energy is eliminated upon a transition to 100 percent WWS.

### 7.3.5 Increasing Energy Efficiency and Reducing Energy Use beyond BAU

Finally, increasing energy efficiency and reducing energy use beyond those occurring in the BAU scenario reduce end-use energy demand further. The BAU projection of energy demand between today and 2050 (Section 7.1) accounts for some end-use energy efficiency improvements and reductions in energy use. However, such improvements are modest. Additional policy-driven energy efficiency measures and incentives can reduce end-use energy demand further. Table 7.3 provides estimated reductions in end-use energy demand that may be achievable for each fuel in each energy sector due to

policy-driven improvements in efficiency beyond what is expected to occur in the BAU case.

Most of the improvements are due to increasing energy efficiency in the residential and commercial sectors. Some additional efficiency improvements arise in the industrial sector due to faster implementation of more advanced equipment than in the BAU case. In the transportation sector, improvements in vehicle design and in lightweight materials relative to BAU may also occur.

### 7.3.6 Overall Reduction in End-Use Demand

Table 7.1 summarizes the final estimated annually averaged end-use power demand, summed over 143 countries in 2050, after a transition from BAU to 100 percent WWS. The table also shows the incremental benefit of the five specific improvements discussed in Section 7.3.1. Table 7.6 provides the 2050 end-use WWS demand, summed over all energy sectors, by country.

Table 7.1 indicates that, of the overall 57.1 percent reduction in end-use power demand between the 2050 BAU case and the 2050 WWS-D case, 38.3 percent is due to the efficiency of using WWS electricity over combustion; 12.1 percent is due to eliminating energy in the mining, transporting, and refining of fossil fuels; and 6.6 percent is due to end-use energy efficiency improvements and reduced energy use beyond those in BAU. Of the 38.3 percent reduction due to the efficiency of WWS electricity, 21.7 percentage points are due to the efficiency of WWS transportation, 3.4 percentage points are due to the efficiency of WWS electricity for industrial heat, and 13.2 percentage points are due to the efficiency of heat pumps.

In sum, transitioning from fossil fuels, biofuels, bioenergy, and uranium to 100 percent WWS has a large potential to reduce power demand worldwide by reducing waste heat, eliminating unnecessary energy in finding and processing fossil fuels, and improving energy efficiency.

## 7.4 Performing a Resource Analysis

The next step in developing a 100 percent WWS roadmap is to perform a resource analysis for each WWS energy-generating technology to determine the maximum WWS renewable resource in each country, state, city, or town that is available to meet demand. Resource analyses can be performed through data analysis, modeling, or both.

Data include local **in situ** (on site) data or **remotely sensed** data. In situ data include, for example, continuous measurements of wind speed with an **anemometer** and solar radiation with a **pyranometer** at a given location.

Remotely sensed data include primarily satellite products. Satellite products are not usually measured directly with an instrument. Instead, they are derived from measurements combined with model predictions. For example, satellite-derived wind speeds can be obtained by combining data from a microwave radiometer, a microwave scatterometer, or an ultraviolet laser with model predictions. The first two of these instruments work best primarily over the ocean. The last can work over the ocean or land.

A **microwave radiometer** on board a satellite detects radiation at millimeter to centimeter (microwave) wavelengths emitted by a surface, such as the ocean surface. When wind speeds are high over the ocean, the winds create strong waves, which, in turn, increase the roughness of the ocean surface. High surface roughness over the ocean increases microwave radiation emissions, and this increase is detected by the radiometer. An algorithm is used to estimate wind speeds 10 m above the ocean surface as a function of the radiation detected.

A **microwave scatterometer** measures the reflectance of a microwave radar wave sent to the Earth. If the reflected beam received back by the instrument is weak, that indicates a smooth ocean surface because less energy is reflected (more is absorbed). If the reflected beam is strong, that indicates a rough surface. A smooth surface indicates a slow wind speed; a rough surface, a fast wind speed. An algorithm estimates wind speed based on reflectance.

In 2018, a **high-power ultraviolet laser** was installed on a satellite to estimate wind speeds. The laser pulses 50 times per second. Light from the laser bounces off the atmosphere at different altitudes and returns to the satellite, where it is retrieved by a large telescope. The light returns with different frequencies than it was sent down with. Wind speed at each altitude is then calculated from the difference in frequency at each altitude.

Whereas in situ data are considered the most accurate, they are available only at a few locations worldwide. Satellite products are less accurate, particularly in the presence of clouds, but satellite data are available over large parts of the world. Two major problems with both in situ and satellite data are (1) they cannot be used to predict WWS energy resources in the future and (2) they cannot be used to predict wind resources in the presence versus absence of wind turbines (Section 6.6.6.4) or wave devices.

Three-dimensional atmospheric models, on the other hand, can cover high spatial resolution, predict the future, and account for competition for limited kinetic energy among wind turbines. However, model predictions have greater uncertainty in their accuracy than do either in situ data or remotely sensed satellite products.

Table 7.4 estimates the annual average power available worldwide from WWS electricity-generating technologies. These results are obtained from modeling (for wind and solar) or a combination of modeling and data analysis. The table contains estimates of both raw resources and resources that can feasibly be extracted in the near term considering cost and location. Finally, it shows the actual power supplied by these WWS technologies in 2017. The table indicates that only wind and solar can provide more annual average power on their own than the world needs for all end uses in 2050 (8.7 TW, Table 7.1).

A resource analysis also necessitates the need to evaluate current and future rooftop areas available for solar PV in each country, province, state, city, or town. Rooftop areas include areas over buildings, parking canopies, parking structures, and road canopies. Rooftop areas suitable for PV exclude areas facing north in the Northern Hemisphere or south in the Southern Hemisphere. They also exclude areas continuously shaded and areas between rows of panels needed for walking and for avoiding shading of one row by another. Table 7.5 provides results from such an analysis for 143 countries.

## 7.5 Selecting a Mix of WWS Energy Generators to Meet Demand

The next step in developing a 100 percent WWS roadmap is to quantify a mix, for each region of interest, of the WWS electricity and heat generators that will supply the end-use all-purpose energy (or power) in the annual average. The penetration of each WWS electricity generator in each country or state is limited by the following constraints: (1) each type of generator cannot draw more power from its renewable resource than exists in the region of interest; (2) the land area taken up among all WWS generators should be no more than a few percent of the land area of the region of interest; (3) no new conventional hydropower dams need to be installed, but existing ones can be improved; (4) wind and solar, which

Table 7.4 **Power available in wind, water, and solar (WWS) energy resources if the energy were used in electricity generation devices worldwide and in likely developable locations. Also shown is the 2017 delivered electricity from WWS resources. All values are before transmission and distribution losses.**

| Energy technology | Power available worldwide (TW) | Power available in likely developable locations (TW) | Power delivered as electricity 2017 (GW) |
|---|---|---|---|
| Wind | 253[a] | 20[b] | 152[c] |
| Wave | 3.4[d] | 0.5[e] | 0.002[f] |
| Geothermal | 45[g] | 0.07–0.14[e] | 9.7[h] |
| Hydroelectric | 1.9[e] | 1.6[e] | 478[i] |
| Tidal | 3.7[e] | 0.02[e] | 0.042[f] |
| Solar PV | 19,400[j] | 1,350[j] | 52.5[k] |
| CSP | 3,960[l] | 200[m] | 1.3[n] |
| Total | | | 694 |

[a] The world wind power available is onshore plus offshore and accounts for competition for available kinetic energy among all wind turbines. Source: Jacobson and Archer (2012) and summarized in Table 6.8.

[b] Locations over land or near the coast, away from remote areas, where the wind speed exceeds 7 m/s.

[c] Source: IEA (2018b).

[d] Source: WEC (2016), who estimate wave power for waves exceeding 5 kW/m and for latitudes less than 66.5 N.

[e] Source: Jacobson (2009) and references therein. This is for tidal power only. Ocean current power is additional.

[f] Source: Marine Energy (2018).

[g] Source: Jacobson and Delucchi (2011) and references therein.

[h] Source: IEA (2018c).

[i] Source: IHA (2018).

[j] Source: Table 5.3.

[k] Source: IEA (2018d).

[l] Estimated from the maximum solar PV potential over land (5,600 TW from Table 5.3) after scaling it by the land area requirements for a CSP plant to that of a PV panel, obtained from Jacobson (2009). This calculation ignores spacing between CSP plants and assumes no cost limitation to putting CSP anywhere.

[m] Estimated assuming CSP can be installed cost effectively only in locations over land with annual solar radiation at the surface greater than or equal to 5 kWh/m²/day.

[n] Estimated from the nameplate capacity of CSP in place in 2017 – 4,952 MW and an estimated capacity factor of 26.6 percent, which was Spain's CSP capacity factor in 2017 (EVWind, 2018).

are complementary in nature, should both be used in roughly equal proportions to the extent possible.

Table 7.6 provides the projected 2050 all-sector annual average, end-use power demand and one initial mix of

10 WWS electricity generators that satisfy the criteria just identified and meet the demand in each of 143 countries. The mix is an "initial mix" because it will be updated when matching demand with supply continuously in Chapter 8. The initial mix of generators provided by country is by no means the only possible mix. It is one of many bounded by the constraints discussed. The numbers in the table assume the generators produce power sufficient to meet the annual average end-use load of the country given in the table. The table ignores cross-border transfers of energy that will occur in reality and does not account for meeting variations in power demand during the year.

Table 7.7 provides a weighted mean of the percentages in Table 7.6 among all 143 countries. It also provides an estimate of the nameplate capacity of existing plus new generators, summed over all countries, to meet the 2050 demand in the 143 countries. Finally, it gives the number of new plants of a specified nameplate capacity needed to meet the demand.

The number of WWS generators required for each country in Table 7.7 is derived starting with the end-use power demand supplied by each generator in each country, calculated from Table 7.6. This is divided by the annually averaged power output from one energy device (e.g., one wind turbine or one solar panel) after transmission, distribution, and maintenance losses have been accounted for. The annual output by device after losses equals the nameplate capacity per device (same for all countries) multiplied by the country-specific annually averaged capacity factor of the device (given in Jacobson et al., 2019), diminished by transmission, distribution, and maintenance losses. For onshore and offshore wind turbines, the capacity factors are calculated country-by-country in 2050 at 100-m hub height from 2050 global model simulations. The simulations account for competition among wind turbines for available kinetic energy; thus, they account for array losses (e.g., Section 6.6.6.4).

The summed nameplate capacities in column (C) of Table 7.7 can provide all energy needs in 143 countries *in the annual average*. However, storage and additional generators are needed to meet energy demand *hour by hour* during the year. In addition, solar and geothermal heat will help meet the heating portion of energy needs. Table 7.7 provides the level of existing solar and geothermal heat in the 143 countries along with an initial estimate of solar heat and CSP with storage needed to help meet energy demand hour by hour in the 143 countries. These numbers are updated in Chapter 8, which provides

Table 7.5 Residential and commercial/government rooftop areas suitable for solar PV panels and potential nameplate capacity of suitable rooftop areas, for 143 countries. An average of about 19.0 percent and 60.1 percent of potential residential and commercial, respectively, rooftop areas from this table are proposed to be installed by no later than 2050 in the roadmaps discussed herein. Potential nameplate capacity is the peak DC (direct current) output among all possible rooftop PV panels in a country.

| Country | Residential roof-top area suitable for PVs in 2012 (km²) | Potential nameplate capacity of suitable area in 2050 (MW) | Commercial/ government rooftop area suitable for PVs in 2012 (km²) | Potential nameplate capacity of suitable area in 2050 (MW) | Country | Residential rooftop area suitable for PVs in 2012 (km²) | Potential nameplate capacity of suitable area in 2050 (MW) | Commercial/ government rooftop area suitable for PVs in 2012 (km²) | Potential nameplate capacity of suitable area in 2050 (MW) |
|---|---|---|---|---|---|---|---|---|---|
| Albania | 25 | 6,004 | 17 | 4,179 | Kyrgyzstan | 77 | 18,454 | 31 | 7,318 |
| Algeria | 686 | 164,115 | 386 | 92,286 | Latvia | 12 | 2,906 | 22 | 5,176 |
| Angola | 758 | 181,328 | 278 | 66,594 | Lebanon | 22 | 5,220 | 11 | 2,713 |
| Argentina | 603 | 144,204 | 419 | 100,166 | Libya | 203 | 48,533 | 113 | 27,115 |
| Armenia | 28 | 6,640 | 16 | 3,817 | Lithuania | 23 | 5,477 | 43 | 10,303 |
| Australia | 907 | 216,896 | 544 | 130,210 | Luxembourg | 2 | 392 | 2 | 388 |
| Austria | 82 | 19,598 | 66 | 15,878 | Macedonia, Rep. of | 23 | 5,428 | 15 | 3,471 |
| Azerbaijan | 139 | 33,206 | 85 | 20,343 | Malaysia | 910 | 217,740 | 348 | 83,310 |
| Bahrain | 10 | 2,465 | 4 | 971 | Malta | 2 | 444 | 1 | 192 |
| Bangladesh | 1,359 | 325,065 | 195 | 46,667 | Mauritius | 22 | 5,313 | 6 | 1,512 |
| Belarus | 35 | 8,376 | 59 | 14,194 | Mexico | 1,940 | 464,094 | 972 | 232,424 |
| Belgium | 22 | 5,276 | 19 | 4,613 | Moldova, Republic of | 16 | 3,882 | 9 | 2,178 |
| Benin | 272 | 65,116 | 42 | 10,071 | Mongolia | 50 | 11,986 | 47 | 11,168 |
| Bolivia | 269 | 64,265 | 105 | 25,120 | Montenegro | 7 | 1,564 | 5 | 1,204 |
| Bosnia & Herzegovina | 39 | 9,383 | 25 | 5,912 | Morocco | 445 | 106,332 | 193 | 46,191 |
| Botswana | 62 | 14,850 | 34 | 8,013 | Mozambique | 717 | 171,550 | 100 | 24,026 |
| Brazil | 3,689 | 882,214 | 1,625 | 388,671 | Myanmar | 1,006 | 240,657 | 225 | 53,888 |
| Brunei Darussalam | 20 | 4,736 | 8 | 1,808 | Namibia | 43 | 10,359 | 21 | 4,955 |
| Bulgaria | 54 | 12,815 | 53 | 12,597 | Nepal | 430 | 102,939 | 55 | 13,112 |
| Cambodia | 350 | 83,804 | 59 | 14,005 | Netherlands | 32 | 7,542 | 53 | 12,676 |

Table 7.5 (*cont.*)

| Country | Residential roof-top area suitable for PVs in 2012 (km²) | Potential nameplate capacity of suitable area in 2050 (MW) | Commercial/government rooftop area suitable for PVs in 2012 (km²) | Potential nameplate capacity of suitable area in 2050 (MW) | Country | Residential rooftop area suitable for PVs in 2012 (km²) | Potential nameplate capacity of suitable area in 2050 (MW) | Commercial/government rooftop area suitable for PVs in 2012 (km²) | Potential nameplate capacity of suitable area in 2050 (MW) |
|---|---|---|---|---|---|---|---|---|---|
| Cameroon | 503 | 120,274 | 114 | 27,250 | Curacao | 2 | 519 | 1 | 213 |
| Canada | 386 | 92,379 | 738 | 176,590 | New Zealand | 81 | 19,433 | 62 | 14,730 |
| Chile | 240 | 57,502 | 158 | 37,756 | Nicaragua | 109 | 26,107 | 33 | 7,785 |
| China | 15,139 | 3,620,836 | 9,211 | 2,203,033 | Niger | 841 | 201,125 | 68 | 16,210 |
| Taiwan | 292 | 69,728 | 127 | 30,384 | Nigeria | 5,005 | 1,196,999 | 1,326 | 317,102 |
| Colombia | 961 | 229,862 | 360 | 86,165 | Norway | 42 | 10,128 | 78 | 18,624 |
| Congo | 202 | 48,304 | 65 | 15,662 | Oman | 103 | 24,586 | 57 | 13,665 |
| Congo, Dem. Republic | 2,153 | 514,979 | 261 | 62,451 | Pakistan | 2,660 | 636,177 | 704 | 168,272 |
| Costa Rica | 67 | 15,973 | 28 | 6,625 | Panama | 110 | 26,395 | 44 | 10,419 |
| Cote d'Ivoire | 538 | 128,697 | 112 | 26,803 | Paraguay | 137 | 32,751 | 59 | 14,048 |
| Croatia | 44 | 10,603 | 35 | 8,327 | Peru | 687 | 164,421 | 271 | 64,899 |
| Cuba | 141 | 33,796 | 67 | 16,017 | Philippines | 2,131 | 509,569 | 532 | 127,334 |
| Cyprus | 30 | 7,294 | 10 | 2,420 | Poland | 203 | 48,538 | 357 | 85,291 |
| Czech Republic | 58 | 13,864 | 59 | 14,227 | Portugal | 139 | 33,348 | 70 | 16,851 |
| Denmark | 24 | 5,713 | 42 | 9,951 | Qatar | 17 | 4,112 | 8 | 1,994 |
| Dominican Republic | 92 | 22,071 | 42 | 10,061 | Romania | 180 | 43,053 | 88 | 21,156 |
| Ecuador | 436 | 104,181 | 140 | 33,473 | Russian Federation | 884 | 211,440 | 1,630 | 389,873 |
| Egypt | 1,963 | 469,527 | 692 | 165,527 | Saudi Arabia | 1,093 | 261,444 | 609 | 145,638 |
| El Salvador | 56 | 13,462 | 20 | 4,794 | Senegal | 325 | 77,673 | 60 | 14,378 |
| Eritrea | 152 | 36,406 | 15 | 3,667 | Serbia | 61 | 14,543 | 61 | 14,628 |
| Estonia | 6 | 1,392 | 11 | 2,654 | Singapore | 28 | 6,744 | 6 | 1,518 |

| Country | | | | |
|---|---|---|---|---|
| Ethiopia | 3,839 | 918,272 | 272 | 64,989 |
| Finland | 30 | 7,088 | 75 | 17,831 |
| France | 542 | 129,641 | 475 | 113,722 |
| Gabon | 92 | 22,051 | 39 | 9,254 |
| Georgia | 38 | 9,057 | 25 | 5,913 |
| Germany | 460 | 110,015 | 502 | 120,097 |
| Ghana | 521 | 124,622 | 124 | 29,632 |
| Gibraltar | 0 | 13 | 0 | 6 |
| Greece | 84 | 20,172 | 73 | 17,492 |
| Guatemala | 298 | 71,360 | 87 | 20,878 |
| Haiti | 93 | 22,271 | 14 | 3,411 |
| Honduras | 158 | 37,825 | 46 | 11,007 |
| Hong Kong, China | 14 | 3,231 | 5 | 1,153 |
| Hungary | 72 | 17,150 | 72 | 17,284 |
| Iceland | 3 | 742 | 6 | 1,422 |
| India | 19,163 | 4,583,398 | 5,075 | 1,213,892 |
| Indonesia | 5,958 | 1,424,942 | 1,885 | 450,757 |
| Iran, Islamic Republic | 1,214 | 290,449 | 729 | 174,348 |
| Iraq | 648 | 154,991 | 362 | 86,531 |
| Ireland | 48 | 11,564 | 55 | 13,271 |
| Israel | 79 | 18,813 | 36 | 8,685 |
| Italy | 708 | 169,446 | 256 | 61,262 |
| Jamaica | 41 | 9,916 | 13 | 3,068 |
| Japan | 716 | 171,213 | 402 | 96,084 |

| Country | | | | |
|---|---|---|---|---|
| Slovak Republic | 42 | 10,094 | 39 | 9,436 |
| Slovenia | 17 | 4,035 | 19 | 4,503 |
| South Africa | 669 | 160,101 | 344 | 82,199 |
| South Sudan | 492 | 117,663 | 54 | 12,995 |
| Spain | 558 | 133,572 | 254 | 60,651 |
| Sri Lanka | 555 | 132,778 | 111 | 26,630 |
| Sudan | 1,554 | 371,671 | 347 | 83,063 |
| Suriname | 23 | 5,502 | 9 | 2,263 |
| Sweden | 48 | 11,472 | 88 | 21,123 |
| Switzerland | 80 | 19,031 | 68 | 16,159 |
| Syrian Arab Republic | 316 | 75,661 | 138 | 33,124 |
| Tajikistan | 114 | 27,290 | 32 | 7,742 |
| Tanzania, United Rep. | 1,067 | 255,127 | 177 | 42,431 |
| Thailand | 1,337 | 319,888 | 485 | 115,995 |
| Togo | 193 | 46,188 | 20 | 4,792 |
| Trinidad and Tobago | 26 | 6,210 | 8 | 1,981 |
| Tunisia | 128 | 30,667 | 69 | 16,467 |
| Turkey | 895 | 214,161 | 616 | 147,450 |
| Turkmenistan | 123 | 29,417 | 80 | 19,118 |
| Ukraine | 222 | 53,207 | 192 | 45,818 |
| United Arab Emirates | 124 | 29,582 | 63 | 15,021 |
| United Kingdom | 194 | 46,350 | 330 | 78,966 |
| United States | 8,259 | 1,975,467 | 5,680 | 1,358,464 |
| Uruguay | 38 | 9,162 | 23 | 5,573 |

Table 7.5 (*cont.*)

| Country | Residential roof-top area suitable for PVs in 2012 (km²) | Potential nameplate capacity of suitable area in 2050 (MW) | Commercial/government rooftop area suitable for PVs in 2012 (km²) | Potential nameplate capacity of suitable area in 2050 (MW) |
|---|---|---|---|---|
| Jordan | 64 | 15,194 | 35 | 8,343 |
| Kazakhstan | 398 | 95,246 | 364 | 87,141 |
| Kenya | 1,234 | 295,198 | 194 | 46,374 |
| Korea, DPR | 146 | 34,863 | 43 | 10,279 |
| Korea, Republic of | 458 | 109,429 | 253 | 60,420 |
| Kosovo | 12 | 2,840 | 7 | 1,732 |
| Kuwait | 28 | 6,769 | 15 | 3,492 |
| Uzbekistan | 326 | 77,918 | 164 | 39,322 |
| Venezuela | 665 | 158,993 | 258 | 61,755 |
| Vietnam | 1,345 | 321,671 | 327 | 78,159 |
| Yemen | 658 | 157,446 | 158 | 37,786 |
| Zambia | 590 | 141,177 | 144 | 34,452 |
| Zimbabwe | 344 | 82,380 | 47 | 11,123 |
| **World total** | 112,000 | 26,759,000 | 46,600 | 11,152,000 |

Source: Jacobson et al. (2019).

Table 7.6 Projected 2050 WWS annually averaged all-sector end-use power demand for 143 countries and initial estimates of one mix of power generators that can meet that annually averaged demand. Annual average power is annual energy (GWh/y) divided by the number of hours per year. The percentages for each country add to 100 percent. Multiply each percent by the end-use demand to get the end-use demand met by each device. Divide the end-use demand for each device by its capacity factor to obtain nameplate capacity of each device needed to meet annual averaged demand. Such nameplate capacities must be updated in order to meet time-dependent demand (Chapter 8). Table 7.7 compares the initial and final nameplate capacities, summed over all countries.

| Country | 2050 end-use demand (GW) | Onshore wind (%) | Offshore wind (%) | Wave (%) | Geo-thermal (%) | Hydro-electric (%) | Tidal (%) | Res PV (%) | Com/gov PV (%) | Utility PV (%) | CSP (%) |
|---|---|---|---|---|---|---|---|---|---|---|---|
| Albania | 1.840 | 21.01 | 6.83 | 0 | 0 | 43.54 | 0.11 | 5.85 | 13.01 | 6.83 | 2.82 |
| Algeria | 39.558 | 42.58 | 2.18 | 0.1 | 0 | 0.29 | 0.01 | 9.37 | 20.82 | 19.68 | 4.98 |
| Angola | 9.589 | 37.54 | 9.64 | 0.89 | 0 | 11.23 | 0.06 | 8.26 | 18.36 | 9.64 | 4.39 |
| Argentina | 57.149 | 38.77 | 9.95 | 0 | 1.4 | 7.9 | 0.02 | 8.53 | 18.95 | 9.95 | 4.53 |
| Armenia | 1.601 | 27.37 | 0 | 0 | 1.25 | 34.72 | 0 | 6.02 | 13.38 | 14.05 | 3.2 |
| Australia | 93.588 | 26.95 | 15.09 | 0.96 | 0.36 | 3.99 | 0.12 | 9.43 | 15.72 | 22.64 | 4.73 |
| Austria | 21.000 | 36.18 | 0 | 0 | 0 | 19.59 | 0 | 10.55 | 9.05 | 24.62 | 0 |
| Azerbaijan | 7.253 | 22.57 | 0 | 0 | 0 | 6.77 | 0 | 12.55 | 27.89 | 29.29 | 0.93 |
| Bahrain | 9.228 | 0.38 | 17 | 0 | 0 | 0 | 0.02 | 3.49 | 1.66 | 72.46 | 5 |
| Bangladesh | 30.916 | 7.08 | 7.12 | 0.62 | 0 | 0.32 | 0.1 | 23.18 | 9.68 | 46.96 | 4.95 |
| Belarus | 12.364 | 44.81 | 0 | 0 | 0.07 | 0.36 | 0 | 7.99 | 7.15 | 39.63 | 0 |
| Belgium | 29.194 | 7.96 | 22.64 | 0 | 0 | 0.19 | 0 | 1.82 | 1.61 | 65.78 | 0 |
| Benin | 2.532 | 33.97 | 13.74 | 1 | 0 | 0.46 | 0.03 | 11.78 | 20.36 | 13.74 | 4.93 |
| Bolivia | 5.640 | 32.66 | 0 | 0 | 18.65 | 4.96 | 0 | 7.18 | 15.97 | 16.76 | 3.82 |
| Bosnia & Herzegovina | 3.457 | 22.32 | 4.79 | 0 | 0 | 27.91 | 0.01 | 8.83 | 19.61 | 15.8 | 0.72 |
| Botswana | 2.155 | 42.75 | 0 | 0 | 0 | 0 | 0 | 9.41 | 20.9 | 21.95 | 5 |
| Brazil | 279.105 | 35.53 | 9.12 | 0.84 | 0 | 16.04 | 0.02 | 7.82 | 17.37 | 9.12 | 4.16 |
| Brunei Darussalam | 1.540 | 2.34 | 25.43 | 1 | 0 | 0 | 0.09 | 21.8 | 18.96 | 25.43 | 4.95 |
| Bulgaria | 9.987 | 40.28 | 11.05 | 0 | 0 | 10.46 | 0.02 | 9.47 | 17.68 | 11.05 | 0 |
| Cambodia | 6.069 | 25.56 | 12.91 | 0 | 0 | 8.34 | 0.04 | 11.07 | 24.59 | 12.91 | 4.58 |
| Cameroon | 4.470 | 39.08 | 3.56 | 0.92 | 0 | 7.61 | 0.05 | 8.6 | 19.11 | 16.5 | 4.57 |

Table 7.6 (cont.)

| Country | 2050 end-use demand (GW) | Onshore wind (%) | Offshore wind (%) | Wave (%) | Geo-thermal (%) | Hydro-electric (%) | Tidal (%) | Res PV (%) | Com/ gov PV (%) | Utility PV (%) | CSP (%) |
|---|---|---|---|---|---|---|---|---|---|---|---|
| Canada | 151.542 | 32.74 | 8.87 | 0.73 | 2.63 | 23.6 | 0.29 | 7.6 | 14.67 | 8.87 | 0 |
| Chile | 35.034 | 36.71 | 10.06 | 0.87 | 3.8 | 9.41 | 0.06 | 8.62 | 16.13 | 10.06 | 4.29 |
| China | 2284.621 | 34.84 | 14.29 | 0.05 | 0.07 | 6.3 | 0.02 | 12.25 | 13.22 | 14.29 | 4.68 |
| Taiwan | 29.285 | 2.21 | 23.92 | 0.15 | 30.38 | 1.05 | 0.01 | 10.01 | 4.94 | 23.92 | 3.42 |
| Colombia | 1.191 | 34.42 | 8.83 | 0.81 | 0 | 18.3 | 0.38 | 7.57 | 16.83 | 8.83 | 4.03 |
| Congo | 15.252 | 41.43 | 10.63 | 0.93 | 0 | 6.91 | 0.09 | 9.11 | 20.25 | 10.63 | 0 |
| Congo, Dem. Republic | 3.965 | 39.8 | 0.86 | 0.07 | 0 | 6.82 | 0 | 9.5 | 17 | 21.3 | 4.66 |
| Costa Rica | 5.685 | 21.46 | 5.51 | 0.51 | 24.74 | 24.43 | 0.13 | 4.72 | 10.49 | 5.51 | 2.51 |
| Cote d'Ivoire | 5.822 | 35.38 | 10.99 | 0.93 | 0 | 6.67 | 0.05 | 9.42 | 20.94 | 10.99 | 4.62 |
| Croatia | 8.063 | 13.31 | 0 | 0 | 0 | 13.96 | 0.24 | 16.25 | 16.6 | 37.92 | 1.72 |
| Cuba | 1.366 | 42.12 | 10.81 | 1 | 0 | 0.35 | 0.13 | 9.27 | 20.59 | 10.81 | 4.93 |
| Cyprus | 1.828 | 8.39 | 22.11 | 1 | 0 | 0 | 0.19 | 18.95 | 22.3 | 22.11 | 4.94 |
| Czech Republic | 17.522 | 43.73 | 0 | 0 | 0 | 2.83 | 0 | 9.59 | 9.39 | 34.47 | 0 |
| Denmark | 9.614 | 44.46 | 22.1 | 1 | 0 | 0.04 | 0.17 | 6.66 | 3.47 | 22.1 | 0 |
| Dominican Republic | 6.233 | 22.1 | 12.74 | 0 | 8.85 | 3.92 | 0.08 | 10.92 | 24.28 | 12.74 | 4.36 |
| Ecuador | 9.744 | 33.38 | 3.09 | 0.79 | 0.33 | 20.23 | 0.56 | 7.34 | 16.32 | 14.05 | 3.9 |
| Egypt | 83.693 | 42.1 | 10.8 | 0 | 0 | 1.52 | 0.01 | 9.26 | 20.58 | 10.8 | 4.92 |
| El Salvador | 2.283 | 22.93 | 5.89 | 0.54 | 36.35 | 9.39 | 0.09 | 5.04 | 11.21 | 5.89 | 2.68 |
| Eritrea | 0.303 | 41.72 | 10.71 | 0 | 0 | 0 | 2.4 | 9.18 | 20.4 | 10.71 | 4.88 |
| Estonia | 2.013 | 44.68 | 19.88 | 0 | 0 | 0.18 | 0.53 | 8.53 | 6.33 | 19.88 | 0 |
| Ethiopia | 17.872 | 35.94 | 0 | 0 | 7.11 | 8.83 | 0 | 7.91 | 17.57 | 18.45 | 4.2 |
| Finland | 21.667 | 41.84 | 22.73 | 0 | 0 | 6.99 | 0.02 | 3.92 | 1.76 | 22.73 | 0 |
| France | 112.365 | 40.28 | 13.15 | 0.92 | 0.03 | 7.51 | 0.2 | 11.27 | 11.65 | 13.15 | 1.83 |
| Gabon | 8.666 | 15.66 | 23.34 | 0.99 | 0 | 1.44 | 0.04 | 20 | 15.2 | 23.34 | 0 |

| Country | | | | | | | | | | | |
|---|---|---|---|---|---|---|---|---|---|---|---|
| Georgia | 3.429 | 25.28 | 6.49 | 0 | 0 | 43.75 | 0.06 | 5.56 | 12.36 | 6.49 | 0 |
| Germany | 155.226 | 41.05 | 20.42 | 0.1 | 0.02 | 1.35 | 0.01 | 8.18 | 8.46 | 20.42 | 0 |
| Ghana | 7.445 | 22.97 | 13.25 | 0.92 | 0 | 8.44 | 0.04 | 11.36 | 25.24 | 13.25 | 4.53 |
| Gibraltar | 1.315 | 1.15 | 96.52 | 0.18 | 0 | 0 | 0.02 | 0.1 | 0.06 | 1.97 | 0 |
| Greece | 13.865 | 37.48 | 9.62 | 0.89 | 2.6 | 8.66 | 0.17 | 8.25 | 18.33 | 9.62 | 4.38 |
| Guatemala | 5.712 | 24.4 | 6.26 | 0.58 | 32.95 | 9.36 | 0.04 | 5.37 | 11.93 | 6.26 | 2.85 |
| Haiti | 1.309 | 20.57 | 17.47 | 0 | 0 | 1.73 | 0.4 | 14.97 | 22.48 | 17.47 | 4.89 |
| Honduras | 3.563 | 35.09 | 9.01 | 0.83 | 11.14 | 5.85 | 0.09 | 7.72 | 17.16 | 9.01 | 4.1 |
| Hong Kong, China | 28.850 | 0.07 | 97.2 | 0.14 | 0 | 0 | 0.01 | 1.09 | 0.5 | 0.98 | 0 |
| Hungary | 11.954 | 15.6 | 0 | 0 | 2.42 | 0.2 | 0 | 17.38 | 14.74 | 49.65 | 0 |
| Iceland | 2.983 | 40.69 | 1.22 | 0.42 | 25.54 | 31.65 | 0.44 | 0 | 0 | 0.04 | 0 |
| India | 926.295 | 36.92 | 6.23 | 0.06 | 0.02 | 2.12 | 0.02 | 12.03 | 15.86 | 21.85 | 4.89 |
| Indonesia | 174.839 | 15.79 | 15.28 | 0.94 | 4.45 | 1.34 | 0.03 | 13.1 | 29.11 | 15.28 | 4.66 |
| Iran, Islamic Republic | 178.193 | 28.92 | 12.56 | 0 | 0 | 2.68 | 0 | 14.67 | 14.63 | 21.67 | 4.87 |
| Iraq | 20.320 | 40.83 | 0.91 | 0 | 0 | 4.49 | 0 | 8.98 | 19.96 | 20.04 | 4.78 |
| Ireland | 7.591 | 43.9 | 17.45 | 0.99 | 0 | 1.4 | 0.07 | 14.96 | 3.78 | 17.45 | 0 |
| Israel | 12.820 | 8.21 | 15.61 | 0 | 0 | 0.03 | 0.02 | 21.26 | 11.01 | 38.87 | 5 |
| Italy | 83.169 | 37.31 | 13.97 | 0.35 | 0.98 | 7.79 | 0.02 | 11.97 | 9.1 | 13.97 | 4.54 |
| Jamaica | 2.268 | 5.4 | 24.03 | 0 | 0 | 0.4 | 0.17 | 20.6 | 20.4 | 24.03 | 4.97 |
| Japan | 177.969 | 10.24 | 31.55 | 0.93 | 0.69 | 5.84 | 0.28 | 11.78 | 7.14 | 31.55 | 0 |
| Jordan | 7.330 | 42.72 | 1.42 | 0 | 0 | 0.07 | 0.01 | 9.77 | 19.63 | 21.39 | 5 |
| Kazakhstan | 35.452 | 43.48 | 0 | 0 | 0 | 3.39 | 0 | 9.56 | 21.25 | 22.32 | 0 |
| Kenya | 10.530 | 35.46 | 9.1 | 0.84 | 12.92 | 3.27 | 0.03 | 7.8 | 17.34 | 9.1 | 4.15 |
| Korea, DPR | 11.620 | 36.13 | 13.74 | 0 | 0 | 17.55 | 1.73 | 11.78 | 4.91 | 13.74 | 0.43 |
| Korea, Republic of | 155.187 | 4.44 | 37.18 | 0 | 0 | 1.98 | 0.15 | 8.9 | 5.27 | 37.18 | 4.89 |
| Kosovo | 1.317 | 22.55 | 0 | 0 | 47.16 | 2.23 | 0 | 4.96 | 11.02 | 11.57 | 0.51 |

Table 7.6 (cont.)

| Country | 2050 end-use demand (GW) | Onshore wind (%) | Offshore wind (%) | Wave (%) | Geo-thermal (%) | Hydro-electric (%) | Tidal (%) | Res PV (%) | Com/gov PV (%) | Utility PV (%) | CSP (%) |
|---|---|---|---|---|---|---|---|---|---|---|---|
| Kuwait | 30.545 | 0.71 | 14.03 | 0 | 0 | 0 | 0.01 | 2.97 | 1.74 | 75.53 | 5 |
| Kyrgyzstan | 3.308 | 26.91 | 0 | 0 | 0 | 37.04 | 0 | 5.92 | 13.16 | 13.82 | 3.15 |
| Latvia | 3.202 | 34.8 | 12.52 | 0 | 0 | 22.58 | 0.09 | 10.73 | 6.77 | 12.52 | 0 |
| Lebanon | 5.762 | 2.62 | 27.44 | 0 | 0 | 1.74 | 0.03 | 14.87 | 7.95 | 40.43 | 4.91 |
| Libya | 10.270 | 42.73 | 10.97 | 0 | 0 | 0 | 0.05 | 9.4 | 20.89 | 10.97 | 5 |
| Lithuania | 4.249 | 44.5 | 14.28 | 0 | 0 | 1.1 | 0.02 | 12.24 | 13.58 | 14.28 | 0 |
| Luxembourg | 2.296 | 7.63 | 0 | 0 | 0 | 0.68 | 0 | 1.66 | 1.74 | 88.3 | 0 |
| Macedonia, Rep. of | 2.024 | 17.5 | 0 | 0 | 0 | 13.69 | 0 | 14.24 | 21.36 | 33.22 | 0 |
| Malaysia | 77.139 | 2.7 | 25.24 | 0.19 | 0 | 3.64 | 0.02 | 21.7 | 16.3 | 25.39 | 4.81 |
| Malta | 1.429 | 1.07 | 43.52 | 0.59 | 0 | 0 | 0.13 | 4.2 | 2.01 | 43.52 | 4.96 |
| Mauritius | 1.790 | 1.68 | 26.76 | 0.98 | 0 | 1.58 | 0.08 | 22.94 | 14.35 | 26.76 | 4.87 |
| Mexico | 131.278 | 39.44 | 10.12 | 0.93 | 2.96 | 3.84 | 0.02 | 8.68 | 19.28 | 10.12 | 4.61 |
| Moldova, Republic of | 2.035 | 44.32 | 0 | 0 | 0 | 1.51 | 0 | 13.45 | 9.32 | 31.39 | 0 |
| Mongolia | 3.340 | 44.87 | 0 | 0 | 0 | 0.28 | 0 | 9.87 | 21.94 | 23.03 | 0 |
| Montenegro | 0.765 | 24.55 | 8.06 | 0 | 0 | 36.18 | 0.24 | 6.91 | 15.36 | 8.06 | 0.64 |
| Morocco | 18.336 | 41.01 | 10.53 | 0.97 | 0 | 3.06 | 0.03 | 9.02 | 20.05 | 10.53 | 4.8 |
| Mozambique | 8.843 | 36.8 | 9.45 | 0.87 | 0 | 10.65 | 2.39 | 8.1 | 17.99 | 9.45 | 4.3 |
| Myanmar | 11.333 | 37.67 | 9.67 | 0.89 | 0 | 10.65 | 0.33 | 8.29 | 18.42 | 9.67 | 4.41 |
| Namibia | 1.939 | 39.59 | 10.16 | 0.94 | 0 | 6.22 | 0.24 | 8.71 | 19.35 | 10.16 | 4.63 |
| Nepal | 6.888 | 25.98 | 0 | 0 | 0 | 4.63 | 0 | 17.29 | 6.98 | 40.34 | 4.77 |
| Netherlands | 40.098 | 10.45 | 43.09 | 0 | 0 | 0.04 | 0.01 | 1.86 | 1.47 | 43.09 | 0 |
| Curacao | 1.366 | 1.31 | 42.9 | 0 | 0 | 0 | 0.16 | 5.38 | 2.36 | 42.9 | 4.99 |
| New Zealand | 17.566 | 32.42 | 10.38 | 0.77 | 9.32 | 13.83 | 0.25 | 8.89 | 9.97 | 10.38 | 3.79 |
| Nicaragua | 1.667 | 32.3 | 8.29 | 0.76 | 20.38 | 3.07 | 0.23 | 7.11 | 15.79 | 8.29 | 3.78 |

| | | | | | | | | | | | |
|---|---|---|---|---|---|---|---|---|---|---|---|
| Niger | 1.409 | | 0 | 0 | 0 | 0 | 0 | 9.41 | 20.9 | 21.95 | 5 |
| Nigeria | 67.956 | 42.75 | 0 | 0.3 | 0 | 1.29 | 0.01 | 14.37 | 31.94 | 33.54 | 4.92 |
| Norway | 20.216 | 13.63 | 5.56 | 0.31 | 0 | 68.83 | 0.39 | 4.77 | 0.87 | 5.56 | 0 |
| Oman | 33.053 | 13.71 | 14.48 | 0.66 | 0 | 0 | 0.02 | 11.11 | 6.94 | 48.87 | 4.97 |
| Pakistan | 82.144 | 12.96 | 10.55 | 0.24 | 0 | 3.65 | 0 | 15.23 | 16.55 | 24.98 | 4.81 |
| Panama | 5.969 | 23.99 | 9.36 | 0.86 | 0 | 13.13 | 0.73 | 8.02 | 17.82 | 9.36 | 4.26 |
| Paraguay | 4.926 | 36.46 | 0 | 0 | 0 | 85.07 | 0 | 1.4 | 3.12 | 3.28 | 0.75 |
| Peru | 17.482 | 6.38 | 0.02 | 0.79 | 6.62 | 14.04 | 0.05 | 7.38 | 16.41 | 17.21 | 3.93 |
| Philippines | 40.458 | 33.56 | 15.13 | 0.6 | 11.24 | 3.95 | 0.27 | 12.97 | 28.82 | 15.13 | 4.2 |
| Poland | 44.642 | 7.68 | 11.46 | 0 | 0.2 | 0.58 | 0.01 | 9.82 | 21.83 | 11.46 | 0 |
| Portugal | 13.064 | 44.64 | 9 | 0.83 | 0.6 | 15.77 | 0.82 | 7.71 | 17.13 | 9 | 4.1 |
| Qatar | 24.422 | 35.05 | 21.02 | 0 | 0 | 0 | 0.01 | 2.31 | 1.32 | 69.77 | 5 |
| Romania | 18.269 | 0.56 | 12.02 | 0 | 0.44 | 16.32 | 0.01 | 10.31 | 11.42 | 12.02 | 0 |
| Russian Federation | 233.033 | 37.45 | 13.06 | 0.5 | 0.17 | 8.96 | 0.03 | 11.19 | 11.88 | 13.06 | 0.9 |
| Saudi Arabia | 174.704 | 40.25 | 3.84 | 0 | 0 | 0 | 0 | 11.28 | 14.65 | 22.48 | 5 |
| Senegal | 3.366 | 42.75 | 10.64 | 0.98 | 0 | 1.95 | 0.08 | 9.12 | 20.27 | 10.64 | 4.85 |
| Serbia | 8.934 | 41.46 | 0 | 0 | 0 | 11.01 | 0 | 16.5 | 20.29 | 38.5 | 0 |
| Singapore | 67.141 | 13.71 | 92.66 | 0 | 4.94 | 0 | 0 | 1.11 | 0.32 | 0.94 | 0 |
| Slovak Republic | 7.684 | 0.02 | 0 | 0 | 0 | 9.97 | 0 | 11.6 | 10.86 | 27.06 | 0 |
| Slovenia | 3.631 | 40.52 | 9.8 | 0 | 2.3 | 16.61 | 0.02 | 9.06 | 14.38 | 11.35 | 0 |
| South Africa | 104.603 | 36.48 | 13.7 | 1 | 0 | 0.3 | 0.01 | 11.74 | 12.43 | 13.7 | 4.93 |
| South Sudan | 0.476 | 42.19 | 0 | 0 | 0 | 0 | 0 | 9.41 | 20.9 | 21.95 | 5 |
| Spain | 65.686 | 42.75 | 11.92 | 0.88 | 0.06 | 11.34 | 0.33 | 10.21 | 11.61 | 11.92 | 4.37 |
| Sri Lanka | 11.835 | 37.36 | 10.18 | 0.94 | 0 | 6.25 | 0.04 | 8.73 | 19.39 | 10.18 | 4.64 |
| Sudan | 11.108 | 39.66 | 10.15 | 0 | 0 | 7.48 | 0.03 | 8.7 | 19.33 | 10.15 | 4.62 |
| Suriname | 0.505 | 39.54 | 8.96 | 0.82 | 0 | 17.08 | 0.48 | 7.68 | 17.06 | 8.96 | 4.08 |
| Sweden | 30.479 | 34.89 | 17.39 | 0 | 0 | 24.93 | 0.07 | 4.45 | 2.03 | 17.39 | 0 |

Table 7.6 (*cont.*)

| Country | 2050 end-use demand (GW) | Onshore wind (%) | Offshore wind (%) | Wave (%) | Geo-thermal (%) | Hydro-electric (%) | Tidal (%) | Res PV (%) | Com/gov PV (%) | Utility PV (%) | CSP (%) |
|---|---|---|---|---|---|---|---|---|---|---|---|
| Switzerland | 16.000 | 26.91 | 0 | 0 | 0 | 39.31 | 0 | 6.56 | 11.9 | 15.32 | 0 |
| Syrian Arab Republic | 6.797 | 38.67 | 9.93 | 0 | 0 | 9.51 | 0.02 | 8.51 | 18.91 | 9.93 | 4.52 |
| Tajikistan | 2.680 | 7.75 | 0 | 0 | 0 | 81.88 | 0 | 1.7 | 3.79 | 3.98 | 0.91 |
| Tanzania, United Rep. | 16.956 | 41.47 | 10.64 | 0.98 | 0 | 1.41 | 0.61 | 9.12 | 20.27 | 10.64 | 4.85 |
| Thailand | 122.975 | 3.75 | 18.81 | 0 | 0.08 | 1.31 | 0.01 | 22.28 | 15.65 | 33.18 | 4.93 |
| Togo | 1.442 | 26.62 | 18.51 | 0.89 | 0 | 1.2 | 0.04 | 15.86 | 13.48 | 18.51 | 4.89 |
| Trinidad and Tobago | 7.375 | 0.45 | 37.95 | 0.5 | 0 | 0 | 0.03 | 13.66 | 4.49 | 37.95 | 4.97 |
| Tunisia | 10.655 | 42.62 | 10.94 | 0 | 0 | 0.27 | 0.04 | 9.38 | 20.84 | 10.94 | 4.98 |
| Turkey | 71.476 | 35.48 | 2.08 | 0 | 1.33 | 15.65 | 0.02 | 7.81 | 17.35 | 16.13 | 4.15 |
| Turkmenistan | 8.725 | 42.75 | 0 | 0 | 0 | 0.01 | 0 | 9.4 | 20.9 | 21.94 | 5 |
| Ukraine | 43.052 | 42.41 | 15.21 | 0 | 0 | 5.73 | 0.02 | 13.04 | 8.39 | 15.21 | 0 |
| United Arab Emirates | 104.716 | 7.1 | 12.26 | 0 | 0 | 0 | 0.01 | 4.3 | 2.45 | 68.88 | 5 |
| United Kingdom | 88.812 | 20.29 | 32.66 | 0.96 | 0 | 0.93 | 2.81 | 5.41 | 4.27 | 32.66 | 0 |
| United States | 939.460 | 31.44 | 16.42 | 0.96 | 0.57 | 3.9 | 0.01 | 10.95 | 14.6 | 16.42 | 4.73 |
| Uruguay | 5.233 | 36.55 | 9.38 | 0.86 | 0 | 13.57 | 0.07 | 8.04 | 17.87 | 9.38 | 4.27 |
| Uzbekistan | 18.840 | 40.94 | 0 | 0 | 0 | 4.23 | 0 | 9.01 | 20.02 | 21.02 | 4.79 |
| Venezuela | 36.234 | 35.91 | 9.22 | 0.18 | 0 | 15.8 | 0.02 | 7.9 | 17.56 | 9.22 | 4.2 |
| Vietnam | 91.272 | 0.72 | 25.19 | 0.57 | 0 | 8.07 | 0.01 | 21.59 | 14.08 | 25.19 | 4.57 |
| Yemen | 2.263 | 13.21 | 16.31 | 0.97 | 3.12 | 0 | 0.26 | 13.98 | 31.06 | 16.31 | 4.78 |
| Zambia | 9.572 | 37.85 | 0 | 0 | 0.73 | 10.74 | 0 | 8.33 | 18.5 | 19.43 | 4.43 |
| Zimbabwe | 5.881 | 39.86 | 0 | 0 | 0 | 6.76 | 0 | 9.58 | 16.8 | 22.34 | 4.66 |
| **World total/ average** | **8.693** | **30.5** | **14.51** | **0.34** | **0.92** | **5.72** | **0.08** | **11.14** | **13.84** | **19.03** | **3.93** |

Source: Jacobson et al. (2019).

Table 7.7 Nameplate capacities needed by generator type for 100 percent WWS. Estimated (C) initial nameplate capacities (meeting the annual average all-purpose end-use power demand) and (D) final nameplate capacities (meeting time-dependent demand) of WWS generators, summed among 143 countries in 24 regions, needed to supply 100 percent of all-purpose energy with WWS in those countries. Also shown are (A) the 143-country-averaged end-use demand estimated to be supplied by the initial nameplate capacity of each generator, (E) the percent of final 2050 nameplate capacity of each generator already installed in 2018, and (F) the final numbers of new devices of specified sizes still needed.

| Energy technology | (A) Nameplate capacity of one plant or device (MW) | (B) 2050 all-purpose annual average demand met by plant/device (%) | (C) Initial nameplate capacity, existing plus new plants or devices to meet annual average demand (GW) | (D) Final nameplate capacity, existing plus new plants or devices to meet time-dependent demand (GW) | (E) Final nameplate capacity already installed 2018 (%) | (F) Number of new plants or devices needed for 143 countries |
|---|---|---|---|---|---|---|
| **Annual average power** | | | | | | |
| Onshore wind | 5 | 30.50 | 8,251 | 11,976 | 4.76 | 2,281,019 |
| Offshore wind | 5 | 14.51 | 3,841 | 3,606 | 0.68 | 716,252 |
| Wave device | 0.75 | 0.34 | 156 | 156 | 0.0001 | 208,313 |
| Geothermal electricity | 100 | 0.92 | 97 | 97 | 13.67 | 837 |
| Hydropower plant[a] | 1,300 | 5.72 | 1,109 | 1,109 | 100.0 | 0 |
| Tidal turbine | 1 | 0.08 | 31 | 31 | 1.76 | 30,075 |
| Res. roof PV[b] | 0.005 | 11.14 | 5,082 | 2,776 | 3.44 | 536,080,000 |
| Com/gov roof PV[b] | 0.1 | 13.84 | 6,705 | 5,121 | 1.87 | 50,250,000 |
| Utility PV plant[b] | 50 | 19.03 | 8,234 | 13,691 | 2.09 | 268,090 |
| Utility CSP plant[b] | 100 | 3.93 | 634 | 1,262 | 0.43 | 12,565 |
| Total for average power | | 100 | 34,138 | 39,842 | 5.53 | 610,045,000 |
| **For peaking/ storage** | | | | | | |
| Additional CSP[c] | 100 | 2.36 | 381 | 0 | 0 | 0 |
| Solar thermal heat[c] | 50 | | 2,573 | 632 | 72.6 | 3,468 |
| Geothermal heat[c] | 50 | | 70.3 | 70.3 | 100.00 | 0 |
| Total peaking/ storage | | 2.36 | 3,024 | 702 | 75.31 | 3,468 |
| **Total all** | | | 37,163 | 40,544 | 6.74 | 610,049,000 |

a more detailed analysis of matching variable demand with variable supply of energy in these countries. Column (D) of Table 7.7 shows the resulting final nameplate capacities, summed among all countries for each generator after the update.

## 7.6 Estimating Avoided Energy, Air Pollution, and Climate Costs

Transitioning to 100 percent WWS will reduce direct energy costs as well as the social (economic) cost of energy. The **social (economic) cost** of energy is the total cost of energy to society. Thus, it is the private energy cost plus all externality costs associated with the energy. The **private energy cost** is the marketplace cost of energy. It does not account for health or climate costs of energy. An **externality cost** is a cost not captured in the market. The most relevant externality costs related to energy are health costs, climate costs, and non-health and non-climate environmental costs. In this section, these costs are discussed.

### 7.6.1 Avoided Energy Costs

The private (non-externality) cost of an energy system includes the costs of electricity generation and storage, heat generation and storage, cold generation and storage, hydrogen generation and storage, transmission and distribution, and the machines and appliances using the electricity, heat, and cold.

With respect to electric power generation costs, two metrics commonly used are the cost per unit energy (e.g., per kWh) generated and the aggregate cost of electricity generation. The aggregate cost is simply the cost per unit energy (per kWh) multiplied by the energy (number of kWh) consumed. Table 7.1 shows that a 100 percent WWS system can reduce annually averaged power (thus energy) demand by about 57 percent. As such, the same cost per kWh between a BAU and WWS system means a 57 percent lower energy bill to a consumer in the WWS case.

Here, the method of calculating cost per unit energy of an electric power generator is described. A cost comparison is then performed between a BAU energy system and a 100 percent WWS system.

The main parameters involved in the private cost of energy for an energy source are the upfront cost, the time-dependent operation and maintenance cost, the time-dependent fuel cost (if any), the decommissioning cost, the discount rate, the time between financing and construction, and energy-generating technology lifetime. For renewable energy generators (e.g., wind, solar, geothermal, hydro, tidal, and wave), the fuel cost is zero. For hydrogen, the fuel cost is the cost of water used to produce the hydrogen by electrolysis.

The **levelized cost of electricity (LCOE)** is the net present value of the total cost of electricity over the life of a generating plant, divided by the net present value of the total energy generated by the plant over its life. The units of LCOE (in USD) are usually either $/MWh or ¢/kWh. LCOE is effectively the average price that an

---

Table 7.7 (cont.)

All values are summed over 143 countries in 24 regions. "Annual average power" is annual average all-purpose energy demand divided by the number of hours per year. The nameplate capacity of each device, shown in column (A), is assumed to be the same for all countries. The percent of annual-average power demand met by each device type, shown in column (B), is a demand-weighted average among the mixes given for 143 countries (Table 7.6) before time-dependent demand calculations were performed. The "initial" nameplate capacity in column (C) is a weighted average among all countries and devices of the total end-use demand for each country from Table 7.6 multiplied by the percentage of demand satisfied by the device from Table 7.6, then divided by the capacity factor of the device. This initial nameplate capacity (meeting average-annual demand), for each grid region, is used at the start of time-dependent simulations for matching power demand with supply and storage (Chapter 8). The part of column (C) labeled "For peaking/storage" is the initial estimate of additional CSP installations and solar thermal heat generators for the start of the time-dependent simulations. Column (D) shows 143-country final nameplate capacities needed to match load, after the time-dependent simulations for each of 24 grid regions encompassing the 143 countries (Table 8.5). Columns (D) and (E) are the fraction of final nameplate capacity already installed as of 2018 end and the remaining number of devices of size specified in column (A) still needed, respectively.

[a] The average capacity factor of hydropower plants is assumed to increase from its current world average of ~42 to 54.8 percent. No increase in the number of dams or in the peak discharge rate of hydropower is assumed.

[b] The solar PV panels used for this calculation are SunPower E20 panels. CSP is assumed to have storage with a maximum charge to discharge rate (storage size to generator size ratio) of 2.62:1.

[c] Additional CSP is estimated CSP plus storage beyond that for annual average power generation proposed to provide peaking power to stabilize the grid. Additional solar thermal and existing geothermal heat are used for direct heat or underground thermal energy storage. "Geothermal heat" is existing geothermal heat, which is assumed not to change in the future (hence the same values in columns (C) and (D)).

Source: Jacobson et al. (2019).

electricity generator must be paid before taxes for it to break even over its lifetime. The LCOE is calculated as

$$LCOE = \frac{\sum_{t=1}^{n} \dfrac{A_{Capital} + A_{O\&M,t} + A_{Fuel,t} + A_{Decomm}}{(1+i)^t}}{\sum_{t=1}^{n} \dfrac{E_t}{(1+i)^t}}$$

(7.1)

where $A_{Capital}$ is the annualized (over the project period) upfront capital cost, $A_{O\&M,t}$ is the annual operation and maintenance (O&M) cost in year $t$, $A_{Fuel,t}$ is the annual fuel cost in year $t$ (zero for non-hydrogen WWS sources), $A_{Decomm}$ is the annualized (over the project period) decommissioning cost at the end of the project, $n$ is the time between financing and end of decommissioning of the plant (thus includes plant construction, operation, and decommissioning times), $i$ is the **discount rate** (fraction per year, assumed constant over $n$ years), and $E_t$ is the energy generated by the plant in year $t$.

The annualized capital cost (or payment on a loan) is

$$A_{Capital} = P \times CRF_{i,\,n}$$ (7.2)

where $P$ is the present value of the investment (upfront capital cost). In addition, CRF is the **capital recovery factor** (fraction of the present value paid each year),

$$CRF_{i,n} = \frac{i(1+i)^n}{(1+i)^n - 1}$$ (7.3)

The annualized decommissioning cost is

$$A_{Decom} = \frac{F}{(1+i)^n} CRF_{i,n}$$ (7.4)

where $F$ is the decommissioning cost in future year $n$, and $F/(1+i)^n$ is the present value of a future cost. For WWS generators, the O&M costs are usually a fixed percent of the upfront capital cost, increasing slightly each year. For thermal power plants, additional O&M costs are generally required, which are a function of the amount of electricity generated each year (e.g., Table 7.8). WWS plants (aside from hydrogen fuel cell plants) do not have fuel costs, but BAU plants, which rely on natural gas, coal, or uranium, do. All O&M costs and fuel costs start to accrue in Equation 7.1 after construction is completed (when electricity is generated); however, upfront capital costs and decommissioning costs are annualized for all years (construction, operation, and decommissioning).

There are two types of discount rate – a private discount rate and a social discount rate. The social discount rate is discussed shortly.

The **private discount rate** (PDR) is the interest rate that banks will charge builders and consumers for taking out loans. Such loans may be used to pay for the construction of a power plant or to build a house. The PDR is also the **opportunity cost of capital**. In other words, it is the rate of return that can be obtained by investing capital in a market. Private discount rates are appropriate only for relatively short-term public projects that, dollar-for-dollar, crowd out private investment (Moore et al., 2004; NCEE, 2014). The private discount rate in 2020 in the United States was between 3 and 6 percent.

## Example 7.1 Estimating an annuity

What is the annual payment on a $100,000 loan at a 5 percent private discount rate over 30 years?

### Solution:
From Equation 7.3, the capital recovery factor is 0.06505/y. Thus, from Equation 7.2, the annual payment is $100,000 × 0.06505/y = $6,505/y.

Table 7.8 gives the unsubsidized LCOEs of several WWS and BAU electricity generation technologies in the United States in 2018. The analysis is a private cost analysis. The table indicates that onshore wind and utility PV were the least expensive forms of new electric power that year. Natural gas combined cycle plants were third, but natural gas peaker plants were the most expensive. This was due to the fact that they are run at full capacity only 10 percent of the time.

The mean cost of nuclear power from Table 7.8 (15.1 ¢/kWh) is 3.8 times the mean cost of thin-film utility PV (4.0 ¢/kWh). However, the cost of nuclear power is

Table 7.8 Components of the unsubsidized levelized cost of electricity (LCOE) for several new WWS and BAU electricity-generating technologies in the United States in 2018. The analysis is a private cost analysis (see footnote). All costs are in 2018 USD.

| Energy generator | Capital cost ($/W) | Fixed O&M ($/kW-y) | Variable O&M $/MWh | Fuel cost ($/MMBtu) | Heat production (Btu/Wh) | Capacity factor (%) | Construction time (months) | Facility life (years) | LCOE (¢/kWh) |
|---|---|---|---|---|---|---|---|---|---|
| **WWS generators** | | | | | | | | | |
| Onshore wind | 1.15–1.55 | 28–36.5 | 0 | 0 | 0 | 55–38 | 12 | 20 | 2.9–5.6 |
| Offshore wind | 2.25–3.80 | 80–110 | 0 | 0 | 0 | 55–45 | 12 | 20 | 6.2–12.1 |
| Geothermal plant | 4.0–6.4 | | 25–35 | 0 | 0 | 90–85 | 36 | 25 | 7.1–11.1 |
| Residential roof PV | 2.95–3.25 | 14.5–25 | 0 | 0 | 0 | 19–13 | 3 | 25 | 16–26.7 |
| Comm./indust. roof PV | 1.90–3.25 | 15–20 | 0 | 0 | 0 | 25–20 | 3 | 25 | 8.1–17 |
| Community rooftop PV | 1.85–3.00 | 12–16 | 0 | 0 | 0 | 25–20 | 4–6 | 30 | 7.3–14.5 |
| Utility PV-crystalline | 0.95–1.25 | 12–9 | 0 | 0 | 0 | 32–21 | 9 | 30 | 4–4.6 |
| Utility PV-thin film | 0.95–1.25 | 12–9 | 0 | 0 | 0 | 34–23 | 9 | 30 | 3.6–4.4 |
| Utility CSP with storage | 3.85–10 | 75–80 | 0 | 0 | 0 | 43–52 | 36 | 35 | 9.8–18.1 |
| Hydrogen fuel cell | 3.3–6.5 | 0 | 30–44 | 3.45 | 8–7.26 | 95 | 3 | 20 | 10.3–15.2 |
| **BAU generators** | | | | | | | | | |
| Gas combined cycle | 0.7–1.3 | 6–5.5 | 3.5–2 | 3.45 | 6.13–6.9 | 80 | 24 | 20 | 4.1–7.4 |
| Gas peaking | 0.7–0.95 | 5–20 | 4.7–10 | 3.45 | 9.80–8 | 10 | 12–18 | 20 | 15.2–20.6 |
| Nuclear | 6.5–12.25 | 115–135 | 0.75 | 0.85 | 10.5 | 90 | 69 | 40 | 11.2–18.9 |
| Coal | 3.0–8.4 | 40–80 | 2–5 | 1.45 | 8.75–12 | 93 | 60–66 | 40 | 6.0–14.3 |

Assumes 60 percent debt at a private discount rate of 8 percent, and 40 percent equity at 12 percent private cost. Assumes fixed O&M costs escalate 2.25 percent per year. Does not include externality costs, such as nuclear waste disposal costs, air pollution costs, or climate costs. For utility solar, the low end represents single-axis tracking; the high end represents fixed tilt. MMBtu = million Btu.
Source: Lazard (2018).

higher than the table indicates because nuclear's cost was derived assuming a construction time of 5.75 years (Lazard, 2018); whereas the only two reactors being built in the United States in 2020 are the Vogtle 3 and 4 plants, which are scheduled to be completed after 8.5 and 9 years of construction, respectively (Section 3.3.1.1). The additional LCOE due to that delay period is about 3.7 ¢/kWh, bringing the total cost of nuclear to 18.8 ¢/kWh, or 4.7 times that of thin-film PV.

---

## Example 7.2 Estimating the cost of energy of a utility-scale thin-film PV system

Estimate the cost per unit energy of a $P_r$ = 100 kW nameplate capacity utility-scale thin-film PV system with a capital cost of $1/W, a capacity factor of $CF$ = 30 percent, an O&M cost of $10/kW-y, and a lifetime of 30 years. Assume the private discount rate is 3 percent, the construction time is 1 year, the decommissioning cost is $10/kW, and decommissioning occurs immediately after the last year of operation. Also assume transmission, distribution, and downtime losses are 10 percent.

### Solution:

Table 7.9 summarizes the results for this problem. The annual energy production of the solar farm can be estimated from Equation 6.24, $E_t = P_r \times CF \times H$ = 262,800 kWh/y. However, accounting for the 10 percent transmission, distribution, and downtime losses reduces this to 236,520 kWh/y. The energy is produced only in years 2 to 31 (Table 7.9), since construction is occurring in year 1. Each year, the annual energy production must be brought forward to the present, as shown in Equation 7.1. This is done in column (d) of Table 7.9.

The upfront capital cost is 100 kW × $1,000/kW = $100,000. With a project lifetime of 31 years, the capital recovery factor from Equation 7.3 is 0.050. From Equation 7.2, the annuity is, therefore, $5,000/y. This annuity must be paid during all 31 years of the project in Table 7.9. Financing could be constructed such that, during construction, the loan is interest only, whereas during operation, the balance after one year is amortized. That would make only a small difference in this case. The O&M cost is $10/kW-y of the nameplate capacity, or $1,000/y total. This cost accrues during only years 2 to 31. The future decommissioning cost is $1,000, or $20/y when that value is annualized with Equation 7.4. This cost accrues during all years of the project. The sum of the three annual costs is then brought forward to the present, as shown in column (h) of Table 7.9.

Finally, dividing the summed present value from column (h) by that from column (d) gives the LCOE of solar PV from Equation 7.1 in this case as 2.65 ¢/kWh.

---

The **social discount rate (SDR)** is the discount rate used in a **social cost analysis**. The **social cost** of an investment is the investment's direct cost plus its externality costs (e.g., health and climate costs). A social discount rate is used primarily when the costs and benefits of a project occur at different times and over more than one generation. Such projects are called **intergenerational projects.**

Social discount rates are smaller than private discount rates, because society, as a whole, cares more about the welfare of distant future generations than does the average consumer or investor, who is generally concerned with near-term impacts during his or her lifetime. As a result, social discount rates appropriately weigh the present value of future impacts higher than do private discount rates. The (incorrect) use of a relatively high private discount rate in the evaluation of long-term climate change mitigation would undervalue future social benefits and thus bias present-day investments away from efforts that provide long-term benefits to society. In order to properly evaluate long-term costs and benefits from the perspective of society, the social discount rate must be used.

Moore et al. (2004) reviewed accepted methods of estimating social discount rates and concluded,

Table 7.9 Calculation of the LCOE of a utility PV power plant. Example 7.2 gives the assumptions and values. Dividing the total from column (h) by the total from column (d) gives the LCOE = 2.65 ¢/kWh, as determined from Equation 7.1.

| (a) Year | (b) Factor to divide annuity by to obtain present value =(1+i)(a) | (c) Energy produced in year (kWh/y) | (d) =(c)/(b) Present value of energy produced in year | (e) Annual cost of upfront capital ($/y) | (f) Annual O&M cost ($/y) | (g) Annualized future decom-missioning cost ($/y) | (h) =(e+f+g)/(b) Present value of annual payments ($/y) |
|---|---|---|---|---|---|---|---|
| 1 | 1.030 | 0 | 0 | 5,000 | 0 | 20 | 4,874 |
| 2 | 1.061 | 236,520 | 222,943 | 5,000 | 1,000 | 20 | 5,674 |
| 3 | 1.093 | 236,520 | 216,449 | 5,000 | 1,000 | 20 | 5,509 |
| 4 | 1.126 | 236,520 | 210,145 | 5,000 | 1,000 | 20 | 5,349 |
| 5 | 1.159 | 236,520 | 204,024 | 5,000 | 1,000 | 20 | 5,193 |
| 6 | 1.194 | 236,520 | 198,082 | 5,000 | 1,000 | 20 | 5,042 |
| 7 | 1.230 | 236,520 | 192,312 | 5,000 | 1,000 | 20 | 4,895 |
| 8 | 1.267 | 236,520 | 186,711 | 5,000 | 1,000 | 20 | 4,752 |
| 9 | 1.305 | 236,520 | 181,273 | 5,000 | 1,000 | 20 | 4,614 |
| 10 | 1.344 | 236,520 | 175,993 | 5,000 | 1,000 | 20 | 4,479 |
| 11 | 1.384 | 236,520 | 170,867 | 5,000 | 1,000 | 20 | 4,349 |
| 12 | 1.426 | 236,520 | 165,890 | 5,000 | 1,000 | 20 | 4,222 |
| 13 | 1.469 | 236,520 | 161,059 | 5,000 | 1,000 | 20 | 4,099 |
| 14 | 1.513 | 236,520 | 156,368 | 5,000 | 1,000 | 20 | 3,980 |
| 15 | 1.558 | 236,520 | 151,813 | 5,000 | 1,000 | 20 | 3,864 |
| 16 | 1.605 | 236,520 | 147,391 | 5,000 | 1,000 | 20 | 3,751 |
| 17 | 1.653 | 236,520 | 143,098 | 5,000 | 1,000 | 20 | 3,642 |
| 18 | 1.702 | 236,520 | 138,931 | 5,000 | 1,000 | 20 | 3,536 |
| 19 | 1.754 | 236,520 | 134,884 | 5,000 | 1,000 | 20 | 3,433 |
| 20 | 1.806 | 236,520 | 130,955 | 5,000 | 1,000 | 20 | 3,333 |
| 21 | 1.860 | 236,520 | 127,141 | 5,000 | 1,000 | 20 | 3,236 |
| 22 | 1.916 | 236,520 | 123,438 | 5,000 | 1,000 | 20 | 3,142 |
| 23 | 1.974 | 236,520 | 119,843 | 5,000 | 1,000 | 20 | 3,050 |
| 24 | 2.033 | 236,520 | 116,352 | 5,000 | 1,000 | 20 | 2,961 |
| 25 | 2.094 | 236,520 | 112,963 | 5,000 | 1,000 | 20 | 2,875 |
| 26 | 2.157 | 236,520 | 109,673 | 5,000 | 1,000 | 20 | 2,791 |
| 27 | 2.221 | 236,520 | 106,479 | 5,000 | 1,000 | 20 | 2,710 |
| 28 | 2.288 | 236,520 | 103,377 | 5,000 | 1,000 | 20 | 2,631 |
| 29 | 2.357 | 236,520 | 100,366 | 5,000 | 1,000 | 20 | 2,555 |
| 30 | 2.427 | 236,520 | 97,443 | 5,000 | 1,000 | 20 | 2,480 |
| 31 | 2.500 | 236,520 | 94,605 | 5,000 | 1,000 | 20 | 2,408 |
| Total | | | 4,500,870 | | | | 119,430 |

*... no matter which method one chooses, the estimates for the social discount rate vary ... between 0 and 3.5 percent for projects with intergenerational impacts.* (p. 809)

Drupp et al. (2015) surveyed 197 experts and similarly found that 92 percent of them believe the social discount rate should be between 1 and 3 percent. OMB (2003) also recommended 1 to 3 percent, which is the range adopted by Jacobson et al. (2017).

The importance of using a social discount rate versus a higher private discount rate for a social cost analysis can be illustrated in the following example. Suppose the emission of 1 tonne of $CO_2$ into the air today results in a cost in 20 years of $750. The present value of that cost is

$$P = \frac{F}{(1 + i)^n} \tag{7.5}$$

where $F$ is the future value. If a private discount rate of, say, 6 percent is used, then the present value of the climate damage is $234. If the social discount rate of 2 percent is used, the present value of the climate cost is $505. Thus, a social discount rate weighs the present value of future damage due to today's actions higher than does a private discount rate. As such, it corrects for the fact that a private discount rate weighs impacts on future generations less than it weighs impacts on the current generation. In other words, the use of a private discount rate for a social cost analysis makes damage to future generations appear small to the present generation compared with what the damages are once the damage hits, thus biases present-day policymakers into thinking there is no need to act on climate or pollution problems today. In sum, with the use of a social cost analysis, policymakers can make better decisions for society about what investments to make today.

---

## Example 7.3 Comparing a social discount rate with a private discount rate for a wind turbine

Estimate the levelized cost of energy of a 1,500-kW nameplate capacity wind turbine with a blade diameter of 77 m in the presence of a mean annual wind speed of 7.5 m/s. Assume the capital cost of the investment is $1,300 per kW. Also assume a discount rate range (representing a social discount rate and a private discount rate) of 2 to 6 percent, respectively. Assume construction time is 1 year and the turbines last 30 years. Also assume an annual operation and maintenance (O&M) cost of $30/kW-y, a net decommissioning cost (after reimbursement for scrapping) of $13/kW, and transmission/distribution/downtime/array losses of 10 percent.

### Solution:

This solution is less precise than that from Example 7.2 because it is obtained here without using a spreadsheet. The error, though, is small because the construction time is short.

The upfront capital cost is 1,500 kW × $1,300/kW = $1.95 million. From Equation 7.3, the capital recovery factor over a period of 31 years ranges from 0.0436 to 0.0718 for the different discount rates. From Equation 7.1, the annuity is $85,000 to $140,000/y. The O&M cost is 30/kW-y, or $45,000/y. The decommissioning cost in year 31 is $19,500. From Equation 7.4, the decommissioning annual cost is $460/y at a 2 percent discount rate and $230/y at a 6 percent discount rate. Thus, the total annual cost is $130,000 to $185,000/y.

The annual energy production of the turbine can be estimated from Equation 6.24, $E_t = P_r \times CF \times H$, where the capacity factor is derived from Equation 6.28 as $CF = 0.087 \times V_m - P_r/D^2$. Plugging the mean annual wind speed, the nameplate capacity, and rotor diameter of the turbine gives a capacity factor of 0.40. Plugging this into Equation 6.24 gives the annual energy output from the turbine as 5.25 million kWh/y. Reducing this by the 10 percent loss rate gives 4.72 million kWh/y. Reducing it further by 1/31, the ratio of the number of years of construction (when no electricity was produced) to the total number of years of the project, gives 4.57 million kWh/y.

Finally, dividing the turbine cost range per year by the energy output per year gives the LCOE range of the turbine as 2.9 to 4.1 ¢/kWh. The only difference between these two numbers is the use of a social discount rate versus a private discount rate.

Table 7.10 compares the 2050 LCOE of BAU electricity (the electricity cost before health and climate costs are accounted for) with that of WWS electricity for 143 countries. The table indicates that a transition to WWS may reduce the LCOE of electricity by about 24 percent on average. In addition, Table 7.1 indicates that WWS reduces all-purpose end-use energy by 57.1 percent, so although the LCOE of WWS is only modestly smaller than that of BAU electricity, the cost to consumers of WWS direct energy is about 32 percent that of BAU electricity.

## Transition highlight

An analysis of the investment and environmental costs of transitioning the energy infrastructures of towns, cities, states, provinces, and countries to 100 percent WWS between today and 2050 is a social cost analysis. As such, it requires the use of a social discount rate of 1 to 3 percent. Not only does such an analysis cover multiple generations of infrastructure, but it also accounts for the health and climate cost avoidance of such a transition. A social discount rate is applied to *all* components of a social cost analysis, including the direct investment costs, because all costs are treated from the perspective of society. Social cost analyses need to use a social discount rate for the entire analysis. Private cost analyses need to use a private discount rate for the whole analysis.

## Transition highlight

When examining a 100 percent WWS system versus a BAU system, it is also important to analyze the relative cost of a transition in other energy sectors aside from the electricity sector. Table 7.11 provides one such analysis. It compares the fuel cost of driving a hydrogen fuel cell (HFC) vehicle with that of driving a gasoline vehicle. It illustrates that an HFC vehicle costs $0.64 to $2.94 to travel the same distance as 1 gallon of gasoline would take a gasoline vehicle. For comparison, the U.S. cost of gasoline has been $2.75 to $4.25 per gallon. As such, the fuel cost of driving an HFC vehicle is equivalent to or less than that driving a gasoline vehicle.

### 7.6.2  Avoided Health Costs from Air Pollution

Transitioning homes, towns, cities, states, provinces, and countries to WWS immediately reduces air pollution

health problems. Fewer health problems save money by reducing hospitalization rates, emergency room visits, lost workdays, lost school days, insurance rates, taxes, worker's compensation rates, and loss of companionship while improving quality of life.

Air pollution causes premature mortality in several ways. It contributes to death from heart disease, stroke, **chronic obstructive pulmonary disease (COPD)**, lower respiratory tract infection, lung cancer, and asthma. Common types of COPD are chronic bronchitis and emphysema. Common types of lower respiratory tract infections are the flu, bronchitis, and pneumonia.

In 2016, 56.9 million people died from all causes worldwide (WHO, 2017a). Table 7.12 shows that air pollution caused between 24 and 45 percent of the deaths for each of 5 out of the 6 leading causes of death. About 4.5 million people died prematurely from outdoor air pollution and 7.1 million, from indoor plus outdoor pollution in 2016 (Table 7.12). **Thus, about 12.5 percent of all deaths worldwide in 2016 were due to indoor plus outdoor air pollution, making it the second leading cause of death after heart disease**. Twenty percent of premature air pollution deaths were of children age five and younger.

The Global Burden of Disease study (GBD, 2015) similarly estimated that about 5.5 (5.1 to 5.9) million deaths worldwide in 2013 were caused by indoor plus outdoor air pollution. Of these, 2.8 to 3.1 million were from outdoor particulate matter smaller than 2.5 μm ($PM_{2.5}$), 0.16 to 0.27 million were from outdoor ozone, and 2.5 to 3.3 million were from indoor air pollution from solid fuel burning.

## Transition highlight

Table 7.13 shows the 2016 mean number of deaths from indoor plus outdoor air pollution by country for 183 out of 195 countries of the world. China and India absorb the brunt of mortalities, a combined total of 2.6 million per year (37 percent of all deaths). In addition, Nigeria, Pakistan, Indonesia, Bangladesh, the Philippines, and Russia all suffer more than 100,000 air pollution deaths per year. The highest per capita air pollution death rates are in North Korea (Korea, DPR), Georgia, Chad, Nigeria, Bosnia and Herzegovina, Somalia, Sierra Leone, the Ivory Coast (Cote d'Ivoire), India, Bulgaria, the Central African Republic, China, Niger, and Montenegro, respectively.

Table 7.10 Mean value of the levelized cost of energy (LCOE) for (a) conventional fuels (BAU) in 2050 in the electricity sector and (b) WWS in all energy sectors (which will be electrified) in 2050 for 143 countries. The LCOE estimates do not include externality costs and, for WWS, are determined based on WWS technologies meeting annual average power demand (not time-dependent demand). Chapter 8 discusses costs when matching time-dependent demand. The 2050 LCOEs are used to calculate (c) the energy cost savings per person per year in each country due to switching from BAU to WWS. (d) 2050 air pollution premature mortalities calculated from Equation 7.7. (e) Avoided costs of energy-related (90 percent of the total) air-pollution mortalities, morbidities, and non-air-pollution effects per unit energy upon a conversion to WWS. (f) Percentage of world $CO_2$ emissions by country in 2017. (g) Avoided climate change costs per unit energy due to transitioning each country to 100 percent WWS. (h) Ratio of the BAU-to-WWS social cost and the total avoided social cost (energy plus health plus climate cost per person per year) due to transitioning are provided. All costs are in 2013 USD and use a social discount rate.

| Country | (a) 2050 LCOE of BAU (¢/kWh-electricity) | (b) 2050 LCOE of WWS (¢/kWh-all-energy) | (c) 2050 Average BAU retail electricity cost savings to country due to switching to WWS electricity ($/person/y) | (d) 2050 Estimated all-cause air pollution premature mortalities/y | (e) 2050 Mean avoided health cost ($2013) ¢/kWh-BAU-all-energy | (f) 2017 Percent of world $CO_2$ emissions | (g) 2050 Mean avoided climate cost ¢/kWh-BAU-all-energy | (h) Ratio of 2050 BAU:WWS social cost per kWh | (i) 2050 Average electricity + country health + world climate cost savings due to switching to WWS in country ($/person/y) |
|---|---|---|---|---|---|---|---|---|---|
| Albania | 6.37 | 6.95 | 69 | 1,766 | 34.1 | 0.014 | 10.4 | 7.3 | 5,968 |
| Algeria | 11.88 | 7.88 | 200 | 10,815 | 5.3 | 0.446 | 15.1 | 4.1 | 6,013 |
| Angola | 8.95 | 5.85 | 25 | 20,206 | 32.8 | 0.086 | 14.1 | 9.5 | 2,675 |
| Argentina | 10.31 | 8.66 | 155 | 12,153 | 6.2 | 0.586 | 13.7 | 3.5 | 5,443 |
| Armenia | 9.4 | 6.66 | 155 | 1,429 | 21.1 | 0.013 | 8.8 | 5.9 | 4,484 |
| Australia | 10.34 | 7.47 | 870 | 3,039 | 1.6 | 1.122 | 17.9 | 4 | 13,506 |
| Austria | 8.67 | 6.53 | 543 | 1,744 | 4.2 | 0.202 | 11.1 | 3.7 | 9,318 |
| Azerbaijan | 11.53 | 6.9 | 271 | 3,755 | 19.1 | 0.091 | 14.2 | 6.5 | 5,527 |
| Bahrain | 11.89 | 6.9 | 2,055 | 172 | 1.2 | 0.1 | 25.6 | 5.6 | 24,305 |
| Bangladesh | 11.8 | 5.91 | 35 | 161,254 | 70.8 | 0.236 | 15.6 | 16.6 | 2,312 |
| Belarus | 11.88 | 6.5 | 569 | 5,004 | 13.9 | 0.174 | 14.9 | 6.3 | 12,599 |
| Belgium | 11.12 | 6.44 | 790 | 2,300 | 3.8 | 0.291 | 11.3 | 4.1 | 10,132 |
| Benin | 11.69 | 6.46 | 10 | 17,112 | 36.4 | 0.02 | 10.2 | 9 | 1,754 |
| Bolivia | 10.34 | 7.02 | 45 | 5,510 | 12.8 | 0.057 | 12 | 5 | 2,503 |
| Bosnia & Herzegovina | 8.49 | 6.78 | 230 | 3,647 | 35.9 | 0.071 | 27.4 | 10.6 | 11,976 |
| Botswana | 9.72 | 7.5 | 132 | 940 | 12.8 | 0.022 | 19.8 | 5.6 | 5,510 |

Table 7.10 (cont.)

| Country | (a) 2050 LCOE of BAU (¢/kWh-electricity) | (b) 2050 LCOE of WWS (¢/kWh-all-energy) | (c) 2050 Average BAU retail electricity cost savings to country due to switching to WWS electricity ($/person/y) | (d) 2050 Estimated all-cause air pollution premature mortalities/y | (e) 2050 Mean avoided health cost ($2013) ¢/kWh-BAU-all-energy | (f) 2017 Percent of world CO$_2$ emissions | (g) 2050 Mean avoided climate cost ¢/kWh-BAU-all-energy | (h) Ratio of 2050 BAU: WWS social cost per kWh | (i) 2050 Average electricity + country health + world climate cost savings due to switching to WWS in country ($/person/y) |
|---|---|---|---|---|---|---|---|---|---|
| Brazil | 8.53 | 6.01 | 161 | 49,584 | 6.1 | 1.375 | 9.2 | 4 | 3,211 |
| Brunei Darussalam | 11.89 | 7.1 | 1,128 | 36 | 1 | 0.019 | 17.7 | 4.3 | 14,685 |
| Bulgaria | 9.51 | 7.03 | 692 | 3,776 | 17 | 0.138 | 19.1 | 6.5 | 16,282 |
| Cambodia | 8.31 | 7.7 | 16 | 12,111 | 26.5 | 0.029 | 9.4 | 5.7 | 2,209 |
| Cameroon | 7.77 | 5.16 | 16 | 26,050 | 43.2 | 0.027 | 8.1 | 11.4 | 2,120 |
| Canada | 8.24 | 7.79 | 539 | 3,768 | 1.1 | 1.722 | 13.8 | 3 | 13,337 |
| Chile | 9.53 | 8.5 | 237 | 4,119 | 5.7 | 0.252 | 14.6 | 3.5 | 6,625 |
| China | 9.27 | 7.96 | 265 | 1,090,410 | 21.4 | 30.342 | 16.4 | 5.9 | 13,100 |
| Taiwan | 9.27 | 8.14 | 745 | 6,670 | 5.3 | 0.738 | 22 | 4.5 | 20,540 |
| Colombia | 8.1 | 5.3 | 90 | 11,703 | 9.3 | 0.209 | 9.9 | 5.2 | 2,485 |
| Congo | 8.95 | 5.43 | 13 | 4,532 | 41.1 | 0.015 | 15.4 | 12.1 | 2,509 |
| Congo, Dem. Republic | 6.38 | 5.05 | 3 | 93,575 | 17.9 | 0.01 | 1.1 | 5 | 509 |
| Costa Rica | 8.24 | 6.81 | 89 | 1,008 | 7.8 | 0.023 | 9.9 | 3.8 | 2,313 |
| Cote d'Ivoire | 11.05 | 5.84 | 34 | 33,708 | 57.1 | 0.035 | 9.8 | 13.3 | 2,773 |
| Croatia | 8.48 | 6.3 | 337 | 1,964 | 14.8 | 0.049 | 10.4 | 5.3 | 8,813 |
| Cuba | 11.98 | 8.26 | 204 | 4,852 | 26.5 | 0.087 | 23 | 7.4 | 7,035 |
| Cyprus | 12.06 | 8.14 | 436 | 280 | 8.7 | 0.02 | 14.8 | 4.4 | 6,650 |
| Czech Republic | 9.9 | 6.73 | 601 | 3,222 | 7.5 | 0.306 | 19.1 | 5.4 | 12,451 |
| Denmark | 12.61 | 8.11 | 606 | 1,004 | 4.6 | 0.094 | 9.8 | 3.3 | 6,461 |
| Dominican Republic | 11.35 | 8.08 | 110 | 3,213 | 15.7 | 0.064 | 18.5 | 5.6 | 3,004 |

| Country | | | | | | | | | |
|---|---|---|---|---|---|---|---|---|---|
| Ecuador | 9.13 | 6.03 | 85 | 2,873 | 6.4 | 0.11 | 16.3 | 5.3 | 2,504 |
| Egypt | 11.49 | 8.66 | 164 | 63,338 | 20.4 | 0.722 | 18.7 | 5.8 | 4,813 |
| El Salvador | 11.32 | 7.4 | 98 | 1,560 | 13.7 | 0.022 | 15.1 | 5.4 | 2,348 |
| Eritrea | 11.9 | 8.34 | 5 | 6,885 | 91.9 | 0.002 | 8.3 | 13.4 | 928 |
| Estonia | 12.52 | 6.87 | 1,460 | 298 | 5.3 | 0.05 | 24.6 | 6.2 | 18,124 |
| Ethiopia | 6.91 | 7.39 | 1 | 152,284 | 31.3 | 0.042 | 2.5 | 5.5 | 845 |
| Finland | 9.74 | 6.97 | 1,067 | 545 | 1.5 | 0.131 | 8.4 | 2.8 | 8,615 |
| France | 9.39 | 7.97 | 363 | 10,528 | 4.7 | 0.943 | 10.1 | 3 | 5,043 |
| Gabon | 9.51 | 6.39 | 86 | 1,059 | 6.4 | 0.018 | 6.5 | 3.5 | 4,870 |
| Georgia | 7.58 | 7.13 | 102 | 4,102 | 36.7 | 0.032 | 11.8 | 7.9 | 9,806 |
| Germany | 10.85 | 7.95 | 583 | 19,077 | 6.2 | 2.222 | 16.4 | 4.2 | 10,735 |
| Ghana | 8.76 | 6.29 | 25 | 25,500 | 46.5 | 0.052 | 13.8 | 11 | 2,429 |
| Gibraltar | 10.84 | 4.85 | 703 | 20 | 0.5 | 0.002 | 0.9 | 2.5 | 23,256 |
| Greece | 10.6 | 7.53 | 482 | 4,605 | 12.8 | 0.201 | 16.3 | 5.3 | 8,976 |
| Guatemala | 9.96 | 7.07 | 38 | 7,226 | 17 | 0.05 | 9.8 | 5.2 | 2,000 |
| Haiti | 11.44 | 8.2 | 3 | 10,487 | 32.3 | 0.01 | 7.4 | 6.2 | 1,312 |
| Honduras | 10.74 | 6.33 | 58 | 3,161 | 12.2 | 0.029 | 12.5 | 5.6 | 1,554 |
| Hong Kong, China | 10.42 | 5.73 | 1,225 | 3,972 | 7.1 | 0.125 | 7.9 | 4.4 | 17,950 |
| Hungary | 10.23 | 6.26 | 396 | 4,162 | 12.5 | 0.142 | 12.4 | 5.6 | 8,342 |
| Iceland | 8.36 | 7.97 | 977 | 36 | 0.8 | 0.011 | 6 | 1.9 | 9,758 |
| India | 9.68 | 7.61 | 79 | 1,444,634 | 48.8 | 6.847 | 19.7 | 10.3 | 6,712 |
| Indonesia | 10.4 | 6.99 | 87 | 155,519 | 24.7 | 1.426 | 16.6 | 7.4 | 5,042 |
| Iran, Islamic Republic | 11.57 | 8.23 | 288 | 21,470 | 4 | 1.873 | 19.3 | 4.2 | 9,177 |
| Iraq | 11.51 | 7.85 | 93 | 12,511 | 17.5 | 0.556 | 47 | 9.7 | 5,414 |
| Ireland | 11.81 | 8.06 | 340 | 782 | 5.6 | 0.109 | 16.6 | 4.2 | 5,780 |
| Israel | 10.9 | 8.04 | 386 | 1,545 | 6.2 | 0.187 | 19.5 | 4.6 | 5,774 |
| Italy | 11.06 | 7.53 | 407 | 18,054 | 8.9 | 1.007 | 12.5 | 4.3 | 7,045 |

Table 7.10 (cont.)

| Country | (a) 2050 LCOE of BAU (¢/kWh-electricity) | (b) 2050 LCOE of WWS (¢/kWh-all-energy) | (c) 2050 Average BAU retail electricity cost savings to country due to switching to WWS electricity ($/person/y) | (d) 2050 Estimated all-cause air pollution premature mortalities/y | (e) 2050 Mean avoided health cost ($2013) ¢/kWh-BAU-all-energy | (f) 2017 Percent of world $CO_2$ emissions | (g) 2050 Mean avoided climate cost ¢/kWh-BAU-all-energy | (h) Ratio of 2050 BAU:WWS social cost per kWh | (i) 2050 Average electricity + country health + world climate cost savings due to switching to WWS in country ($/person/y) |
|---|---|---|---|---|---|---|---|---|---|
| Jamaica | 11.85 | 8.31 | 74 | 697 | 7.4 | 0.021 | 17 | 4.4 | 2,902 |
| Japan | 10.78 | 7.81 | 501 | 27,181 | 7.2 | 3.684 | 21.2 | 5 | 9,121 |
| Jordan | 11.88 | 8.23 | 198 | 1,857 | 7.3 | 0.069 | 19.4 | 4.7 | 3,501 |
| Kazakhstan | 9.8 | 7.39 | 410 | 7,774 | 7.7 | 0.743 | 19.4 | 5 | 13,344 |
| Kenya | 10.65 | 7.38 | 14 | 17,789 | 12.9 | 0.052 | 6.8 | 4.1 | 917 |
| Korea, DPR | 7.39 | 7.47 | 17 | 37,704 | 42.3 | 0.105 | 26.6 | 10.2 | 4,433 |
| Korea, Republic of | 10.14 | 6.82 | 1,048 | 8,990 | 3.4 | 1.878 | 17.7 | 4.6 | 14,515 |
| Kosovo | 9.57 | 8.06 | 212 | 266 | 5.8 | 0.024 | 25.5 | 5.1 | 5,385 |
| Kuwait | 11.89 | 6.56 | 2,940 | 888 | 1.9 | 0.271 | 18.4 | 4.9 | 33,434 |
| Kyrgyzstan | 6.89 | 6.88 | 48 | 3,791 | 22.3 | 0.031 | 13.4 | 6.2 | 2,835 |
| Latvia | 10.35 | 6.7 | 478 | 877 | 12.3 | 0.022 | 8.6 | 4.7 | 10,310 |
| Lebanon | 11.74 | 7.7 | 524 | 1,297 | 7.8 | 0.064 | 24.2 | 5.7 | 8,578 |
| Libya | 11.89 | 8.55 | 192 | 2,935 | 6.7 | 0.161 | 25.7 | 5.2 | 8,115 |
| Lithuania | 12.05 | 7.4 | 516 | 1,340 | 13.6 | 0.043 | 11.1 | 5 | 9,982 |
| Luxembourg | 11.96 | 4.82 | 1,157 | 103 | 2.7 | 0.027 | 11.9 | 5.5 | 11,874 |
| Macedonia, Rep. of | 8.81 | 6.79 | 264 | 1,486 | 33.5 | 0.022 | 17.6 | 8.8 | 9,329 |
| Malaysia | 10.44 | 6.61 | 424 | 9,353 | 5.7 | 0.722 | 21.1 | 5.6 | 9,757 |
| Malta | 12.03 | 5.61 | 909 | 104 | 2.8 | 0.005 | 3.5 | 3.3 | 7,537 |
| Mauritius | 11.13 | 8.27 | 247 | 418 | 7.4 | 0.011 | 11.8 | 3.7 | 5,609 |
| Mexico | 11.1 | 7.31 | 237 | 29,995 | 7.9 | 1.415 | 18.1 | 5.1 | 5,110 |
| Moldova, Republic of | 11.61 | 7.39 | 264 | 1,384 | 13.6 | 0.023 | 13.2 | 5.2 | 6,006 |

| Country | | | | | | | | |
|---|---|---|---|---|---|---|---|---|
| Mongolia | 9.86 | 7.58 | 119 | 2,600 | 21.9 | 0.072 | 41.9 | 9.7 | 11,111 |
| Montenegro | 8.01 | 6.93 | 279 | 480 | 30.4 | 0.012 | 24.6 | 9.1 | 13,172 |
| Morocco | 10.4 | 8.53 | 68 | 10,344 | 11.9 | 0.172 | 19.2 | 4.9 | 2,879 |
| Mozambique | 7.12 | 7.33 | 3 | 24,816 | 12.8 | 0.022 | 4.4 | 3.3 | 616 |
| Myanmar | 8.6 | 6.93 | 15 | 50,419 | 55.6 | 0.079 | 10.9 | 10.8 | 3,002 |
| Namibia | 6.49 | 8.18 | -3 | 965 | 10.7 | 0.012 | 11 | 3.5 | 4,678 |
| Nepal | 6.38 | 6.46 | 3 | 38,210 | 39.7 | 0.023 | 4.5 | 7.8 | 2,158 |
| Netherlands | 11.15 | 7.92 | 510 | 3,352 | 3.9 | 0.488 | 12.5 | 3.5 | 8,963 |
| Curacao | 11.15 | 5.51 | 209 | 10 | 0.1 | 0.013 | 7.9 | 3.5 | 11,921 |
| New Zealand | 9.2 | 7.93 | 474 | 444 | 1.6 | 0.103 | 10.8 | 2.7 | 7,271 |
| Nicaragua | 12.37 | 6.44 | 71 | 1,908 | 17.7 | 0.017 | 13.2 | 6.7 | 1,851 |
| Niger | 10.97 | 7.55 | 3 | 52,062 | 111.3 | 0.007 | 5.8 | 17 | 1,076 |
| Nigeria | 10.88 | 6.34 | 12 | 417,695 | 73 | 0.265 | 4.7 | 14 | 4,676 |
| Norway | 6.61 | 6.58 | 577 | 569 | 1.7 | 0.131 | 7.5 | 2.4 | 8,215 |
| Oman | 11.89 | 7.78 | 811 | 752 | 1.4 | 0.219 | 16.2 | 3.8 | 18,122 |
| Pakistan | 10.07 | 7.68 | 33 | 205,431 | 40.2 | 0.55 | 13.5 | 8.3 | 3,311 |
| Panama | 8.47 | 5.82 | 141 | 782 | 3.6 | 0.034 | 7.3 | 3.3 | 3,627 |
| Paraguay | 6.37 | 5.97 | 45 | 2,511 | 11.8 | 0.018 | 6.5 | 4.1 | 1,984 |
| Peru | 9.23 | 6.11 | 87 | 13,130 | 16.9 | 0.156 | 12.7 | 6.3 | 3,357 |
| Philippines | 10.59 | 7.95 | 53 | 126,709 | 76 | 0.383 | 21 | 13.5 | 4,539 |
| Poland | 10.25 | 7.7 | 384 | 14,363 | 11.3 | 0.89 | 20.2 | 5.4 | 10,624 |
| Portugal | 10.89 | 7.97 | 364 | 1,654 | 5.3 | 0.158 | 14.2 | 3.8 | 5,552 |
| Qatar | 11.89 | 6.83 | 2,357 | 203 | 0.5 | 0.273 | 17.4 | 4.4 | 45,522 |
| Romania | 9.69 | 7.52 | 232 | 13,080 | 30.2 | 0.226 | 14.9 | 7.3 | 10,727 |
| Russian Federation | 10.21 | 7.95 | 674 | 55,075 | 8 | 4.923 | 17.5 | 4.5 | 15,677 |
| Saudi Arabia | 11.89 | 8.05 | 1,076 | 9,804 | 3.8 | 1.782 | 23.9 | 4.9 | 21,257 |
| Senegal | 11.44 | 7.66 | 20 | 12,993 | 33 | 0.027 | 14.8 | 7.7 | 1,380 |
| Serbia | 8.78 | 6.48 | 468 | 4,208 | 20.2 | 0.162 | 26.9 | 8.6 | 13,866 |

Table 7.10 (cont.)

| Country | (a) 2050 LCOE of BAU (¢/kWh-electricity) | (b) 2050 LCOE of WWS (¢/kWh-all-energy) | (c) 2050 Average BAU retail electricity cost savings to country due to switching to WWS electricity ($/person/y) | (d) 2050 Estimated all-cause air pollution premature mortalities/y | (e) 2050 Mean avoided health cost ($2013) ¢/kWh-BAU-all-energy | (f) 2017 Percent of world $CO_2$ emissions | (g) 2050 Mean avoided climate cost ¢/kWh-BAU-all-energy | (h) Ratio of 2050 BAU:WWS social cost per kWh | (i) 2050 Average electricity + country health + world climate cost savings due to switching to WWS in country ($/person/y) |
|---|---|---|---|---|---|---|---|---|---|
| Singapore | 11.89 | 5.16 | 1,204 | 2,107 | 1.6 | 0.153 | 3.5 | 3.3 | 12,441 |
| Slovak Republic | 9.47 | 6.92 | 373 | 1,731 | 8.5 | 0.106 | 14.4 | 4.7 | 8,433 |
| Slovenia | 8.88 | 7.06 | 459 | 534 | 6.7 | 0.042 | 14.4 | 4.2 | 9,689 |
| South Africa | 9.69 | 8.49 | 332 | 18,139 | 5.2 | 1.305 | 27.4 | 5 | 13,708 |
| South Sudan | 11.9 | 7.38 | 3 | 19,104 | 183.2 | 0.003 | 8.6 | 27.6 | 1,185 |
| Spain | 10.84 | 7.76 | 369 | 8,585 | 5.5 | 0.786 | 12.9 | 3.8 | 5,433 |
| Sri Lanka | 8.69 | 7.38 | 38 | 13,636 | 33.4 | 0.067 | 11.6 | 7.3 | 4,540 |
| Sudan | 7.8 | 8.1 | 6 | 65,754 | 68.6 | 0.059 | 8.9 | 10.5 | 2,235 |
| Suriname | 8.57 | 5.9 | 166 | 225 | 13.1 | 0.006 | 19.3 | 7 | 4,973 |
| Sweden | 8.7 | 7 | 714 | 981 | 2 | 0.142 | 6.6 | 2.5 | 5,559 |
| Switzerland | 7.79 | 6.33 | 371 | 1,089 | 4.2 | 0.111 | 8.9 | 3.3 | 5,676 |
| Syrian Arab Republic | 11.76 | 8.11 | 59 | 9,262 | 32.7 | 0.079 | 24.1 | 8.5 | 2,240 |
| Tajikistan | 6.42 | 6.16 | 31 | 5,315 | 39.2 | 0.016 | 9.9 | 9 | 1,840 |
| Tanzania, United Rep. | 10.05 | 6.75 | 11 | 31,301 | 14.8 | 0.041 | 3.9 | 4.3 | 1,260 |
| Thailand | 11.41 | 7.44 | 360 | 35,599 | 11.1 | 0.779 | 14.6 | 5 | 8,966 |
| Togo | 8.19 | 6.48 | 6 | 12,450 | 33.9 | 0.008 | 7.1 | 7.6 | 1,188 |
| Trinidad and Tobago | 11.89 | 6.75 | 1,108 | 271 | 1.4 | 0.105 | 20.7 | 5 | 37,598 |
| Tunisia | 11.91 | 7.98 | 203 | 4,211 | 7.8 | 0.088 | 12.9 | 4.1 | 5,154 |
| Turkey | 9.94 | 7.4 | 147 | 28,480 | 14.4 | 1.198 | 19.9 | 6 | 4,994 |
| Turkmenistan | 11.89 | 7.68 | 275 | 2,073 | 5 | 0.202 | 15.5 | 4.2 | 11,488 |

| Country | | | | | | | | | |
|---|---|---|---|---|---|---|---|---|---|
| Ukraine | 9.55 | 8.15 | 273 | 26,830 | 17.1 | 0.574 | 16.6 | 5.3 | 9,895 |
| United Arab Emirates | 11.89 | 6.86 | 2,347 | 797 | 0.6 | 0.566 | 13 | 3.7 | 31,334 |
| United Kingdom | 11.16 | 7.98 | 368 | 13,823 | 6.7 | 1.058 | 12.2 | 3.8 | 5,819 |
| United States | 10.43 | 7.81 | 600 | 62,676 | 3.7 | 14.247 | 15.2 | 3.8 | 9,612 |
| Uruguay | 9.11 | 7.55 | 193 | 675 | 5.3 | 0.019 | 7.3 | 2.9 | 3,377 |
| Uzbekistan | 10.65 | 7.58 | 115 | 11,609 | 13 | 0.266 | 15.7 | 5.2 | 3,954 |
| Venezuela | 8.37 | 5.89 | 115 | 7,249 | 5.5 | 0.407 | 16.7 | 5.2 | 4,585 |
| Vietnam | 9.2 | 8.04 | 87 | 44,139 | 12.9 | 0.61 | 18.4 | 5 | 4,165 |
| Yemen | 11.89 | 8.65 | 15 | 26,192 | 162.2 | 0.035 | 28 | 23.4 | 2,048 |
| Zambia | 6.54 | 6.29 | 10 | 15,969 | 25.2 | 0.014 | 3.4 | 5.6 | 1,311 |
| Zimbabwe | 8.02 | 6.9 | 20 | 10,769 | 9.4 | 0.034 | 8.1 | 3.7 | 1,250 |
| **World total/ average** | **9.99** | **7.61** | **171** | **5,285,036** | **16.9** | **99.74** | **16** | **5.6** | **6,052** |

a) The 2050 LCOE cost of retail electricity for BAU fuels in each country combines the percentage mix of BAU electricity generators in 2050 with 2050 mean LCOEs for each generator, derived herein. Such costs include all-distance transmission, pipeline, and distribution costs, but they exclude health and climate externality costs. The 2050 BAU mix includes some existing WWS (mostly hydropower) plus future increases in WWS electricity and future energy efficiency improvements.

b) The 2050 LCOE of WWS in the country combines the 2050 mix of WWS generators among all energy sectors from Table 7.6 with the 2050 mean LCOEs for each WWS generator. The LCOEs of individual generators account for all-distance transmission and distribution.

c) The 2050 average BAU retail electricity sectors cost savings per capita per year due to switching to WWS is calculated as the cost of electricity use in the electricity sector in the BAU case (the product of BAU electricity use and the 2050 BAU LCOE) less the annualized cost of the assumed efficiency improvements in the WWS case beyond BAU improvements and less the total cost of BAU retail electricity converted to WWS (product of WWS electricity use replacing BAU electricity and the 2050 WWS LCOE), all divided by 2050 population.

d) Avoided premature mortalities in each country in 2050 are estimated by projecting country-specific indoor plus outdoor air pollution mortality estimates for 2016 from WHO (2017b) to 2050 with Equation 7.7. Table 7.12 indicates that the mean number of such deaths in 2016 was about 7.1 million. The present table suggests that the number of deaths may drop somewhat, but not substantially, by 2050 in a BAU economy because lower emissions due to improvements in emission controls will have a greater impact than the higher population exposed to air pollution.

e) The total avoided damage cost of indoor plus outdoor air pollution caused by BAU fuels (fossil-fuel and biofuel combustion and evaporative emissions) in a country is the sum of avoided mortality costs, morbidity costs, and non-health costs (e.g., lost visibility and agricultural output) in the country. It is calculated with Equation 7.6. The resulting avoided damage cost per kWh-BAU-all-energy is determined by dividing the total air pollution damage cost per year in 2050 per country by the kWh from all energy in the BAU case per year produced by the country.

f) Percentage of 2017 world anthropogenic $CO_2$ emissions by country (European Commission, 2019). The total worldwide (218 countries) was 35,849 gigatonnes (GT)-$CO_2$ and in the 143 countries treated here was 35,756 GT-$CO_2$. These numbers are estimated to rise by 2050 in the BAU scenario to 57,103 GT-$CO_2$ for 218 countries and 56,955 GT-$CO_2$ for 143 countries.

g) Product of the $CO_2$ emissions rate per country and the mid-value of the social cost of carbon (SCC) from Table 7.16 ($500/tonne-$CO_2$e), divided by the kWh from all energy in the BAU case per year produced by the country.

h) The ratio of the BAU to WWS social cost is roughly columns (a) + (e) + (h), divided by column (c).

i) The per capita electricity plus climate cost savings is calculated as columns (e) + (h), multiplied by the BAU kWh/yr per country and divided by the population of the country. The result is then added to column (c).

Source: Jacobson et al. (2019).

Table 7.11 **Calculation of the cost of producing, compressing, and using hydrogen for an HFC passenger vehicle and comparison with the fuel cost of driving a gasoline vehicle.**

| Parameter | Low case | High case |
|---|---|---|
| **1. Hydrogen electrolyzer cost** | | |
| a) Energy to make $H_2$ by electrolysis (kWh/kg-$H_2$) | 53.4 | 53.4 |
| b) Electrolyzer capital cost ($/kW) | 300 | 400 |
| c) Electrolyzer lifetime (y) | 19 | 10 |
| d) Social (low) / private (high) discount rate (percent) | 2 | 5 |
| e) Capital recovery factor (Equation 7.3) (fraction/y) | 0.0638 | 0.1295 |
| f) Installation factor to multiply capital cost by (–) | 1.2 | 1.3 |
| g) Operation & maintenance (fraction of capital cost/y) | 0.015 | 0.015 |
| h) Electrolyzer use factor (fraction of time it is in use)[a] | 0.50 | 0.95 |
| i) Hours per year | 8,760 | 8,760 |
| **j) Overall electrolyzer cost ($/kg-$H_2$) = ab(ef+g)/(hi)** | **0.34** | **0.53** |
| **2. Hydrogen compressor cost** | | |
| a) Compressor capital cost ($) | 400,000 | 500,000 |
| b) Compression rate (kg-$H_2$/h) | 33 | 33 |
| c) Compressor lifetime (y) | 19 | 10 |
| d) Social (low) / private (high) discount rate (percent) | 2 | 5 |
| e) Capital recovery factor (Equation 7.3) (fraction/y) | 0.0638 | 0.1295 |
| f) Installation factor to multiply capital cost by (–) | 1.2 | 1.3 |
| g) Operation & maintenance (fraction of capital cost/y) | 0.015 | 0.015 |
| h) Compressor use factor (fraction of time it is in use)[a] | 0.50 | 0.95 |
| i) Hours per year | 8,760 | 8,760 |
| **j) Overall compressor cost ($/kg-$H_2$) = a(ef+g)/(bhi)** | **0.25** | **0.34** |
| **3. Hydrogen storage tank cost** | | |
| a) Storage tank capital cost ($/kg-$H_2$-stored) | 450 | 550 |
| b) Storage tank lifetime (y) | 50 | 30 |
| c) Social (low) / private (high) discount rate (percent) | 2 | 5 |
| d) Capital recovery factor (Equation 7.3) (fraction/y) | 0.0318 | 0.0651 |
| e) Installation factor to multiply capital cost by (–) | 1.2 | 1.3 |
| f) Max kg-$H_2$ stored as a fraction of kg-$H_2$/y produced | 0.0137 | 0.0137 |
| g) Operation & maintenance (fraction of capital cost/y) | 0.015 | **0.015** |
| **h) Overall storage tank cost ($/kg-$H_2$) = af(de+g)** | **0.33** | **0.75** |
| **4. Wind electricity cost for electrolysis & compression** | | |
| a) Electricity required to electrolyze $H_2$ (kWh/kg-$H_2$) | 53.4 | 53.4 |
| b) Electricity required to compress $H_2$ (kWh/kg-$H_2$) | 5.64 | 5.64 |
| c) Electrolysis + compression electricity (kWh/kg-$H_2$) = a+b | 59.0 | 59.0 |
| d) Cost of electricity from wind ($/kWh) (Table 7.8) | 0.029 | 0.056 |
| e) $H_2$ leakage rate (fraction of total $H_2$ produced) | 0.003 | 0.01 |
| **f) Overall electricity cost ($/kWh) = dc(1–e)** | **1.72** | **3.34** |
| **5. Cost of water for electrolysis** | | |
| a) Molecular weight of $H_2$ (g-$H_2$/mol) | 2.0158 | 2.0158 |
| b) Molecular weight of $H_2O$ (g-$H_2O$/mol) | 18.015 | 18.015 |
| c) Water required for electrolyzer (kg-$H_2O$/kg-$H_2$) = b/a | 8.94 | 8.94 |
| d) Density of liquid water (kg-$H_2O$/m³) | 1,000 | 1,000 |
| e) Gallons per cubic meter | 264.2 | 264.2 |
| f) Water required for electrolyzer (gal-$H_2O$/kg-$H_2$) | 2.36 | 2.36 |
| g) Cost of water per gallon ($/gal) | 0.002 | 0.004 |
| **h) Cost of water per kg-$H_2$ produced ($/kg-$H_2$)** | **0.0047** | **0.0094** |
| **6. Overall $H_2$ fuel cell cost ($/kg-$H_2$) = 1j+2j +3h+4f+5h** | **2.64** | **4.97** |

### 7. Gasoline-equivalent cost of H₂

| | | |
|---|---|---|
| a) Lower heating value of gasoline (MJ/kg-gasoline) | 44 | 44 |
| b) Tank-to-wheel efficiency of gasoline vehicle (fraction) | 0.17 | 0.20 |
| c) Gasoline density (kg-gasoline/m³) | 750 | 750 |
| d) Gallons per cubic meter | 264.2 | 264.2 |
| e) Energy in gasoline (MJ/gal) = ac/d | 125 | 125 |
| f) Lower heating value of H₂ (MJ/kg-H₂) | 119.96 | 119.96 |
| g) Fuel cell+wiring+motor efficiency (Table 7.2) (fraction) | 0.41 | 0.67 |
| h) H₂ leakage rate (fraction of total H₂ produced) | 0.003 | 0.01 |
| i) kg-H₂/gal-gasoline = eb/(fg(1−h)) | 0.27 | 0.51 |
| j) Gasoline equiv. H₂ cost in HFC vehicle ($/gal-gas)[b] = 6×i | 0.70 | 2.54 |
| k) Comparative cost of gasoline in United States ($/gal-gasoline) | 2.75 | 4.00 |

[a] The electrolyzer and compressor lifetimes are inversely proportional to the use factors since, for example, an electrolyzer that is used half the time should last twice as long before it needs to be replaced.

[b] This is the cost to produce and use H₂ in an HFC vehicle to go the same distance that 1 gallon of gasoline would take a gasoline vehicle. See Jacobson et al. (2005) for a similar analysis.

More recently, Burnett et al. (2018) calculated 8.9 (7.5 to 10.3) million deaths per year worldwide in 2015 due to indoor plus outdoor air pollution. They hypothesized that the additional deaths from air pollution may have been because previous studies considered only a limited number of categories of death that air pollution contributes to.

The air pollution deaths in Tables 7.12 and 7.13 are almost all due to combustion products of fossil fuels, biofuels, bioenergy, open biomass burning, and human-caused wildfires. The indoor mortalities are additionally due to the indoor burning of bioenergy (e.g., wood, dung, waste), coal, and gas for home heating and cooking, primarily in developing countries. A 100 percent WWS world will eliminate about 90 percent of the outdoor plus indoor air pollution deaths. Controlling open biomass burning and human-caused wildfires will address most of the rest (Section 2.9.1).

Because premature mortalities arising from a BAU energy infrastructure result in a social health cost to society, it is important to quantify the avoided cost of reducing such mortalities, related morbidities, and non-health costs due to air pollution. This is done next.

The total annual damage cost ($/y USD) of air pollution due to conventional fuels (fossil-fuel and biofuel combustion and evaporative emissions) in a country is estimated as

$$AP_{C,Y} = D_{C,Y} VOSL_{C,Y} F_1 F_2 \tag{7.6}$$

(Jacobson et al., 2017), where $D_{C,Y}$ is the air pollution premature mortality rate (deaths/y) in country $C$ in target year $Y$, $VOSL_{C,Y}$ is the value of statistical life ($/death) in the country and for the target year, $F_1$ is the ratio of mortality plus **morbidity** (illness) costs to mortality costs alone, and $F_2$ is the ratio of health cost (mortality plus morbidity costs) plus non-health costs to health costs alone. **Non-health costs** include costs due to animal health impacts, lost visibility, reduced agricultural output, and corrosion to building materials and works of art. Table 7.14 gives low, medium, and high estimates of $F_1$ and $F_2$. These are held constant for all countries and years.

The premature mortality rate in a country in target year $Y$ is projected from a base year ($BYD$) during which death rates from air pollution are available, with

$$D_{C,Y} = D_{C,BYD}\left(e^{\Delta A_C[Y-BYD]}\right)\left(\frac{P_{C,Y}}{P_{C,BYD}}\right)^{\kappa} \tag{7.7}$$

where $\Delta A_C$ (Table 7.14) is the fractional rate of change per year in the air pollution death rate in country $C$ due to emission controls, $P$ is population, and $\kappa$ is the change in exposed population per unit change in population (Table 7.14). Figure 7.1 shows the application of Equation 7.7 to 24 world regions encompassing 143 countries. These countries were home to 96 percent of 2016 indoor plus outdoor air pollution mortalities worldwide. The figure indicates that BAU mortalities may be less in 2050 than in 2016 in almost all regions despite higher populations in 2050. The reason is that improvements in BAU emission-reduction technologies between 2016 and 2050 outpace population growth. The only exception is in Africa, where population growth outpaces technology improvements, resulting in higher air pollution mortality in 2050 than in 2016.

Figure 7.1 indicates that BAU mortalities may be, on average, about 22 percent less in 2050 than in 2016.

**Table 7.12** Leading causes of death worldwide in 2016. Also shown are the percentage and number of deaths in each category due to indoor plus outdoor air pollution and, separately, outdoor air pollution alone.

| Cause of Death | Total All-Cause[a] Number of deaths/y (millions) | Indoor Plus Outdoor Air Pollution Percent of all-cause deaths[b] | Number of deaths/y (millions) | Outdoor Air Pollution Only Percent of all-cause deaths[c] | Number of deaths/y (millions) |
|---|---|---|---|---|---|
| 1. Ischemic heart disease (coronary artery disease) | 9.43 | 25 | 2.36 | 17 | 1.60 |
| 2. Stroke | 5.78 | 24 | 1.39 | 16 | 0.81 |
| 3. COPD (chronic bronchitis, emphysema)[d] | 3.04 | 43 | 1.31 | 25 | 0.76 |
| 4. Lower respiratory infection (flu, bronchitis, pneumonia) | 2.96 | 45 | 1.32 | 26 | 0.77 |
| 5. Alzheimer's disease/dementia | 2.00 | 0 | 0 | 0 | 0 |
| 6. Trachea, bronchus, lung cancers | 1.71 | 29 | 0.50 | 16 | 0.27 |
| 7. Diabetes | 1.60 | 0 | 0 | 0 | 0 |
| 8. Road accidents | 1.40 | 0 | 0 | 0 | 0 |
| 9. Diarrheal disease (cholera, dysentery) | 1.38 | 0 | 0 | 0 | 0 |
| 10. Tuberculosis | 1.29 | 0 | 0 | 0 | 0 |
| 11. Asthma | 0.42 | 43 | 0.18 | 25 | 0.10 |
| **Total number of deaths worldwide** | **56.9** | **12.5** | **7.1** | **7.9** | **4.5** |

[a] Source: WHO (2017a).

[b] Source: WHO (2017b), except that the percentage of lower respiratory infection deaths that are due to indoor plus outdoor air pollution is estimated as the percentage of respiratory deaths that are from outdoor air pollution from WHO (2017b) multiplied by the ratio of the percentage of deaths from outdoor plus indoor to outdoor air pollution for COPD. The asthma percentage is assumed to be the same as the COPD percentage.

[c] Source: WHO (2017b), except that the percentage of stroke deaths that are due to outdoor air pollution is estimated as the percentage of stroke deaths that are from indoor plus outdoor air pollution from WHO (2017b) multiplied by the ratio of the percentage of outdoor to indoor plus outdoor air pollution for ischemic heart disease. The asthma percentage is assumed to be the same as the COPD percentage.

[d] Chronic obstructive pulmonary disease (COPD) deaths are due to smoking and air pollution. They exclude asthma deaths, which are added separately.

Reductions occur in almost all world regions despite higher populations in all regions in 2050. The reason is that improvements in BAU emission-reduction technologies between 2016 and 2050 outpace population growth in almost all regions. The only exception is in Africa, where population growth is so high, it outpaces technology improvements, resulting in higher air pollution mortality in 2050 than in 2016.

The **value of statistical life** is a widely used metric determined by economists to assign the cost of reducing mortality risk. It is the value of reducing 1 statistical mortality in a population. For example, if the average person in a city of 100,000 is willing to pay $75 to reduce her or his mortality risk by 1/100,000th, the statistical value of reducing 1 mortality is $7.5 million. The value of statistical life

is also determined from how much more employers pay their workers who have a higher risk of dying on the job.

The VOSL varies with time and country. An estimate of the variation of VOSL ($million per death in 2013 USD) in country $C$ during year $Y$ is

$$VOSL_{C,Y} = VOSL_{US,Y}\left(T + [1-T]\left[\frac{G_{C,Y}}{G_{US,Y}}\right]^{\gamma_{GDP,US,BYV}\left(\frac{G_{C,Y}}{G_{US,BYV}}\right)^{\gamma_{GDP}}}\right)$$

(7.8)

where

$VOSL_{US,Y}$ is the VOSL in the United States in year $Y$ (given in Table 7.14 for $Y = 2050$);

$T$ is the fraction of the country's VOSL that is held constant at the U.S. VOSL for that year;

**Table 7.13** 2016 Mean number of indoor plus outdoor air pollution (AP) deaths, deaths per 100,000 population (WHO, 2017c), and population by country, for 183 countries of the world. Ranked from highest to lowest number of deaths.

| | Country | 2016 AP deaths | 2016 AP deaths per 100,000 | 2016 Population | | Country | 2016 AP deaths | 2016 AP deaths per 100,000 | 2016 Population |
|---|---|---|---|---|---|---|---|---|---|
| 1 | China | 1,912,570 | 140 | 1,366,119,400 | 93 | Tajikistan | 5,830 | 70 | 8,328,400 |
| 2 | India | 1,785,870 | 141 | 1,266,575,400 | 94 | Azerbaijan | 5,430 | 55 | 9,866,000 |
| 3 | Nigeria | 299,500 | 159 | 188,363,000 | 95 | Canada | 5,300 | 15 | 35,357,400 |
| 4 | Pakistan | 228,270 | 113 | 202,012,600 | 96 | Netherlands | 5,270 | 31 | 17,014,400 |
| 5 | Indonesia | 209,070 | 81 | 258,113,600 | 97 | Kyrgyzstan | 4,430 | 74 | 5,993,200 |
| 6 | Bangladesh | 176,940 | 103 | 171,788,200 | 98 | Belgium | 4,080 | 39 | 10,456,200 |
| 7 | Philippines | 130,520 | 117 | 111,558,600 | 99 | Dominican Rep. | 4,030 | 38 | 10,605,000 |
| 8 | Russian Fed. | 116,320 | 86 | 135,256,400 | 100 | Australia | 3,910 | 17 | 22,988,600 |
| 9 | Ethiopia | 87,400 | 82 | 106,591,200 | 101 | Croatia | 3,830 | 86 | 4,457,400 |
| 10 | Congo, DR of | 82,160 | 101 | 81,350,000 | 102 | Moldova | 3,760 | 107 | 3,510,400 |
| 11 | United States | 78,060 | 24 | 325,264,000 | 103 | Congo | 3,570 | 73 | 4,892,800 |
| 12 | Brazil | 66,460 | 31 | 214,398,400 | 104 | Liberia | 3,570 | 83 | 4,302,200 |
| 13 | Myanmar | 65,980 | 116 | 56,881,200 | 105 | Ecuador | 3,540 | 22 | 16,075,400 |
| 14 | Egypt | 65,730 | 73 | 90,041,600 | 106 | Honduras | 3,470 | 39 | 8,890,600 |
| 15 | Vietnam | 61,900 | 65 | 95,223,400 | 107 | Mongolia | 3,260 | 97 | 3,361,400 |
| 16 | Ukraine | 59,900 | 137 | 43,719,400 | 108 | Mauritania | 3,240 | 88 | 3,678,600 |
| 17 | Thailand | 58,150 | 85 | 68,406,800 | 109 | Slovak Republic | 3,240 | 59 | 5,495,600 |
| 18 | Korea, DPR | 58,020 | 231 | 25,115,000 | 110 | Austria | 3,210 | 39 | 8,223,200 |
| 19 | Japan | 54,450 | 43 | 126,637,400 | 111 | Albania | 3,190 | 105 | 3,038,200 |
| 20 | Sudan | 53,640 | 105 | 51,082,400 | 112 | Paraguay | 3,160 | 46 | 6,864,800 |
| 21 | Nepal | 42,670 | 133 | 32,082,600 | 113 | Libya | 3,120 | 43 | 7,257,400 |
| 22 | Mexico | 39,560 | 33 | 119,882,000 | 114 | Portugal | 3,030 | 28 | 10,828,400 |
| 23 | Turkey | 38,350 | 46 | 83,369,800 | 115 | Lithuania | 2,860 | 82 | 3,483,000 |
| 24 | Germany | 36,320 | 45 | 80,715,200 | 116 | Turkmenistan | 2,700 | 51 | 5,290,600 |
| 25 | Tanzania | 35,170 | 75 | 46,896,200 | 117 | Macedonia | 2,620 | 125 | 2,099,400 |
| 26 | Cote d'Ivoire | 34,180 | 144 | 23,736,800 | 118 | El Salvador | 2,590 | 42 | 6,156,200 |
| 27 | Afghanistan | 31,710 | 95 | 33,380,000 | 119 | Nicaragua | 2,570 | 43 | 5,967,000 |
| 28 | Uganda | 30,700 | 74 | 41,491,000 | 120 | Armenia | 2,420 | 81 | 2,990,600 |
| 29 | Italy | 30,360 | 49 | 61,964,600 | 121 | Singapore | 2,250 | 39 | 5,781,200 |
| 30 | South Africa | 29,480 | 61 | 48,334,800 | 122 | Lesotho | 2,210 | 113 | 1,952,200 |

Table 7.13 (*cont.*)

| | Country | 2016 AP deaths | 2016 AP deaths per 100,000 | 2016 Population | | Country | 2016 AP deaths | 2016 AP deaths per 100,000 | 2016 Population |
|---|---|---|---|---|---|---|---|---|---|
| 31 | Poland | 29,060 | 76 | 38,231,400 | 123 | Lebanon | 2,170 | 52 | 4,169,400 |
| 32 | Iran | 28,970 | 35 | 82,767,800 | 124 | Latvia | 2,090 | 98 | 2,137,000 |
| 33 | Niger | 27,690 | 140 | 19,777,000 | 125 | Gambia | 1,950 | 97 | 2,009,200 |
| 34 | Ghana | 27,300 | 101 | 27,024,800 | 126 | Switzerland | 1,930 | 25 | 7,708,600 |
| 35 | Romania | 26,560 | 123 | 21,593,400 | 127 | Guinea-Bissau | 1,900 | 108 | 1,759,400 |
| 36 | Cameroon | 25,730 | 118 | 21,803,800 | 128 | Israel | 1,850 | 23 | 8,043,800 |
| 37 | Yemen | 24,550 | 90 | 27,279,000 | 129 | Jordan | 1,760 | 26 | 6,754,000 |
| 38 | Chad | 21,460 | 181 | 11,856,000 | 130 | Denmark | 1,680 | 30 | 5,594,000 |
| 39 | United Kingdom | 20,620 | 32 | 64,422,600 | 131 | Namibia | 1,670 | 75 | 2,222,200 |
| 40 | Madagascar | 20,320 | 80 | 25,395,600 | 132 | Sweden | 1,650 | 18 | 9,171,400 |
| 41 | Sri Lanka | 19,780 | 89 | 22,220,200 | 133 | Costa Rica | 1,320 | 27 | 4,870,800 |
| 42 | Kenya | 18,680 | 40 | 46,711,600 | 134 | Botswana | 1,170 | 53 | 2,208,800 |
| 43 | Burkina Faso | 18,170 | 93 | 19,541,200 | 135 | Slovenia | 1,130 | 57 | 1,976,600 |
| 44 | Peru | 17,830 | 58 | 30,739,000 | 136 | Uruguay | 1,070 | 32 | 3,351,200 |
| 45 | Mali | 17,280 | 107 | 16,152,400 | 137 | Kuwait | 1,050 | 37 | 2,830,000 |
| 46 | Korea, Rep. of | 17,210 | 35 | 49,164,400 | 138 | Finland | 1,000 | 19 | 5,271,200 |
| 47 | France | 16,640 | 25 | 66,544,400 | 139 | Swaziland | 1,000 | 69 | 1,451,400 |
| 48 | Mozambique | 16,620 | 64 | 25,963,000 | 140 | Timor-Leste | 1,000 | 77 | 1,295,400 |
| 49 | Somalia | 16,480 | 152 | 10,844,200 | 141 | Ireland | 990 | 20 | 4,949,000 |
| 50 | Argentina | 16,210 | 37 | 43,821,400 | 142 | Panama | 960 | 26 | 3,704,400 |
| 51 | Colombia | 16,050 | 34 | 47,206,600 | 143 | Jamaica | 950 | 32 | 2,970,200 |
| 52 | Uzbekistan | 15,920 | 54 | 29,473,000 | 144 | United Arab Emirates | 950 | 16 | 5,923,000 |
| 53 | Guinea | 15,380 | 127 | 12,108,000 | 145 | Norway | 910 | 19 | 4,770,400 |
| 54 | Algeria | 14,810 | 40 | 37,030,800 | 146 | Montenegro | 900 | 140 | 645,400 |
| 55 | Cambodia | 13,880 | 87 | 15,952,600 | 147 | Gabon | 890 | 51 | 1,739,400 |
| 56 | South Sudan | 13,680 | 109 | 12,546,200 | 148 | Djibouti | 840 | 99 | 846,800 |
| 57 | Angola | 13,530 | 67 | 20,196,800 | 149 | Equatorial Guinea | 760 | 100 | 760,000 |
| 58 | Morocco | 13,460 | 40 | 33,649,600 | 150 | Comoros | 750 | 94 | 794,000 |
| 59 | Spain | 13,100 | 27 | 48,520,000 | 151 | Estonia | 740 | 60 | 1,239,800 |
| 60 | Haiti | 12,990 | 127 | 10,226,600 | 152 | Oman | 740 | 22 | 3,356,600 |

| 61 | Benin | 12,790 | 119 | 10,750,400 | 153 | Fiji | 690 | 76 | 914,400 |
|---|---|---|---|---|---|---|---|---|---|
| 62 | Burundi | 11,950 | 100 | 11,945,000 | 154 | Bhutan | 660 | 88 | 750,000 |
| 63 | Iraq | 11,910 | 35 | 34,025,800 | 155 | Mauritius | 650 | 48 | 1,347,800 |
| 64 | Saudi Arabia | 10,980 | 39 | 28,165,400 | 156 | New Zealand | 630 | 14 | 4,473,400 |
| 65 | Malaysia | 10,830 | 35 | 30,941,600 | 157 | Guyana | 560 | 76 | 742,000 |
| 66 | Senegal | 10,600 | 74 | 14,328,000 | 158 | Trinidad & Tobago | 550 | 45 | 1,219,400 |
| 67 | Kazakhstan | 10,460 | 57 | 18,344,000 | 159 | Solomon Islands | 430 | 67 | 634,600 |
| 68 | Belarus | 10,340 | 110 | 9,401,000 | 160 | Cyprus | 400 | 33 | 1,204,600 |
| 69 | Syria | 10,230 | 44 | 23,252,000 | 161 | Cape Verde | 380 | 69 | 553,400 |
| 70 | Zambia | 10,160 | 63 | 16,128,200 | 162 | Suriname | 300 | 51 | 586,000 |
| 71 | Cuba | 9,910 | 90 | 11,011,200 | 163 | Qatar | 290 | 13 | 2,244,800 |
| 72 | Malawi | 9,830 | 54 | 18,212,800 | 164 | Bahrain | 210 | 15 | 1,378,600 |
| 73 | Zimbabwe | 9,750 | 67 | 14,550,400 | 165 | Malta | 180 | 44 | 415,000 |
| 74 | Bulgaria | 9,600 | 141 | 6,807,400 | 166 | Vanuatu | 180 | 76 | 239,000 |
| 75 | Togo | 8,930 | 115 | 7,763,200 | 167 | Barbados | 170 | 57 | 291,800 |
| 76 | Sierra Leone | 8,920 | 148 | 6,029,000 | 168 | Sao Tome +Principe | 160 | 82 | 197,400 |
| 77 | Serbia | 8,720 | 122 | 7,144,000 | 169 | Belize | 120 | 35 | 353,600 |
| 78 | Venezuela | 8,610 | 29 | 29,675,200 | 170 | Luxembourg | 120 | 23 | 532,800 |
| 79 | Georgia | 8,290 | 184 | 4,508,000 | 171 | Samoa | 120 | 62 | 199,200 |
| 80 | Greece | 8,290 | 77 | 10,769,200 | 172 | Micronesia | 100 | 93 | 104,400 |
| 81 | Hungary | 8,190 | 83 | 9,872,800 | 173 | Kiribati | 90 | 88 | 107,200 |
| 82 | Central African Republic | 7,770 | 141 | 5,511,800 | 174 | Bahamas | 70 | 22 | 327,600 |
| 83 | Laos | 7,720 | 110 | 7,019,000 | 175 | Saint Lucia | 60 | 38 | 164,400 |
| 84 | Rwanda | 7,670 | 59 | 12,995,000 | 176 | Tonga | 60 | 57 | 106,800 |
| 85 | Guatemala | 7,590 | 50 | 15,188,000 | 177 | Grenada | 50 | 44 | 111,400 |
| 86 | Bosnia & Herzegovina | 7,330 | 159 | 4,612,800 | 178 | Iceland | 50 | 17 | 321,000 |
| 87 | Czech Republic | 6,470 | 64 | 10,106,600 | 179 | Maldives | 50 | 14 | 392,800 |
| 88 | Eritrea | 6,340 | 95 | 6,674,400 | 180 | St. Vinc. & Grenad. | 50 | 48 | 102,600 |
| 89 | Tunisia | 6,340 | 57 | 11,128,400 | 181 | Seychelles | 50 | 56 | 92,800 |
| 90 | Chile | 6,150 | 35 | 17,559,600 | 182 | Brunei Darussalam | 40 | 9 | 436,800 |

Table 7.13 (*cont.*)

| | Country | 2016 AP deaths | 2016 AP deaths per 100,000 | 2016 Population | | Country | 2016 AP deaths | 2016 AP deaths per 100,000 | 2016 Population |
|---|---|---|---|---|---|---|---|---|---|
| 91 | Papua New Guinea | 6,110 | 90 | 6,789,400 | 183 | Antigua & Barbuda | 30 | 28 | 93,200 |
| 92 | Bolivia | 6,030 | 55 | 10,968,800 | | | | | |
| | | | | | | Total | 7.01 million | 95.6 | 7.33 billion |

$G_{C,Y}$ is the **gross domestic product (GDP)** per capita in the country in year $Y$;

$G_{US,Y}$ is the U.S. GDP per capita in year $Y$ (estimated in Table 7.14 for $Y = 2050$);

$G_{US,BYV}$ is the U.S. GDP per capita in base year $BYV$ for calculating VOSL (Table 7.14 for BYV = 2006);

$\gamma_{GDP,US,BYV}$ is the elasticity of the GDP per capita in base year $BYV$ (Table 7.14 for BYV = 2006); and

$\gamma_{GDP}$ is the elasticity of the GDP per capita for all years (Table 7.14).

The GDP per capita in Equation 7.8 is the GDP at **purchasing power parity (PPP)**. The GDP at PPP means that the GDP is determined by equalizing the value of a basket of goods in one country versus another, taking into account the currency exchange rate. For example, suppose countries A and B have a normal GDP per capita of $20 and $40, respectively, but a cup of coffee costs $5 in country A and $20 in country B. In that case, a consumer in country B can purchase only one-fourth the goods that a consumer in country A can for the same amount of money. If we make country A the reference country, then its GDP at PPP is still $20 per person, but the GDP at PPP of the second country is $40 per capita × ($5 per cup in country A / $20 per cup in country B) = $10 per person. So, although the normal GDP per capita is higher in country B, the GDP at PPP is higher in country A.

Equation 7.8 and the corresponding values of $T$ in Table 7.14 indicate that a small portion of the VOSL is assumed to be constant across all countries. This constant portion is the fraction of the VOSL that is independent of relative wealth, productivity or consumption. The equation also indicates that the VOSL is a function of change in income. In addition, the elasticity of the GDP per capita is itself a function of the GDP per capita ratio between the country and the United States.

## Example 7.4  Estimating the value of statistical life

Estimate the medium number of premature air pollution deaths in country $C$ in 2050 if the number of deaths in base year 2018 is 10,000/y, the population in the base year is 30 million, and the population in 2050 is 50 million. Also, estimate the VOSL and cost of air pollution in 2050 in the country assuming the GDP per capita in 2050 in the country is $40,000/person.

### Solution:

From Equation 7.7, the number of premature deaths in the country is estimated to increase to about 10,909/y. Thus, the impact of the increase in population is offset partly by the impact of better emission controls. From Equation 7.8, the value of statistical life in the country in 2050 is $6.59 million in 2013 USD. Finally, from Equation 7.6, the air pollution mortality, morbidity, and non-health effects cost is $90.9 billion/y.

Equation 7.7 requires a premature mortality rate from indoor plus outdoor air pollution in a base year in each country. Such an estimate can be obtained from data, such as those data shown in Table 7.12, but for individual countries. Such data include the number of actual deaths occurring in each country due to a specific cause, as

Table 7.14 **Parameters in the calculation of the value of statistical life over time and by country.**

| Parameter | LCHB | Middle | HCLB |
|---|---|---|---|
| U.S. VOSL in base year 2006 ($VOSL_{US,BYV}$) ($mil/death USD 2006) | 9.00 | 7.00 | 5.00 |
| U.S. VOSL in target year 2050 ($VOSL_{US,Y}$) ($mil/death USD 2013) | 15.37 | 10.40 | 6.47 |
| 2006 global average VOSL ($mil/death USD 2006) | 4.00 | 3.48 | 3.43 |
| 2050 global average VOSL ($mil/death USD 2013) | 8.15 | 7.09 | 6.99 |
| U.S. GDP per capita in 2006 ($G_{US,BYV}$) ($/person 2006 USD) | 52,275 | 52,275 | 52,275 |
| U.S. GDP per capita target year 2050 ($G_{US,Y}$) ($/person USD 2013) | 96,093 | 96,093 | 96,093 |
| Multiplier for morbidity impacts ($F_1$) | 1.25 | 1.15 | 1.05 |
| Multiplier for non-health impacts ($F_2$) | 1.10 | 1.10 | 1.05 |
| Fractional reduction in mortalities per year ($\Delta A_c$) | −0.014 | −0.015 | −0.016 |
| Exponent giving change in mortality with population change ($\kappa$) | 1.14 | 1.11 | 1.08 |
| Fraction of country's VOSL fixed at U.S. TY value ($T$) | 0.10 | 0.00 | 0.00 |
| GDP/capita elasticity ($\gamma_{GDP, US,BYV}$) of VOSL, U.S. base year 2006 | 0.75 | 0.50 | 0.25 |
| GDP/capita elasticity ($\gamma_{GDP}$) of VOSL, all years | −0.15 | −0.15 | −0.15 |

LCHB = low cost, high benefit. HCLB = high cost, low benefit. VOSL = value of statistical life. GDP = gross domestic product at purchasing power parity (PPP).
Source: Jacobson et al. (2017), except that the low and high fraction reduction in mortalities per year are updated here.

denoted on death certificates, and an approximation of how many deaths by each cause were due to air pollution. Table 7.10 estimates the avoided premature mortalities from air pollution by country in 2050 upon a conversion to 100 percent WWS. The mortalities are determined from Equation 7.7 using base-year death rates from 2016. Base-year indoor plus outdoor air pollution death rates are determined for each country by multiplying the country-specific total number of air pollution mortalities per 100,000 population from WHO (2017b) by the population of the country.

Table 7.10 also shows the resulting avoided health cost (from Equation 7.6) per unit of BAU energy eliminated due to switching to WWS. Transitioning to 100 percent WWS will avoid millions of premature deaths each year and save the equivalent of 16.9 ¢/kWh (USD 2013), which is more than the direct cost of energy. The cost savings per unit energy translates to an aggregate avoided air pollution mortality, morbidity, and non-health cost in 2050 of about $30 trillion/y.

A second way to determine premature air pollution mortalities by region or country is with a health effects equation. The equation combines concentrations of $PM_{2.5}$ and $O_3$ with estimates of the population exposed to those concentrations and with relative risk estimates of premature mortality as a function of $PM_{2.5}$ and $O_3$ concentrations.

The **health effects equation** gives the death rate (e.g., deaths per year), cancer rate, hospitalization rate, etc., due to exposure to a pollutant as

$$y = y_0 P(1 - \exp[-\beta \times \max(x - x_{th}, 0)]) \qquad (7.9)$$

where $x$ is the average concentration or mixing ratio of the pollutant, $x_{th}$ is the threshold concentration or mixing ratio below which no health effect occurs, $\beta$ is the **relative risk**, or fractional increase in the risk of the health effect occurring per unit concentration $x$, $y_0$ is the baseline health effect rate per unit population (e.g., all-cause deaths per year per 100,000 population), and $P$ is the population (e.g., Jacobson, 2010b).

The concentrations are obtained from measurements or computer model simulation. The advantage of using measurements is that they are fairly accurate. However, such data are usually scattered sparsely throughout a country, and concentrations measured in one location

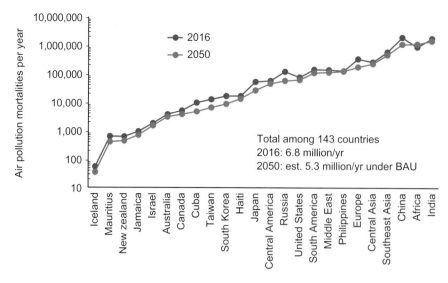

**Figure 7.1** Estimated air pollution mortalities in 2016 and 2050 by world region. These are indoor plus outdoor air pollution mortalities per year in 24 world regions encompassing 143 countries (see Table 8.5 for a list of countries in each region). Year 2016 data are obtained by multiplying country-specific indoor plus outdoor air pollution deaths per 100,000 population from WHO (2017c) by 2016 country population. 2050 estimates are obtained from Equation 7.7. BAU energy is responsible for about 90 percent of the mortalities. Most of the rest are from open biomass burning, wildfires, and dust.

may not be representative of concentrations nearby. The advantage of using a model is that it has complete horizontal and vertical coverage of a country. The disadvantage is that the model estimates are less accurate than are the measurements.

Table 7.15 gives relative risks and threshold values for $PM_{2.5}$, $O_3$, and carcinogens.

Equation 7.9 can be applied over small regions, such as a neighborhood, or a whole country. When it is applied to each of many small regions adjacent to each other, results are summed over all regions to obtain a total number of premature mortalities per year, for example.

---

## Example 7.5 Estimating the number of premature mortalities due to ozone

Estimate the number of premature mortalities in the United States per year due to short-term ozone exposure if the entire population of 300 million were exposed to 40 parts-per-billion-volume (ppbv) ozone over a year. Assume the all-cause death rate is approximately $y_0 = 833$ deaths per year per 100,000 population.

### Solution:

Substituting $y_0 = 0.00833$ deaths per person, $P = 300,000,000$ people, $\beta = 0.0004$ per ppbv, $x = 40$ ppbv, and $x_{th} = 35$ ppbv into Equation 7.9 gives $y \approx 5,000$ additional premature deaths per year due to ozone in the United States.

---

### 7.6.3 Avoided Climate Change Damage Costs

The damage that carbon dioxide equivalent ($CO_2e$) emissions (Section 3.1) cause to the global economy through their impacts on climate is the **social cost of carbon (SCC)**. Units of the SCC are U.S. dollars per metric tonne-$CO_2e$ emissions. Climate change damage costs include costs arising from higher sea levels (coastal

infrastructure losses), reduced crop yields for certain crops, more intense hurricanes, more droughts and floods, more wildfires and air pollution, more migration due to crop losses and famine, more heat stress and heat stroke, more disease of certain types, fishery and coral reef losses, and greater air cooling requirements, among other impacts. Only a portion of these costs is offset by

Table 7.15 Relative risks ($\beta$) of premature mortality due to $PM_{2.5}$ and $O_3$, and of cancers over 70 years for various carcinogens.

| | Low | Medium | High |
|---|---|---|---|
| **Long-term $PM_{2.5}$ exposure[a]** | | | |
| All-cause mortality | 0.0035 | 0.0055 | 0.0076 |
| Cardiopulmonary mortality | 0.0100 | 0.0129 | 0.0159 |
| Ischemic heart disease mortality | 0.0175 | 0.0217 | 0.0259 |
| Lung cancer mortality | 0.0055 | 0.0129 | 0.0203 |
| | | | |
| **Short-term $O_3$ exposure[b]** | | | |
| All-cause mortality (1-h max $O_3$) | 0.0002 | 0.0004 | 0.0006 |
| All-cause mortality (8-h max $O_3$) | 0.00027 | 0.00053 | 0.0008 |
| All-cause mortality (24-h $O_3$) | 0.0005 | 0.001 | 0.0015 |
| **Cancers over 70 years[c]** | U.S. EPA CURES | | OEHHA CURES |
| Formaldehyde | $1.3 \times 10^{-5}$ | | $6.0 \times 10^{-6}$ |
| Acetaldehyde | $2.2 \times 10^{-6}$ | | $2.7 \times 10^{-6}$ |
| Butadiene | $3.0 \times 10^{-5}$ | | $1.7 \times 10^{-4}$ |
| Benzene | $5.0 \times 10^{-6}$ | | $2.9 \times 10^{-5}$ |

[a] The relative risks due to the long-term effects of particulate matter smaller than 2.5 μm ($PM_{2.5}$) are the fractional increases in the cause of death specified per μg/m$^3$ increase in annual average outdoor $PM_{2.5}$ concentration (Krewski et al., 2009), based on data for 1999 to 2000 (CARB, 2010). The relative risks apply only above a threshold concentration of $x_{th}$ = 5.8 μg/m$^3$, the lowest annual averaged $PM_{2.5}$ concentration measured in Krewski et al. (2009). The relative risks apply only for people older than 30 years. The low threshold due to health problems from $PM_{2.5}$ is 0 μg/m$^3$, but the relative risk down to zero is uncertain. Jacobson (2010b) estimates the relative risk down to zero as one-fourth that above $x_{th}$. The low and high values in the table represent 95 percent confidence intervals. Ischemic heart disease is a subset of cardiopulmonary causes of death. Those two and lung cancer are a subset of the all-cause death rate.

[b] The relative risks (for all ages) due to the short-term effects of $O_3$ are the fractional increase in all-cause daily mortality that is due to short-term exposure to a 1 ppbv (parts-per-billion-volume) increase in the highest 1-hour average ozone level during a day, the highest 8-hour average ozone level during a day, or the 24-hour average ozone during a day (Ostro et al., 2006). The low threshold for ozone health effects is $x_{th}$ = 35 ppmv.

lower heating requirements and higher yields of some crops.

The SCC of emissions is likely to increase over time as $CO_2$ and other warming agents accumulate in the atmosphere and temperatures continue to rise. The accumulation will accelerate as more people rise out of poverty in developing countries and consume energy faster than their predecessors did. As such, the SCC is tied to the gross domestic product (GDP) of a country.

Van den Bergh and Botzen (2014) argue that the lower bound of the SCC (in 2014) should be at least $125 per tonne-$CO_2$e. Moore and Diaz (2015) found that incorporating the effect of climate change on the rate of economic growth can increase the SCC to between $200 and $1,000 per tonne-$CO_2$e. Burke et al. (2015) similarly found that accounting for the long-term effects of temperature rise on economic productivity results in climate change damage estimates that are 2.5 to 100 times higher than those from earlier studies. Table 7.16 provides estimates of the SCC in 2050 derived from these studies plus the estimated growth rates of the SCC. The mid-value from the range, $500 per metric tonne-$CO_2$e in 2013 USD (Jacobson et al., 2017), is used to derive the avoided climate change cost in Table 7.10.

### 7.6.4 Summary of Avoided Energy, Health, and Climate Damage Costs

Table 7.10 indicates that the world-average 2050 levelized cost of BAU electricity (in 2013 USD) may be about 10 ¢/kWh-BAU-electricity; the world-average air pollution damage cost due to BAU energy will be about 16.9 ¢/kWh-BAU-all-energy; and the world-average climate damage cost due to BAU energy will be about 16 ¢/kWh-BAU-all-energy. Thus, the total social (economic) cost of BAU electricity will be about 42.9 ¢/kWh-BAU-electricity.

[c] CURES are cancer unit risk estimates. They are 70-year cancer risks per μg/m$^3$ sustained concentration change of a carcinogen. Thus, divide the CURES by 70 years for use in Equation 7.9 to obtain the number of new cancers (not necessarily cancer deaths) per year due to exposure to the carcinogen. No low thresholds apply. The two sources of CURES are the U.S. Environmental Protection Agency (U.S. EPA) and the California Office of Environmental Health Hazard Assessment (OEHHA) (Jacobson, 2010b).

Table 7.16 **Low, mid-, and high estimates of the social cost of carbon (SCC).**

| Parameter | Low estimate | Mid-estimate | High estimate |
|---|---|---|---|
| 2010 global SCC (2007 USD) | 125 | 250 | 600 |
| Annual percentage increase in SCC | 1.8 | 1.5 | 1.2 |
| 2050 global SCC (2013 USD) | 282 | 500 | 1,063 |

Units of the SCC are USD per metric tonne-$CO_2$e.
Source: Jacobson et al. (2017).

In comparison, the world-average WWS social cost will simply be its direct LCOE of 7.6 ¢/kWh-WWS-all-energy, which is all electricity (Table 7.10). As such, the ratio of the BAU social cost (per kWh) to the WWS social cost of electricity is about 5.6:1. However, this WWS cost does not account for matching power demand with supply continuously. That WWS cost will be determined in Chapter 8.

In addition, a WWS energy system needs 57.1 percent less energy than does a BAU system (Table 7.1). As such, the ratio of the BAU social cost to the WWS social cost (to meet annual average power demand), in terms of aggregate amount of money, is about 13:1. In other words, a BAU system costs the world about 13 times the social cost as a WWS system in terms of the aggregate amount of money society needs to pay for each.

## 7.7 Summary

This chapter discussed the steps in developing 100 percent clean, renewable WWS roadmaps for individual countries, states, provinces, cities, or towns. The process involves first projecting BAU end-use energy demand for all fuels in all energy sectors to a future year. The next step is to convert each technology for each BAU fuel in each energy sector to a technology run on electricity or direct heat. This calculation results in a reduction in end-use energy demand. A resource analysis is then performed. Based on the resource analysis and land use and other constraints, a mix of WWS generators is proposed to match power demand in the annual average. The mix is updated and combined with storage and demand response in Chapter 8 to match demand continuously. Based on the WWS mix, the footprint and spacing areas of the WWS energy system are estimated. The energy cost of the WWS and BAU systems are then compared. The comparison is of the energy, health, and climate costs of each. Such a calculation among 143 countries worldwide finds that a transition to 100 percent WWS among all energy sectors in those countries reduces end-use energy demand significantly, reduces the private cost of energy, and reduces the social (economic) cost of energy (private plus air pollution health and climate cost). Whereas the numbers in this chapter are based on matching annual average power demand, those in the next chapter are based on matching power demand continuously with 100 percent WWS supply.

## Further Reading

Delucchi, M. Z., and M. Z. Jacobson, 2011. Providing all global energy with wind, water, and solar power, Part II: reliability, system and transmission costs, and policies, *Energy Policy*, **39**, 1170–1190, doi:10.1016/j.enpol.2010.11.045.
Jacobson, M. Z., and M. A. Delucchi, November 2009. A path to sustainable energy by 2030, *Scientific American*.
Jacobson, M. Z., and M. A. Delucchi, 2011. Providing all global energy with wind, water, and solar power, Part I: technologies, energy resources, quantities and areas of infrastructure, and materials, *Energy Policy*, **39**, 1154–1169, doi:10.1016/j.enpol.2010.11.040.

Jacobson, M. Z., R. W. Howarth, M. A. Delucchi, S. R. Scobies, J. M. Barth, M. J. Dvorak, et al., 2013. Examining the feasibility of converting New York State's all-purpose energy infrastructure to one using wind, water, and sunlight, *Energy Policy*, **57**, 585–601.

Jacobson, M. Z., M. A. Delucchi, G. Bazouin, Z. A. F. Bauer, C. C. Heavey, E. Fisher, et al., 2015. 100 percent clean and renewable wind, water, sunlight (WWS) all-sector energy roadmaps for the 50 United States, *Energy Environ. Sci.*, **8**, 2093–2117, doi:10.1039/C5EE01283J.

Jacobson, M. Z., M. A. Delucchi, Z. A. F. Bauer, S. C. Goodman, W. E. Chapman, M. A. Cameron, et al., 2017. 100 percent clean and renewable wind, water, and sunlight (WWS) all-sector energy roadmaps for 139 countries of the world, *Joule*, **1**, 108–121, doi:10.1016/j.joule.2017.07.005.

# 7.8 Problems and Exercises

7.1.  Explain the difference between primary energy and end-use energy.

7.2.  Explain the difference between power and energy. What is the difference between instantaneous power and annual average power?

7.3.  Identify four energy sectors.

7.4.  Identify four ways that electrifying all energy and providing the electricity with 100 percent wind, water, and solar reduces end-use power demand.

7.5.  Explain why the conversion from an internal combustion engine gasoline vehicle to an electric vehicle reduces end-use power demand.

7.6.  Why might we want to use battery-electric passenger vehicles as much as possible before using hydrogen fuel cell passenger vehicles?

7.7.  Which two WWS energy resources are in significant enough abundance to power the entire world for all purposes at reasonably low cost?

7.8.  What is the annual payment on a $1,000,000 loan at a 3.6 percent private discount rate over 30 years?

7.9.  What is the future value of the loan in Problem 7.8 after year 30 if it accrues interest over 30 years?

7.10. If a wind farm produces $50,000/y in annual revenue for 30 years, what is the present value of that income, assuming a private discount rate of 4 percent?

7.11. Estimate the cost per unit energy of a $P_r = 5$ MW nameplate capacity wind turbine with a capital cost of $1,200/kW, a capacity factor of $CF = 34$ percent, an O&M cost of $30/kW-y, and a lifetime of 30 years. Assume the discount rate is 3.5 percent, the construction time is 2 years, and the decommissioning cost is $13/kW. Also assume transmission, distribution, downtime, and array losses are 10 percent. *Hint*: Construct a table like Table 7.9.

7.12. Estimate the middle number of premature air pollution deaths in a country in 2050 if the number of deaths in year 2020 is 100,000, the population in the base year is 200 million, and the estimated population in 2050 is 250 million. Also estimate the VOSL and cost of air pollution in 2050 in the country, assuming the 2050 GDP per capita in the country is $50,000.

7.13. Estimate the medium number of premature mortalities per year in a country due to long-term $PM_{2.5}$ exposure if the population of the country older than 30 years old exposed to $PM_{2.5}$ pollution is 100 million. Assume they are exposed to 15 $\mu g/m^3$ $PM_{2.5}$ over multiple years. Assume also that the all-cause death rate is $y_0 = 1,000$ deaths per year per 100,000 population and the relative risk below the low threshold of 5.8 $\mu g/m^3$ is 1/4th that above the threshold.

# 8 Matching Electricity, Heat, Cold, and Hydrogen Demand Continuously with 100 Percent WWS Supply, Storage, and Demand Response

One of the greatest concerns facing the implementation of a 100 percent clean, renewable energy and storage system for all purposes worldwide is whether electricity, heat, cold, and hydrogen will be available when needed. In other words, can a 100 percent system avoid blackouts, which occur when the electric power grid fails because not enough electricity is available to meet demand at a given moment? Similarly, will a 100 percent system always have enough heat, cold, and hydrogen at the times needed?

The electric power grid in a 100 percent WWS world will be much different from the grid today. Today, electricity comprises about 20 percent of all end-use energy (or 40 percent of primary energy). However, in a 100 percent world, electricity will comprise close to 100 percent of all end-use energy, which itself will equal primary energy less transmission and distribution losses. The rest of end-use energy will come from direct geothermal or solar heat. The sectors that will be electrified (transportation, building heating and cooling, industry, agriculture/forestry/fishing, and the military) will use more energy-efficient technologies than their fossil-fuel counterparts. Such technologies include electric vehicles, hydrogen fuel cell vehicles, and heat pumps, among others. The reduction in energy use will help reduce electric power demand. Demand will also decrease because no more energy will be used to mine, transport, or process fossil fuels, biofuels, bioenergy, or uranium. End-use energy efficiency will increase, and measures will be put in place

to reduce energy use. A future electric grid will also be coupled with electricity, heat, cold, and hydrogen storage. Finally, a future grid will have more long-distance, high voltage direct current (HVDC) electricity transmission instead of fossil-fuel pipelines.

Electricity in the future grid will be used to produce heat, cold, hydrogen, and electricity. As such, **the main challenge in a future grid will be one of matching electricity, heat, cold, and hydrogen demand with 100 percent WWS electricity and heat supply plus storage, while using demand response.**

This chapter discusses matching such demand with supply, storage, while using demand response both on the short timescale (seconds to minutes) and long timescale (months to seasons to years). It starts by describing methods of meeting energy demand continuously with supply (Section 8.1). It then provides a case study example of how to calculate costs and match demand with supply over time (Section 8.2). It then quantifies the footprint and spacing areas needed for a worldwide transition to 100 percent WWS (Section 8.3) and the changes in jobs due to a transition (Section 8.4).

## 8.1 Methods of Meeting Energy Demand Continuously

Winds, waves, and sunlight produce power that varies over short timescales (seconds to minutes) as well as over long timescales (months to seasons to years) due to continuous variations in the weather and climate. Thus, these energy sources are referred to as **variable WWS resources**. Another term commonly used to describe variability is **intermittency**. However, all energy resources are intermittent due to scheduled and unscheduled maintenance, whereas variable resources are those whose energy outputs vary with the weather in addition to being affected by maintenance.

One concern with the use of variable WWS resources is whether such resources can provide electric power, heat, cold, and hydrogen when energy is needed. Any electricity system, in particular, must respond to changes in demand (load) over periods of seconds, minutes, hours, seasons, and years, and accommodate unanticipated changes in the availability of generation. It is not possible to control the weather; thus, a sudden change in demand often cannot be met by a variable WWS resource unless either storage, demand response, or long-distance transmission or all three are coupled with generation.

The concern about matching demand with supply, though, applies to all energy sources, not just to the variable ones. For example, because geothermal, tidal, coal, and nuclear power can provide relatively constant (**baseload**) supplies of electricity during the day, they do not match power demand, which varies significantly over short periods. In addition, all baseload supplies of electricity are down for maintenance anywhere from 3 to 12 percent of all hours in a year. As a result, gap-filling resources, such as natural gas and hydroelectricity, are currently used to meet peaks in demand in the presence of baseload resources.

Even nuclear power, which has been designed in some countries to ramp slowly, does not match load. In France, for example, nuclear power ramps by 1 to 5 percent per minute, but this is 20 to 100 times slower than hydropower, pumped hydropower storage, or a battery and 4 to 20 times slower than an open cycle gas turbine (see Table 2.1).

Although concern exists about stability of the current grid, it generally works, increasing the desire of grid operators to maintain the status quo. In several places worldwide today, wind energy alone has supplied between 50 and 110 percent of a state or country's power for a period without the lights going out.

For example, during the first six months of 2019, wind provided twice the electricity needed to power all the homes in Scotland (population, 5.4 million), which is a partially independent country that is part of the United Kingdom (Cockburn, 2019).

Similarly, during the first six months of 2019, the Coquimbo region of Chile (population, 742,000) produced 99.7 percent of its electricity from WWS, with 74.9, 20.8, and 4 percent coming from wind, solar PV, and hydropower, respectively (Renewables Now, 2019).

On May 5, 2019, renewables alone provided 96 percent of California's electric power from 3 to 3:30 PM. The breakdown of renewables was 48 percentage points from solar, 22 from hydropower, 19 from wind, and 5 from geothermal. During the same day, 91 percent of electricity in the state from 12:30 to 5:30 PM and 65 percent from the full day was from renewables. Although several environmental issues exist with hydropower, it provides both stored electric power and immediate peaking power (with a ramp rate of 100 percent per minute from Table 2.1). However, hydropower storage levels vary during the year because rainfall runoff and energy consumption vary during the year. As such, if hydropower is the only WWS resource available in a country, electricity shortages may arise from time to time. For example, although Tajikistan supplies far more hydropower electricity than the country needs in the annual average (exporting the difference), it has, until recently, had to require its citizens to reduce electricity use during winter due to low water levels in existing reservoirs. Tajikistan addressed this problem recently by building additional reservoirs.

## Transition highlight

Impressively, **by 2019, at least 9 countries had obtained 95 to 100 percent of their electric power from wind-water-solar resources.** These include Iceland, Norway, Costa Rica, Albania, Paraguay, Uruguay, Bhutan, Tajikistan, and Kenya. Table 8.1 shows the breakdown of WWS electricity generation in most of these countries. The dominant renewable in most of them is hydropower. In Kenya, it is geothermal. Paraguay and Uruguay produce >280 percent and 113 percent, respectively, of their electricity from WWS. The excess in both cases is exported to Argentina and Brazil. Bhutan produces ~300 percent of its electricity from WWS. The excess is exported to India.

Scotland is an example of a country with a high penetration of WWS (68.4 percent in 2018), but with wind (53.6 percent of all electricity) rather than hydropower as

Table 8.1 **Percent electric power consumed within selected countries that is generated by wind-water-solar (WWS) clean, renewable sources in the year for which the latest data were available. Also shown is the population of each country.**

| | Iceland 2018 | Norway 2018 | Costa Rica 2018 | Albania 2018 | Paraguay 2019 | Uruguay 2019 | Bhutan 2019 | Tajikistan 2018 | Scotland 2018 | Sweden 2017 |
|---|---|---|---|---|---|---|---|---|---|---|
| Pop (mil.) 2019 | 0.36 | 5.33 | 4.91 | 2.88 | 6.81 | 3.46 | 0.81 | 8.92 | 5.44 | 10.1 |
| Hydropower | 69.66 | 95.0 | 72.24 | 95 | >280 | 67.55 | >300 | >100 | 13.9 | 45.68 |
| Wind | 0.02 | 2.66 | 16.14 | 0 | 0 | 41.76 | 0.02 | 0 | 53.6 | 12.45 |
| Geothermal | 30.31 | 0 | 8.92 | 0 | 0 | 0 | 0 | 0 | 0 | 0.163 |
| Solar PV | 0 | 0 | 0.09 | 0 | 0 | 3.68 | 0 | 0 | 0.918 | 0 |
| Tidal/wave | 0 | 0 | 0 | 0 | 0 | 0 | 0 | 0 | 0.022 | 0 |
| Total | 99.99 | 97.7 | 97.4 | 95 | >280 | 113 | >300 | >100 | 68.4 | 58.29 |

its dominant source of electricity. Kenya (not shown in Table 8.1) has installed sufficient nameplate capacity in 2020 (47 percent geothermal, 11.7 percent wind, and 30 percent hydropower) to produce 90 to 100 percent of its electricity from WWS. In the United States, the states of North Dakota, Kansas, and Iowa similarly produced 53.5, 47.1, and 43.2 percent, respectively, of their electricity from wind plus solar during 2018 (Puiu, 2019). In all four locations, the grid remained stable.

There are several steps in designing an integrated energy system over a region in which electricity, heat, cold, and hydrogen demand are matched with 100 percent WWS electricity and heat supply plus storage, demand response, and transmission. These are discussed below. The overall strategy, though, is as follows:

- Electrify or use direct heat for all energy.
- Reduce energy use and implement energy efficiency measures.
- Generate all electricity and direct heat with WWS.
- Interconnect WWS electricity generators through the electric transmission and distribution system.
- Use WWS electricity to produce cold and low-temperature heat with heat pumps.
- Use WWS electricity to produce hydrogen with electrolyzers.
- Store electricity, heat, cold, or hydrogen that is not used immediately.
- Use demand response to shift peak electricity, heat, cold, or $H_2$ demand that can't be met to a later time.
- Use storage to satisfy unmet peaks in electricity, heat, cold, or $H_2$ demand.

This basic strategy is divided below into two categories. The first category (I) of steps relates to designing and creating the infrastructure needed to provide all-sector energy with WWS. The second category (II) relates to matching demand with supply using the infrastructure. The steps in Category I are as follows:

## I. Steps in Creating a 100 Percent WWS Infrastructure

A. Transition transportation to run on batteries and hydrogen fuel cells, where the hydrogen is produced from electricity (Section 2.2). The hydrogen that is produced can be stored and used in vehicles immediately or stored in stationary hydrogen storage tanks for later use.

B. Transition all building air and water heat to be provided by electric heat pumps (Section 2.3), direct solar and geothermal heat (Section 2.3), and district heat (Section 2.8.2). Also, ensure that district heat is produced by electric heat pumps, direct solar heat, and/or direct geothermal heat. Either use district heat immediately or store it in water tanks (Section 2.8.1) or underground. The types of underground storage include borehole thermal energy storage (BTES, Section 2.8.3.1), pit thermal energy storage (PTES, Section 2.8.3.2), and aquifer thermal energy storage (ATES, Section 2.8.3.3).

C. Transition building cooling to run on the same heat pumps as used for building heating (Section 2.3) or to run on a district cooling loop that is part of a fourth-generation or higher district heating system (Section 2.8.2). The district cooling system should include storage of cold in water tanks or in ice.

D. Electrify high-, medium-, and low-temperature heat used in industry with electric arc furnaces (Section 2.4.1), induction furnaces (Section 2.4.2), resistance furnaces (Section 2.4.3), dielectric heaters (Section 2.4.4), electron beam heaters (Section 2.4.5), concentrated solar power steam (Section 2.4.6), and/or heat pump steam (Section 2.4.6).

E. Transition all remaining combustion appliances, machines, and processes to electric ones. These include moving to electric induction cookers (Section 2.5.1), electric fireplaces (Section 2.5.2), electric leaf blowers (Section 2.5.3), and electric lawnmowers (Section 2.5.4), for example.

F. Increase energy efficiency in buildings by moving to LED lights, using more energy-efficient appliances, reducing air leaks in buildings, and increasing insulation in walls and windows (Section 2.6). Improve passive heating and cooling techniques in buildings, such as using thermal mass to store heat and cold, using ventilated façades, using window blinds and films, and using night ventilation (Section 2.8.4).

G. Provide all the electricity for a WWS world with onshore and offshore wind, solar PV on rooftops and in power plants, concentrated solar power, geothermal power, tidal power, wave power, and hydroelectric power.

H. Store WWS electricity in either storage associated with concentrated solar power, pumped hydropower storage, stationary or vehicle batteries, flywheels, compressed air energy storage, and gravitational storage with solid masses.

I. Build sufficient short- and long-distance HVAC transmission, long-distance HVDC transmission, and AC distribution to interconnect WWS supply and demand centers.

Figure 8.1 summarizes the main components of a 100 percent WWS system. It includes WWS electricity and heat generation, storage, and end use along with transmission. The second category of steps relates to matching instantaneous electricity, heat, cold, and hydrogen demand with WWS electricity and heat supply, storage, and demand response. **Demand response**, or **demand response management (DRM)**, is the use of financial incentives to shift the time of a flexible demand (load) to a future time, when more energy is available to satisfy the demand. The incentive can be either a lower price for electricity during times of low demand versus high demand or a direct payment to the customer in exchange for the customer agreeing not to use electricity during the time of peak demand. The demand that is shifted can be either an electricity, heat, cold, or hydrogen demand.

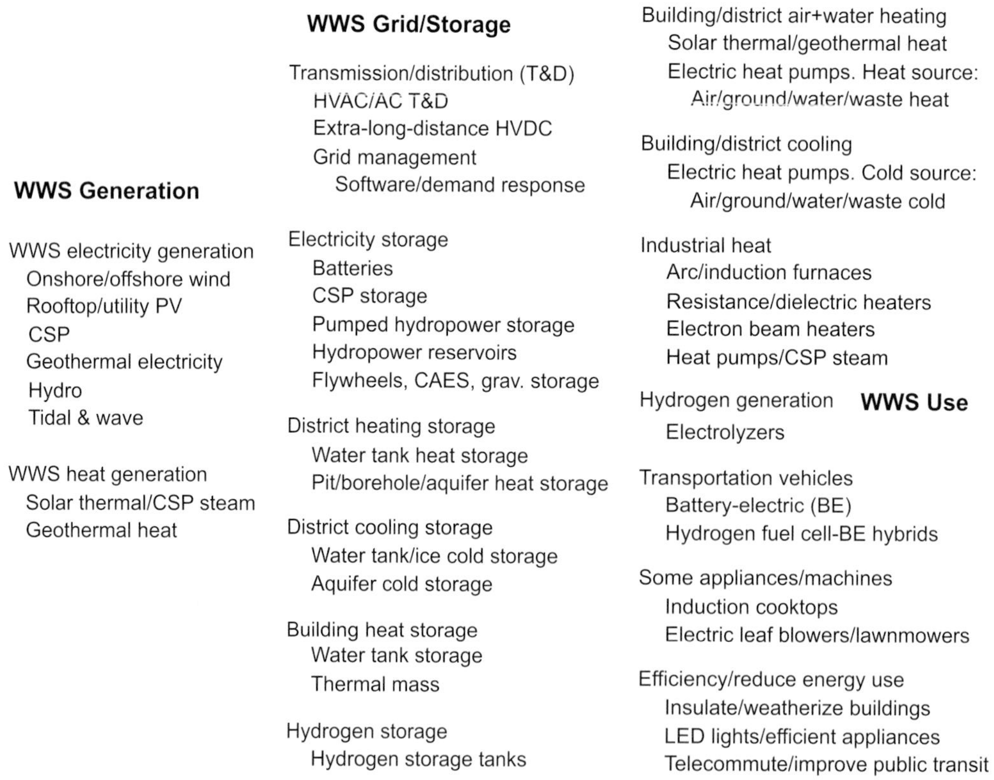

**WWS Grid/Storage**

Transmission/distribution (T&D)
   HVAC/AC T&D
   Extra-long-distance HVDC
   Grid management
      Software/demand response

Electricity storage
   Batteries
   CSP storage
   Pumped hydropower storage
   Hydropower reservoirs
   Flywheels, CAES, grav. storage

District heating storage
   Water tank heat storage
   Pit/borehole/aquifer heat storage

District cooling storage
   Water tank/ice cold storage
   Aquifer cold storage

Building heat storage
   Water tank storage
   Thermal mass

Hydrogen storage
   Hydrogen storage tanks

**WWS Generation**

WWS electricity generation
   Onshore/offshore wind
   Rooftop/utility PV
   CSP
   Geothermal electricity
   Hydro
   Tidal & wave

WWS heat generation
   Solar thermal/CSP steam
   Geothermal heat

Building/district air+water heating
   Solar thermal/geothermal heat
   Electric heat pumps. Heat source:
      Air/ground/water/waste heat

Building/district cooling
   Electric heat pumps. Cold source:
      Air/ground/water/waste cold

Industrial heat
   Arc/induction furnaces
   Resistance/dielectric heaters
   Electron beam heaters
   Heat pumps/CSP steam

Hydrogen generation   **WWS Use**
   Electrolyzers

Transportation vehicles
   Battery-electric (BE)
   Hydrogen fuel cell-BE hybrids

Some appliances/machines
   Induction cooktops
   Electric leaf blowers/lawnmowers

Efficiency/reduce energy use
   Insulate/weatherize buildings
   LED lights/efficient appliances
   Telecommute/improve public transit

**Figure 8.1 Main generation, transmission, storage, and use components of a 100 percent WWS system to power the world for all purposes.**

**Flexible demands or loads** are demands for electricity, heat, cold, or hydrogen that can be shifted forward in time during a day or over several days by demand response.

For instance, charging an electric vehicle is a flexible load because the time of charging can be shifted to a non-peak time of grid electricity use through a financial incentive, such as a lower nighttime electricity rate. In fact, any electricity load that can be shifted in time is a flexible load. Examples are electricity used for (a) a wastewater treatment plant, (b) charging a stationary battery, (c) pumping water to an upper reservoir in a pumped hydropower system, (d) storing energy in a flywheel, (e) heating water in a hot water tank with a heat pump, (f) producing ice for later use, or (g) producing hydrogen for later use. Similarly, electricity demand for high-temperature industrial heat is a flexible load if the heating can be scheduled during times of a day that are not times of peak electricity demand.

Thus, for example, a utility may establish an agreement with a flexible load wastewater treatment plant for the plant to use electricity only during certain hours of the day in exchange for a better electricity rate. In this way, the utility can shift the time of demand to a time when more supply is available.

Similarly, the demand for electricity to charge BE vehicles is a flexible load because such vehicles are generally charged at night, and it is not critical which hours of the night the electricity is supplied so long as the full power is provided sometime during the night. In this case, a utility can use a smart meter to provide electricity for the BE vehicle when WWS availability is high and reduce the power supplied when it is low. Utility customers would sign up their BE vehicles under a plan by which the utility controlled the nighttime (primarily) or daytime supply of power to the vehicles.

**Inflexible loads**, on the other hand, are electricity demands that cannot be shifted in time, such as electricity use for computers, lighting, and electric induction cooktops.

Whereas the electricity stored in an onboard battery used to drive an electric vehicle or to power an electric leaf blower is an inflexible load, the electricity used to charge the batteries in both cases is a flexible load. Similarly, whereas low-temperature heat and cold demands for building heating and cooling are inflexible loads, the electricity or direct heat used to provide heat or cold for the storage media – water tanks, borehole fields, water pits, and aquifers – are flexible loads. Finally, whereas the use of hydrogen in a hydrogen fuel cell vehicle is an inflexible load, the electricity used to produce hydrogen for a stationary hydrogen storage tank is a flexible load.

The steps for meeting demand in Category II are as follows:

### II. Steps for Matching Flexible and Inflexible Demand with WWS Supply, Storage, and Demand Response

A. Interconnect, through the transmission system, geographically dispersed variable wind, wave, and solar resources to turn some of their variable supply into more steady supply while reducing overall transmission requirements. Also, interconnect geographically dispersed or co-located wind and solar together to take advantage of their complementary nature, thereby reducing overall variability.

B. Determine the annual average end-use demand in each energy sector in the year of interest. Size WWS electricity and heat generation to meet the annual average all-sector end-use demand in each region.

C. Estimate the additional generation, storage, and demand response needed to meet demand continuously over time rather than just in the annual average. Do this by first quantifying (1) electric and direct heat demands for low-temperature heating that must be supplied; (2) electric demands for cooling and refrigeration that must be supplied; (3) electric demands for producing, compressing, and storing hydrogen for fuel cells used in transportation that must be supplied; and (4) all other electric demands (including industrial heat demands). Of the demands in each of these categories, identify which ones are inflexible and which ones are flexible, thus subject to demand response.

D. Once the generation, storage, and demand response system is designed, carry out the following measures to match instantaneous demand with supply.

E. When the current WWS electricity or direct (solar and geothermal) heat supply exceeds the current inflexible electricity or heat demand, use the supply to satisfy the demand.

F. Use the remaining instantaneous WWS electricity or heat supply to satisfy as much existing flexible electric or heat demand as possible.

G. Use any excess electricity after that to fill electricity, heat, cold, or hydrogen storage. Prioritize which type of storage is filled first.

H. Use any excess geothermal or solar heat after satisfying inflexible and flexible heat demand to fill heat storage.

I. When the current inflexible plus flexible electricity load exceeds the current WWS electricity supply from the grid, use electricity storage and electricity transmitted in from far away to fill in the gap in supply. Use the electricity to supply inflexible loads first.

J. If electricity storage is depleted and flexible load remains, use demand response to shift the flexible load to a future hour.

K. If inflexible plus flexible building heat demand exceeds current supply, use stored heat. If stored heat is exhausted, shift flexible heat demands to a future hour with demand response.

L. If inflexible plus flexible building cold demand exceeds current supply, use stored cold. If stored cold is exhausted, shift flexible cold demands to a future hour with demand response.

M. Provide hydrogen demands with current hydrogen supply, where the current hydrogen is produced by current electricity. If current hydrogen supply is exhausted, use stored hydrogen.

N. If the measures above are insufficient to match power demand with supply continuously, either (a) increase the nameplate capacity of the electric power generators; (b) increase electricity, heat, cold, or hydrogen storage nameplate capacity; (c) increase the nameplate capacity of the transmission system to allow more renewables that are far away, thus subject to different weather patterns, help fill in gaps in supply; (d) use electricity stored in vehicle batteries to help fill in gaps in supply on the grid; and/or (e) integrate weather forecasts into system operations to improve the efficiency of the grid.

Below, these steps are discussed in more detail.

## 8.1.1 Interconnecting Geographically Dispersed Generators

Interconnecting geographically dispersed wind, wave, or solar farms to a common transmission grid smoothens out electricity supply significantly (Kahn, 1979). For wind alone, interconnection over regions as small as a few hundred kilometers apart can eliminate hours of zero power, accumulated over all wind farms.

Figure 8.2 shows the impact of interconnecting three and eight geographically dispersed wind sites versus not interconnecting at one site. All locations are within a 550 km × 700 km region in the Great Plains of the United States. The figure shows that, with just one location, low wind speeds resulted in zero power output from wind turbines (with a 3 m/s cut-in wind speed) 7.6 percent of the hours during the year between 12:00 and 15:00 in the

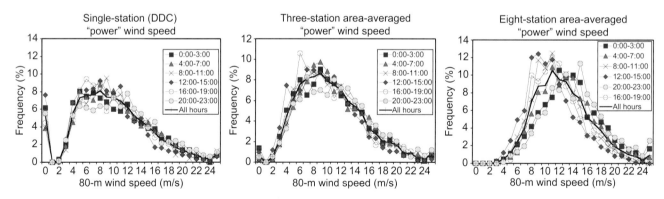

Figure 8.2 Frequency distribution of the "power" wind speed 80 m above ground level at different times of day (given in 4-hour blocks) integrated over one, three, and eight wind sites. The one-site station was DDC (a three-letter acronym) in Kansas. The three-site stations were DDC, RSL, and GCK, over a 160 km × 120 km area in Kansas. The eight-site stations were those plus AMA, GDP, CSM, HBR, and CAO, over a 550 km × 700 km area in Kansas, New Mexico, Texas, and Oklahoma. The "power" wind speed is obtained by determining the power at a given wind speed at each station, summing and averaging the resulting power over all stations in each area considered, then back-calculating the wind speed that gives that average power. The 80-m wind speeds are determined by extrapolating 10-m hourly wind speeds at the stations vertically using interpolated vertical profiles from data gathered from several sounding stations surrounding each surface site.
From Archer and Jacobson (2003).

afternoon. However, with three interconnected wind sites, the number of zero-power hours during that time decreased to 2.6 percent. With eight wind sites, the number decreased to 0 percent. In other words, **interconnecting geographically dispersed wind farms eliminates times of the year when no wind power output occurs over a region** (Archer and Jacobson, 2003).

Interconnecting farms also narrows the frequency distribution of wind during the year, turning a Rayleigh distribution (Section 6.6.2) to a narrower Gaussian distribution. This narrowing means that wind, aggregated geographically, acts more like a single power plant with steadier, closer-to-baseload power production. In fact, when 19 geographically dispersed wind sites in the Midwest, over a region 850 km × 850 km, were hypothetically interconnected, between 33 and 47 percent of yearly averaged wind power was calculated to be available at the same baseload reliability as that of a coal-fired power plant (Archer and Jacobson, 2007). In addition, the amount of power guaranteed over the year by having the wind farms dispersed over 19 sites was four times greater than the amount of power guaranteed by having one wind farm. There was no saturation of the benefits. In other words, the benefits of interconnecting increased with more and more interconnected sites, although with diminishing returns.

Co-locating a wind and a wave farm has similar impacts as interconnecting geographically dispersed wind farms. Namely, it reduces the impact of local power variations (Stoutenburg et al., 2010) and reduces transmission requirements (Stoutenburg and Jacobson, 2011). Figure 8.3, for example, shows that combining wind and wave power at the same location results in lower variability and fewer hours of zero power output than a wind farm or wave farm acting alone.

The idea of interconnecting geographically dispersed wind and co-located wind and wave power extends to interconnecting geographically dispersed solar farms or co-locating wind and solar farms. **Wind and solar, for example, are complementary in nature.** During the day, when the wind is not blowing, the sun is often shining and vice versa. This occurs physically because, in a high-pressure system, descending air evaporates clouds, increasing sunlight to the surface. In a high-pressure system, pressure gradients are also generally weak, causing winds to be slow. Conversely, in a low-pressure system, rising air increases cloudiness, reducing sunlight penetrating to the surface. Pressure gradients in low-pressure systems are generally strong, causing winds to be fast (Section 6.8.2.4). Thus, co-locating wind and solar reduces the variability of the overall power output of wind alone or solar alone.

## Transition highlight
One more benefit of interconnecting is that it reduces total transmission requirements with little power loss. For example, connecting the 19 wind farms just described to a common point and then connecting that point to a far-away city allows the long-distance portion of transmission nameplate capacity to be reduced by, for example, 20 percent with only a 1.6 percent loss of energy (Archer and Jacobson, 2007).

Figure 8.4 illustrates how the combined use of wind (variable), solar rooftop PV (variable), CSP (solar thermal) with storage (variable), geothermal (baseload), and hydroelectric (dispatchable) can be used together to match hourly power demand, while accounting for transmission and distribution losses. It shows computer model results

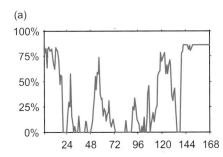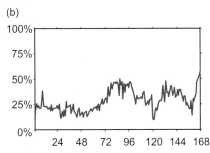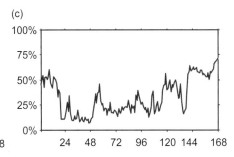

**Figure 8.3 Hourly capacity factors (hourly electric power production as a percentage of nameplate capacity) for 7 days of (a) wind turbines running alone, (b) wave devices running alone, and (c) a mix of 50 percent wind and 50 percent wave.** From Stoutenburg et al. (2010).

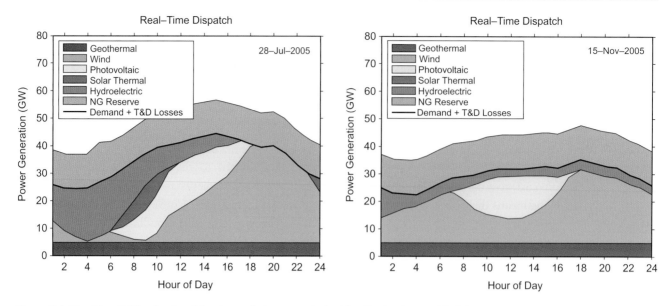

**Figure 8.4** Matching California electricity demand plus transmission/distribution losses (black line) with 100 percent clean, renewable supply based on a least-cost optimization model calculation for two days in 2005. Natural gas was available for backup but was not needed during these days. From Hart and Jacobson (2011).

System nameplate capacities: wind, 73.5 GW; CSP (solar thermal) with 3 hours' of storage, 26.4 GW; photovoltaics, 28.2 GW; geothermal, 4.8 GW; hydroelectric, 20.8 GW; and natural gas (NG) reserves, 24.8 GW. Transmission and distribution losses were 7 percent of the demand. The least-cost optimization accounted for the day-ahead forecast of hourly resources, carbon emissions, wind curtailment, and thermal storage at CSP facilities, and it allowed for the nighttime production of electricity from CSP storage. The hydroelectric supply was based on historical reservoir discharge data and 2005 imported generation from the Pacific Northwest. The wind and solar supplies were obtained by aggregating hourly wind and solar power at several sites in California estimated from wind speed and solar irradiance data for those hours applied to a specific turbine power curve, a specific concentrated solar plant configuration (parabolic trough collectors on single-axis trackers), and specific rooftop PV characteristics. California's developable geothermal resources limited the geothermal supply.

for two days in California in 2005. The geothermal power installed was increased over 2005 levels but was limited by California's geothermal resources. The daily hydroelectric generation was determined by estimating the historical generation on those days from reservoir discharge data. Wind and solar capacities were increased substantially over 2005 levels, but they did not exceed maximum levels determined by prior land and resource availability studies.

Figure 8.4 illustrates the potential for matching hourly power demand with WWS supply on a contemporary electricity grid. Although results for only two days are shown, results for all hours of all days of 2005 and 2006 (730 days total) found that 99.8 percent of delivered energy during those days could be produced carbon free from WWS technology in the absence of most storage (Hart and Jacobson, 2011). The only sources of storage were water stored in hydropower reservoirs and 3 hours of CSP storage. Transportation, building heat, and industrial heat were not electrified. In addition, no demand response or hydrogen was used.

## 8.1.2 Determining Annual Average Demands and Sizing WWS Generation to Meet Them

The second step in the process of matching demand with supply is to estimate future (e.g., 2050) all-sector end-use annual average demands, then to propose a mix of WWS electricity and heat generators to match those demands in the annual average. Table 7.6 provides an estimate of 2050 annual average end-use all-sector demand for 143 countries. The table also proposes one mix of WWS generators for each country to meet each country's demand. Dividing the annual average demand supplied by each generator in each country by the average capacity factor (fraction of nameplate capacity that is actually produced during the year) of the generator in the country gives an estimate of the nameplate capacity of each generator needed (Section 7.5). Summing the nameplate capacities of each generator among all countries gives the 143-country nameplate capacity of each generator, shown in column (c) of Table 7.7. All generators taken together in that column of the table can theoretically power these 143 countries for all purposes in the annual average in 2050.

Whereas the nameplate capacities given in column (c) of Table 7.7 are sufficient to match annual average all-purpose power demand, they are not sufficient to meet demand every minute. Demand varies continuously (e.g., Figure 8.5) as does electricity production from variable WWS generators. Whereas WWS generators provide more than enough electricity during some minutes of a year, they provide less electricity than needed during others. As such, additional methods must be used to match demand with supply. One method is oversizing the nameplate capacity of generators (Section 8.1.5.1). A second is using storage. A third is using demand response (Section 8.1.4.2). Interconnecting geographically dispersed renewables also helps (Section 8.1.1).

Oversizing the nameplate capacities of variable electric power generators while simultaneously interconnecting them on the grid (Section 8.1.1) increases the minimum power output available at any given instant on the grid.

This helps to ensure (but does not guarantee) that the highest peak demand (about 700 GW in Figure 8.5) will always be met because it increases the chance that the minimum electricity supplied always exceeds the peak demand, regardless of when that demand occurs during the year.

However, in order to guarantee demand will be met every minute of a year without any storage or demand response, it may be necessary to oversize the nameplate capacity of WWS generation by a factor of 5 to 10 or more. This, of course, results in an exorbitant cost of energy, particularly because, in the annual average, the WWS generators will produce 5 to 10 times more electricity than is needed, solely to ensure that the peak demand that occurs (the largest demand in the bottom part of Figure 8.5) is met. The excess electricity will be curtailed (shed), thus wasted, because no electricity, heat, cold, or hydrogen storage is included in this scenario to put the excess electricity into. When the electricity is wasted, the cost per kWh produced goes up, in this case, by a factor of 5 to 10.

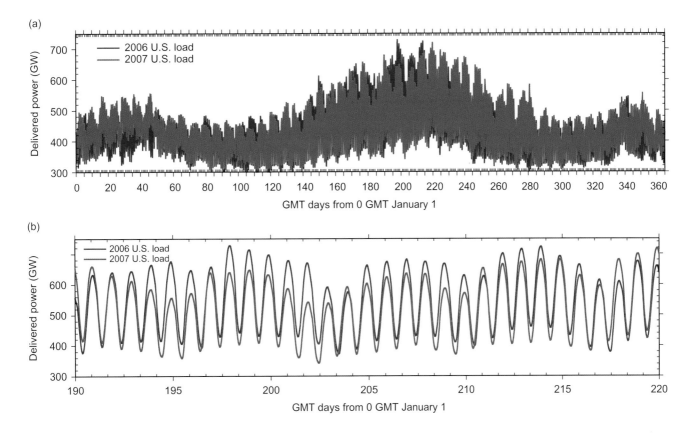

Figure 8.5 (a) Time-dependent (GMT time) U.S. end-use electric loads (demands) by hour in 2006 and 2007. The average load in 2006 was 437.51 GW and that in 2007 was 450.14 GW. (b) Same as (a), but for a peak 20-day period.
From Corcoran et al. (2012) and Jacobson et al. (2015b).

### 8.1.3 Sizing Additional Generation, Storage, and Demand Response

No future proposal calls for solely oversizing nameplate capacity to meet peak demand. Future proposals call for combining some oversizing with storage and demand response. The first step in sizing the additional generation, storage, and demand response needed to meet demands continuously (e.g., minute-by-minute) over time is to separate annual average total demand into (1) electricity and heat loads needed for low-temperature heating; (2) electric loads needed for cooling and refrigeration; (3) electricity loads needed to produce, compress, and store hydrogen for fuel cells used for transportation; and (4) all other electricity loads (including industrial heat loads).

Each of these loads is then divided further into flexible and inflexible loads. Flexible loads include electricity and heat loads that can be used to fill cold and low-temperature heat storage, all electricity used to produce hydrogen (since all hydrogen can be stored), and remaining electricity and heat loads subject to demand response. Inflexible loads are all loads that are not flexible. The flexible loads may be shifted forward in time with demand response. The inflexible loads must be met immediately.

#### 8.1.3.1 Estimating Heat, Cold, Hydrogen, and Electricity Loads

Electricity and heat needed for low-temperature building air and water heat are collectively referred to as **low-temperature heat loads**, or just **heat loads**. Ideally time-dependent heat load data are available from a recent year for the grid region or country being analyzed. If time-dependent data are not available, then they can be approximated as follows: First, the annual average low-temperature heat load (GW) across all energy sectors in a given region is

$$L_{heat} = L_{heat,r} + L_{heat,c} + L_{heat,i} \qquad (8.1)$$

where

$$L_{heat,r} = (F_{ah,r} + F_{wh,r})L_r \qquad (8.2)$$

$$L_{heat,c} = (F_{ah,c} + F_{wh,c})L_c \qquad (8.3)$$

$$L_{heat,i} = F_{ah,i}L_i \qquad (8.4)$$

are the low-temperature heat loads in the residential, commercial, and industrial sectors, respectively. In these equations, $F_{ah,r}$ and $F_{wh,r}$ are the fractions of the residential heat load ($L_r$) that are for air and water heating, respectively; $F_{ah,c}$ and $F_{wh,c}$ are the fractions of the commercial heat load ($L_c$) that are for air and water heating, respectively; and $F_{ah,i}$ is the fraction of the industrial load ($L_i$) that is for low-temperature air heating. The last parameter is estimated with

$$F_{ah,i} = F_{hvac}H/(C + H) \qquad (8.5)$$

where $F_{hvac}$ is the fraction of total industrial load that is for the sum of air heating, air cooling, and refrigeration in the industrial sector (0.0624 from U.S. data, Jacobson et al., 2018a); and H and C are the average number of heating and cooling degree days, respectively, in a year.

**Heating degree days** (HDDs) are the number of degrees that the outside air temperature must be raised to reach an indoor comfort-level reference temperature, summed over all days of a month or year. HDDs are a proxy for the heating requirement of a building. HDDs are more specifically calculated as the number of outdoor air temperature degrees below a reference temperature in a day, summed over all days during some period. If the air temperature is above the reference temperature for 24 hours of a day, the number of HDDs is zero for that day. The reference temperature varies depending on the country, but it is typically 18.33 °C (65 °F). Whereas the number of HDDs is a good proxy for air heating requirements, it is less accurate for water heating requirements. For example, even if the number of HDDs is zero for a day, a building may still need hot water for dishwashing, clothes washing, showers, cooking, and cleaning.

**Cooling degree days** (CDDs) are a proxy for the cooling requirements of a building. They are the number of degrees that the outside air temperature must be cooled to reach the same reference temperature as for HDDs, summed over all days during a month or year. If the outdoor temperature for 24 hours of a given day is below the reference temperature, then the number of CDDs on that day is zero. Whereas the number of CDDs is a good proxy for air cooling requirements, it does not help so much for refrigeration requirements in a home or building.

Figure 8.6 shows the average number of HDDs and CDDs per year for the 24 world regions listed in Table 8.5. Countries or regions at higher latitudes experience more HDDs, thus heating requirements, than do tropical (low-latitude) countries. Conversely, countries or regions at lower latitudes experience more CDDs, thus cooling requirements, than do higher-latitude countries.

The number of CDDs for Iceland, summed over 2013 and 2014, was 0 °C, and the number of HDDs was

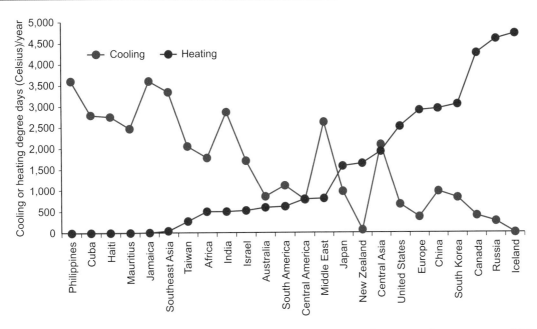

**Figure 8.6** Number of cooling degree days (CDDs) and heating degree days (HDDs), in °C, averaged over two years (either 2013 and 2014 or 2017 and 2018) for 24 world regions, each defined in Table 8.5. The reference temperature was 18.33 °C (65 °F). From Bizee (2019). For individual countries, the values are from one location in the country. For grid regions, they are a weighted average of all countries in the region, where the weighting is based on the end-use power demand in the country.

4,720 °C, suggesting significant air heating requirements but hardly any air conditioning requirements. On the other hand, the numbers of HDD and CDD days in the Haiti region were 2,754 °C and 1 °C, respectively, suggesting significant air conditioning requirements but virtually no air heating requirements.

---

## Example 8.1  Calculating heating and cooling degree days

Calculate the number of heating degree days and cooling degree days over a four-day period relative to a reference temperature of 18 °C if the outdoor air temperatures on each day are 10 °C, 16 °C, 20 °C, and 25 °C, respectively.

### Solution:

HDDs are calculated only for outdoor air temperatures less than 18 °C; CDDs are calculated for the remaining days. As such, the number of HDDs = (18 °C – 10 °C) + (18 °C – 16 °C) = 10 °C and the number of CDDs = (20 °C –18 °C) + (25 °C –18 °C) = 9 °C.

---

The fraction of total load in a sector that is low-temperature heat load is

$$F_h = F_{ah} + F_{wh} \qquad (8.6)$$

Table 8.2 gives values of $F_h$, which apply to both the residential and commercial sectors, for several countries and world regions. The table indicates that, worldwide, about 79 percent of all energy consumed in residential and commercial buildings is used for air and water heating. If total residential or commercial energy demand is known in a region, it can be multiplied, as a first estimate, by $F_h$ from Table 8.2 to obtain an estimate of the low-temperature (air plus water) heat load in buildings.

Based on U.S. data, the fraction of residential load needed for water heating is related to that needed for air heating by $F_{wh,r} = 0.4265F_{ah,r}$. Combining this relationship with Equation 8.6 gives $F_{ah,r} = 0.701F_{h,r}$, and $F_{wh,r} = 0.299F_{h,r}$. In the commercial sector, $F_{wh,c} = 0.2118 \times F_{ah,c}$, giving $F_{ah,c} = 0.8252 \times F_{h,c}$ and $F_{wh,c} = 0.1748 \times$

Table 8.2 Fraction of 2010 annual average residential or commercial total energy (electricity plus heat) load that is heat load (the rest is electricity load) and fraction of annual average industrial sector total energy load that is high-temperature heat load (the rest is electricity plus low-temperature heat load), by country or region. Heat load includes load for both air and water heating.

| Country or region | Fraction of total load in the residential or commercial sector that is low-temperature heat load ($F_h$) | Fraction of total load in the industrial sector that is high-temperature heat load ($F_{ht}$) |
|---|---|---|
| Asia other | 0.816 | 0.643 |
| Australia | 0.649 | 0.623 |
| Brazil | 0.660 | 0.658 |
| Canada | 0.723 | 0.640 |
| China | 0.857 | 0.637 |
| Russia | 0.881 | 0.721 |
| France | 0.757 | 0.578 |
| Germany | 0.804 | 0.588 |
| India | 0.856 | 0.705 |
| Italy | 0.816 | 0.581 |
| Japan | 0.665 | 0.594 |
| LAM other | 0.756 | 0.623 |
| MEA other | 0.743 | 0.709 |
| Nigeria | 0.963 | 0.836 |
| OECD other | 0.748 | 0.590 |
| Poland | 0.865 | 0.666 |
| RE other | 0.811 | 0.634 |
| South Africa | 0.746 | 0.503 |
| United Kingdom | 0.805 | 0.588 |
| United States | 0.689 | 0.643 |
| World average | 0.787 | 0.647 |

Asia other = Asia other than China and India; LAM other = Latin America other than Brazil; MEA other = Middle East and Africa other than South Africa and Nigeria; OECD other = countries in the Organisation for Economic Co-operation and Development other than Australia, France, Germany, Japan, Italy, United Kingdom, United States; RE other = reforming economies in Eastern Europe and the former Soviet Union other than Poland and Russia. Source: Jacobson et al. (2018a), derived from data in De Stercke (2014).

$F_{h,c}$ (Jacobson et al., 2018a), where $F_{h,r} = F_{h,c} = F_h$ (given in Table 8.2).

Table 8.2 also gives the fraction ($F_{ht}$) of total industrial load ($L_i$) that is high-temperature industrial heat load. The high-temperature heat load, $L_{temp,i}$, is thus

$$L_{htemp,i} = F_{ht}L_i \tag{8.7}$$

Table 8.2 indicates that worldwide, about 65 percent of industrial energy is used for high-temperature heat. The rest is used for low-temperature air heat, air conditioning, refrigeration, transportation, and normal electricity.

Table 8.3 provides values for $F_{ah}$, $F_{wh}$, and $F_{ht}$ for the residential, commercial, and/or industrial sectors for 24 world regions.

**Cold loads** are loads in each sector for air conditioning and commercial refrigeration. The total cold load across all energy sectors is

$$L_{cold} = L_{cold,r} + L_{cold,c} + L_{cold,i} \tag{8.8}$$

where

$$L_{cold,r} = F_{ac,r}L_r \tag{8.9}$$

$$L_{cold,c} = (F_{ac,c} + F_{rf,c})L_c \qquad (8.10)$$

$$L_{cold,i} = (F_{ac,i} + F_{rf,i})L_i \qquad (8.11)$$

In these equations, $F_{ac,r}$, $F_{ac,c}$, and $F_{ac,i}$ are the fractions of the total residential, commercial, and industrial loads, respectively, that are for air conditioning; and $F_{rf,c}$ and $F_{rf,i}$ are the fractions of the commercial and industrial loads, respectively, that are for refrigeration.

The fractions of total load in the residential and commercial sectors that are for air conditioning are estimated to be the smaller of the air heating load multiplied by the ratio of cooling to heating degree days and a maximum allowable fraction of building electric load used for air cooling:

$$F_{ac,r} = \min(F_{ah,r}C/H, F_{e,r}F_{max}) \qquad (8.12)$$

$$F_{ac,c} = \min(F_{ah,c}C/H, F_{e,c}F_{max}) \qquad (8.13)$$

$$F_{ac,i} = \min(F_{ah,i}C/H, F_{e,i}F_{max}) \qquad (8.14)$$

respectively, where $F_{e,r} = 1 - F_{h,r}$, $F_{e,c} = 1 - F_{h,c}$, and $F_{e,I} = 1 - F_{ht}$ are the fractions of total load in the residential and commercial sectors, respectively, that are non-high-temperature electric loads ($F_{h,r}$ and $F_{h,c}$ are the fractions of total load in the residential and commercial sectors, respectively, that are low-temperature heat loads, and $F_{ht}$ is the fraction of total industrial sector load that is high-temperature heat load from Table 8.2). Finally, $F_{max}$ is the maximum allowable fraction of building electric load that is for air conditioning (set to 0.4) (Jacobson et al., 2018a).

The fraction of total load in the commercial sector that is refrigeration load is then estimated as

$$F_{rf,c} = 0.7383 F_{ac,c} \qquad (8.15)$$

Lastly, $F_{rf,i}$ is assumed to be 0.024 for all world regions based on U.S. data (Jacobson et al., 2018a). Table 8.3 provides 2050 estimates of $F_{ac}$ and $F_{rf}$ for the residential, commercial, and industrial sectors in 24 world regions (defined in Table 8.5).

### 8.1.3.2 Estimating Loads Subject to Storage and Demand Response

The next step is to determine the loads subject to heat, cold, hydrogen, and electricity storage; the load subject to demand response; and the inflexible load.

The **load subject to heat storage, which can be charged flexibly,** is

$$L_{heat,stor} = F_{dh}L_{heat} + (F_{flx,wh} - F_{dh})(F_{wh,r}L_r + F_{wh,c}L_c) \quad (8.16)$$

where $F_{dh}$ is the fraction of all low-temperature heat and cold load in each region that is provided by district heating and cooling. Table 8.3 gives values for 24 world regions. The **district heating and cooling fraction** is important, because cold and low-temperature heat energy provided by district heating and cooling is stored in water tanks, borehole fields, water pits, aquifers, or ice. Stored heat or cold is always produced hours to months before it is used, so the electricity or heat charging the storage is a flexible load. Thus, for example, an electric heat pump may produce heat or cold for storage whenever excess electricity is available on the grid. Conversely, if an electricity shortage occurs on the grid, heat pumps can be shut off, freeing electricity for the rest of the grid.

In Equation 8.16, $F_{flx,wh} = 0.95$ is the fraction of water heating that occurs in water storage tanks and that can be charged flexibly. This term accounts for the fact that building hot water tanks that are *not* in district heating systems can be charged flexibly. Hot water tanks that are part of district heating systems are accounted for in the first term on the right side of Equation 8.16.

The **load subject to cold storage, all of which can be charged flexibly,** is

$$L_{cold,\ stor} = F_{dh}L_{cold} \qquad (8.17)$$

Thus, only cold loads subject to district cooling can be charged flexibly. Such cold loads are stored in water, ice, and aquifers, all of which can be charged flexibly.

Hydrogen will be used primarily for transportation. The **load subject to hydrogen storage** is

$$L_{H2,stor} = F_{H2}L_t \qquad (8.18)$$

Where $F_{H2}$ is the fraction of the transportation load ($L_t$) needed to produce, compress, and store hydrogen. Table 8.2 provides estimates of $F_{H2}$ for 24 world regions. Loads for producing hydrogen are all flexible because all hydrogen produced can be stored before use so long as hydrogen storage is sized correctly. As such, loads for producing hydrogen are subject to demand response. However, hydrogen is separated from other loads subject to demand response since the quantity of stored hydrogen must be tracked.

**Loads, aside from hydrogen loads, subject to demand response** are estimated with

$$L_{DR} = L_{DR,r} + L_{DR,c} + L_{DR,i} + L_{DR,t} + L_{DR,a} + L_{DR,o} \qquad (8.19)$$

Table 8.3 Parameters for estimating thermal energy use in different world regions. $F_{dh}$ is the fraction of total 2050 air heating, water heating, air conditioning, plus refrigeration load that is subject to district heating, thus thermal energy storage. $F_{H2}$ is the fraction of total 2050 all-sector end-use demand (from Table 7.6) needed to produce, compress, and store hydrogen for transportation. The average across all regions is 6.01 percent, which represents 37.1 percent of the transportation load. The remaining values are the fractions of either residential, commercial, or industrial 2050 load (from Table 7.6) that are required for either air heating ($F_{ah}$), water heating ($F_{wh}$), air cooling ($F_{ac}$), refrigeration ($F_{rf}$), or high-temperature industrial processes ($F_{ht}$) in each region.

| Region | $F_{dh}$ | $F_{H2}$ | Residential | | | Commercial | | | | | Industrial | | |
| --- | --- | --- | --- | --- | --- | --- | --- | --- | --- | --- | --- | --- | --- |
| | | | $F_{ah}$ | $F_{wh}$ | $F_{ac}$ | $F_{ah}$ | $F_{wh}$ | $F_{ac}$ | $F_{rf}$ | $F_{ht}$ | $F_{ah}$ | $F_{ac}$ | $F_{rf}$ |
| Africa | 0.1 | 0.084 | 0.56 | 0.24 | 0.08 | 0.63 | 0.13 | 0.09 | 0.07 | 0.66 | 0.014 | 0.049 | 0.024 |
| Australia | 0.1 | 0.073 | 0.46 | 0.19 | 0.14 | 0.54 | 0.11 | 0.14 | 0.10 | 0.62 | 0.026 | 0.037 | 0.024 |
| Canada | 0.2 | 0.056 | 0.51 | 0.22 | 0.01 | 0.60 | 0.13 | 0.01 | 0.01 | 0.64 | 0.061 | 0.001 | 0.024 |
| Central America | 0.1 | 0.102 | 0.53 | 0.23 | 0.10 | 0.62 | 0.13 | 0.10 | 0.07 | 0.62 | 0.031 | 0.031 | 0.024 |
| Central Asia | 0.01 | 0.050 | 0.57 | 0.24 | 0.07 | 0.67 | 0.14 | 0.07 | 0.05 | 0.64 | 0.030 | 0.033 | 0.024 |
| China | 0.3 | 0.031 | 0.60 | 0.26 | 0.06 | 0.71 | 0.15 | 0.06 | 0.04 | 0.64 | 0.047 | 0.016 | 0.024 |
| Cuba | 0.15 | 0.035 | 0.53 | 0.23 | 0.10 | 0.62 | 0.13 | 0.10 | 0.07 | 0.62 | 0.000 | 0.062 | 0.024 |
| Europe | 0.5 | 0.067 | 0.54 | 0.23 | 0.07 | 0.64 | 0.14 | 0.08 | 0.06 | 0.59 | 0.055 | 0.007 | 0.024 |
| Haiti | 0.05 | 0.095 | 0.53 | 0.23 | 0.10 | 0.62 | 0.13 | 0.10 | 0.07 | 0.62 | 0.000 | 0.062 | 0.024 |
| Iceland | 0.92 | 0.033 | 0.52 | 0.22 | 0.00 | 0.62 | 0.13 | 0.00 | 0.00 | 0.59 | 0.062 | 0.000 | 0.024 |
| India | 0.1 | 0.049 | 0.60 | 0.26 | 0.06 | 0.71 | 0.15 | 0.06 | 0.04 | 0.71 | 0.009 | 0.053 | 0.024 |
| Israel | 0.2 | 0.079 | 0.46 | 0.19 | 0.14 | 0.54 | 0.11 | 0.14 | 0.10 | 0.62 | 0.015 | 0.048 | 0.024 |
| Jamaica | 0 | 0.105 | 0.53 | 0.23 | 0.10 | 0.62 | 0.13 | 0.10 | 0.07 | 0.62 | 0.000 | 0.062 | 0.024 |
| Japan | 0.1 | 0.054 | 0.47 | 0.20 | 0.13 | 0.55 | 0.12 | 0.13 | 0.10 | 0.59 | 0.039 | 0.024 | 0.024 |
| Mauritius | 0.05 | 0.187 | 0.52 | 0.22 | 0.10 | 0.61 | 0.13 | 0.10 | 0.08 | 0.71 | 0.000 | 0.062 | 0.024 |
| Middle East | 0.05 | 0.069 | 0.52 | 0.22 | 0.10 | 0.61 | 0.13 | 0.10 | 0.08 | 0.71 | 0.015 | 0.048 | 0.024 |
| New Zealand | 0.05 | 0.064 | 0.46 | 0.19 | 0.02 | 0.54 | 0.11 | 0.02 | 0.02 | 0.62 | 0.060 | 0.003 | 0.024 |
| Philippines | 0.05 | 0.102 | 0.57 | 0.24 | 0.07 | 0.67 | 0.14 | 0.07 | 0.05 | 0.64 | 0.000 | 0.062 | 0.024 |
| Russia | 0.5 | 0.051 | 0.62 | 0.26 | 0.04 | 0.73 | 0.15 | 0.04 | 0.03 | 0.72 | 0.059 | 0.003 | 0.024 |
| South America | 0.1 | 0.073 | 0.50 | 0.21 | 0.12 | 0.58 | 0.12 | 0.12 | 0.09 | 0.64 | 0.022 | 0.040 | 0.024 |
| Southeast Asia | 0.1 | 0.099 | 0.57 | 0.24 | 0.07 | 0.67 | 0.14 | 0.07 | 0.05 | 0.64 | 0.001 | 0.062 | 0.024 |
| South Korea | 0.15 | 0.054 | 0.52 | 0.22 | 0.10 | 0.62 | 0.13 | 0.10 | 0.07 | 0.59 | 0.049 | 0.013 | 0.024 |
| Taiwan | 0.15 | 0.059 | 0.57 | 0.24 | 0.07 | 0.67 | 0.14 | 0.07 | 0.05 | 0.64 | 0.007 | 0.055 | 0.024 |
| United States | 0.2 | 0.083 | 0.48 | 0.21 | 0.12 | 0.57 | 0.12 | 0.12 | 0.09 | 0.64 | 0.049 | 0.013 | 0.024 |

Source: Jacobson et al. (2019).

where

$$L_{DR,r} = F_{flx,r}[L_r - (L_{heat,r} - L_{cold,r})F_{dh}] \quad (8.20)$$

$$L_{DR,c} = F_{flx,c}[L_c - (L_{heat,c} - L_{cold,c})F_{dh}] \quad (8.21)$$

$$L_{DR,i} = F_{flx,hti}L_{htemp,i} + F_{flx,tri}L_{trans,i} + F_{flx,oi}$$
$$[L_i - L_{htemp,i} - L_{trans,i} - (L_{heat,i} - L_{cold,i})F_{dh}] \quad (8.22)$$

$$L_{DR,t} = F_{flx,t}L_t \quad (8.23)$$

$$L_{DR,a} = F_{flx,a}L_a \quad (8.24)$$

$$L_{DR,o} = F_{flx,o}L_o \quad (8.25)$$

are the loads subject to demand response in the residential ($L_r$), commercial ($L_i$), industrial ($L_i$), transportation ($L_t$), agriculture/forestry/fishing ($L_a$), and military/other ($L_o$) sectors, respectively. In these equations, $F_{flx}$ is the fraction of a given load that is flexible, thus subject to demand response. For example, $F_{flx,r} = F_{flx,c} = F_{flx,oi} = F_{flx,a} = 0.15$ is the fraction of the residential, commercial, and agriculture/forestry/fishing non-heating, non-cooling, non-transportation, and non-high-temperature loads that is flexible. $F_{flx,hti} = 0.70$ is the fraction of the high-temperature industrial load that is flexible. Many high-temperature industrial loads are subject to demand response, thus flexible (Section 8.1.6). $F_{flx,t} = F_{flx,tri} = 0.85$ is the fraction of the transportation-sector transportation load and the industrial-sector transportation load that is flexible. $F_{flx,o} = 0.75$ is the fraction of other-sector loads that is flexible. In Equation 8.22,

$$L_{trans,i} = F_{tr,i}L_i \quad (8.26)$$

is the industrial-sector transportation load, where $F_{tr,i} = 0.0072$ is the fraction of the industrial-sector load that is for transportation, based on U.S. data (Jacobson et al., 2018a).

Finally, the **inflexible load** ($F_{inflex}$) is the all-sector total load ($L_{total} = L_r + L_c + L_i + L_t + L_a + L_o$) minus the loads subject to heat storage, cold storage, hydrogen storage, and demand response. Thus,

$$F_{inflex} = L_{total} - L_{heat, stor} - L_{cold,stor} - L_{H2,stor} - L_{DR} \quad (8.27)$$

Inflexible loads need to be satisfied immediately. However, if the cost of an inflexible load is high enough, many individuals will forgo the loads to save money. **As such, no load is truly inflexible.** They are only inflexible up to a certain cost. Beyond that cost, such loads can be shifted with demand response.

Figure 8.7 shows the resulting annual average end-use 2050 WWS load by sector and separated into inflexible and flexible loads for 24 world regions. Flexible loads are separated into loads subject to low-temperature heat storage, loads subject to low-temperature cold storage, loads subject to hydrogen storage, and remaining flexible loads, which are all subject to demand response. The hydrogen load is the load needed to produce, compress, and store hydrogen. It accounts for hydrogen leakage as well. Hydrogen is used only in fuel cells for long-distance, heavy transport in trucks, trains, ships, and aircraft in scenarios discussed in this chapter.

### 8.1.3.3 Estimating Daily and Hourly Loads from Annual Loads

Next, each region's annually averaged total heating load ($L_{heat}$) is distributed into daily heating loads with the use of heating degree day data as follows:

$$L_{heat,day} = L_{heat}H_{day}D_y/H \quad (8.28)$$

where $H_{day}$ is the number of heating degree days on a given day of the year, $D_y = 365$ (or 366 for leap years) is the number of days per year, and H is the number of heating degree days per year. The total heat load is then partitioned into a daily heat load subject to storage and not subject to storage, respectively, with

$$L_{heat,stor,day} = L_{heat,stor}L_{heat,day}/L_{heat} \quad (8.29)$$

$$L_{heat,nostor,day} = L_{heat,day} - L_{heat,stor,day} \quad (8.30)$$

The latter parameter, the heat load not subject to storage, may be treated as inflexible or partly subject to demand response. Here, 15 percent ($= F_{flx,r} = F_{flx,c} = F_{flx,oi} = F_{flx,a}$) is treated as flexible and subject to demand response in all sectors. This flexible term is already included as an annual average term in the equations for loads subject to demand response (Equations 8.20 to 8.25). In Equation 8.20, for example, the term is $F_{flx,r}L_{heat,r}(1 - F_{dh})$. This and similar terms must be removed and replaced with daily terms, such as $F_{flx,r}L_{heat,r}(1 - F_{dh})L_{heat,day}/L_{heat}$, to obtain the daily load subject to demand response. Equations 8.20 to 8.25 are modified further for other daily or hourly terms, as discussed shortly.

Similarly, each region's annually averaged total cooling load ($L_{cool}$) is distributed into daily cooling loads with the use of cooling degree day data as follows,

$$L_{cool,day} = L_{cool}C_{day}D_y/C \quad (8.31)$$

where $C_{day}$ is the number of cooling degree days on a given day of the year and C is the number of cooling

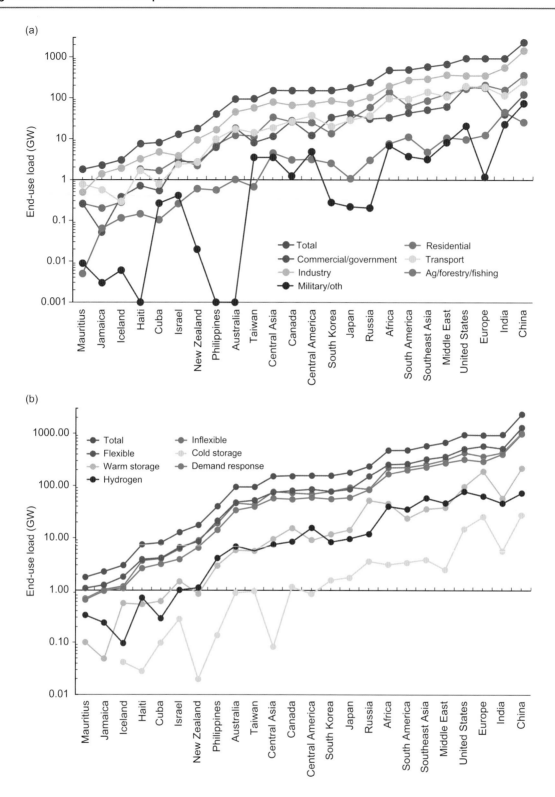

**Figure 8.7** (a) Annual average end-use 2050 WWS total load (GW) in 24 world regions, broken down by energy-use sector. (b) Annual average end-use 2050 WWS total load broken into inflexible and flexible load, with flexible load broken into low-temperature heat load subject to storage (warm storage), cold load subject to storage (cold load), hydrogen load, and load subject to demand response.

degree days per year. The total cold load is then partitioned into a daily cold load subject to storage and not subject to storage, respectively, with

$$L_{cold,stor,day} = L_{cold,stor}L_{cold,day}/L_{heat} \qquad (8.32)$$

$$L_{cold,nostor,day} = L_{cold,day} - L_{cold,stor,day} \qquad (8.33)$$

The cold load not subject to storage may be treated as inflexible or as partly subject to demand response. As with the heat load not subject to storage, 15 percent is treated as flexible and subject to demand response in all sectors. This flexible term is already included as an annual average term in the equations for loads subject to demand response (Equations 8.20 to 8.25). In Equation 8.20, for example, the term is $F_{flx,r}L_{cold,r}(1 - F_{dh})$. This and similar terms must be removed and replaced with daily terms, such as $F_{flx,r}L_{cold,r}(1 - F_{dh})L_{cold,day}/L_{cold}$, to obtain the daily load subject to demand response.

Because hydrogen loads are flexible and hydrogen will be needed every day for long-distance heavy transport, the annual average hydrogen load can initially be spread evenly each hour of the year. Demand response will adjust the actual timing of hydrogen production.

Once daily inflexible heat and cold loads are calculated, they are added to hourly inflexible loads, which are all remaining loads aside from those used to produce hydrogen, subject to heat or cold storage, or subject to demand response.

Hourly 2050 inflexible electricity loads are obtained from contemporary hourly BAU load data (e.g., ENTSO-E, 2016; Neocarbon Energy, 2016). Figure 8.5 illustrates hourly load data for the United States as a whole. Electricity load data include cold loads, since air conditioning and refrigeration currently run on electricity. On the other hand, they do not include many low-temperature heat loads, because air and water today are heated mostly with natural gas, fuel oil, and wood pellets rather than electricity.

The contemporary hourly electricity BAU loads are extrapolated to 2050 and converted to WWS total loads (flexible plus inflexible loads) using the annual average ratio of the total 2050 WWS load to the total contemporary BAU load (e.g., Sections 7.1 and 7.2). From the resulting hourly 2050 WWS load, hourly heat, cold, and hydrogen loads subject to storage are subtracted out to give the remaining electric load. The electric load is then partitioned into hourly flexible and inflexible loads. The hourly flexible loads are subject to demand response and

replace the corresponding annual average flexible loads subject to demand response in Equations 8.20 to 8.25.

### 8.1.3.4 Sizing Storage and Additional Generation

The generator nameplate capacities provided in Table 7.7 were calculated based on meeting annual average electricity, heat, cold, and hydrogen loads. However, to match power demand with supply over time, additional generators and storage are needed. The next step in analyzing grid stability is to estimate the sizes of additional generation nameplate capacity and the storage needed to match supply with demand over time rather than in the annual average alone.

Because decades of time and billions to trillions of dollars of investment are required to change a country's energy infrastructure, estimates of the amount of excess generation, storage, and demand response should be modeled before a complete infrastructure is implemented. Such modeling requires taking a first guess at a possible infrastructure, then performing reliability and cost calculations to determine whether demand can be met by supply, storage, and demand response over time at low cost. If not, the proposed energy infrastructure should be modified until demand can be met. Table 7.7 gives one such first guess of additional generators needed, summed over 143 countries.

Similarly, a first guess of the demand response time limit is needed. The **demand response (DR) time limit** is the maximum allowable number of hours a load can be delayed by being shifted forward in time before it must be met. It is an adjustable parameter, but it is usually limited to 8 hours. During each grid integration model timestep, if a flexible load subject to demand response cannot be met, the load is shifted forward one timestep (e.g., 30 seconds). If a portion of the load subject to demand response still cannot be met during that timestep, that portion is shifted forward again, and so forth. If, after 8 hours, the remaining load still cannot be met, it immediately becomes an inflexible load and must be met; otherwise, the model stops. Each load that is shifted forward in time in this way is tracked independently of loads that are shifted forward starting at different times to ensure no load is shifted forward more than the DR time limit allows.

A first guess of the electricity, heat, cold, and hydrogen storage size is also needed. Three important characteristics of storage are its peak charge rate, peak discharge

Table 8.4 **Present value of mean 2019 to 2050 lifecycle cost of new storage capacity and round-trip efficiency of several storage technologies.**

| Storage Technology | Present Value of Lifecycle Cost of New Storage ($/kWh-Max Energy Storage Capacity) | | | Round-Trip Charge/store/Discharge Efficiency (%) |
|---|---|---|---|---|
| | Middle | Low | High | |
| **Electricity** | | | | |
| PHS | 14 | 12 | 16 | 80 |
| CSP-PCM | 20 | 15 | 23 | 99 |
| LI batteries | 60 | 30 | 90 | 85 |
| **Cold** | | | | |
| CW-STES | 6.5 | 0.13 | 12.9 | 84.7 |
| ICE | 36.7 | 12.9 | 64.5 | 82.5 |
| **Heat** | | | | |
| HW-STES | 6.5 | 0.13 | 12.9 | 83 |
| UTES | 0.90 | 0.071 | 1.71 | 56 |

Source: Jacobson et al. (2018a), except with 2019 to 2050 mean battery costs updated from BloombergNEF (2019).
PHS = pumped hydropower storage; CSP-PCM = concentrated solar power with phase change material for storage; LI batteries = lithium-ion batteries; CW-STES = cold water sensible-heat thermal energy storage; ICE = ice storage; HW-STES = hot water sensible-heat thermal energy storage; UTES = underground thermal energy storage (borehole). PHS efficiency is the ratio of electricity delivered to the sum of electricity delivered and electricity used to pump the water.
2050 storage lifetimes are 17 (12 to 22) years for batteries and 32.5 (25 to 40) years for all other storage.
The CSP-PCM cost is for the PCM material and storage tanks. The CSP-PCM efficiency is the ratio of the heat available for the steam turbine after storage to the heat from the solar collector that goes into storage. The additional energy losses due to reflection and absorption by the CSP mirrors (45 percent of incident solar energy is lost to reflection) and due to converting CSP heat to electricity (71.3 percent of heat is wasted and only 28.7 percent is converted to electricity) are accounted for in the CSP efficiency without storage. Battery efficiency is the ratio of electricity delivered to electricity put into the battery. CW-STES and HW-STES efficiencies are the ratios of the energy returned as cooling and heating, respectively, after storage, to the electricity input into storage. The UTES efficiency is the fraction of heated fluid entering underground storage that is ultimately returned during the year (either short or long term) as air or water heat for a building.

rate, and maximum capacity. The **peak charge rate** (kW) is the maximum amount of energy per unit time that can be added to storage. The **peak discharge rate** (kW) is the maximum amount of energy per unit time that can be removed from storage. The **peak storage capacity** (kWh) is the maximum amount of energy that can be stored. It equals the peak discharge rate multiplied by the number of hours of storage at the maximum discharge rate. Another parameter is the number of full charge and discharge cycles the storage can go through before wearing down.

For example, a Tesla Powerwall 1 battery has a peak charge and discharge rate of 3.3 kW and a storage capacity of 6.4 kWh. Thus, it can continuously discharge 3.3 kW for 1.94 hours before emptying out.

Whereas the peak discharge rate of a hydropower reservoir is limited by the turbine nameplate capacity and the maximum water flow rate to the turbine, the hydropower charge rate is unpredictable because it is controlled by natural rain and stream flow, which are intermittent. The energy storage capacity of a hydropower reservoir equals the peak discharge rate (kW) multiplied by the number of hours required for the reservoir to empty at the peak discharge rate.

Storage size is an issue because storage is more expensive the faster the peak discharge rate and the larger the peak storage capacity are. Whereas heat, cold, and hydrogen storage have relatively low or modest costs per unit peak storage capacity, the most convenient electricity storage, battery storage, is still relatively expensive in 2020, and its high cost (relative to other forms of electricity storage) may persist into the future (Table 8.4). Other forms of electricity storage (CSP storage, pumped hydropower storage) are less expensive (Table 8.4) but are limited to certain areas or face other hurdles that limit the pace at which they are adopted in comparison with batteries.

If the cost of electricity storage were not a barrier, enough battery storage would be built immediately, not only to fill short-term gaps in electricity supply, but also to store large amounts of electricity that could be used months later. However, this is not done yet because of the relatively high cost of battery storage (Table 8.4). On the other hand, hydropower currently serves the dual purpose of providing electricity over both minutes and seasons. Conversely, hydropower is limited by how many new dams can be built, the current peak discharge rate of the turbines and generators in each hydropower plant, and the current peak storage capacity of each reservoir. Most other electricity storage operates over timescales of minutes to days to a couple of weeks at most.

Underground thermal energy storage in boreholes, water pits, and aquifers similarly holds heat and cold over periods of minutes to seasons. The cost of such storage is relatively low (Table 8.4), so it can and has been used seasonally. Hydrogen storage tanks can also store hydrogen over minutes to seasons.

### 8.1.4 Solutions to Instantaneous Over and Under Generation

Once electricity and heat generator sizes, storage sizes, and demand response characteristics are defined, an energy system is ready for testing. The test can be successful only if the energy system matches instantaneous demand with instantaneous generation plus storage plus demand response over time. This section discusses such matching in two parts. One part concerns what to do with excess instantaneous generation when it exceeds instantaneous demand. The second discusses how to meet instantaneous demand when it exceeds instantaneous supply.

#### 8.1.4.1 Solutions When Instantaneous WWS Electricity or Heat Supply Exceeds Instantaneous Demand

This section discusses the steps taken when the current (instantaneous) supply of WWS electricity or heat exceeds the current electricity or heat demand (load). The total load, whether for electricity or heat, consists of flexible and inflexible loads. Whereas flexible loads

may be shifted forward in time with demand response, inflexible loads must be met immediately.

If WWS instantaneous electricity or heat supply exceeds the instantaneous inflexible electricity or heat load, then the supply is used to satisfy that load. The excess WWS is then used to satisfy as much current flexible electric or heat load as possible.

If any excess electricity exists after inflexible and current flexible loads are met, the excess electricity is sent to fill electricity, heat, cold, or hydrogen storage. Electricity storage is filled first. Excess CSP energy goes to CSP electricity storage. Remaining instantaneous CSP electricity and excess electricity from other sources are used next to fill pumped hydropower storage followed by battery storage, flywheels, gravitational masses, compressed air energy storage, cold water storage, ice storage, hot water tank storage, and underground thermal energy storage. Remaining electricity is used to produce hydrogen. Any residual after that is shed.

Heat and cold storage are filled by using the excess electricity to run a heat pump to move heat or cold from the air, water, or ground to the thermal storage medium. Hydrogen storage is filled by using electricity in an electrolyzer to produce hydrogen and in a compressor to compress the hydrogen, which is then moved to a storage tank.

If any excess direct geothermal or solar heat exists after it is used to satisfy inflexible and flexible heat loads, it is used to fill either district heat storage (water tank and underground heat storage) or an individual home's hot water tank heat storage.

#### 8.1.4.2 Solutions When Instantaneous Load Exceeds Instantaneous WWS Electricity or Heat Supply

When current inflexible plus flexible electricity load exceeds the current WWS electricity supply from the grid, the first step is to use electricity storage (CSP, pumped hydro, hydropower, battery, flywheel, gravitational mass, and compressed air energy storage) to fill in the gap in supply. The electricity is used to supply the inflexible load first, followed by the flexible load.

If electricity storage becomes depleted and flexible load persists, demand response is used to shift the flexible load to a future hour.

If the inflexible plus flexible heat load subject to storage exceeds WWS direct heat supply, then stored district heat (in water tanks and underground storage) is used to satisfy district heat loads subject to storage, and stored building heat (in hot water tanks) is used to satisfy building water heat loads. If stored heat becomes exhausted, then any remaining low-temperature air or water heat load becomes either an inflexible load (85 percent) that must be met immediately with electricity or a flexible load (15 percent) that can be met with electricity but can be shifted forward in time with demand response.

Similarly, if the inflexible plus flexible cold load subject to storage exceeds cold storage (in ice or water), excess cold load becomes either an inflexible load (85 percent) that must be met immediately with electricity or a flexible load (15 percent) that can be met with electricity but can be shifted forward in time with demand response.

Finally, if the current hydrogen load depletes hydrogen storage, the remaining hydrogen load becomes an inflexible electrical load that must be met immediately with current electricity.

## 8.1.5 Measures Needed When Instantaneous Load Cannot Be Met with Instantaneous Supply or Storage

If the measures just described are insufficient for matching power demand with supply continuously, the system, whether a real-life system or a computer-modeled system, must be modified to ensure reliability. The standard for reliability is a **loss of load expectation (LOLE)** of no more than 1 day (24 hours) in 10 years. Loss of load arises due to the failure to match instantaneous load with instantaneous supply or storage. It can arise due to a too severe lull in wind speed and solar radiation over a long period, substantial scheduled or unscheduled maintenance of many energy generators, unexpected load due to a heat wave or a cold spell, or transmission line congestion.

If supply does not match load consistently, it will be necessary to either (a) increase the nameplate capacity of the electric power generators in locations that are most peak coincident with the load; (b) increase the electricity, heat, cold, or hydrogen nameplate storage capacity; (c) increase the nameplate capacity of the transmission system to allow more renewables that are far away, thus subject to different weather patterns, to help fill in gaps in supply; (d) use electricity stored in vehicle batteries to help fill in gaps in supply on the grid; and/or (e) integrate weather forecasts into system operations to improve efficiency of the grid. Some of these measures also help to increase the

efficiency of the grid, even when supply matches demand continuously. These techniques are discussed next.

### 8.1.5.1 Oversizing Wind, Water, and Solar Generation to Help Meet Demand

Oversizing the nameplate capacity of WWS installations so that annual average WWS supply exceeds annual average energy demand can help to reduce the number of times during a year that available WWS power supply is below inflexible demand. This solution, therefore, reduces the number of outages that occur during a year. Whereas oversizing helps to meet hours of peak demand, it also results in excess supply during many other hours. Ideally, the excess supply is put into electricity storage or used to produce heat, cold, and/or hydrogen, which are either stored or used immediately. Using excess WWS in these ways avoids WWS curtailment and thus reduces overall system cost.

### 8.1.5.2 Oversizing Storage to Help Meet Peaks in Demand

Oversizing storage is a second way to improve matching power demand with supply and storage on the grid. Because wind, solar, and wave power are variable WWS resources, they often overproduce electricity. During times of overproduction, they often fill electricity storage, and remaining excess electricity is used to produce and store heat, cold, or hydrogen. However, if more electricity storage capacity were available, more excess electricity would be stored and used to meet peaks in electricity demand over a longer period, reducing the frequency of outages. Heat, cold, and hydrogen storage capacities can similarly be increased if these storages are also depleted during peaks of heat, cold, or hydrogen demand.

### 8.1.5.3 Increasing Transmission Nameplate Capacity to Help Meet Demand

As discussed in Section 8.1.1, interconnecting geographically dispersed variable renewable resources such as wind, solar, and wave power smoothens the aggregate power supply from these generators. Similarly, interconnecting wind and solar, which are complementary in nature, smoothens out overall supply. As such, an additional solution to keeping the grid stable is to import faraway, variable renewable electricity through an upgraded transmission system. The long-distance transmission lines should be HVDC lines to reduce line losses, thus reducing system costs. Whereas each country will ideally produce its own energy, interconnecting renewables among nearby countries will likely reduce overall costs as well. The reason is that some countries have better wind and hydropower resources than others, whereas others have better solar resources. Interconnecting nearby countries to take advantage of the plentiful resources in each country will reduce times that no power is available (Section 8.1.1) and allow storage to be replenished more regularly. Interconnecting also allows aggregate transmission requirements to be reduced with little loss in annual energy output (Section 8.1.1).

### 8.1.5.4 Helping to Balance Demand with Vehicle-to-Grid

Another method to help match demand with supply is to store electric power in the batteries of BE vehicles, and then to withdraw such power when needed to supply electricity back to the grid. This concept is referred to as **vehicle-to-grid (V2G)** (Kempton and Tomic, 2005a). With V2G, a utility operator would enter into a contract with an individual BE vehicle owner to allow electricity transfers back to the grid at any time during an agreed-upon period in exchange for a lower electricity price.

V2G is expected to wear down batteries faster by increasing the number of charge and discharge cycles the battery goes through during a year. On the other hand, one study suggests that, under idealized conditions, it may be possible for V2G to increase battery life (Uddin et al., 2017).

In either case, only a small percentage of vehicles needs to participate in V2G to help balance the grid. One study suggests only 3.2 percent of vehicles in the United States is needed to smoothen out U.S. electricity demand when 50 percent of demand is supplied by wind (Kempton and Tomic, 2005b). Through computer modeling, Budischak et al. (2013) and Child et al. (2018) further found that using V2G would help to match demand with supply and storage hourly on the grid in the Eastern United States and in the Aland Islands (Baltic Sea), respectively. An alternative to V2G is simply to avoid charging BE vehicles when electric power is in short supply. Such an action is a form of demand response.

### 8.1.5.5 Using Weather Forecasts to Plan for and Reduce Backup Requirements

Forecasting the weather (winds, sunlight, waves, tides, and precipitation) gives grid operators more time to plan ahead for a backup energy supply when a variable electricity source might produce less electricity than anticipated. Good forecast accuracy can also allow spinning reserves to be shut down more frequently, increasing the overall grid's efficiency (Hart and Jacobson, 2011). Forecasting is done with either a numerical weather prediction model, the best of which can produce minute-by-minute predictions 1 to 4 days in advance with good accuracy, or statistical analyses of local measurements. The use of forecasting reduces uncertainty and makes planning more dependable, thus reducing the impacts of variability.

---

### Example 8.2  Matching supply with demand and storage

Suppose the instantaneous inflexible load in a system is 4 MW, the flexible load is 3 MW, and the charge and discharge rate of total storage (electricity, heat, cold, plus hydrogen storage) are both 2 MW. Describe how to meet load with supply in four cases: when the supply is (a) 10 MW, (b) 5 MW, (c) 3 MW, and (d) 1 MW. Assume storage is empty in the first case and full in the remaining cases.

#### Solution:
a. Use 7 MW of supply to satisfy the flexible plus inflexible loads. Use 2 MW to charge storage. Shed the remaining 1 MW.
b. Use 4 MW of supply to satisfy the inflexible load. Use 1 MW of supply and 2 MW of storage to satisfy the inflexible load.
c. Use 3 MW of supply and 1 MW of storage to satisfy the inflexible load. Use 1 MW of storage to supply 1 MW of flexible load. Shift the remaining 2 MW of flexible load forward in time with demand response.
d. Use 1 MW of supply and 2 MW of storage to satisfy the inflexible load. Since 1 MW of inflexible load is not satisfied, it is necessary to modify the system by oversizing generation or storage, increasing transmission capacity, using vehicle-to-grid, or improving weather forecasting.

## 8.1.6 Ancillary Services: Load Following, Regulation, Reserves, and Voltage Control

An electric power system needs to meet two major requirements. One is to match current load with current generation plus storage (Sections 8.1.4 and 8.1.5). The second is to manage power flows between and among individual transmission facilities. This section discusses both issues with respect to practical grid operations.

Meeting instantaneous load is difficult because load and supply continuously vary in time. Load can vary over seconds to minutes due to the random turning on and off of millions of individual inflexible loads. Load can also vary seasonally due to changes in heating and cooling demand over the year. WWS electricity and heat generator output also varies minutely to seasonally.

**Ancillary services** are services performed by equipment and people that are, according to the U.S. Federal Energy Regulatory Commission (FERC), "necessary to support the transmission of electric power from seller to purchaser given the obligations of control areas and transmitting utilities within those control areas to maintain reliable operations of the interconnected transmission system" (Kirby, 2004).

The main ancillary services include load following, regulation, spinning reserves, supplemental reserves, replacement control, and voltage control. These are discussed, in turn.

### 8.1.6.1 Load Following

**Load following** is the use of power generators or storage to follow the rate of changes of load (e.g., MW per minute), averaged over a 5- to 15-minute period. Load following can also be accomplished by reducing the load with demand response. Load following can be scheduled automatically or manually. Load following, generators or storage can have a relatively modest ramp rate. Thus, in a WWS world, hydropower, a portion of CSP output, or a portion of solar PV output can be used for load following, as can CSP storage or any electricity storage. In fact, any WWS generator can load follow, since if a generator projects (e.g., due to wind or solar forecasting) that it will have electricity available over a coming period, it can compete in a real-time electricity market to sell the electricity for load following.

Load following can similarly be accomplished by reducing power consumption in residences and commercial buildings. For example, demand response can help push the time of battery charging for electric vehicles to late at night. It can also push the time of heating water in a hot water tank or of running a dishwasher, washing machine, or dryer to a time of low electricity demand on the grid. Similarly, demand response can push the time of running a wastewater treatment plant or hydrogen production by electrolysis to a time of low demand.

### Transition highlight

Load following can also be accomplished by rapidly reducing the power consumption of large industrial customers. Some industrial loads that can be shifted include air liquefaction; induction and ladle metallurgy; water pumping with variable speed drives; and production by electrolysis of aluminum, chlor-alkali, potassium hydroxide, magnesium, sodium chlorate, and copper (Kirby, 2004). Indeed, NRC (2010) states, "The ability of industry to cut peak electric loads is a motivator for utilities to incentivize demand response (shifting loads to off-peak periods) in industry. ... In combination with peak-load pricing for electricity, energy efficiency and demand response can be a lucrative enterprise for industrial customers."

### 8.1.6.2 Regulation

**Regulation** is the automatic (through electronic grid controls) use of pre-contracted power generators, storage, or load reduction to fill in small gaps between the actual load and the load that results from load following. Over a 5- to 15-minute period, the actual load varies subminutely due to hundreds to millions of people turning loads on or off. Load following meets only the average change in load over a period, not the minute-by-minute change in load. Minute-by-minute regulation can be met with generators, storage, and load reduction methods that have ramp rates 5 to 10 times the minimum ramp rates required for load following.

In a WWS world, regulation will be performed primarily with hydropower, batteries, pumped hydropower storage, flywheel storage, compressed air storage, and gravitational storage with solid masses, all of which have fast ramp rates (see Table 2.1). The best electricity producers for regulation are generally fast-responding storage technologies that can repeatedly operate through many cycles without degrading. Batteries, for example, respond quickly (within milliseconds), have output that can be controlled precisely, and can operate through many cycles. As such, they are ideal for regulation.

When intermittent WWS generators, such as wind, solar, or wave power, are aggregated together over a large geographic region, their overall variability is reduced (Section 8.1.1). This reduces the regulation requirement for, thus cost of, wind, solar, and wave power (Kirby, 2004). In other words, aggregation makes it easier for battery storage, for example, to meet the difference between instantaneous fluctuation in demand and WWS generation.

Because of the fast response time required for regulation services, such services are not obtained from market signals, which respond too slowly. Instead, generators, storage, and load reduction used for regulation are contracted ahead of time.

### 8.1.6.3 Frequency Regulation

Regulation helps to maintain the frequency on the grid. The AC frequency on the grid is required to be close to 60 Hz in the United States and some other countries and 50 Hz in Europe and remaining countries (Section 4.1.3). The deviation allowed from this mean frequency in the United States is $\pm 0.5$ percent, or from 59.7 Hz to 60.3 Hz. In most of Europe, it is $\pm 0.4$ percent, or from 49.8 Hz to 50.2 Hz. In some places in Eastern Europe, it is broader, $\pm 6$ percent, or from 47 Hz to 53 Hz. Frequency is maintained within a narrow range because too high or too low a frequency can destroy electrical equipment on the grid and end-use electrical appliances and devices.

When the frequency on the grid falls outside the range allowed, averaged over a minute, grid operators reduce the frequency down into the permissible range. They do this by reducing electricity added to the grid from generation or storage or by increasing load on the grid. In other words, a frequency that is too high can be reduced by adding slightly less electricity to the grid than is needed. Conversely, adding more electricity to the grid than is needed can push a low frequency into the normal range. The increase or decrease of generation, storage, or load to regulate frequency on the grid is called **frequency regulation**.

### Transition highlight

Traditionally, grid operators have increased frequency by increasing generation from gas, coal, oil, nuclear, or hydroelectric plants. In a WWS world, operators will increase frequency by increasing hydroelectric, geothermal, wind, and solar PV output as well as output

from CSP storage, batteries, pumped hydropower storage, flywheels, compressed air storage, and gravitational storage with solid masses. Conversely, grid operators will reduce frequencies that are too high by reducing generation and storage or turning on artificial loads.

In the case of wind turbines, an operator can increase generation by running some turbines in partial load mode such that, when a decrease in grid frequency occurs, the operator increases turbine output by varying blade pitch. Alternatively, the operator may control short releases of electricity from wind turbines to the grid (Erlich and Wilch, 2010). Similarly, some solar PV plants can be run at less than full output such that, when a decrease in grid frequency occurs, output is increased. Inverters can also be optimized to provide frequency control to the grid (Roselund, 2019).

### 8.1.6.4 Spinning, Supplemental, and Replacement Reserves and Voltage Control

Three ancillary services – spinning reserves, supplemental reserves, and replacement reserves – are used to supply load in the event that a large power generator or transmission line goes down.

**Spinning reserves** are electricity generators or storage media, online and connected to the grid, that can increase output immediately in response to a major generator or transmission outage. Spinning reserves are required to reach full output within 10 minutes.

**Supplemental reserves** are the same as spinning reserves, but they are not required to respond immediately. They can be offline but, when turned online, they must reach full output within 10 minutes.

**Replacement reserves** are the same as supplemental reserves, but they must reach full output within 30 minutes. They are also used to replenish spinning and supplemental reserves to their pre-contingency status.

**Voltage control** is the injection or absorption of reactive power to maintain transmission-system voltages within required ranges (Section 4.9). Such control is needed on the timescale of seconds. System operators obtain voltage control by installing transmission equipment or obtaining voltage support from local generators. Such equipment includes tap changers, capacitors, reactors, and static Volt-Ampere-reactive (VAr) compensators (Kirby, 2004).

## 8.2 Case Study of Meeting Demand with 100 Percent WWS

Nine countries of the world currently meet 95 to 100 percent of their annual average electricity from WWS (see Table 8.1). Other countries, such as Scotland, are approaching that threshold rapidly as well. In most of the countries, though, hydropower is the major WWS generator. The major generator in Scotland is wind. Most countries don't have the luxury of having so much hydropower. In addition, electricity is only 20 percent of all end-use energy in the global average, and a transition in all other energy sectors (transportation, building heating/cooling, industrial heat, agriculture/forestry/fishing, and the military) to electricity and direct heat from WWS is needed as well. Because no country has a 100 percent all-sector WWS energy system in place, it is necessary to model such a system to help build them optimally. In this section, past modeling work with 100 percent or near 100 percent WWS systems is first described. Then, results from a recent study of meeting all-sector energy demand with supply, storage, and demand response among 24 world regions are discussed.

### 8.2.1 Previous Studies of Matching Demand with or near 100 Percent WWS

Many studies support the possibility of matching power demand with near 100 percent WWS supply in one or more energy sectors. Whereas several of these studies examine matching demand with supply in the annual average, dozens of others examine it at time intervals ranging from 30 seconds to hours. Both types of studies are summarized below.

Sorensen (1975) examined the supply of wind and solar in Denmark and suggested that enough in combination might be available to supply all building heat, transportation, industrial, and electrical demand in the annual average by the year 2050. Sorensen (1996) then estimated the annual average quantity of biomass, biofuels, and biogas energy together with WWS (solar, wind, and hydroelectric) energy that might be needed to provide all world energy.

Jacobson and Masters (2001) estimated the number of wind turbines needed, in the annual average, and their cost to replace 60 percent of coal in the United States to satisfy the Kyoto Protocol.

Jacobson et al. (2005) then modeled the health and climate benefits of transitioning 100 percent of the U.S. vehicle fleet to hydrogen fuel cell vehicles, where the hydrogen was obtained from either electrolysis (with the electricity from wind), steam reforming of natural gas, or coal gasification. The result was that wind electrolysis is the cleanest among all options for producing hydrogen for transportation. Thus, this was an early study examining a 100 percent WWS system for the transportation sector.

Archer and Jacobson (2005) used data from 7,753 surface stations and 446 sounding stations worldwide to map and quantify the land and coastal wind potential for every inhabited continent worldwide. They concluded that enough wind power is available in high-wind-speed locations (where the mean annual wind speed exceeds 6.9 m/s) at 80-m hub height to power the entire world for all purposes several times over.

Czisch (2005) and Czisch and Giebel (2007) developed a least-cost optimization model that simulated the electric power grid over Europe, North Africa, and the Middle East divided into 19 regions interconnected by HVDC transmission. They found that a renewable electricity supply of wind, solar photovoltaics, concentrated solar power, bioenergy, and hydroelectric power could keep the grid stable at low cost every three hours for one year.

Lund (2006) examined optimized mixes of wind, solar PV, and wave power that resulted in meeting 100 percent of electricity demand in Denmark. Such mixes resulted in the least excess production, thus the least curtailment, of these resources. He concluded that an optimized mix for Denmark's electric power sector may be about 50 percent wind, 30 percent wave, and 20 percent PV.

Hoste et al. (2009), whose results were used in Jacobson (2009) and Jacobson and Delucchi (2009), found that by combining at least four renewables, including baseload geothermal, intermittent wind, intermittent solar, and gap-filling hydroelectric, and by assuming an expanded transmission network, California might meet 100 percent of its monthly averaged hour-by-hour electricity demand in the two months tested, April and July 2020. Hydropower, constrained by its monthly availability, was used to fill gaps in supply. No other gap filling was used. Thus, whereas wind and solar are intermittent, treating them as a bundle with baseload geothermal and gap-filling hydropower increased the ability of all resources to meet peak demand.

Jacobson (2009) evaluated 9 electric power sources and 12 options for producing energy for vehicles in terms of 11 impact categories (e.g., on climate, air pollution, land use, water supply, reliability, impacts on wildlife, energy security, and more). The study concluded that WWS resources (onshore and offshore wind, solar PV, CSP, geothermal, hydro, tidal and ocean current, and wave power) may be the best in terms of minimizing such impacts and should be used to advance solutions to global warming, air pollution, and energy security.

Jacobson and Delucchi (2009) subsequently estimated the WWS resources needed to provide the world's 2030 end-use power demand for all purposes (electricity, transportation, heating/cooling, and industry, which itself included agriculture, forestry, fishing, and the military). The main barriers were social and political, not technical or economic. Material requirements presented some challenges but not limitations. The study considered grid stability, albeit simplistically, and land requirements.

Results were obtained without nuclear power, fossil fuels with carbon capture, direct air capture, biomass, biofuels, or biogas. As such, the study was the first to propose a worldwide all-sector WWS energy system that eliminated energy-related air pollution, climate-relevant emissions, and energy insecurity simultaneously. The previous global study of Sorensen (1996) had assumed the use of about 50 percent bioenergy (biomass, biofuels, and biogas); thus, it was not a WWS system and allowed for the continuation of combustion air pollution and large amounts of land (e.g., Figure 3.2). The Archer and Jacobson (2005) study did include a 100 percent WWS system with wind alone but did not consider the practicality of a wind-alone system in terms of cost, materials, land requirements, or grid stability.

A methodology similar to that in Jacobson and Delucchi (2009) was used to develop more detailed roadmaps to meet annually averaged power demand for both the world and the United States in Jacobson and Delucchi (2011) and Delucchi and Jacobson (2011). Subsequent roadmaps of the same type were developed for New York (Jacobson et al., 2013), California (Jacobson et al., 2014b), Washington State (Jacobson et al., 2016), the 50 U.S. states (Jacobson et al., 2015a), 139 countries (Jacobson et al., 2017), 143 countries (Jacobson et al., 2019), and 53 towns and cities in North America (Jacobson et al., 2018b).

Lund and Mathiesen (2009) modeled the hour-by-hour matching of electricity demand with wind, wave, solar, and biomass supply in Denmark for the year 2050 and determined that such a system can remain reliable.

Mason et al. (2010) modeled the electricity system in New Zealand with 53 to 60 percent hydropower, 22 to 25 percent wind, 12 to 14 percent geothermal, 1 percent biomass, 0 to 12 percent added hydropower to provide peaking power to fill in gaps in supply, and assumptions about demand response. They found it possible to match demand with supply with this system (near 100 percent WWS) replacing the 2005 to 2007 electricity system in New Zealand.

Hart and Jacobson (2011) modeled the California electric power grid in 2005 to 2006 as if the grid were transitioned to renewables. They found that it was possible to match power demand with WWS supply such that 99 percent of the electricity supplied was from non-carbon sources, even before using demand response or large-scale storage. Gaps in supply were filled primarily by hydropower (Figure 8.4).

Between 2011 and 2019, at least 38 additional studies among 12 independent groups and 73 scientists subsequently supported the earlier results. They have found that matching 100 percent or near 100 percent time-dependent power demand with time-dependent supply and storage in one or more energy sectors is feasible at low cost. Such studies include Connolly et al. (2011, 2016); Mathiesen et al. (2011, 2015); Hart and Jacobson (2012); Hart et al. (2012); Elliston et al. (2012, 2013, 2014); Rasmussen et al. (2012); Budischak et al. (2013); Steinke et al. (2013); Becker et al. (2014, 2015); Connolly and Mathiesen (2014); Jacobson et al. (2015b, 2018a, 2019); Bogdanov and Breyer (2016); Child and Breyer (2016); Aghahosseini et al. (2016, 2019); Blakers et al. (2017, 2018); Barbosa et al. (2017); Lu et al. (2017); Gulagi et al. (2017a, 2017b); Esteban et al. (2018); Zapata et al. (2018); Child et al. (2018); Sadiqa et al. (2018); Barasa et al. (2018); Caldera and Breyer (2018); Liu et al. (2018); Teske et al. (2019); Ram et al. (2019); Hansen et al. (2019b); and Bogdanov et al. (2019).

Two reviews of several of these studies, by Brown et al. (2018) and Diesendorf and Elliston (2018), have similarly found that the methods used in the studies were rigorous. Hansen et al. (2019a) discuss the growth in the number of 100 percent renewable energy papers.

## 8.2.2 Types of Models for Meeting Demand

Three main types of computer models that simulate matching power demand with supply, storage, and/or demand response on an electric power grid have been developed. These include power flow/load flow models, optimization models, and trial-and-error simulation models. Such models are discussed next.

### 8.2.2.1 Power Flow or Load Flow Models

One type of computer model of the electric grid is called a **power flow** or **load flow model**. Such a model treats thousands of individual transmission lines connected to electricity generators and load centers. Inputs are the real power and voltage magnitude of each generator and the real and reactive power (Section 4.9) of each load. At one arbitrary generator (called the slack bus) the voltage phase angle (Section 4.8) is also known. Outputs are the voltage magnitude of all the loads and the voltage phase angles of both the generators (except the slack bus, where the phase angle is already known) and loads. The resulting set of nonlinear equations does not depend on time. In other words, the equations represent an equilibrium state, and the solution to the equations is an equilibrium (independent of time) solution. The solution is not an optimized solution. Instead, it is the only real solution to a set of $N$ equations and $N$ unknowns. One method of solving the equations is with an iterative Newton–Raphson technique. Because power flow models are equilibrium models that represent one snapshot in time, they are not used to examine matching power demand with supply over time. Their requirement for iteration, thus heavy computational cost, also makes them difficult to use to match power demand with supply over time, even if modified to do so. On the other hand, they are the only type of model currently available that can be used to simulate flows through individual transmission lines.

### 8.2.2.2 Optimization Models

A second type of model is a **least-cost or least-carbon optimization model**. Most models used for the studies described in Section 8.2.1 are optimization models. Such models either assume perfect transmission (no representation of individual transmission lines) or include just a few main transmission lines between generators and load centers. As such, they do not treat the transmission system in the same detail as a power flow model does.

However, unlike power flow models, optimization models treat multiple time intervals simultaneously.

With an optimization model, a problem is set up to calculate the least-cost portfolio of generators or the lowest-carbon-emitting portfolio of generators to meet load (e.g., Hart and Jacobson, 2011). Cost usually includes capital cost, fixed and variable operation and maintenance (O&M) cost, fuel cost, and decommissioning cost of each generator and storage technology. Carbon emissions include those from each generator, if any.

The minimization in each case is subject to a power balance constraint that ensures that the generation, summed over all generators, minus transmission and distribution losses and plus or minus changes from storage, equals the load on the grid each time period (usually an hour). The inputs into the calculation may include the cost (or carbon emissions) of each generator as a function of nameplate capacity; the peak charge rates, discharge rates, and capacities of storage; hourly electric power generation per unit nameplate capacity from each generator; transmission loss rates; constraints for hourly removals from or additions to storage; and hourly loads. A set of equations is derived that requires balancing load each hour over a specified time period. The equations are solved among all hours simultaneously to minimize cost or carbon. Since the cost and carbon emissions are a function of the nameplate capacity of each generator, the resulting nameplate capacities of each generator are then known.

The equations are solved with an optimization solver, of which many exist. The main problem with an optimization model is that, the greater the numbers of hours and variables (e.g., different types of generators and storage) that are solved for, the more difficult it is to converge the resulting set of equations in a reasonable amount of computing time, even with a large computer. As such, many optimization models have been unable to solve for a full year at an hour resolution. Instead, such models have solved over a reduced number of hours per year, usually representing blocks of days in different seasons (see Table 1 of Frew and Jacobson, 2016, for examples). Models that do solve every hour generally simplify the system by reducing the types of storage or generators or by omitting demand response, hydrogen loads, heat loads, and/or cold loads.

The problem with not treating at least every consecutive hour of a year with an optimization model is that storage becomes impossible to track correctly. The amount of electricity, heat, cold, or hydrogen in storage at a given hour can be determined only from knowing how much was in storage during the last hour and how much was added to or removed from storage during the current hour. If a model is optimizing over a few consecutive hours in one month and a few consecutive hours in a second month, the model has no knowledge at the beginning of the second month of how much storage is currently available. As such, an optimization model that treats storage cannot provide accurate information about whether load can be met unless it treats all hours consecutively.

Finally, because optimization models are so time consuming, even at a 1-hour resolution, it is difficult, given current computing resources, to solve systems of equations with them over multiple years at 1 minute or less time resolution.

### 8.2.2.3 Trial-and-Error Simulation Models

The trial-and-error simulation model (Jacobson et al., 2015b, 2018a, 2019) was designed to overcome the main problems of the least-cost or least-carbon optimization model. Namely, it allows for many more types of energy generation and storage than an optimization model while taking orders of magnitude smaller timesteps *and* orders of magnitude less computing time to cover the same period. For example, a trial-and-error simulation assuming perfect transmission over 5.3 million 30-s timesteps (5 years) requires about 2.6 minutes on a single 3.2-GHz computer processor (Jacobson et al., 2018a). This is 1/500th to 1/100,000th the computer time of an optimization model for the same number of timesteps.

On the other hand, the trial-and-error simulation model does not necessarily find the least-cost solution. Instead, it finds multiple low-cost solutions with zero load loss. The lowest-cost solution among this set is selected.

The trial-and-error simulation model works by a user running multiple simulations, one after the other. Each simulation marches forward for several (e.g., 2 to 50) years, one timestep at a time, just as the real world does. As such, unlike an optimization model, which solves among all timesteps simultaneously, a trial-and-error model does not know what the weather will be during the next timestep. Timesteps can be of any size (e.g., seconds, minutes, hours). Results for the simulations shown in the next section are calculated with a 30-s timestep.

The trial-and-error simulation model starts from some set of initial conditions (e.g., nameplate capacities of generators and storage) and marches forward in time. Its main constraint is that electricity, heat, cold, and hydrogen load, adjusted by demand response, must match energy supply and storage during every timestep for an entire simulation period. If load is not met during a single timestep, the simulation stops. A new simulation is then restarted from the beginning with an adjustment to either the nameplate capacity of one or more generators, the characteristics of storage (peak charge rate, peak discharge rate, peak storage capacity), or characteristics of demand response. New simulations are run until load is met every timestep of the simulation period. After load is met once, additional simulations are run with further-adjusted inputs to generate a set of solutions that matches load every timestep. The lowest cost solution in this set is then selected.

Because the model does not permit load loss at any time, it is designed to exceed the utility industry standard of load loss once every 10 years. Other aspects of planning and operating the grid, such as frequency regulation (Section 8.1.6.3) and voltage control (Section 8.1.6.4), are not treated here because the model is not simulating individual lines.

Model inputs are as follows: (1) time-dependent electricity produced from onshore and offshore wind turbines, wave devices, tidal turbines, rooftop PV, utility PV, CSP, and geothermal plants; (2) a hydropower plant peak discharge rate (nameplate capacity), mean recharge rate (from rainfall), and annual average electricity output; (3) time-dependent geothermal heat and solar-thermal heat generation rates; (4) specifications of **hot water and chilled water sensible-heat thermal energy storage** (HW-STES and CW-STES) (peak charge rate, peak discharge rate, peak storage capacity, and energy losses during charging and discharging); (5) specifications of **underground thermal energy storage** (UTES), including borehole, water pit, and aquifer storage; (6) specifications of **ice storage** (ICE); (7) specifications of electricity storage in pumped hydropower storage (PHS), phase change materials coupled with concentrated solar power plants (CSP-PCM), and batteries; (8) specifications of hydrogen electrolysis, compression, and storage equipment; (9) specifications of electric heat pumps for air and water heating and cooling; (10) specifications of a demand response system; (11) losses along short- and long-distance transmission and distribution lines; (12)

scheduled and unscheduled maintenance downtimes for generators, storage, and transmission; and (13) time-dependent electricity, heat, cold, and hydrogen loads.

Cost inputs are also needed for electricity, heat, and cold generators and storage; transmission and distribution; and hydrogen electrolysis, compression, and storage. These costs include capital costs, technology lifetimes, fixed and variable operation and maintenance costs, decommissioning costs, and a social discount rate. WWS generators have no fuel costs, except the water cost for hydrogen production by electrolysis.

A trial-and-error model can be run assuming perfect transmission (without representing individual transmission lines), or it can be modified to treat some major transmission lines. In both cases, costs and power losses during transmission and distribution are calculated, but in the former case, power flows through individual transmission lines or substations are not treated.

### 8.2.3 Matching Demand with WWS Supply, Storage, and Demand Response in 24 World Regions

Most of the studies described in Section 8.2.1 discuss meeting load in specific parts of the world. An important question is whether load can be met among all energy sectors in all regions of the world with 100 percent WWS together with storage and demand response. This section discusses that issue based on results from the trial-and-error model developed in Jacobson et al. (2015b, 2018a, 2019). In the first study, the model (LOADMATCH) modeled matching demand with supply, storage, and demand response among all energy sectors in the 48 contiguous U.S. states. In the second study, it was updated and applied, with three different storage scenarios to 20 world regions encompassing 139 countries. This section summarizes results from the third study, where the model was applied to study matching demand with 100 percent WWS supply and storage in 24 world regions encompassing 143 countries.

Table 8.5 summarizes the countries included in each of the 24 world regions. Generation and load were assumed to be interconnected perfectly by transmission among all countries in each region. Wind and solar fields were predicted with a global climate model every 30 seconds for 3 years worldwide. The time-dependent wind and solar fields were used to calculate the time-dependent

electric power output from onshore and offshore wind turbines, wave devices, solar PV panels on rooftops, utility-scale solar PV panels, CSP plants, and solar thermal heat collectors every 30 seconds in each of 143 countries. The wind power estimates from the climate model accounted for competition among wind turbines for available kinetic energy. Figure 6.12 shows an example of the spatial distribution of the loss in power extracted due to such competition. The time-dependent wind and solar power generation from the climate model was input into the LOADMATCH model.

LOADMATCH also required specifications of tidal, geothermal, and hydropower electricity production; geothermal heat production; electricity, heat, cold, and hydrogen storage; transmission; electricity, heat, cold, and hydrogen demand; and demand response. Most such data are summarized in Table 7.7, Section 8.1.3, and the present section.

Ten general types of storage were treated: two for building air and water heating: hot water tank storage (HW-STES) and underground thermal energy storage (UTES); two for air conditioning and refrigeration: cold water tank storage (CW-STES) and ice storage (ICE); four for electric power storage: batteries, pumped hydropower storage (PHS), conventional hydropower storage, and phase change materials coupled with concentrated solar power plants (CSP-PCM); and two for vehicle transport (hydrogen and batteries). Conventional hydropower is a type of electricity storage because its source of power is water stored in a reservoir. But, because reservoirs can be recharged only naturally, hydropower is not artificially rechargeable unless it is turned into a PHS facility.

Demand response was treated by allowing certain flexible loads (e.g., those described in Equations 8.18 and 8.19) to be shifted forward to the next 30-s timestep for up to a maximum of 8 hours. If a flexible load was not satisfied by the 8th hour, it was converted into an inflexible load that must be satisfied immediately, or the code would fail.

With these initial inputs, the model was run for each of the 24 world regions. Heat pumps provided all building heating and cooling. Some building heating and cooling was subject to district heating. No turbines were added to existing hydropower plants (Table 8.6). Adding such turbines would allow the hydropower peak discharge rate to increase without increasing the annual average water consumed by the hydropower facility.

Table 8.5 **The 24 world regions and 143 countries within them treated in this section.**

| Region | Country(ies) within each region |
|---|---|
| Africa | Algeria, Angola, Benin, Botswana, Cameroon, Congo, Democratic Republic of the Congo, Ivory Coast, Egypt Eritrea, Ethiopia, Gabon, Ghana, Kenya, Libya, Morocco, Mozambique, Namibia, Niger, Nigeria, Senegal, South Africa, South Sudan, Sudan, Tanzania, Togo, Tunisia, Zambia, Zimbabwe |
| Australia | Australia |
| Canada | Canada |
| Central America | Costa Rica, El Salvador, Guatemala, Honduras, Mexico, Nicaragua, Panama |
| Central Asia | Kazakhstan, Kyrgyz Republic, Pakistan, Tajikistan, Turkmenistan, Uzbekistan |
| China | China, Hong Kong, Democratic Republic of Korea, Mongolia |
| Cuba | Cuba |
| Europe | Albania, Austria, Belarus, Belgium, Bosnia & Herzegovina, Bulgaria, Croatia, Cyprus, Czech Republic, Denmark, Estonia, Finland, France, Germany, Gibraltar, Greece, Hungary, Ireland, Italy, Kosovo, Latvia, Lithuania, Luxembourg, Macedonia, Malta, Moldova Republic, Montenegro, Netherlands, Norway, Poland, Portugal, Romania, Serbia, Slovakia, Slovenia, Spain, Sweden, Switzerland, Ukraine, United Kingdom |
| Haiti | Haiti, Dominican Republic |
| Iceland | Iceland |
| India | India, Nepal, Sri Lanka |
| Israel | Israel |
| Jamaica | Jamaica |
| Japan | Japan |
| Mauritius | Mauritius |
| Mideast | Armenia, Azerbaijan, Bahrain, Iran, Iraq, Israel, Jordan, Kuwait, Lebanon, Oman, Qatar, Saudi Arabia, Syrian Arab Republic, Turkey, United Arab Emirates, Yemen |
| New Zealand | New Zealand |
| Philippines | Philippines |
| Russia | Georgia, Russia |
| South America | Argentina, Bolivia, Brazil, Chile, Colombia, Ecuador, Dutch Antilles, Paraguay, Peru, Trinidad and Tobago, Uruguay, Venezuela |
| Southeast Asia | Bangladesh, Brunei Darussalam, Cambodia, Indonesia, Malaysia, Myanmar, Singapore, Thailand, Vietnam |
| South Korea | South Korea |
| Taiwan | Taiwan |
| United States | United States |

Table 8.7 shows the peak instantaneous charge rates, discharge rates, and storage capacities for six of the 24 regions simulated.

In each of the 24 world regions, the model was run forward in time, generally following the proposed priorities for matching load outlined in Section 8.1.4. Simulations were run for 3 years (3.15 million 30-s timesteps) from 2050 to 2052. For each scenario, the initial inputs were adjusted until a zero-load-loss solution was found among all timesteps, typically within 10 simulations. The model was then run another 4 to 20 times to find lower-cost solutions. Thus, multiple zero-load-loss solutions were obtained for each region, but only the lowest-cost solution is presented.

Table 8.6 **Several of the main characteristics of the model simulations discussed here for matching supply with demand in 24 different world regions among 143 countries.**

| Parameter | Case |
| --- | --- |
| Onshore and offshore wind electricity | Yes |
| Residential, comm./govt. PV electricity | Yes |
| Utility PV electricity | Yes |
| CSP electricity | Yes |
| Geothermal electricity | Yes |
| Tidal and wave electricity | Yes |
| Direct solar and geothermal heat | Yes |
| Battery storage | Yes |
| CSP storage | Yes |
| Pumped hydropower storage | Yes |
| Existing hydropower dam storage | Yes |
| Added hydropower turbines | No |
| Heat storage (water, underground) | Yes |
| Cold storage (water, ice) | Yes |
| Hydrogen storage | Yes |
| Hydrogen fuel cells for transportation | Yes |
| Battery-electric vehicles | Yes |
| District heating | Some |
| Electric heat pumps | Yes |
| Electrified industry | Yes |
| Losses from T&D, storage, and downtime | Yes |
| Wind array losses | Yes |
| Perfect transmission interconnections | Yes |
| Costs of generation, T&D, and storage | Yes |
| Avoided cost of air pollution damage | Yes |
| Avoided cost of climate damage | Yes |
| Land requirements | Yes |
| Changes in job numbers | Yes |

T&D = transmission and distribution

Supply matched total load (end-use load plus changes in storage plus losses plus shedding) every 30 s for 3 years in all 24 regions encompassing the 143 countries. Results are summarized for select regions in Figure 8.8 and Tables 8.8 to 8.10. Table 8.9 summarizes energy costs among all world regions.

The WWS cost per unit energy in each case includes the costs of new electricity and heat generation, short-distance transmission, long-distance transmission, distribution, heat storage, cold storage, electricity storage, and hydrogen production/compression/storage (Table 8.8).

In a 2050 WWS world, WWS energy private costs (costs of energy alone) are assumed to equal WWS energy social costs (energy private costs plus health and climate damage costs due to energy). The reason is that, in 2050, WWS generators, storage, and transmission will result in zero emissions while in use. Further, their production and decommissioning will also be free of energy-related emissions. The health and climate costs of zero emissions are zero.

The 2050 all-energy WWS mean social cost per unit energy, when weighted by generation among all 24 regions, is 8.96 ¢/kWh-all-energy (USD 2013) (Table 8.8). However, Figure 8.9 shows that the individual regional averages range from 6.54 ¢/kWh-all-energy (Iceland) to 13.1 ¢/kWh-all-energy (Israel). The largest portion of cost is generation (which includes the costs of capital, operation, maintenance, and decommissioning), followed by transmission and distribution, electricity storage, hydrogen production, and thermal energy storage, respectively.

Figure 8.9 also indicates that the overall upfront capital cost of transitioning 143 countries while keeping the grid stable, spent between today and 2050, may be about $72.8 trillion (USD 2013). Individual regional ranges are $2.6 billion for Iceland to $16.6 trillion for China. The U.S. cost is about $7.8 trillion, and the Europe cost, about $6.2 trillion.

However, the **upfront capital cost** is not the most relevant metric. A more useful metric is the aggregate annual energy cost in comparison with business-as-usual (BAU). An even more relevant metric is the WWS-to-BAU aggregate social cost ratio, derived shortly. Figure 8.10 illustrates these parameters.

Figure 8.10a provides the 2050 cost per unit energy for each region in the 100 percent WWS cases. It also shows, for comparison, the estimated cost (per unit energy) of energy, air pollution, and climate damage from the BAU case. The figure indicates that, whereas the private costs per unit energy of WWS and BAU energy alone are similar, the WWS cost per unit energy of energy plus health damage plus climate damage is only about 21 percent that of BAU. Multiplying the private cost per unit energy in Figure 8.10a by the end-use energy consumed per year (or by the annual average power) in the WWS and BAU cases, respectively, gives the **aggregate annual**

**Table 8.7** Aggregate (among all storage devices in a country or region) maximum instantaneous charge rates, maximum instantaneous discharge rates, and maximum energy storage capacities of the different types of electricity storage (PHS, CSP-PCM, batteries, and hydropower), cold storage (CW-STES and ICE), and heat storage (HW-STES and UTES) technologies treated in the simulations here, for six world regions.

| Storage technology | Max charge rate (GW) | Max discharge rate (GW) | Max storage capacity (TWh) | Max charge rate (GW) | Max discharge rate (GW) | Max storage capacity (TWh) | Max charge rate (GW) | Max discharge rate (GW) | Max storage capacity (TWh) |
|---|---|---|---|---|---|---|---|---|---|
| | United States | | | China | | | Europe | | |
| PHS | 95.8 | 95.8 | 1.342 | 116 | 116 | 1.62 | 197 | 197 | 2.76 |
| CSP-elec. | 92.9 | 92.9 | – | 296 | 296 | – | 21.1 | 21.1 | – |
| CSP-PCM | 149.8 | – | 2.098 | 478 | – | 6.69 | 34.0 | – | 0.475 |
| Batteries | 3,300 | 3,300 | 6.402 | 2,600 | 2,600 | 5.04 | 1,200 | 1,200 | 2.33 |
| Hydropower | 36.7 | 80.1 | 321 | 146 | 318 | 1,279 | 75.8 | 167.4 | 664 |
| CW-STES | 6.0 | 6.0 | 0.084 | 11.0 | 11.0 | 0.154 | 10.1 | 10.1 | 0.142 |
| ICE | 9.0 | 9.0 | 0.126 | 16.5 | 16.5 | 0.232 | 15.2 | 15.2 | 0.213 |
| HW-STES | 324 | 324 | 2.59 | 712 | 712 | 3.56 | 507 | 507 | 3.04 |
| UTES-heat | 18.3 | 324 | 15.5 | 351 | 712 | 137 | 168 | 507 | 122 |
| UTES-electricity | 972 | – | – | 1,425 | – | – | 760 | – | – |
| | Jamaica | | | Australia | | | Africa | | |
| PHS | 3.00 | 3.00 | 0.042 | 10.7 | 10.7 | 0.150 | 27.8 | 27.8 | 0.389 |
| CSP-electricity | 0.28 | 0.28 | – | 13.0 | 13.0 | – | 45.9 | 45.9 | – |
| CSP-PCM | 0.45 | – | 0.006 | 21.0 | – | 0.294 | 74.1 | – | 1.037 |
| Batteries | 15.0 | 15.0 | 0.029 | 500 | 500 | 0.970 | 1,300 | 1,300 | 2.52 |
| Hydropower | 0.01 | 0.02 | 0.080 | 3.7 | 8.1 | 32.7 | 12.4 | 29.3 | 109 |
| CW-STES | 0 | 0 | 0 | 0.4 | 0.4 | 0.005 | 1.2 | 1.2 | 0.018 |
| ICE | 0 | 0 | 0 | 0.5 | 0.5 | 0.007 | 1.9 | 1.9 | 0.026 |
| HW-STES | 0.44 | 0.44 | 0.004 | 23.2 | 23.2 | 0.186 | 90.8 | 90.8 | 0.727 |
| UTES-heat | 0.00 | 0.44 | 0.011 | 6.6 | 23.2 | 2.78 | 2.0 | 90.8 | 21.8 |
| UTES-electricity | 0.13 | – | – | 69.6 | – | – | 272.5 | – | – |

PHS = pumped hydropower storage; PCM = phase change material; CSP = concentrated solar power; CW-STES = chilled water sensible-heat thermal energy storage; HW-STES = hot water sensible-heat thermal energy storage; and UTES = underground thermal energy storage (either boreholes, water pits, or aquifers). The peak energy storage capacity equals the maximum discharge rate multiplied by the maximum number of hours of storage at the maximum discharge rate. Storage times for PHS and ICE were 14 hours; for CSP times were 22.6 hours; for batteries times were 1.94 hours (patterned after the Tesla Powerwall 1, which has 3.3-kW maximum charge and discharge rates and 6.4 kWh of maximum storage); for CW-STES and HW-STES times were 8 hours, except in China, where they were 5 hours and Europe, where they were 6 hours; and for UTES times were 1 day in Jamaica, 2 days in the United States, 5 days in Australia, 8 days in China, and 10 days in Africa and Europe.

Heat captured by CSP solar collectors can either be used immediately to produce electricity, put in storage, or both. The maximum direct CSP electricity production rate (CSP-elec.) equals the maximum electricity discharge rate, which equals the nameplate capacity of the generator. The maximum charge rate of CSP phase change material storage (CSP-PCM) is set to 1.612 multiplied by the maximum electricity discharge rate, which allows more energy to be

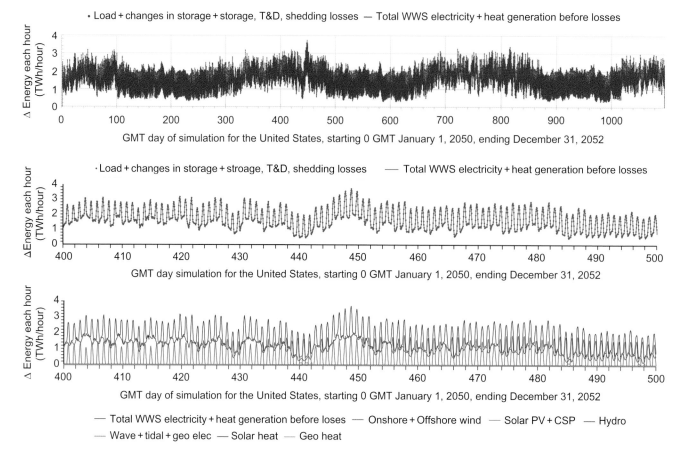

Figure 8.8 **Time-series comparison, from 2050 to 2052, for the United States, the China region (China–Hong Kong–Mongolia–North Korea), and Jamaica. First row: modeled total WWS power generation versus total load plus losses plus changes in storage plus shedding over all three years. Second row: same as first row, but for a window of days 400 to 500 during the 3-year period. Third row: a breakdown of WWS power generation by source during the window. The model was run at 30-s resolution. Results are shown hourly. No load loss occurred during any 30-s interval. From Jacobson et al. (2019).**

**private energy cost** in each case. The result is shown in Figure 8.10b. The figure indicates that, among 143 countries, the aggregate annual private energy cost in the WWS case is $6.83 trillion/y and in the BAU case is $17.7 trillion/y. The main difference is the 57.1 percent lower end-use energy consumption in the WWS case (Table 7.7).

---

**Table 8.7 (*cont.*)**

collected than discharged directly. Thus, the maximum overall simultaneous direct electricity plus storage CSP production rate is 2.612 multiplied by the discharge rate. The maximum energy storage capacity equals the maximum electricity discharge rate multiplied by the maximum number of hours of storage at full discharge, set to 22.6 hours, or 1.612 multiplied by the 14 hours required for CSP storage to charge when charging at its maximum rate.

Hydropower can be charged only naturally, but its annual-average charge rate must equal at least its annual energy output divided by the number of hours per year. It is assumed simplistically here that hydro is recharged at that rate, where its annual energy output in 2050 is close to its current value. Hydropower's maximum discharge rate in 2050 is its 2018 nameplate capacity. The maximum storage capacity is set equal to the 2050 annual energy output of hydro.

The CW-STES charge/discharge rate is set equal to 40 percent of the maximum daily averaged cold load subject to storage, which itself is calculated as the maximum of Equation 8.32 during the period of simulation. The ICE storage charge/discharge rate is set to 60 percent of the same peak cold load subject to storage.

The HW-STES charge and discharge rates are set equal to the maximum daily averaged heat load subject to storage, calculated as the maximum value during the period of simulation from Equation 8.29.

UTES heat stored in underground soil can be charged by either solar or geothermal heat or excess electricity. The maximum charge rate of heat to UTES storage (UTES-heat) is set to the nameplate capacity of the solar thermal collectors. In several regions, no solar thermal collectors are used. Instead, UTES is charged only with excess grid electricity. The maximum charge rate of excess grid electricity converted to heat stored in UTES (UTES-elec.) is set by trial and error for each country. The maximum UTES heat discharge rate is set to that of HW-STES storage, which is limited by the warm storage load.

Source: Jacobson et al. (2019).

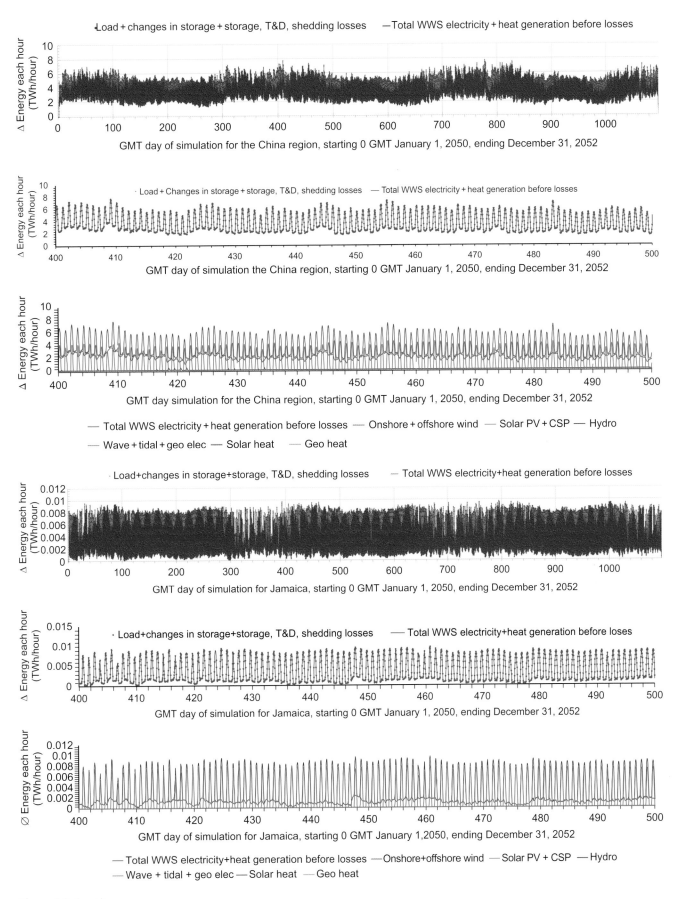

**Figure 8.8** (*cont.*)

**Table 8.8** Summary of 2050 mean capital costs of new electricity plus heat generators and storage ($trillion in 2013 USD) and mean levelized private costs of energy (LCOE) (¢/kWh-all-energy or ¢/kWh-electricity-replacing-BAU-electricity) averaged over the 3-year simulations for six world regions (defined in Table 8.5) and among all 24 world regions. Also shown are the energy consumed per year in each case and the resulting aggregate annual energy cost to the region.

| | United States | China | Europe | Jamaica | Australia | Africa | 24 world regions |
|---|---|---|---|---|---|---|---|
| Capital cost new generators only ($trillion) | 6.85 | 14.83 | 5.16 | 0.022 | 0.71 | 3.33 | **63.7** |
| Capital cost new generators + storage ($trillion) | 7.819 | 16.58 | 6.15 | 0.025 | 0.82 | 3.77 | **72.8** |
| *Components of total LCOE (¢/kWh-all-energy)* | | | | | | | |
| Short-distance transmission (¢/kWh-all-energy) | 1.05 | 1.05 | 1.05 | 1.05 | 1.05 | 1.05 | **1.05** |
| Long-distance transmission | 0.121 | 0.121 | 0.121 | 0.000 | 0.122 | 0.122 | **0.116** |
| Distribution | 2.375 | 2.375 | 2.375 | 2.375 | 2.375 | 2.375 | **2.375** |
| Electricity generators | 5.109 | 4.324 | 3.984 | 5.484 | 4.590 | 4.307 | **4.747** |
| Additional hydro turbines | 0 | 0 | 0 | 0 | 0 | 0 | **0** |
| Solar thermal collectors | 0.056 | 0.169 | 0.229 | 0.000 | 0.072 | 0.005 | **0.09** |
| CSP-PCM+PHS+battery storage | 0.436 | 0.168 | 0.171 | 0.952 | 0.657 | 0.339 | **0.388** |
| CW-STES+ICE storage | 0.004 | 0.003 | 0.006 | 0.000 | 0.002 | 0.001 | **0.003** |
| HW-STES storage | 0.012 | 0.007 | 0.014 | 0.007 | 0.008 | 0.006 | **0.01** |
| UTES storage | 0.010 | 0.035 | 0.076 | 0.003 | 0.018 | 0.027 | **0.044** |
| $H_2$ production/compression/storage | 0.157 | 0.082 | 0.158 | 0.247 | 0.149 | 0.146 | **0.141** |
| **Total LCOE (¢/kWh-all-energy)** | **9.33** | **8.33** | **8.18** | **10.12** | **9.04** | **8.38** | **8.964** |
| LCOE (¢/kWh-replacing BAU electricity) | 9.14 | 8.21 | 7.93 | 9.86 | 8.86 | 8.19 | **8.762** |
| GW annual avg. end-use demand (Table 8.10) | 939.5 | 2,328.4 | 939.7 | 2.3 | 93.6 | 481.7 | **8,693** |
| TWh/y end-use demand (GW × 8,760 h/y) | 8,230 | 20,397 | 8,232 | 20 | 820 | 4,220 | **76,151** |
| Annual energy cost ($billion/y) | **767.6** | **1,699.5** | **673.5** | **2.0** | **74.1** | **353.6** | **6,825** |

The LCOEs are derived from capital costs assuming a social discount rate for an intergenerational project (Section 7.6.1) of 2.0 (1 to 3) percent and lifetimes, annual O&M, and end-of-life decommissioning costs that vary by technology, all divided by the total annualized end-use demand met, given in the present table. Capital costs are an estimated average of those between 2015 and 2050 and are a mean (in USD 1 million/MW) of 1.27 for onshore wind, 3.06 for offshore wind, 2.97 for residential rooftop PV, 2.06 for commercial/government PV, 1.32 for utility PV, 4.33 for CSP with storage, 3.83 for geothermal electricity and heat, 2.81 for hydropower, 3.57 for tidal, 4.01 for wave, and 1.22 for solar thermal for heat.

Since the total end-use load includes heat, cold, hydrogen, and electricity loads (all energy), the "Electricity generators" cost, for example, is a cost per unit all energy rather than per unit electricity alone. The "Total LCOE" gives the overall cost of energy, and the "Electricity LCOE" gives the cost of energy for the electricity portion of load replacing BAU electricity end use. It is the total LCOE less the costs for UTES and HW-STES storage, $H_2$, and less the portion of long-distance transmission associated with $H_2$.

Short-distance transmission costs are $0.0105 (0.01–0.011)/kWh.

Distribution costs are $0.02375 (0.023–0.0245)/kWh.

Long-distance transmission costs are $0.00406 (0.00152–0.00903)/kWh (in USD 2013) (Table S28 of Jacobson et al., 2017 brought from 2012 to 2013), which assumes 1,200- to 2,000-km lines. It is assumed that 30 percent of all annually averaged electricity generated is subject to long-distance transmission in all regions except Cuba, Haiti, Iceland, Israel, Jamaica, Mauritius, South Korea, and Taiwan (0 percent); New Zealand (15 percent); and Central America, Japan, and the Philippines (20 percent).

Storage costs are the product of the storage peak energy capacity and the levelized cost of storage per unit of storage peak energy capacity of each storage technology (Table 8.4), annualized with the same social discount rate as for power generators, but with 2050 storage lifetimes of 17 (12–22) years for batteries and 32.5 (25–40) years for non-battery storage, all divided by the annual average end-use load met, given in the table.

$H_2$ costs are derived as in Table 7.12. These costs exclude electricity costs, which are included separately in the present table.

Source: Jacobson et al. (2019).

Table 8.9 2050 143-country (24-world-region) WWS versus BAU average social cost per unit energy. A social cost includes energy, health, and climate costs for BAU and energy costs only for WWS, since WWS eliminates virtually all health and climate costs. Also shown is the WWS-to-BAU aggregate social cost ratio and the components of its derivation (Equation 8.34).

| | |
|---|---|
| a) BAU electricity private cost per unit energy (¢/kWh)[1] | 9.99 |
| b) BAU health cost per unit energy (¢/kWh) | 16.9 |
| c) BAU climate cost per unit energy (¢/kWh) | 16.0 |
| d) BAU social cost per unit energy (¢/kWh) (a+b+c) | 42.9 |
| e) WWS private and social cost per unit energy (¢/kWh)[1] | 8.96 |
| f) BAU end-use power demand (GW)[2] | 20,255 |
| g) WWS end-use power demand (GW)[2] | 8.693 |
| h) BAU electricity sector aggregate annual energy private cost ($tril/y) (af) | 17.7 |
| i) BAU health cost ($tril/y) (bf) | 30.0 |
| j) BAU climate cost ($tril/y) (cf) | 28.4 |
| k) BAU social cost ($tril/y) (df) | 76.1 |
| l) WWS private and social cost ($tril/y) (eg) | 6.82 |
| m) WWS-to-BAU energy private cost/kWh ratio ($R_{WWS:BAU-E}$) (e/a) | 0.90 |
| n) BAU-energy-private-cost/kWh-to-BAU-social-cost/kWh ratio ($R_{BAU-S:E}$) (a/d) | 0.23 |
| o) WWS-kWh-used-to-BAU-kWh-used ratio ($R_{WWS:BAU-C}$) (g/f) | 0.43 |
| WWS-to-BAU aggregate social cost ratio ($R_{ASC}$) (mno) | 0.09 |
| WWS-to-BAU aggregate private cost ratio ($R_{APC}$) (mo) | 0.39 |
| WWS-to-BAU social cost per unit energy ratio ($R_{SCE}$) (mn) | 0.21 |

[1] This is the BAU electricity-sector cost of energy per unit energy. It is assumed to equal the BAU cost of all energy per unit energy. The WWS cost per unit energy is already for all energy, which is almost all electricity (plus a small amount of direct heat).
[2] Multiply GW by 8,760 hr/y to obtain GWh/y.
Source: Jacobson et al. (2019).

A social cost of an energy source is the source's private energy cost plus the health and climate costs of the chemical emissions from the source. The **WWS aggregate annual social cost** ($A_{WWS}$) is effectively the same as the WWS aggregate annual private energy cost, $6.83 trillion/y, because WWS has virtually no health or climate impacts. The **BAU aggregate annual social cost** ($A_{BAU}$), on the other hand, is $76.1 trillion/y.

The ratio of the two social costs is the **WWS-to-BAU aggregate social cost ratio** ($R_{ASC}$),

$$R_{ASC} = A_{WWS}/A_{BAU} = R_{WWS:BAU-E} \times R_{BAU-S:E} \times R_{WWS:BAU-C} \tag{8.34}$$

where $R_{WWS:BAU-E}$ is the WWS-to-BAU levelized cost of energy (cost per kWh) ratio, $R_{BAU-S:E}$ is the BAU-energy-cost-per-kWh-to-social-cost-per-kWh ratio, and $R_{WWS:BAU-C}$ is the WWS-to-BAU end-use energy consumption (kWh/y or annual average kW) ratio. Based on the aggregate annual social costs in Figure 8.10, $R_{ASC}$ is 9 percent (Figure 8.10c). Thus, in terms of aggregate social costs (private plus health plus climate costs of energy), a WWS system costs only 9 percent of a BAU system each year.

Table 8.9 illustrates how to calculate the WWS-to-BAU aggregate annual social cost ratio using the second definition in Equation 8.34. The result is the same. The annual energy plus health plus climate cost in a 100 percent WWS region is 91 percent less than when the region is under BAU.

A related parameter is the **WWS-to-BAU aggregate private cost ratio**,

$$R_{AEC} = R_{WWS:BAU-E} \times R_{WWS:BAU-C} \tag{8.35}$$

which gives an indication of the aggregate private energy cost per year in a region in a WWS versus a BAU case. The 143-country aggregate private cost ratio from Table 8.9 is 39 percent. In other words, people in a region with 100 percent WWS pay 61 percent less for energy per year than when BAU energy is used.

A third parameter is the **WWS-to-BAU social cost per unit energy ratio**,

$$R_{SCE} = R_{WWS:BAU-E} \times R_{BAU-S:E} \tag{8.36}$$

which gives an indication of the energy plus health plus climate cost per unit energy in a WWS case versus a BAU case. In the present example for 143 countries, the social cost per unit energy is 79 percent less in the WWS case than in the BAU case.

Figure 8.9 indicates that the 2050 WWS cost of energy per unit energy is relatively low for big regions (e.g.,

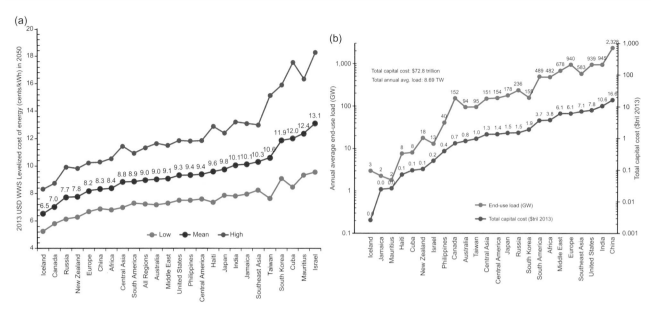

**Figure 8.9** Energy private costs, capital costs, and load by world region. (a) Low, mean, and high modeled levelized private costs of all energy in 24 world regions encompassing 143 countries upon converting all countries for all energy purposes to 100 percent WWS. (b) Annual average all-purpose end-use load and overall capital cost for a 100 percent WWS system.

Canada, Russia, Africa, China, Europe, the United States) and small countries with good WWS resources (e.g., Iceland, New Zealand). The large geographical areas of the big regions allow for the effective aggregation of wind and solar power, resulting in less intermittency of these resources (Section 8.1.1). These regions also have a good balance of solar and wind, which are complementary in nature seasonally. Finally, they have substantial existing hydropower resources that can provide peaking power. Iceland has substantial resources of hydropower, geothermal, and wind.

## Transition highlight

In island countries, such as in the Caribbean, the actual direct price paid for BAU electricity is a mean of about 33 ¢/kWh. This price reflects, among other factors, the cost of transporting fuels to the islands and price hikes due to frequent supply shortages. Such high costs should not occur when WWS electricity, which is produced locally, is combined with storage.

The current BAU energy cost per unit energy (33 ¢/kWh) is 2.8 to 3.4 times the modeled cost of WWS energy replacing BAU energy for Haiti–Dominican Republic (9.62 ¢/kWh), Cuba (~12.0 ¢/kWh), and Jamaica (10.1 ¢/kWh) (Figure 8.9). In addition, the WWS-to-BAU energy consumption ratios in those three regions are 0.41, 0.56, and 0.48, respectively (Figure 8.10c), so the aggregate private cost people will pay in a 100 percent WWS world will be 12 to 20 percent what they pay currently.

## Example 8.3 Aggregate social cost ratio

Calculate the WWS-to-BAU aggregate social cost ratio, aggregate private cost ratio, and social cost per unit energy ratio in a country if the BAU energy, air pollution, and climate costs are 10, 15, and 15 ¢/kWh, respectively, the WWS energy cost is 10 ¢/kWh, and the end-use energy consumption in the WWS case is 50 percent that in the BAU case.

### Solution:

In this example, $R_{WWS:BAU-E}$ = 10 ¢/kWh / 10 ¢/kWh = 1, $R_{BAU-S:E}$ = 10 ¢/kWh / (10+15+15) ¢/kWh = 0.25, and $R_{WWS:BAU-C}$ = 0.5.

   Thus, the aggregate social cost ratio is $R_{ASC}$ = 1 × 0.25 × 0.5 = 0.125

   The aggregate private cost ratio is $R_{AEC}$ = 1 × 0.5 = 0.5

   The social cost per unit energy ratio is $R_{SCE}$ = 1 × 0.25 = 0.25

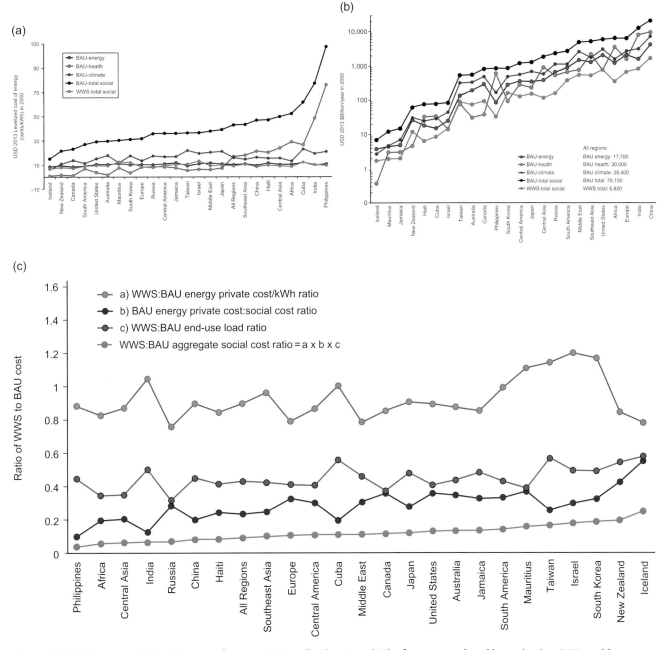

**Figure 8.10** WWS versus BAU social costs of energy. (a) Levelized cost per kWh of energy produced by region in a BAU world versus a WWS world. BAU costs include energy, health, and climate costs. WWS costs include only energy costs, because energy externality costs are approximately zero. (b) Same as (a), except for the annual aggregate cost per year, obtained by multiplying the cost per unit energy in (a) by the end-use energy consumption per year in the BAU or WWS case, respectively. Note the resulting spread in the BAU versus WWS energy cost. (c) The WWS-to-BAU aggregate social cost ratio and the component factors: the WWS-to-BAU cost per unit energy ratio (obtained from panel (a)), the BAU energy private-cost-to-social-cost ratio (obtained from panel (b)), and the WWS-to-BAU end-use load ratio (e.g., from Table 7.7, but for each region). Ninety percent of all air pollution mortalities are ascribed to BAU energy. Most of the rest are ascribed to open biomass burning, wildfires, and dust. All costs are in 2013 USD. From Jacobson et al. (2019).

Costs are highest in small countries with high population densities (Taiwan, Cuba, South Korea, Mauritius, and Israel). Nevertheless, the WWS private energy cost per year in all five regions is 43 to 65 percent that of BAU, indicating that a transition to WWS reduces energy costs even under the least favorable circumstances (Jacobson et al., 2019).

Table 8.10 summarizes the energy budget (total end-use load, supply, changes in storage, and losses) for six of the world regions. Statistics among all 24 world regions

Table 8.10 **Summary of energy requirements met, energy losses, energy supplies, and changes in storage, during the 3-year (26,291.5-hour) simulations for 6 world regions. All units are TWh over the 3-year simulation. Table 8.5 identifies the countries within each region.**

| | United States | China | Europe | Jamaica | Australia | Africa |
|---|---|---|---|---|---|---|
| **A1. Total end-use demand** | **24,696** | **61,207** | **24,705** | **60** | **2,461** | **12,665** |
| Electricity for electricity inflexible demand | 11,465 | 27,306 | 10,186 | 26 | 1,227 | 5,926 |
| Electricity for electricity, heat, cold storage + DR | 11,180 | 31,998 | 12,854 | 27 | 1,054 | 5,673 |
| Electricity for $H_2$ direct use + $H_2$ storage | 2,051 | 1,903 | 1,664 | 6 | 180 | 1,066 |
| **A2. Total end-use demand** | **24,696** | **61,207** | **24,705** | **60** | **2,461** | **12,665** |
| Electricity for direct use, electricity storage, + $H_2$ | 22,027 | 55,335 | 19,675 | 58 | 2,302 | 11,438 |
| Low-T heat load met by heat storage | 2,514 | 5,607.6 | 4,757 | 1.3 | 150.9 | 1196.0 |
| Cold load met by cold storage | 155 | 264 | 272 | 0 | 8 | 32 |
| **A3. Total end-use demand** | 24,696 | **61,207** | **24,705** | **60** | **2,461** | **12,665** |
| Electricity for direct use, electricity storage, DR | 19,720 | 52,834 | 17,346 | 52 | 2,105 | 10,310 |
| Electricity for $H_2$ direct use + $H_2$ storage | 2,051 | 1,903 | 1,664 | 6 | 180 | 1,066 |
| Electricity + heat for heat subject to storage | 2,529 | 5,746 | 5,029 | 1 | 153 | 1,207 |
| Electricity for cold load subject to storage | 395 | 725 | 666 | 0 | 23 | 82 |
| | | | | | | |
| **B. Total losses** | 12,111 | **23,522** | **5,710** | **27.6** | **1010** | **2596** |
| Transmission, distribution, downtime losses | 2,587 | 5,798.30 | 2,031.65 | 4.65 | 230 | 962 |
| Losses CSP storage | 4.00 | 14.28 | 1.09 | 0.04 | 1.49 | 4.00 |
| Losses PHS storage | 69.4 | 69.7 | 172.8 | 2.3 | 20.1 | 38.6 |
| Losses battery storage | 27.1 | 17.5 | 9.5 | 0.0 | 11.6 | 35.6 |
| Losses CW-STES + ICE storage | 28.0 | 47.69 | 49.21 | 0.00 | 1.39 | 5.73 |
| Losses HW-STES storage | 464 | 813.8 | 818.1 | 0.3 | 20.7 | 158.7 |
| Losses UTES storage | 102 | 1,043 | 415 | 0 | 36 | 328 |
| Losses from shedding | 8,828 | 15,717.6 | 2,212.9 | 20.4 | 689 | 1,063 |
| **Net end-use demand plus losses (A1 + B)** | *36,807* | *84,728* | *30,415* | *87* | *3,470* | *15,261* |
| | | | | | | |
| **C. Total WWS supply before T&D losses** | **36,807** | **84,741** | **30,427** | **87.4** | **3,471** | **15,256** |
| Onshore + offshore wind electricity | 22,124 | 49,831.8 | 18,554 | 27.8 | 1,281 | 8,613 |
| Rooftop + utility PV+ CSP electricity | 12,978 | 30,071.5 | 8,949 | 59.3 | 2,008 | 6,073 |
| Hydropower electricity | 1,170 | 4,375.88 | 2,409 | 0.247 | 139.5 | 415 |
| Wave electricity | 255 | 31.78 | 96.68 | 0.000 | 25.41 | 63.6 |
| Geothermal electricity | 153 | 43.83 | 71.71 | 0.000 | 9.51 | 76.8 |
| Tidal electricity | 2.24 | 18.69 | 93.15 | 0.111 | 3.25 | 11.7 |
| Solar heat | 13.85 | 253.41 | 110.61 | 0.000 | 4.81 | 1.49 |
| Geothermal heat | 111 | 114.41 | 142.77 | 0.000 | 0.10 | 0.88 |

Table 8.10 (cont.)

| | United States | China | Europe | Jamaica | Australia | Africa |
|---|---|---|---|---|---|---|
| **D. Net taken from (+) or added to (–) storage** | **–0.26** | **–12.80** | **–12.52** | **–0.007** | **–1.04** | **5.58** |
| CSP storage | –0.21 | –0.27 | –0.05 | 0.003 | 0.02 | 0.34 |
| PHS storage | –0.13 | –0.16 | –0.28 | –0.003 | –0.04 | 0.21 |
| Battery storage | –0.64 | –0.50 | –0.23 | –0.003 | –0.24 | 0.21 |
| CW-STES+ICE storage | –0.021 | 0 | 0 | 0.000 | 0 | 0 |
| HW-STES storage | 2.331 | 2.076 | 0.311 | 0.000 | –0.046 | 0.654 |
| UTES storage | –1.344 | –13.678 | –12.164 | –0.001 | –0.696 | 3.253 |
| $H_2$ storage | –0.244 | –0.23 | –0.077 | –0.002 | –0.049 | 0.876 |
| **Energy supplied plus taken from storage (C+D)** | **36,807** | **84,728** | **30,415** | **87.4** | **3,470** | **15,261** |

Transmission/distribution/maintenance losses are 7.5 (5–10) percent for all technologies except for rooftop PV, which are 1.5 (1–2) percent and solar thermal for heat, which are 3 (2–4) percent. Table 8.4 gives round-trip storage efficiencies. Electricity generation is shed when it exceeds electricity demand, cold storage capacity, heat storage capacity, and $H_2$ storage capacity. Onshore and offshore wind turbines in the climate model are assumed to be Senvion (formerly Repower) 5-MW turbines with 126-m-diameter rotors, 100-m hub heights, a cut-in wind speed of 3.5 m/s, and a cut-out wind speed of 30 m/s. Rooftop PV panels in GATOR-GCMOM were modeled as fixed-tilt panels at the optimal tilt angle of the country they resided in; utility PV panels were modeled as half fixed optimal tilt and half single-axis horizontal tracking (Jacobson and Jadhav, 2018). All panels were assumed to have a nameplate capacity of 390 W and a panel area of 1.629668 m$^2$, which gives a 2050 panel efficiency (watts of power output per watt of solar radiation incident on the panel) of 23.9 percent, which is an increase from the 2015 value of 20.1 percent. Each CSP plant before storage is assumed to have the mirror and land characteristics of the Ivanpah solar plant, which has 646,457 m$^2$ of mirrors and 2.17 km$^2$ of land per 100-MW nameplate capacity and a CSP efficiency (fraction of incident solar radiation that is converted to electricity) of 15.796 percent, calculated as the product of the reflection efficiency of 55 percent and the steam plant efficiency of 28.72 percent. The efficiency of the solar heat hot fluid collector (energy in fluid divided by incident radiation) is 34 percent.

Of all end-use load, 9.65 percent was electricity or direct heat used for low-temperature heat subject to storage, 1.13 percent was electricity used for cold subject to storage, 6.01 percent was electricity used for hydrogen that was either stored or used immediately, and the rest (83.21 percent) was remaining electricity used to meet immediate inflexible demand or subject to demand response.

Source: Jacobson et al. (2019).

suggest the following: 72.2 percent of all energy produced or supplied from storage was used for end-use load; 6.7 percent was lost during short- and long-distance transmission, distribution, and downtime; 2.7 percent was lost during transfer in and out of storage; and 18.4 percent was shed. Most storage losses occurred with UTES storage.

Of all energy generated (for end-use load plus losses plus shedding plus changes in storage), among the 24 regions, 49.5 percent was generated by onshore plus offshore wind, 44.1 percent was generated by utility plus rooftop PV plus CSP, 5.02 percent was generated by hydropower, 0.71 percent was generated by geothermal electricity, 0.26 percent was generated by wave electricity, 0.06 percent was generated by tidal electricity, 0.14 percent was generated by solar heat, and 0.14 percent was generated by geothermal heat.

## 8.3 Estimating Footprint and Spacing Areas of WWS Generators

Once the WWS generation system has been sized to ensure demand can be met with WWS supply and storage, it is possible to estimate the land and water area

requirements of the system. Footprint is the physical area on the top surface of soil or water needed for each energy device. It does not include areas of underground structures (Section 6.7). Spacing is the area between some devices, such as wind turbines, wave devices, and tidal turbines, needed to minimize interference of the wake of one turbine with downwind turbines. Spacing area can be used for multiple purposes, including rangeland, ranching land, forest land, solar panel installation, open space, and open water. Table 8.11 provides estimated footprint and spacing areas per MW of nameplate capacity of WWS energy generation technologies considered here.

Applying the installed power densities in Table 8.11 to the nameplate capacities needed to provide grid stability in the 143 countries aggregated among 24 world regions gives the total land footprint and spacing areas required in these regions. New land footprint arises only for solar PV plants, CSP plants, onshore wind turbines, geothermal plants, and solar thermal plants. Offshore wind, wave, and tidal generators are in water, so they do not take up new land, and rooftop PV does not take up new land. The footprint area of a wind turbine is relatively trivial. It is primarily the area of the cement around the base of the turbine tower that is

Table 8.11 **Footprint and spacing areas per MW of nameplate capacity and installed power densities for WWS electricity or heat generation technologies.**

| WWS technology | Footprint (m²/MW) | Spacing (km²/MW) | Installed power density (MW/km²) |
|---|---|---|---|
| Onshore wind | 3.22 | 0.051 | 19.8 |
| Offshore wind | 3.22 | 0.139 | 7.2 |
| Wave device | 700 | 0.033 | 30.3 |
| Geothermal plant | 3,290 | – | 304 |
| Hydropower plant | 502,380 | – | 2.0 |
| Tidal turbine | 290 | 0.004 | 250 |
| Residential rooftop PV | 5,230 | – | 191 |
| Commercial/ government rooftop PV | 5,230 | – | 191 |
| Solar PV plant | 12,220 | – | 81.8 |
| Utility CSP plant | 29,350 | – | 34.1 |
| Solar thermal for heat | 1,430 | – | 700 |

Source: Jacobson et al. (2017), except that spacing areas for onshore and offshore wind are calculated assuming 20 MW/km² for onshore wind and 7.2 MW/km² for offshore wind, based on data for Europe and outside of Europe (Enevoldsen and Jacobson, 2020) from Table 6.4, and the footprint area for solar thermal for heat is from Jacobson et al. (2015b). The installed power density is the inverse of the spacing, except when the spacing has no value, in which case it is the inverse of the footprint.

visible above the ground. Paved roads associated with wind farms are also considered footprint, but unpaved roads are not.

The total new land area for footprint required with 100 percent WWS is about 0.17 percent of the 143-country land area (Figure 8.11). Almost all of that is for utility PV and CSP. WWS has no footprint associated with mining fuels to run equipment, but both WWS and BAU energy infrastructures require one-time mining for raw materials for new equipment.

The only spacing area over land needed in a 100 percent WWS world is the land area between onshore wind turbines. Figure 8.11 indicates that the spacing area for onshore wind to power 143 countries is about 0.48 percent of the 143-country land area.

Together, the new land footprint and spacing areas for 100 percent WWS across all energy sectors are 0.65 percent of the 143-country land area (Figure 8.11), and most of this land area is multipurpose spacing land. This is equivalent to about 1.85 percent of California's land area for virtually all world energy. In comparison, about 37.4 percent of the world's land was agricultural land in 2016 and 2.5 percent was urban area in 2010 (World Bank, 2017).

In another comparison, Table 3.3 indicates that the land area required for the fossil-fuel infrastructure in the United States is about 1.3 percent of the U.S. land area. Thus, replacing fossil fuels with 100 percent WWS should reduce land area substantially.

In addition, solar PV panels can be installed on some of the space between wind turbines. As such, some footprint and spacing areas overlap, and the land plus footprint area in Figure 8.11 of 0.65 percent is an upper limit. The lower limit is the spacing area alone, which is 0.48 percent.

## 8.4 Estimating Jobs Created and Lost as Part of a Transition

A final metric relevant to policy decision-making is net job creation and loss. A transition to WWS reduces fossil-fuel, biofuel, bioenergy, and nuclear jobs. Such jobs include those in the mining, transporting, and processing of fuels as well as in the generation of electricity. Transitioning also reduces jobs in the building of internal combustion engines, gas water and air heaters, gas stoves, gas turbines, coal plants, pipelines, gas stations, gas storage facilities, and refineries.

However, a transition also creates jobs in the manufacture and installation of solar PV panels, CSP plants, wind turbines, geothermal plants, tidal devices, and wave devices. It also creates jobs in the electricity, heat, cold, and hydrogen storage industries. More transmission lines will be needed in a WWS world, so jobs increase in that sector as well. Jobs are also needed to install battery-charging stations and build electric vehicles, electric heat pumps, electric induction cooktops, electric lawnmowers, electric arc furnaces, electric induction furnaces, and so on. Electric vehicles that are needed include on-road passenger vehicles, non-road vehicles, trucks, buses, trains, ships, aircraft, agricultural machines, construction machines, and military vehicles.

Common tools for estimating the number of jobs produced due to new electric power generation or

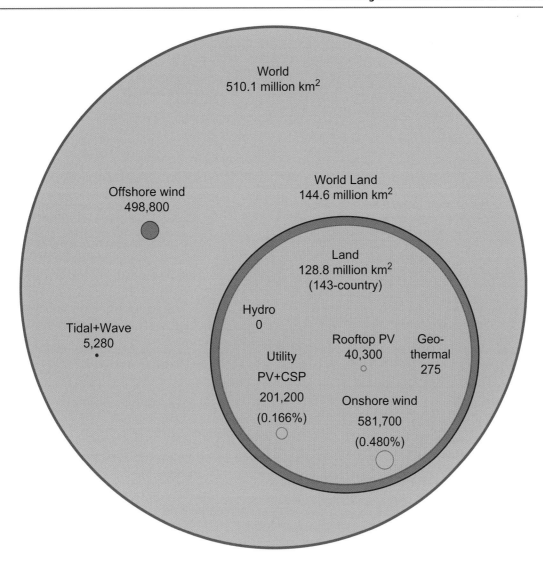

**Figure 8.11** Footprint plus spacing areas (km²) required beyond existing 2018 installations, to power all energy sectors (electricity, transportation, heating/cooling, industry, agriculture/forestry/fishing, the military) in 143 countries in 2050 with 100 percent WWS. For hydropower, the new footprint plus spacing area is zero, since no new installations are proposed. For rooftop PV, the circle represents the additional area of 2050 rooftops that needs to be covered (thus does not represent new land). From Jacobson et al. (2019).

transmission are the Jobs and Economic Development Impact (JEDI) models (NREL, 2017). These models estimate the number of construction and operation jobs plus earnings due to building an electric power generator or transmission line. The models treat direct jobs, indirect jobs, and induced jobs.

**Direct jobs** are jobs for project development, onsite construction, onsite operation, and onsite maintenance of the electricity-generating facility. **Indirect jobs** are revenue and supply chain jobs. They include jobs associated with construction material and component suppliers; analysts and attorneys who assess project feasibility and negotiate agreements; banks financing the project; all

equipment manufacturers; and manufacturers of blades and replacement parts. The number of indirect manufacturing jobs is included in the number of construction jobs. **Induced jobs** result from the reinvestment and spending of earnings from direct and indirect jobs. They include jobs resulting from increased business at local restaurants, hotels, and retail stores and for childcare providers, for example.

Specific output from the JEDI models for each new electric power generator includes temporary construction jobs, permanent operation jobs, and earnings, all per unit nameplate capacity. A **temporary construction job** is defined as a full-time equivalent job required for building infrastructure for one year. A **full-time equivalent (FTE) job** is a job that provides 2,080 hours per year of work. **Permanent operation jobs** are full-time jobs that last as long as the energy facility lasts and that are needed to manage, operate, and maintain an energy generation facility. In a 100 percent WWS system, permanent jobs are effectively indefinite because, once a plant is decommissioned, another one must be built to replace it. The new plant requires additional construction and operation jobs.

The number of temporary construction jobs is converted to a number of permanent construction jobs as follows: One permanent construction job is defined as the number of consecutive one-year construction jobs for $L$ years to replace $1/L$ of the total nameplate capacity of an energy device every year, all divided by $L$ years, where $L$ is the average facility life. In other words, suppose 40 GW of nameplate capacity of an energy technology must be installed over 40 years, which is also the lifetime of the technology. Also, suppose the installation of 1 MW creates 40 one-year construction jobs (direct, indirect, and induced jobs). In that case, 1 GW of wind is installed each year and 40,000 one-year construction jobs are required each year. Thus, over 40 years, 1.6 million one-year jobs are required. This is equivalent to 40,000 forty-year jobs. After the technology life of 40 years, 40,000 more one-year jobs are needed continuously each year in the future. As such, the 40,000 construction jobs are permanent jobs.

Table 8.12 provides an estimated number of permanent, full-time construction and operation jobs per MW of nameplate capacity for several electricity-generating and storage technologies and for transmission and distribution expansion. The total number of jobs produced in a region is simply the new nameplate capacity of each electricity generator multiplied by the number of construction-plus-operation jobs per MW from the table.

The jobs per unit nameplate capacity in Table 8.12 are for the United States. The same metric can be derived for other countries as a function of energy output and GDP per capita as described in Jacobson et al. (2017). As energy output increases, the number of jobs per MW of nameplate capacity decreases slightly because of economies of scale and efficiencies in the use of labor. Similarly, as GDP per capita increases, the number of jobs per MW of nameplate capacity decreases slightly because of a substitution of capital for labor.

Job losses due to a transition to WWS will include losses in the mining, transport, processing, and use of fossil fuels, biofuels, bioenergy, and uranium. Jobs will also be lost in the BAU electricity generation industry and in the manufacturing of appliances that use combustion fuels. In addition, when comparing the number of jobs in a BAU versus WWS system, jobs are lost due to *not* constructing BAU electricity generation plants, petroleum refineries, and oil and gas pipelines.

Table 8.13 estimates the number of permanent, full-time jobs created and lost due to a transition in 143 countries to 100 percent WWS by 2050. The table accounts for jobs in the electricity, heat, cold, and hydrogen generation, storage, and transmission (including HVDC transmission for long distances) industries. However, it does not account for job creation in manufacturing of electric appliances, vehicles, or machines.

In terms of job losses, the table accounts for losses due to eliminating mining, transporting, processing, and consuming fossil fuels, biofuels, and uranium. However, it does not account for jobs lost in the manufacture of combustion appliances, vehicles, or machines. It does account for the retention of non-energy petroleum industry jobs (e.g., for lubricants, asphalt, petrochemical feedstock, and petroleum coke) and for jobs lost due to not building BAU infrastructure.

## Transition highlight

Table 8.13 indicates that, whereas a transition may reduce the world workforce by almost 1 percent due to job losses in the fossil-fuel, biofuel, and nuclear industries, it will more than make up for those losses with job increases in WWS generation, storage, and transmission. In fact, a transition may increase the net number of long-term, full-time jobs created over those lost by about 28.6 million among the 143 countries considered. As such, a transition to 100 percent WWS is expected not only to save consumers money by reducing energy, health, and climate costs, but also to create many more long-term, full-time jobs worldwide than lost.

Table 8.12 **Estimated mean number of permanent, full-time construction and operation jobs per MW nameplate capacity of different electric power sources and storage types in the United States. A full-time job is a job that requires 2,080 hours per year of work. The job numbers include direct, indirect, and induced jobs.**

| Electric power generator | Construction jobs/MW or jobs/km | Operation jobs/MW or jobs/km |
|---|---|---|
| Onshore wind electricity | 0.24 | 0.37 |
| Offshore wind electricity | 0.31 | 0.63 |
| Wave electricity | 0.15 | 0.57 |
| Geothermal electricity | 0.71 | 0.46 |
| Hydropower electricity | 0.14 | 0.30 |
| Tidal electricity | 0.16 | 0.61 |
| Residential rooftop PV | 0.88 | 0.32 |
| Commercial/government rooftop PV | 0.65 | 0.16 |
| Utility PV electricity | 0.24 | 0.85 |
| CSP electricity | 0.31 | 0.86 |
| Solar thermal for heat | 0.71 | 0.85 |
| Geothermal heat | 0.14 | 0.46 |
| Pumped hydro storage (PHS) | 0.77 | 0.3 |
| CSP storage (CSP-PCM) | 0.62 | 0.3 |
| Battery storage | 0.092 | 0.2 |
| Chilled water storage (CW-STES) | 0.15 | 0.3 |
| Ice storage (ICE) | 0.15 | 0.3 |
| Hot water storage (HW-STES) | 0.15 | 0.3 |
| Underground heat storage (UTES) | 0.15 | 0.3 |
| Hydrogen production and storage | 0.32 | 0.3 |
| AC transmission (jobs/km) | 0.073 | 0.062 |
| AC distribution (jobs/km) | 0.033 | 0.028 |
| HVDC transmission (jobs/km) | 0.088 | 0.082 |

Source: Jacobson et al. (2019). The number of long-term, full-time construction jobs is the number of 1-year jobs divided by the lifetime (in years) of the device (Table 3.5, Footnote b), for electricity generation. For transmission, the lifetime was assumed to be 70 y.

Table 8.13 **Estimated 143-country jobs created and lost due to transitioning from BAU energy to WWS across all energy sectors. The job creation accounts for new jobs in the electricity, heat, cold, and hydrogen generation, storage, and transmission (including HVDC transmission) industries. However, it does not account for changes in jobs in the production of electric appliances, vehicles, and machines. Construction jobs are for new WWS devices only. Operation jobs are for new and existing devices. The losses are due to eliminating jobs for mining, transporting, processing, and using fossil fuels, biofuels, bioenergy, and uranium. Also shown is the percentage of total jobs in each sector that are lost. Fossil-fuel jobs due to non-energy uses of petroleum, such as lubricants, asphalt, petrochemical feedstock, and petroleum coke, are retained. For transportation sectors, the jobs lost are those due to transporting fossil fuels (e.g., through truck, train, barge, ship, or pipeline); the jobs not lost are those for transporting other goods. The table does not account for jobs lost in the manufacture of combustion appliances, vehicles, or machines.**

| Energy sector | Number of jobs produced or lost | Percent of jobs in sector that are lost |
|---|---|---|
| **WWS jobs created** | | |
| *Construction* | 24,389,000 | |
| Generation | 15,285,000 | |
| Storage | 8,159,000 | |
| Transmission | 945,000 | |
| *Operation* | 30,151,000 | |
| Generation | 21,709,000 | |
| Storage | 7,697,000 | |
| Transmission | 745,000 | |
| **Total jobs produced** | 54,540,000 | |
| **BAU jobs lost** | | |
| Oil and gas extraction | 2,217,000 | 89 |
| Coal mining | 1,257,000 | 96 |
| Uranium mining | 85,100 | 100 |
| Support for oil and gas | 3,329,000 | 89 |
| Oil and gas pipeline construction | 1,506,000 | 89 |
| Mining & oil/gas machinery | 1,074,000 | 89 |
| Petroleum refining | 651,000 | 94 |
| Asphalt paving and roofing materials | 0 | 0 |

Table 8.13 (*cont.*)

| Energy sector | Number of jobs produced or lost | Percent of jobs in sector that are lost |
|---|---|---|
| Gas stations with stores | 1,775,000 | 30 |
| Other gas stations | 420,000 | 50 |
| Fossil electric power generation utilities | 992,000 | 100 |
| Fossil electric power generation non-utilities | 302,000 | 100 |
| Nuclear and other power generation | 1,150,000 | 100 |
| Natural gas distribution | 1,169,000 | 100 |
| Auto oil change shops/other repair | 59,100 | 10 |
| Rail transportation of fossil fuels | 584,100 | 52 |
| Water transportation of fossil fuels | 279,000 | 23 |
| Truck transportation of fossil fuels | 863,000 | 8 |
| Biofuel except electricity | 6,682,000 | 100 |
| Jobs lost from not increasing fossil-fuel use[a] | 1,488,000 | |
| Total jobs lost | 25,892,000 | 0.70[b] |
| Net jobs produced minus jobs lost | 28,648,000 | |

[a] Jobs lost from not expanding fossil-fuel production are additional refinery and electric power construction and operation jobs that would have accrued by 2050 if BAU instead of WWS continued.
[b] The total world labor force in 2018 was 3.47 billion.
Source: Jacobson et al. (2019).

## 8.5 Summary

One of the greatest barriers facing a 100 percent WWS system worldwide is the concern that intermittent wind, water, and solar resources may make matching intermittent demand for electricity, heat, cold, and hydrogen expensive. However, combining intermittent generation with storage and demand response helps to solve that problem. In addition, interconnecting geographically dispersed intermittent wind, solar, and wave power smoothens out the supply of these generators. Oversizing the nameplate capacity of generators, oversizing storage, oversizing transmission, using vehicle-to-grid technology, and using weather forecasting also help to solve the problem.

The grid in a 100 percent WWS world will look completely different from today. The future grid will contain many more flexible loads due to the electrification of transportation, building heat, and industrial heat, and the expansion of district heating. Most all transportation loads (for battery-electric vehicles and hydrogen fuel cell vehicles) are flexible, as are all district air and water heating and cooling loads and building water heating loads. In addition, many industrial loads are flexible. Flexible loads are subject to demand response so can either be met ahead of time by storing heat in water or in soil underground or storing electricity in batteries or hydrogen or shifting demand forward in time (such as with industrial heat and car charging loads). The significant increase in flexible loads upon electrifying everything makes matching power demand on the grid much easier while reducing costs.

A crucial step in creating the most efficient future grid possible is to model the impacts of different types and amounts of generation, storage, and demand response. Three categories of models – power flow (load flow), optimization, and trial-and-error models – have been developed to attack this problem. However, only the latter two can be used practically to address it. Solutions with both of these types of models suggest it is possible to match power demand with supply, storage, and demand response at low cost while creating jobs and minimizing land requirements worldwide with 100 percent WWS.

# Further Reading

Archer, C. L., and M. Z. Jacobson, 2007. Supplying baseload power and reducing transmission requirements by interconnecting wind farms, *J. Applied Meteorol. Climatol.*, *46*, 1701–1717, doi:10.1175/2007JAMC1538.1.

Bogdanov, D., J. Farfan, K. Sadovskaia, A. Aghahosseini, M. Child, A. Gulagi, et al., 2019. Radical transformation pathway towards sustainable electricity via evolutionary steps, *Nat. Commun.*, *10*, 1077, doi: 10.1038/s41467-019-08855-1.

Brown, T. W., T. Bischof-Niemz, K. Blok, C. Breyer, H. Lund, and B. V. Mathiesen, 2018. Response to "Burden of proof: a comprehensive review of the feasibility of 100% renewable electricity systems," *Renew. Sust. Energ. Rev.*, *92*, 834–847.

Budischak, C., D. Sewell, H. Thompson, L. Mach, D. E. Veron, and W. Kempton, 2013. Cost-minimized combinations of wind power, solar power, and electrochemical storage, powering the grid up to 99.9% of the time, *J. Power Sources*, *225*, 60–74.

Diesendorf, M., and B. Elliston, 2018. The feasibility of 100% renewable electricity systems: a response to critics, *Renew. Sust. Energ. Rev.*, *93*, 318–330.

Elliston, B., I. MacGill, and M. Diesendorf, 2013. Least cost 100% renewable electricity scenarios in the Australian National Electricity Market, *Energy Policy*, *59*, 270–282.

Hansen, K., C. Breyer, and H. Lund, 2019. Status and perspectives on 100% renewable energy systems, *Energy*, *175*, 471–480.

Hart, E. K., and M. Z. Jacobson, 2011. A Monte Carlo approach to generator portfolio planning and carbon emissions assessments of systems with large penetrations of variable renewables, *Renew. Energy*, *36*, 2278–2286, doi:10.1016/j.renene.2011.01.015.

Jacobson, M. Z., M. A. Delucchi, M. A. Cameron, and B. A. Frew, 2015. A low-cost solution to the grid reliability problem with 100 percent penetration of intermittent wind, water, and solar for all purposes, *Proc. Nat. Acad. Sci.*, *112*(49), 15,060–15,065, doi: 10.1073/pnas.1510028112.

Jacobson, M. Z., M. A. Delucchi, M. A. Cameron, S. J. Coughlin, C. Hay, I. P. Manogaran, et al., 2019. Impacts of Green-New-Deal energy plans on grid stability, costs, jobs, health, and climate in 143 countries, *One Earth*, *1*, 449–463, doi:10.1016/j.oneear.2019.12.003.

Kempton, W., C. L. Archer, A. Dhanju, R. W. Garvine, and M. Z. Jacobson, 2007. Large $CO_2$ reductions via offshore wind power matched to inherent storage in energy end-uses, *Geophys. Res. Lett.*, *34*, L02817, doi:10.1029/2006GL028016.

Mathiesen, B. V., H. Lund, D. Connolly, H. Wenzel, P. Z. Ostergaard, B. Moller, et al., 2015. Smart energy systems for coherent 100% renewable energy and transport solutions, *Appl. Energ.*, *145*, 139–154.

Ram, M., D. Bogdanov, A. Aghahosseini, A. Gulagi, A. S. Oyewo, M. Child, et al., 2019. *Global Energy System Based on 100% Renewable Energy – Power, Heat, Transport, and Desalination Sectors*, Lappeenranta University of Technology Research Reports 91, ISSN: 2243-3376, Lappeenranta, Berlin, http://energywatchgroup.org/wp-content/uploads/EWG_LUT_100RE_All_Sectors_Global_Report_2019.pdf (accessed September 6, 2019).

Teske, S., D. Giurco, T. Morris, K. Nagrath, F. Mey, C. Briggs, et al., 2019. Achieving the Paris Climate Agreement, https://oneearth.app.box.com/s/hctp4qlk34ygd0mw3yjdtctsymsdtaqs (accessed September 6, 2019).

## 8.6 Problems and Exercises

8.1. What is the main challenge with a 100 percent WWS system?

8.2. What are the basic strategic steps to match electricity, heat, cold, and hydrogen demand with 100 percent WWS supply over time?

8.3. Identify two ways that interconnecting geographically dispersed variable (intermittent) WWS resources benefits a 100 percent WWS electric power grid.

8.4. List four methods of changing the electric power system when instantaneous electric power generation plus storage cannot meet instantaneous demand during some times of the year.

8.5. Do you expect a solar PV farm alone, a wind farm alone, or a solar farm co-located with a wind farm to produce the smoothest time-dependent electricity supply to the grid? Why?

8.6. Why should bundling of wind, solar, geothermal, and hydroelectric power significantly help to match power demand on the grid with power supply?

8.7. Calculate the number of heating degree days and cooling degree days over a four-day period relative to a reference temperature of 18 °C if the outdoor air temperatures on each day are –5 °C, 0 °C, 30 °C, and 35 °C, respectively.

8.8. What is the difference between load following and regulation?

8.9. Which WWS generators can help with load following?

8.10. What WWS technologies can be used to perform regulation services?

8.11. List three industrial loads and three home loads that are subject to demand response.

8.12. What are the differences among spinning reserves, supplemental reserves, and replacement reserves?

8.13. What is one advantage and one disadvantage of a power flow model, an optimization model, and a trial-and-error model?

8.14. Calculate the U.S. WWS-to-BAU social cost per unit energy ratio, the WWS-to-BAU aggregate social cost ratio, and the WWS-to-BAU aggregate private cost ratio given the following information: The WWS-to-BAU energy use ratio is 0.408; the WWS private and social costs per unit energy are 9.81 ¢/kWh-all-energy. The BAU private, health, and climate costs are 10.4, 8.05, and 15.2 ¢/kWh-all-energy, respectively.

8.15. Calculate the WWS-to-BAU aggregate social cost ratio, aggregate private cost ratio, and social cost per unit energy ratio in a country if the BAU energy, air pollution, and climate costs are 10, 12, and 17 ¢/kWh, respectively; the WWS energy cost is 9 ¢/kWh; and the end-use energy consumption in the WWS case is 45 percent that of the BAU case.

# 9 Evolution of the 100 Percent Movement and Policies Needed for a WWS Solution

The solution to global warming, air pollution, and energy security requires not only a technical and economic roadmap but also popular support and political will. In fact, the main limitations of a transition to 100 percent clean, renewable energy and storage are neither technical nor economic; instead, they are social and political. People need to believe that a solution is possible, to understand what changes they can make in their own lives to solve the problems, to make such changes, and to support policymakers who can pass laws speeding a transition. Policymakers, themselves, need to take bold steps in affecting a transition. Thus, one of the most important factors leading to a change is education about what is possible and why it is possible. This textbook aims to contribute toward that education.

Since the early 2000s, substantial progress has been made toward raising awareness and affecting a transition. A movement toward 100 percent clean, renewable energy has taken shape. Popular support for such a transition has grown, as have the numbers of laws and commitments toward that goal. Such commitments have been made not only by towns, cities, states, and countries, but also by international businesses, nonprofits, community groups, policymakers, and individuals.

This chapter breaks away from the scientific and engineering aspects of a transition to 100 percent clean, renewable energy. It moves onto a personal historical story, progress made to date on a transition, and policies needed to complete the transition. It starts with me reflecting on my personal journey to developing and implementing 100 percent WWS roadmaps, and how these roadmaps and collaborations with other scientists, cultural influencers, business leaders, and community

leaders shaped the **100 percent WWS movement** (Section 9.1). It also tracks the town, city, state, country, and corporate laws and commitments made toward 100 percent to date. It then discusses the necessary timeline for a transition (Section 9.2), obstacles in the way of meeting the timeline (Section 9.3), and policies that are needed to overcome the obstacles to meet the 100 percent goal on time (Section 9.4).

## 9.1 Personal Journey to 100 Percent WWS

This section discusses my personal journey toward developing and disseminating science-based 100 percent WWS roadmaps, co-founding the nonprofit Solutions Project, and helping to build and work with a coalition of science, business, cultural, and community leaders to start the 100 percent WWS movement. The roadmaps evolved from my life goal of trying to understand and help solve large-scale air pollution and climate problems. This goal originated as follows.

### 9.1.1 First Exposure to Severe Air Pollution

During the summer of 1978, as a 13-year-old tennis player I traveled to San Diego from my home in Northern California to play in a tennis tournament. What struck me was the air pollution visible on the freeways and off. I could not only see it, but also taste and smell it. The pollution hung like a morbid pall. It was one thing to sit inactively in a car in the middle of this pollution. It was another to run through it for hours while playing tennis. During play, my throat and lungs became irritated, and my eyes became scratchy. Taking deep breaths while lunging for tennis balls was a chore. If this soup of pollution was hurting me after only a few minutes, I imagined the damage it caused people who lived in it. Indeed, living in such pollution is equivalent to smoking two to three packs of cigarettes per day.

I took additional trips to Los Angeles and San Diego during the next few years. After only the second trip, I realized the smog was not just a one-time event. After the third trip, I concluded this pollution was a way of life in these cities. I then began to ask myself, *Why should anyone live like this?* and *Isn't there a solution to this problem?* I decided then and there, that when I grew up, I wanted to understand and try to solve this avoidable air pollution problem, which affects so many people. I knew what I wanted to do with my career.

Later in my teens, I learned from popular magazine articles about the emissions of greenhouse gases since the Industrial Revolution creating a blanket over the Earth, trapping heat and increasing globally averaged temperatures. I also learned about acid rain and its impact on forests and lakes. I realized that these problems were intimately connected to the air pollution problem. The main source of the chemicals that cause air pollution, namely fossil-fuel and biomass combustion, also produced chemicals that caused global warming and acid rain. **If combustion was the problem, then eliminating combustion must be the solution.**

Yet, at that time, I was busy trying to finish high school academically while playing tennis. I had neither the time nor the skills to help solve the problem. I did figure, though, that if I kept studying and learned as much math and science (and later, at university, engineering and economics) as possible, I would equip myself as best I could to help solve these problems.

## 9.1.2 Hungry for Knowledge

When I started my undergraduate career at Stanford University during the autumn of 1983, it was not obvious to me what courses or major to focus on. Indeed, there was no major available to study air pollution, climate, or acid rain. There was not even a general environmental major. The closest degree (not even an undergraduate degree) was a master's degree in Environmental Engineering, which involves the study of groundwater pollution.

Because it was not evident to me what undergraduate major to focus on, I took five engineering courses in different engineering majors during the spring and autumn of 1984. These included courses in material science, electrical engineering, aeronautics and astronautics, statics, and environmental science and technology. The last course broadly covered water pollution, urban air pollution, acid rain, global climate, and energy. Professor Gil Masters was the instructor. His engaging style consistently won him many teaching awards. This course was by far the most interesting to me among the five. It was taught through the Civil Engineering Department. I decided then to major in Civil Engineering, although there were few other related courses available in the department or the university.

One additional course I took that was related was on small-scale energy systems and also taught by Professor Masters. In that course, he taught about the efficiencies and engineering characteristics of solar panels, wind turbines, and other types of renewable energy systems. I took that course during the winter of 1985, in the middle of my sophomore year. Although I was interested in the renewable energy information in that course, I was still more interested at the time in understanding air pollution, climate, and acid rain problems before focusing on solutions. As such, the information in that course stayed dormant inside me until the year 2000, when I dusted off some of the materials from that course to come up with an *aha* moment.

In the meantime, I had taken a basic economics course during my freshman year. Since I had several credits from high school that I could transfer to Stanford, I had room to consider a second major. Although I was not so excited about taking economics courses, I felt it was important to do so because I knew that if I were to try to understand and solve large-scale air pollution, climate, and acid rain problems, I would need to understand costs, financing, and economics. As a result, I selected Economics as a second major.

During the rest of my undergraduate career, I built up skills in engineering and economics. However, I was disappointed by the fact that there were simply no other courses that I could take that focused on my specific interests in air pollution, climate, and acid rain. Toward graduation, I thought deeply about what my next step was. I learned about a program at Stanford to complete a master's degree that would start while I finished my undergraduate courses. I decided to apply to the program with the goal of working toward an Environmental Engineering master's degree through the Department of

Civil Engineering. This program focused on groundwater pollution, which was still not my main interest. However, it was the closest I could come to a program that moved me toward my goal. Since the program took only one more year, I used the time as an opportunity to broaden my knowledge about the environment and obtain some research skills.

### 9.1.3 Lessons for Life

After completing my MS degree at the end of winter 1988, I contemplated my next move. I knew I wanted to enter a PhD program to study the atmosphere. However, I had also been playing tennis continuously for about twelve years, including four years on the Stanford University tennis team, and I felt that I should try to go onto the professional circuit for a period. I gave myself a year and a half, at which point, during the autumn of 1989, I would enter a PhD program. While I always hoped to break through the tennis ranks and rise to the top, I was under no illusion I was good enough to do that, so I planned my obsolescence ahead of time. Otherwise, like many of my tennis-player friends, I could have stayed out on the tennis circuit for years, traveling and playing, but eventually having to come back to real life. I also felt that the longer I procrastinated trying to understand the problems that I was interested in solving, the longer I would need to come to a solution. I felt passionate about finding a solution. As such, I compromised by giving myself a limited time to play.

Unfortunately, halfway through the one-and-a-half years of travel, I developed a bone fracture in my right knee that resulted in a piece of bone breaking loose and floating around my knee joint. This required surgery to screw the bone chip back in and suture up my meniscus, which the bone chip had shredded. During the surgery, the doctor mistakenly sutured my right peroneal nerve down. That nerve runs between the knee and foot. By compressing the nerve, he paralyzed my foot, giving me a drop foot. He realized this the next morning and took me back in for another surgery to undo the suture. Even though the nerve was not cut, it was compressed so severely that my foot remained paralyzed for two years and did not recover fully for seven years. Feeling gradually returned to my leg and foot at the rate of one inch per month down my leg, starting at my knee. Needless to say, this was the end of my competitive tennis career, although I tried to use a prosthetic to run and play tennis with a drop foot. However, while I could play, the loss of speed made it almost impossible to win matches.

I never regret playing tennis. It taught me many lessons that I still use today.

One is to focus, regardless of all the distractions around you. Always keep your eye on the ball. Don't let outside noise or movement disturb you. I have applied this in my research countless times. There are many distractions that can take the focus away from research on solutions.

Second, there is no substitute for practice and hard work. The harder one trains and practices tennis, the less chance there is that a loss will be due to not being in shape or not training sufficiently. The more one studies and practices a research subject, the deeper one's knowledge of the subject becomes and the chance of making an error lessens.

Third is switching gears. While playing tennis through high school and university, I simultaneously took heavy course loads. This was difficult, but it taught me to be efficient in my current work, where I need to teach, research, advise students, write proposals, respond to emails, review applications, attend committee meetings, and travel for meetings. Even though switching gears to work on different tasks has been stressful at times, I learned how to focus on each task based on years of training.

Fourth is being accurate and truthful. There is nothing more important than trying to be as accurate as possible in research and in reviewing other people's work. In most tennis tournaments I participated in while growing up, players would call their own lines. Thus, if someone cheated, it would be known quickly. As such, most players tried to be honest despite the temptation to get an extra point here or there. Honesty is critical in science as well. It is incredibly important to be accurate not only in one's own work but also when describing another's work. Too often, scientists criticize other studies either without fully understanding them, because they don't want to spend the time to understand them, or, in some cases, because they simply want to make other researchers look bad to make their own work look better in comparison. These scientists would have benefited from learning tennis etiquette.

### 9.1.4 Building a Coupled Regional Air Pollution–Weather Prediction Computer Model

During my hiatus from school playing tennis, I looked at PhD programs at the University of Washington and

UCLA, both of which had strong Atmospheric Science programs. I ended up at UCLA because I found an advisor there, Professor Richard Turco, who had a project available on the main topic I wanted to study, urban air pollution. In addition, I felt that if I wanted to understand and solve air pollution, I needed to live inside it for a while, and Los Angeles was ground zero for air pollution in the United States.

I started at UCLA during the autumn of 1989 and completed my studies in June of 1994 with an MS and PhD in Atmospheric Sciences. There, I learned how to develop physical equations describing phenomena in the atmosphere, how to solve the equations, how to write computer programs to represent the equations and their solutions, and how to compare model results with data. My project was to build a coupled urban air pollution–weather prediction model and apply it to study air pollution in Los Angeles. I loved my project from beginning to end.

An **air pollution computer model** is really a mathematical representation of atmospheric processes, including gas, particle, transport, radiation, and surface processes, among others. At the time, air pollution models were decoupled from weather prediction models. Instead, the sources of wind speed and direction data for moving gases and particles from place to place in the air pollution model were either winds interpolated from observations or winds from a file, where the file was produced by running a weather prediction model independently. No regional air pollution model in the world at the time had a built-in weather prediction model, where the winds drove pollution movement *and* the pollutants themselves fed back to modify the weather.

In 1990, I began my computer modeling research by focusing on building gas chemistry and particle microphysics and chemistry computer algorithms. Some of the codes I eventually developed for the air pollution model included three algorithms that simulated gas-phase photochemistry and others that simulated particle condensation, dissolution, coagulation, and chemistry. In particular, I focused on the details of treating pollution particles of different sizes and composition.

I then coupled these gas and particle algorithms with existing transport and radiation algorithms developed by Dr. Owen Toon and my advisor, Professor Turco. Dr. Toon was a colleague of Professor Turco. I worked during the summer of 1990 in Dr. Toon's lab at NASA Ames Research Center in Mountain View, California. I ultimately updated the radiation model to treat the optical properties of the gases and complex particles I was adding to the model. Later, I added a different transport algorithm suitable for transporting chemicals with sharp peaks and valleys in concentration over distance.

The resulting combination of gas, particle, radiation, and transport algorithms was an air pollution model. This model predicted gas and particle concentrations in space and time in the Los Angeles basin, but it was also applicable anywhere that data were available. At the time, though, the only source of winds for the model was a file with interpolated observations. This was not ideal because the observations were far apart in space and time. In our research group at UCLA, another graduate student, Rong Lu, was building a **weather prediction model**. Such a model predicts winds, temperatures, pressures, and humidity.

In 1993, I coupled Rong's weather model with my air pollution model in such a way that heating rates from the air pollution model, which are a function of gas and aerosol concentration, were now fed back to the weather prediction model to determine temperatures. Conversely, winds from the weather prediction model now drove the transport of gases and particles in the air pollution model. Also, temperatures and air pressures from the weather prediction model now affected gas and particle temperature- and pressure-dependent chemical reactions and microphysics. Little did I know at the time, but this was the first air pollution–weather prediction model, coupled with feedback for gases and aerosol particles, developed worldwide to study urban or regional air pollution. I gave a copy of the resulting model back to Rong for him to use as well, which he did. Sadly, Rong passed away from a brain tumor just a few years after he graduated.

I named the coupled model GATOR-MMTD (Gas, Aerosol, TransPOrt, Radiation-Mesoscale Meteorological, and Tracer Dispersion model). I subsequently used the model to simulate gas, aerosol, radiative, and meteorological parameters and to compare results with hourly data in different locations in the Los Angeles basin (Jacobson, 1994, 1997; Jacobson et al., 1996).

### 9.1.5 Expanding from the Regional to the Global Scale

During 1994, before graduating from UCLA, I was fortunate to land a job in what became the Department of Civil and Environmental Engineering at Stanford University. Because I was also interested in understanding global climate, I expanded the regional air pollution model to the global scale, essentially building a souped-up (due to the intense air pollution chemistry, particle physics, and radiation) global climate model. To do that, I expanded the regional model to treat the globe. For the global version, I needed to predict the winds, temperatures, and pressures globally rather than just regionally. Fortunately, UCLA scientists specialized in this, and in 1994, I coupled, with interactive feedback, my air pollution model to the dynamical portion of the UCLA General Circulation Model (GCM), developed primarily by Professor Akio Arakawa. The dynamical portion provided winds, temperatures, pressures, and moisture. I performed the same type of interactive coupling as I did previously with the regional dynamical model. The result was the first coupled global air quality–meteorological model to treat the feedback of gases and size- and composition-resolved particles to weather and climate and vice versa.

A regional air pollution model simulation requires inputs at its horizontal boundaries. Such inputs include wind speeds and direction, gas concentrations, and particle concentrations. These values must come from the larger scale (e.g., the global scale). At the time, they were coming from interpolated winds primarily. Since I now had both a global and regional model, I realized I could improve the regional model by coupling the global model to it and feeding in boundary conditions from the global model, not only horizontally, but also vertically at the regional model top. I did this coupling and ensured atmospheric processes on all scales were solved consistently. The result, completed in 1998, was a nested global-through-regional model.

A **nested model** starts at the global scale to produce meteorological, gas, and aerosol fields. These fields are then fed into one or more smaller, more finely resolved domains nested anywhere within the global domain. Within each smaller domain lies one or more even smaller, more finely resolved nested domains that receive boundary conditions from the next-larger domain. In this way, air pollution or weather can be modeled anywhere in the world at high resolution while receiving boundary conditions that continuously vary in time and space. The nested model was called GATOR-GCMOM (Gas, Aerosol, TransPort, Radiation-General

Circulation, Mesoscale, and Ocean Model) (Jacobson, 2001b; Jacobson et al., 2007). It was the first model to nest gases, particles, radiation, and meteorology from the global to local scale. Since then, many models have made efforts to nest pollution, weather, and climate in a similar manner.

### 9.1.6 Black Carbon, the Kyoto Protocol, and Wind versus Coal

I used the models over the years to study several phenomena, including the impacts of aerosol particles on ultraviolet radiation, temperatures, and ozone (Jacobson, 1998, 1999) and the effects of black carbon on climate. One hypothesis I made was that black carbon, the main component of soot, may be the second leading component of global warming after carbon dioxide in terms of direct radiative forcing (Jacobson, 2000, 2001a). I further hypothesized that controlling the emissions of black carbon may be the fastest method of slowing global warming (Jacobson, 2002, 2010a). This finding was subsequently supported by several studies including the review of Bond et al. (2013).

Up to the year 2000, I had focused on understanding air pollution and climate problems. In 2000, I started to delve into solutions. In particular, I became concerned about whether the United States would ratify the Kyoto Protocol. The **Kyoto Protocol** was an international climate agreement adopted by the United Nations Framework Convention on Climate Change (UNFCCC) on December 11, 1997. The protocol called for developed countries to reduce greenhouse gas emissions but did not call for reductions by less developed countries. Between the adoption of the protocol and 2019, 191 countries and the European Union ratified it. The only countries not to ratify it were the United States, Sudan, and Afghanistan. Canada renounced its ratification in 2012.

In 2000, I wondered what it would take for the United States to satisfy its share of the Kyoto Protocol, which was to reduce greenhouse gas emissions 7 percent below 1990 levels. I first estimated that this could be accomplished by reducing about 59 percent of 1999 U.S. coal emissions at the time. I then wondered how many wind turbines this would require. I went back to my notes from the Environmental Science and Technology course I took in 1984 from Professor Gil Masters. I vaguely remembered an equation that he derived that could help answer this question. The equation (Equation 6.28) estimated the capacity factor of almost any wind turbine given just three parameters: mean wind speed, rated power of the turbine, and blade diameter. At the time, the equation didn't exist anywhere except in those course notes. I discussed my goal with him, and we used the equation along with information about a new, efficient 1.5-MW wind turbine to calculate the number of such turbines that might be needed to replace enough coal in order for the United States to satisfy the Kyoto Protocol (Jacobson and Masters, 2001).

Despite this effort, the U.S. Senate never ratified the Kyoto Protocol. The United States signed the Kyoto Protocol on November 12, 1998, but the treaty was not submitted to the U.S. Senate for ratification. Shortly after George W. Bush became U.S. president, on June 11, 2001, he stated his reasons for not submitting the Kyoto Protocol for ratification. He stated in a speech that the protocol would have "a negative economic impact, with layoffs of workers and price increases for consumers" and that "Kyoto also failed to address two major pollutants that have an impact on warming, black soot and tropospheric ozone. Both are proven health hazards. Reducing both would not only address climate change, but also dramatically improve people's health" (*New York Times*, 2001).

Both of these justifications were not true. As discussed in Chapters 7 and 8, transitioning to 100 percent WWS would create more long-term, full-time jobs than would be lost and would reduce both the direct cost and social cost of energy.

The second reason offered by President Bush ironically originated from the Jacobson (2002) paper just discussed. On May 18, 2001, three weeks before President Bush's speech, the president's office requested, through the Environmental Protection Agency, and received an early draft of that paper from me. The main conclusion of that paper was this: "Reductions in BC+OM (black carbon plus organic matter) emissions from fossil-fuel sources will not only slow global warming, but also improve health." Instead of stating that the Kyoto Protocol could be improved by including black carbon (and ozone), the president used this conclusion as a reason to pull out of the Kyoto Protocol. He claimed, "The Kyoto Protocol was fatally flawed in two fundamental ways" – one for not including black carbon and ozone (*New York Times*, 2001). Needless to say, it was not a satisfying feeling seeing the president of the United States use correct conclusions from a paper to draw an incorrect conclusion about the Kyoto Protocol.

## 9.1.7 Wind Energy Analysis and Comparing Impacts of Energy Technologies

The short (three-quarters of one page) 2001 paper on wind versus coal generated so much feedback from readers that it motivated me to examine wind energy in more detail. I asked a graduate student, Cristina Archer, who had substantial experience with meteorology, if she wanted to work on a wind-mapping project for the United States. She agreed to work on this along with her main project, which was to analyze a rotational wind (eddy) around the Monterey Bay that she discovered while visiting the Santa Cruz beach. Over the next few years, she developed the world's first wind maps from data alone at 80 m above ground level, one for the United States and one for the world as a whole (Archer and Jacobson, 2003, 2005). The first of these studies along with another study (Archer and Jacobson, 2007) also showed that interconnecting geographically dispersed wind could turn completely intermittent wind power into partial baseload power.

Another student, Mike Dvorak, then performed high-resolution modeling and compared results with data to map the wind resources offshore of California and of the East Coast of the United States (Dvorak et al., 2010, 2012, 2013). These studies found not only substantial offshore resources but also that peak offshore winds often coincided in time with the time of peak energy demand.

In the meantime, I was using the GATOR-GCMOM model to compare the impacts on air pollution and climate of different energy technologies. I and colleagues compared the effects of diesel versus gasoline on air pollution concentrations (Jacobson et al., 2004); the effects of gasoline versus hydrogen fuel cell versus gasoline-electric hybrid vehicles on U.S. air quality and climate (Colella et al., 2005; Jacobson et al., 2005); the effects of hydrogen fuel cell versus gasoline vehicles on the global ozone layer and global climate (Jacobson, 2008b); and the effects of ethanol versus gasoline vehicles on air pollution mortality (Jacobson, 2007; Ginnebaugh et al, 2010; Ginnebaugh and Jacobson, 2012). These studies formed the basis of some of the analysis in Section 3.5.

Following the wind analyses and the comparisons of different fuels and electricity sources, I began to think that it would be useful to review systematically different proposed solutions to global warming, air pollution, and energy security. At the time, several technologies were being proposed as solutions for electricity and others were being proposed as solutions for transportation. Aside from renewables, major proposed technologies for electricity included nuclear power and coal with carbon capture. For transportation, battery-electric vehicles, hydrogen fuel cell vehicles, and biofuel vehicles were being proposed.

Given our previous work on evaluating some of these technologies, I felt I was ready to carry out such a review. The study (Jacobson, 2009) compared 11 different proposed energy solutions to global warming, air pollution, and energy security in terms of 11 different impact categories. Such categories included carbon dioxide equivalent ($CO_2e$) emissions, air pollution mortality, land footprint on the ground, spacing area, water consumption, resource abundance, effects on wildlife, thermal pollution, water chemical pollution/radioactive waste, risk of energy supply disruption, and normal operating reliability.

With respect to reliability, in 2008, I asked a graduate student, Graeme Hoste, to see if it were possible to match California's hourly electricity demand with only wind, solar, geothermal, and hydroelectric power supply. Geothermal would provide constant power each hour; solar and wind would provide intermittent power; and hydroelectric power would fill in the gaps. Graeme completed a report on this topic (Hoste et al., 2009). He found that, in theory, California could meet 100 percent of its monthly averaged hour-by-hour power demand in the two months tested, April and July 2020. The upshot of this was that, **when treated as a bundle with geothermal and hydroelectric, wind and solar are more reliable for meeting peak demand than they are when treated individually.** This result was borne out in dozens of subsequent independent studies (Section 8.2.1) and was used in my 2009 review paper.

The overall conclusion of the review paper was that the best electric power-generating technologies for minimizing the 11 impacts were onshore and offshore wind, solar PV, CSP, geothermal, tidal, wave, and hydroelectric power. These technologies were all **wind, water, and solar (WWS)** technologies. The paper also proposed the use of battery-electric and hydrogen fuel cell vehicles for transportation. All such technologies were introduced in Chapter 2.

## 9.1.8 100 Percent Wind-Water-Solar and the TED Debate

The review paper garnered substantial interest in the climate and energy community. Shortly after the paper was published, I was approached by *Scientific American* to consider writing a follow-up paper about the feasibility of powering the world with the best technologies identified in the review. At that time, I asked a colleague of mine, Dr. Mark Delucchi, who was a research scientist at the University of California at Davis Institute of Transportation Studies at the time (subsequently at UC Berkeley), whether he would be interested in partnering with me on such a study. He agreed. Together, we analyzed the technical and economic feasibility of transitioning the world's all-purpose energy to 100 percent WWS. We quantified the change in energy demand upon a conversion to WWS, the numbers of WWS devices needed, the land areas required for footprint and spacing, the materials needed, reliability, and costs. The study concluded that, with aggressive policies, a transition to 100 percent WWS by 2030 was technically and economically possible, but for social and political reasons, a more likely transition end-goal was around 2050.

The paper was published in November 2009. It was immediately attacked as pie-in-the-sky and an impossible dream. Yet, within 10 years, the 100 percent goal outlined in the paper was much further along and a movement had started around it. Many 100 percent laws had been put in place in cities, states, countries, and businesses. WWS costs had come down substantially, and the public was overwhelmingly supportive of the goal. Nevertheless, as of 2020, the solution is still a long way from being fully realized.

As a result of the review paper and the *Scientific American* article, I was asked to take part in a debate at a TED (Technology, Entertainment, Design) conference in Long Beach, California, on June 10, 2010. The debate was with Stewart Brand, former editor of the *Whole Earth Catalog* and now a nuclear advocate. The debate was on nuclear power versus renewables. I had given lots of talks but had not experienced this debate format before, where we each had 6 minutes to lay out a case and a few minutes each for rebuttals.

At the beginning of the debate, the moderator, Mr. Chris Anderson, asked the audience of about 2,000 if they favored or opposed nuclear energy. To my surprise, about

75 percent favored it. Given the strong support for Mr. Brand's position from the start and the facts that he was an experienced speaker at TED events, had such a charming personality, and was the first to speak, I thought I was doomed. The only advantage I had was that the data I was about to show were based substantially on new, raw research our group had performed, and he had not seen them before, including data from wind mapping studies and comparisons of different energy sources. He was using information exclusively from third-party sources.

After his 6 minutes of speaking, I felt more in a hole because he presented lots of information with a compelling style. I felt I needed to get my words out correctly from the get-go; otherwise, I would lose the audience. I focused on my first few words and managed to start strongly. I then reeled off statistics and my own graphs. While I was presenting my 6 minutes of information, I could see out of the corner of my eye the concern growing on his face, along with the audience resonating with the information I was presenting. In fact, a few times during the talk, the audience started cheering me on. A vote after the debate indicated that I had switched about 10 percent of the audience, or 200 people, such that in the end, 65 instead of 75 percent favored nuclear.

This debate, coupled with the review paper that found that nuclear was better than some technologies but not as good as WWS technologies, rendered me a target for nuclear advocates. Most never understood or did not want to understand that my goal has always been focused on solving air pollution, climate, and energy security problems with the technologies that result in the fastest solution, least cost, and greatest benefit. Nuclear power, while beneficial for some processes on some timescales, is not so good as other solutions. Thus, it isn't that I am *against* nuclear power. Instead, from a scientific point of view, nuclear is not so good as other technologies, and I needed to state that honestly, as I have been trained to do. The fact that nuclear has problems has consistently been borne out over the 11 years since the review paper, as described in Section 3.3.

Subsequent to the *Scientific American* paper, Dr. Delucchi and I embarked on writing more detailed versions of the global 100 percent WWS roadmap and also a roadmap for the United States as a whole. These were published in the journal *Energy Policy* (Jacobson and Delucchi, 2011; Delucchi and Jacobson, 2011). Concurrently, I engaged another PhD student, Elaine Hart, to

develop an optimization model (Section 8.2.2.2) to simulate rigorously the matching of electricity demand with a bundle of geothermal, wind, solar, and hydropower supply in California. She completed that work (Hart and Jacobson, 2011, 2012), confirming in more detail what Graeme Hoste had found. Another student, Bethany Frew, followed that work with an optimization model study that involved combining a bundle of renewables to increase electricity grid reliability across the United States (Frew et al., 2016). On a related topic, an additional PhD student, Eric Stoutenburg, examined the impact of combining wind and wave power to increase reliability of both (Stoutenburg et al., 2010).

## 9.1.9 The Solutions Project

In the midst of the flurry of research activity, I was invited to a dinner at the Axis Café and Gallery in San Francisco on July 10, 2011, to discuss the economic viability of renewable energy alternatives for the state of New York. The dinner was hosted by my now good friend Marco Krapels. At the time, Marco worked in the banking business. He had also started a nonprofit, Empowered by Light, whose goal was to bring solar power to remote communities worldwide.

Marco had invited several others to this dinner, from both the local area and from New York. In particular, he invited Mark Ruffalo and Josh Fox. Mark is an actor and activist who had been asked by the governor of New York, Andrew Cuomo, to participate in a New York renewable energy task force. Josh directs documentaries and was coming off the success of his 2010 documentary *Gasland*, which exposed the problem of **natural gas hydrofracturing**, or **fracking**, to a worldwide audience. The movie received an Academy Award nomination.

I had been invited to the dinner because Marco and several others had seen our *Scientific American* article about powering the world for all purposes with renewable energy. My contribution to the dinner would be to offer ideas of what New York could do to obtain its energy from something other than fracked natural gas. At the time, fracking was not legal in New York, but the governor was under significant pressure to legalize it. Fracking was legal in nearby Pennsylvania, and thousands of wells had already been drilled there, causing damage and upheaval in many communities.

At the dinner, Marco, Josh, Mark, and I bonded together. It was as if we had known each other for years.

We were all passionate about finding a sustainable solution to New York's energy issue and about eliminating air pollution and climate problems in general. The interesting thing was that we all had different backgrounds. I was a scientist. Marco was a businessperson. Mark and Josh were cultural heroes – artists and entertainers. Later, this combination of science-business-culture would prove to be a powerful combination much stronger than any of the individual parts.

At the dinner, we discussed the fracking issue in New York and whether there was an alternative energy solution to it. I speculated that New York could be powered entirely by WWS for all purposes. This prompted the question of whether I would be interested in developing a clean, renewable energy plan for New York. My immediate impulse was that this would be great to do, but I knew how much work it would take, so I waivered. Put on the spot, I told them I would be willing to write a one-paragraph summary, but that it would be better for someone else to take that and turn it into a real roadmap given how much time it would take.

Josh Fox then suggested I speak to two Cornell University professors, Tony Ingraffea and Robert Howarth, to get their perspective on a New York plan, to which I had only committed a paragraph. Tony was a professor in Civil and Environmental Engineering who had worked a great deal on methane leaks from cement casings in natural gas fracking wells. Robert was a professor in Ecology and Evolutionary Biology who had just published a seminal paper with Tony on methane leakage rates from the fracking of shale rock (Howarth et al., 2011). I spoke with them and Josh Fox on the phone almost two weeks later. We discussed different renewable energy resources in New York.

After the call, I sent them a wind resource analysis of New York that my PhD student Mike Dvorak had put together. Josh Fox, Robert Howarth, Tony Ingraffea, and I then scheduled a follow-up call with a larger group, adding Gianluca Signorelli, a work associate of Marco Krapels; Stan Scobies, a retired researcher in New York familiar with the energy landscape; and Marcia Calicchia, a policy expert and researcher at Cornell University.

On the follow-up call, which was at 4 PM Pacific time on September 13, 2011, I was asked again if I could put together a 100 percent renewable energy plan for New York. Feeling the stress of a lot of other work commitments, I waivered again. I repeated that I could write only a brief one-paragraph summary to help guide someone else do the rest.

Later that same night, when I started writing, I began to think more deeply about a state plan for New York, and something in me clicked. I thought to myself, *If we really want to solve the problems of air pollution, global warming, and energy security, we need granular state- and country-level plans.* I also realized that a New York plan would be a natural extension of the global and U.S. roadmaps Mark Delucchi and I had previously developed.

Inspired, I worked into the night. I found wind, solar, hydroelectric, and geothermal resource and air pollution mortality data for New York. I then did an energy, air pollution, and climate data cost analysis of fossil fuels versus WWS. Finally, I identified a set of policy mechanisms to implement 100 percent WWS in the state.

In the morning, I woke up from my trance and happily emailed the paragraph-turned-14-page, single-spaced draft of the 100 percent New York energy roadmap to the group. They were as shocked as I was. Stan quipped, "I figure about three more inspiring conference calls ought to have the whole thing done." We now had a starting point to change the New York energy infrastructure.

The first draft catalyzed a flurry of activity and edits among several in the group we had just formed. Ultimately, the New York roadmap went through forty drafts before it was completed and published in the journal *Energy Policy* 18 months later, on March 13, 2013 (Jacobson et al., 2013).

At the time, though, the immediate goal was to develop a draft paper that was thorough enough to be presented to New York Governor Cuomo and his staff. I worked feverishly on this, engaging students and incorporating comments by several in the group. The group as a whole also started holding regular phone conversations. Marco, Mark, Josh, and I also began talking more, taking part in several middle-of-the-night phone conversations. We became close as we shared a common passion and goal to solve major problems.

The communication and written material from our overall group had become substantial enough that we decided we needed a name. We wanted the name to represent something positive because we felt it is better to be for rather than against something. Since this all-volunteer group was focused on solutions for New York's energy future, we settled on **The Solutions Project**. On December 15, 2011, Marcia sent the group its first brochure, which summarized The Solutions Project as a collaboration among

*… scientists, renewable energy industry pioneers, experts in renewable energy financing and investment, mission-oriented*

*investors, businesses, labor organizations, inner city community groups, farmers, environmentalists, and many celebrities and cultural figures.*

Its initial main goals, laid out in a February 25, 2012, brochure, were to

- Raise mass awareness about the viability of renewable energy,
- Raise awareness about the 100 percent WWS roadmap we were developing for New York, and
- Leverage private capital to demonstrate the success and viability of some specific renewable energy projects.

On January 20, 2012, The Solutions Project, still behind the scenes, began recruiting members for an advisory council. Ultimately, the advisory council consisted of scientists, business leaders, and celebrities. Some of the celebrities who agreed to take part included Deepak Chopra, Leonardo DiCaprio, Jesse Eisenberg, Woody Harrelson, Ethan Hawke, Scarlett Johansson, Sean Lennon, Julianne Moore, Leilani Munter, Elon Musk, Edward Norton, Yoko Ono, Robert Redford, Eileen Rockefeller, Antonio St. Lorenzo, Wendy Schmidt, Martha Stewart, Channing Tatum, Michelle Williams, and Deborah Winger, among others. The purpose of the advisory council was not only to provide feedback, but also to help The Solutions Project meet its main goals, which were to bring awareness to large numbers of people about the potential of a 100 percent transition to clean, renewable WWS energy. Several members of the advisory council proved helpful in disseminating information.

During early March 2012, a new integral member came on board The Solutions Project. Jon Wank had been working in advertising and learned about The Solutions Project from Marco Krapels. Jon's passion for making a difference was so strong that he quit his job and volunteered to work with The Solutions Project on media and social engagement. It was Jon's ingenuity that led to the development of the state, country, and city infographics, such as the one in Figure 9.1. Each infographic summarizes a 100 percent roadmap. The infographics have been used worldwide and are still available online (Solutions Project, 2019).

### 9.1.10 Effects of New York State Roadmap on New York Policies

The original focus of The Solutions Project was to bring the science-based 100 percent New York energy roadmap to the governor of New York. The hope was that the governor might see that a WWS plan is the best solution for the state's future and to translate the plan into law. Such an action would also mean there would be no need for hydraulic fracturing.

To that end, The Solutions Project began to engage other groups, including the National Resources Defense Council (NRDC), which had a major presence in New York. After some discussion with the NRDC, they committed, on January 2, 2012, to support The Solutions Project goal of bringing 100 percent WWS to New York. This was an important first step.

The Solutions Project held its first all-hands meeting at Cornell University on May 21, 2012. It was there that many of the people who had been involved in numerous phone calls could finally meet each other and focus further on a path forward. Such a path included the idea of getting information out to large numbers of people. Some of the ideas that came forth were to hold a concert, go on a speaking tour, hold rallies, and develop media content. Marcia introduced finger puppets that we could wave at each other to lighten the mood.

Shortly after that meeting, on June 20, 2012, Mark Ruffalo, Marco Krapels, and I met at Stanford University, California, where we spoke with students, some of whom were helping develop the New York energy transition roadmap. We then went over to Google in Mountain View, where all three of us gave a joint talk that was also being recorded (Ruffalo et al., 2012a) to a packed house. We offered our experiences, through the lenses of culture (Mark Ruffalo), business (Marco Krapels), and science (me), about how it is possible and necessary to transition from fossil fuels to 100 percent WWS.

We then met with the Google energy and sustainability team led by its director, Rick Needham. Google had been making investments in renewable energy for several years. In 2007, they installed a 1.6-MW PV array on their Mountain View buildings. In 2010, they purchased two wind farms in North Dakota and contracted for additional wind in Iowa. Our meeting with them was a meeting of like minds. We discussed with them the importance of companies to work together with scientists, cultural and community leaders, and policymakers to effect a transition. Ultimately, in 2017, Google became the first company in the world to provide at least 100 percent of its annual average power from WWS.

On the same day, we visited Facebook and their sustainability and energy efficiency team, led by Bill Weihl.

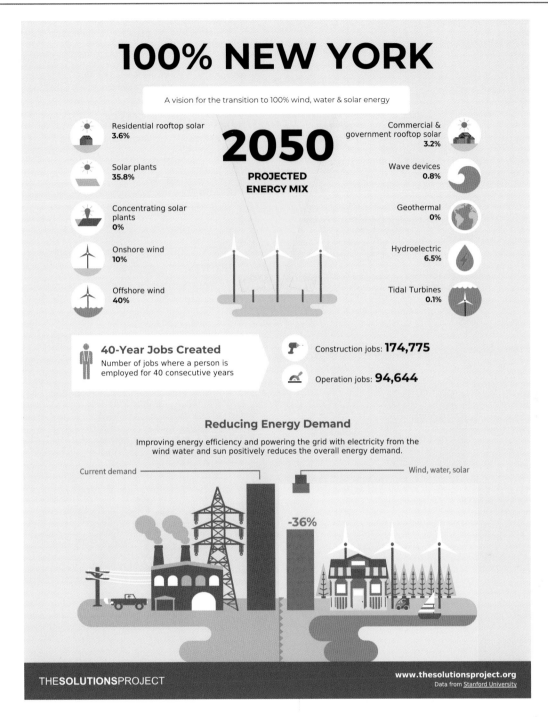

Figure 9.1  Part of an infographic summarizing the 100 percent all-sector WWS roadmap for New York, updated in Jacobson et al. (2015a). From Solutions Project (2019), with graphics developed by Jon Wank.

We similarly discussed the need for businesses to participate in a 100 percent transition.

The talk at Google was received so well, we were asked to come again as a trio to a conference on Nantucket island, Massachusetts, on October 6, 2012 (Ruffalo et al., 2012b). There, policymakers from across political parties and media were gathered. Some included Secretary of State John Kerry; President of Americans for Tax Reform, Grover Norquist; presidential advisor David Gergen; U.S. Treasury Secretary Larry Summers; and MSNBC commentator Chris Matthews.

This talk led to an invitation to Chris Matthews' home on February 27, 2013, where Mark, Marco, and I presented The Solutions Project vision to U.S. Senator

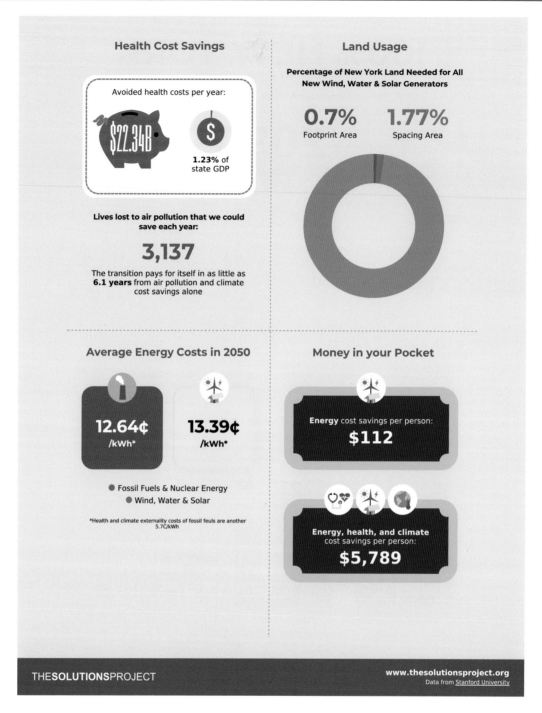

**Figure 9.1** (*cont.*)

Kirsten Gillibrand, Secretary Kerry, some members of the U.S. House of Representatives, and some members of President Obama's staff. Whereas these meetings didn't lead to direct policies at the time, they provided new information to policymakers, increased the familiarity of The Solutions Project name, and allowed us to hone our explanations and learn about the concerns of others.

In a parallel effort to raise awareness on a large scale, Jon Wank created an animated cartoon, *Tommy and the Professor*, during August 2012 (Wank et al., 2012). Mark Ruffalo came up with this idea at our May 12, 2012, retreat at Cornell. He and I were brainstorming outside, and he suggested a brilliantly crazy idea to produce a cartoon where he was an annoying kid and I was a pedantic professor trying to teach him about renewable energy. Months later, Jon Wank brought this cartoon idea to life. Tommy (voiced by Mark Ruffalo) was the annoying student who hung out with a girl (voiced by Zoe Saldana). After seeing that the Professor's house was the only one on the block with its lights on after a blackout, the two came to a class on renewable energy taught by the Professor (voiced by me). There, they learned about energy from wind, water, and sunlight. After the lecture, Tommy came to an epiphany about what 100 percent WWS meant and wanted to inform the world about it.

In the meantime, during the rest of 2012, several students at Stanford plus Tony Ingraffea, Bob Howarth, Stan Scobies, economist Dr. Jannette Barth, and I continued to revise the New York energy roadmap. I also engaged Dr. Mark Delucchi to work on the New York paper. He has been a coauthor on all the roadmaps to date.

On November 17, 2012, Mark Ruffalo, Marco Krapels, Josh Fox, Jon Wank, and I brought the latest version of the roadmap with us to a meeting with the NRDC in New York City. The goal of the meeting was to discuss how the NRDC and The Solutions Project could bring 100 percent WWS to the state. On February 5, 2013, the NRDC and The Solutions Project wrote a joint letter to the governor of New York, informing him about our New York State 100 percent WWS energy roadmap. The letter, signed by Frances Beinecke (president of the NRDC), Mark Ruffalo, and me, stated in part,

*We are writing today to let you know about a clean energy framework for New York that has been developed by scientists, financial specialists, business leaders, and policy experts under the Solutions Project and reviewed and updated for practical implementation in New York State by the Natural Resources Defense Council . . .*

The letter then went on to propose several first steps that the governor could take to reach 100 percent WWS. These steps included increasing large-scale wind and distributed solar, scaling up energy efficiency, deploying distributed renewables, removing barriers to the adoption of electric vehicles, creating a **green bank** (which leverages public and private financing to advance clean energy projects), and expanding the existing solar rooftop program.

Soon afterward (during March 2013) our New York roadmap paper was published and received attention in the press, particularly in New York. The roadmap spread like wildfire throughout New York. I was asked to give a talk at an anti-fracking rally about the potential of New York to go to 100 percent WWS. The rally was on the footsteps of the governor's office in Albany. I spoke on June 17, 2013, to a crowd of thousands. This was the first and only rally I ever spoke at. It was gratifying because the crowd was on board with transitioning to 100 percent renewables. The 100 percent WWS roadmap for New York gave people hope that there was an alternative solution to natural gas.

Ultimately, on December 17, 2014, Governor Cuomo did ban hydraulic fracturing in New York because of concerns over its health risks. A decision on fracking had been held in abeyance for six years while health effects data were gathered. Grassroots organizations throughout the state ensured that the health risks were considered. The ban was also facilitated by the fact that an alternative to fracking now existed: WWS.

On July 18, 2019, Governor Cuomo went further, signing a law requiring New York to reach 70 percent WWS by 2030 and up to 100 percent by 2040 in the electric power sector. The law also required emission reductions in other sectors sufficient for overall greenhouse gas emissions in the state to be reduced 85 percent by 2050. It was gratifying to see science (through the published New York roadmap), business, culture, and community could come together to motivate the passage of a law mandating a clean, renewable energy solution.

## Transition highlight

To that end, with the fracking issue behind him, Governor Cuomo submitted to the New York Public Service Commission a proposal for New York to obtain half of its electric power from renewable sources by 2030. On August 1, 2016, the Public Service Commission approved the proposal, giving rise to a mandatory and enforceable 50 percent WWS law by 2030 for the state of New York called the **Clean Energy Standard**. The law required half of the state's electricity to come from onshore and offshore wind, solar, and hydroelectric power. Our New York roadmap called for 80 percent WWS in all energy sectors by 2030 (and 100 percent by 2050), but the Clean Energy Standard requirement of 50 percent WWS in the electric power sector was a good first step. It was the first piece of proposed legislation that the 100 percent WWS roadmaps provided a scientific basis for.

## 9.1.11 How the California Roadmap Led to Transitioning Towns and Cities

In the meantime, by September 2012, I had a desire to develop a 100 percent all-sector WWS roadmap for California. Aside from the fact that I live in the state, California is the most populated U.S. state, it has one of the largest economies in the world (5th largest in 2020), and it has an affinity for renewable energy. In addition, our research group had done previous work in analyzing offshore wind and the grid in California. I formalized a Solutions Project student research group at Stanford and engaged over 20 new students to help develop the California roadmap. I also worked with several of the same scientists who had helped with the New York roadmap on the California plan.

We finished an early rough draft of the California roadmap on January 8, 2013. I sent a copy that day to the Sierra Club to review for practical implementation. Jodie Van Horn, at the Sierra Club, became the point person, and she obtained internal suggestions for improvement. This was the first of many iterations of the California roadmap over the next year and a half.

In early 2013, The Solutions Project was still an all-volunteer group. However, the group was thinking about becoming a formal nonprofit. Before deciding to take the leap, we considered other options. Given that we were now interacting more with the Sierra Club, it was logical to discuss possibilities with them as well. On February 6, 2013, Mark Ruffalo approached Michael Brune, the executive director of the Sierra Club, about the possibility of The Solutions Project and Sierra Club becoming "one diverse coalition around science, business, and culture advancing 100 percent renewables."

The Sierra Club was strongly interested and proposed to incorporate The Solutions Project as a campaign inside the Sierra Club and to move resources from their Beyond Coal Campaign to a 100 Percent WWS Campaign led by The Solutions Project. The Sierra Club was interested because The Solutions Project offered raw science, the media power of celebrity, and additional creativity. The offer was tempting to The Solutions Project because it negated the need for us to go through the pains of forming a nonprofit. However, it also would have reduced the autonomy of The Solutions Project, whose success to date was based on being free to pursue ideas without gaining the consensus of a large board. After lengthy deliberations, including a meeting in New York City on April 20, 2013 (Figure 9.2), The Solutions Project chose to become its own nonprofit. Its focus remained, for the time being, on amplifying and educating the public and policymakers about the 100 percent roadmaps. The Sierra Club saw this as an important mission as well but shifted to focus on what it does best, grassroots campaigning. In this case, the campaign centered on getting cities across America to commit to 100 percent WWS. This effort, called the Ready for 100 Campaign, started in earnest in late 2015.

On October 29, 2015, Jodie Van Horn from the Sierra Club contacted me again. We talked about the Sierra Club's pending cities campaign, the goal of which was to get 100 U.S. cities to commit to 100 percent WWS within three years. She asked if I could prepare 100 percent WWS roadmaps for some targeted cities. On the one hand, I knew that each roadmap took a lot of work. On the other hand, I knew they were important for analyzing the ability of cities, in this case, to transition. I agreed but told her it would take some time. Indeed, the roadmap paper we finally completed for 53 towns and cities in North America was not published until two-and-a-half years later, on June 30, 2018 (Jacobson et al., 2018b).

The Solutions Project finally became an independent nonprofit in mid-2013. Much of the rest of the year was spent hiring an executive director, Sarah Hope, and staff, setting up the organization, and preparing a launch event. The launch event finally took place on June 19, 2015, in New York City. Leonardo DiCaprio joined the rest of The Solutions Project team at the event.

## Transition highlight

In the meantime, the Sierra Club used the 100 percent state roadmaps and resulting infographics (Solutions Project, 2019) to help convince communities and town and city leaders that 100 percent was possible. When the town and city roadmaps finally dribbled in, they became helpful as well. However, by the time the town and city roadmaps were published in 2018, the Sierra Club had already obtained commitments from over 100 cities and counties (Table 9.1), meeting their goal of at least 100 within three years. The 100 percent movement had really accelerated with many groups working together.

(a)

(b)

(c)

**Figure 9.2 (a) Meeting of The Solutions Project and Sierra Club to consider merging on April 20, 2013, in New York City. From left to right: Marco Krapels, Jodie Van Horn (Sierra Club), Jon Wank, Mark Z. Jacobson, Mark Ruffalo, and Josh Fox.** © Jon Wank.
**(b) Backstage with Mark Ruffalo before appearing on the *Late Show with David Letterman* on September 23, 2013.** © Peter Sullivan.
**(c) On the *Late Show with David Letterman* the same day.** © Mark Z. Jacobson.

During May 2013, while still updating the California roadmap, I embarked on an individual roadmap for Washington State. Washington was chosen due to the substantial renewable resources it had (wind and existing hydroelectric in particular). I thought that Washington might be a state that could reach 100 percent quickly with respect to electric power. During the autumn of 2013, I recruited a new set of over 20 students to work more intensely on the roadmap. The Washington State roadmap was accepted for publication August 1, 2015 (Jacobson et al., 2016).

### 9.1.12 The *Late Show with David Letterman*

On September 25, 2013, a television producer, Mike Buckiewicz, invited me to appear on the *Late Show with David Letterman*. He said that he had invited me because he had seen my interview in Josh Fox's documentary *Gasland Part II*. Josh interviewed me on July 12, 2011, the day after we first met in San Francisco. His film is about the damage done by natural gas fracking. My short interview was about how it was possible to abandon gas, coal, and oil and move completely to renewables. The producer liked the positive message I related. He told me

Table 9.1 Countries and U.S. states, districts, territories, counties, cities, and towns that have committed to or reached 100 or near 100 percent clean, renewable electricity or all energy as of January 16, 2020. The country, state, and district commitments are just in the electricity sector. The county, town, and city commitments are either in the electricity sector (in most cases) or in all sectors, and they are beyond just town/city/county operations. Cities in blue have a population of at least 190,000 as of January 1, 2020. Countries having reached 95 to 100 percent WWS are primarily from Table 8.1. REN21 (2019b) lists 250 cities internationally that had committed to 100 percent renewables in the electric power sector by the end of 2018. Fifty of these cities had committed to 100 percent renewables in multiple sectors (e.g., electricity, transport, and building heating/cooling). Country commitments are from Table R6 of REN21 (2019a) (whose data are for 2018 end) for all countries except Spain and Portugal, which committed during 2018. County, town, and city commitments are from Sierra Club (2019).

**10 countries that have reached 95 to more than 100 percent WWS electricity**

| | | | | | |
|---|---|---|---|---|---|
| Albania | Costa Rica | Kenya[a] | Paraguay | Tajikistan | |
| Bhutan | Iceland | Norway | Scotland[b] | Uruguay | |

**61 countries that have laws requiring 100 percent renewable electricity**

| | | | | | |
|---|---|---|---|---|---|
| Afghanistan | Cook Islands[c] | Guatemala | Morocco | Senegal | Tunisia |
| Aruba[c] | Costa Rica[e] | Haiti | Nepal | Solomon Islands[e] | Tuvalu[c] |
| Bangladesh | Denmark[g] | Honduras | Niger | South Sudan | Scotland[c] |
| Barbados | Djibouti[c] | Kenya | Niue[c] | Spain | Vanuatu[e] |
| Bhutan | Dominica | Kiribati | Palau | Sri Lanka | Vietnam |
| Burkina Faso | Dominican Rep. | Lebanon | Palestine | St. Lucia | Yemen |
| Cabo Verde[d] | Ethiopia | Madagascar | Papua N. Guinea[e] | Sudan | |
| Cambodia[d] | Fiji[e] | Malawi | Philippines | Sweden[f] | |
| Colombia | Gambia | Maldives | Portugal | Tanzania | |
| Comoros | Ghana | Marshall Islands | Rwanda | Timor-Leste | |
| Congo, DR | Grenada | Mongolia | Samoa[e] | Tokelau | |

**14 U.S. states, districts, and territories that have laws or executive orders requiring 100 or up to 100 percent electric power from renewables**

| | | | | |
|---|---|---|---|---|
| California | Maine | New Mexico | Rhode Island | Washington State |
| Connecticut | Nevada | New York | Virginia | Wisconsin |
| Hawaii | New Jersey | Puerto Rico | Washington, DC | |

**13 U.S counties that have committed to 100 percent renewables**

| | | | | |
|---|---|---|---|---|
| Buncombe, NC | Multnomah, OR | Salt Lake, UT | Taos, NM | Whatcom, WA |
| Floyd, VA | Orange, NC | Summit, CO | Ventura, CA | |
| Grand, UT | Pueblo, CO | Summit, UT | Wake, NC | |

6 U.S. towns and cities that have reached 100 percent renewable electricity

| | | | | | |
|---|---|---|---|---|---|
| Aspen, CO | Burlington, VT | Georgetown, TX | Greensburg, KS | Kodiak Island, AK | Rock Port, MO |

153 additional U.S. towns and cities that have committed to 100 percent renewable electricity

| | | | | | |
|---|---|---|---|---|---|
| Abita Springs, LA | Conshohocken, PA | Golden, CO | Minneapolis, MN | Portland, OR | Spokane, WA |
| Ambler, PA | Cornish, NH | Goleta, CA | Missoula, MT | Portola Valley, CA | Springdale, UT / Springfield, PA |
| Amherst, MA | Cottonwood Hts, UT | Hanover, NH | Moab, UT | Pueblo, CO | St. Louis, MO |
| Angel Fire, NM | Culver City, CA | Haverford, PA | Monona, WI | Questa, NM | St. Louis Park, MN |
| Apex, NC | Del Mar, CA | Hillsborough, NC | Monterey, CA | Radnor, PA | St. Paul, MN |
| Arlington, VA | Denton, TX | Holladay, UT / Ivins, UT | Narberth, PA | Reading, PA | St. Petersburg, FL |
| Athens, GA | Denver, CO | Kamas, UT | Nederland, CO | Red River, NM | State College, PA |
| Atlanta, GA | Downingtown, PA | Kansas City, MO | Nevada City, CA | Rolling Hills Est., CA | Tallahassee, FL |
| Augusta, GA | Dunedin, FL | Kearns, UT | New Brunswick, NJ | Safety Harbor, FL | Taos, NM |
| Berkeley, CA | Durango, CO | Keene, NH | Norman, OK | Salt Lake City, UT | Taos, Ski Valley, NM |
| Blacksburg, VA / Bluffdale UT | Eagle Nest, NM | Kennett, PA | Norristown, PA | San Diego, CA | Thousand Oaks, CA |
| Boise, ID | East Bradford, PA | La Crosse, WI | Northampton, MA | San Francisco, CA | Traverse City, MI |
| Boulder, CO | East Hampton, NY | La Mesa, CA | Oakley, UT | San Jose, CA | Tredyffrin, PA |
| Breckenridge, CO | East Pikeland, PA | Lafayette, CO | Ogden, UT | San Luis Obispo, CA | Truckee, CA |
| Cambridge, MA / Castle Valley, UT | Eau Claire, WI | Lakewood, OH | Ojai, CA | Santa Barbara, CA | Upper Merion, PA |
| Chapel Hill, NC | | Largo, FL | Orlando, FL | Santa Monica, CA | Uwchlan, PA |

Table 9.1 (*cont.*)

| | | | | | |
|---|---|---|---|---|---|
| Cheltenham, PA | Edmonds, WA / Emigration, UT | Longmont, CO | Oxnard, CA | Sarasota, FL | Ventura, CA |
| Chicago, IL | Encinitas, CA | Los Angeles, CA | Palo Alto, CA | Satellite Beach, FL | |
| Chula Vista, CA | Eureka, CA | Lowell, MA | Park City, UT | Schuylkill, PA | West Chester, PA |
| Cincinnati, OH | Evanston, IL | Madison, WI | Petoskey, MI | Solana Beach, CA | West Hollywood, CA |
| Clarkston, GA | Fayetteville, AR | Menlo Park, CA | Philadelphia, PA | South Lake Tahoe, CA | Whitemarsh, PA |
| Cleveland, OH | Fort Collins, CO | Middleton, WI | Phoenixville, PA | South Miami, FL | Windsor, MA |
| Coalville, UT / Columbia, SC | Francis, UT | Millcreek, UT | Plainfield, NH | South Pasadena, CA | |
| Concord, NH | Gainesville, FL | Milwaukie, OR | Plymouth, PA | Southampton, NY | |

[a] Kenya's nameplate capacity in mid-2019 was 88.7 percent WWS, sufficient to produce 90 to 100 percent of its annual average electricity from WWS (Section 8.1). [b] Scotland is included based on a preliminary estimate that it can produce 100 percent of its annual average power consumption from WWS by 2020, which is its commitment. Country commitments are all for 2050, except [c] 2020; [d] 2025; [e] 2030, [f] 2030. [g] Denmark's law is for all energy.

that when he showed the clip to David Letterman, the latter asked, "Why haven't we had that guy on before?" David Letterman himself had a serious concern about our future in the face of climate change, so he felt hopeful there may be a possible solution.

I was humbled about the thought of being on the *Late Show*. Only a handful of scientists had appeared on a late-night show on a major network. One whom I deeply admired was the late Professor Stephen Schneider, who was a climate scientist and ironically had an office a few doors down from me at Stanford University. He had appeared on Johnny Carson's *The Tonight Show* in 1977 to talk about climate science. Ever since I was a young scientist and heard about Stephen Schneider doing this, I thought that being able to speak on such a stage about science would be impossible but amazing if it could happen. When I was offered an appearance on the show, I felt a weight and responsibility. I needed to communicate clearly and accurately, given that Stephen had been such an effective communicator, and I had nothing like his skills.

I appeared on the show on the evening of October 9, 2013 (Letterman, 2013). I spent the whole day training in New York City with Mark Ruffalo and others. Mark came with me backstage to the Green Room to help prepare me further (Figure 9.2). I thought I had everything under control. My time slot was right after Lucy Liu. I was given 11 minutes to converse with Mr. Letterman. This was a long time slot for an interview on a show like this.

As I was walking toward the stage with about 30 seconds to go, the music was playing and the lights were blinding. Even though I had been teaching for over two decades and had given hundreds of talks and had prepared for this all day, I panicked. I was overthinking what I wanted to say, and all of a sudden, I had too much running through my head. Then, the thought *I'm going to be speaking to millions of people* crept into my head. With seconds ticking by, all I could think about was, *This is going to be a disaster*. I kept walking toward the stage in a panic, as if I were walking to my death.

Then, with about four seconds to go, I remembered one of my tennis lessons. *Keep your eye on the ball.* I thought to myself, *Okay, relax. Take a breath. Just focus on one thing to say. Say something about why I want to change the energy system.* Because my concern about air pollution is the reason that I started this career, I immediately knew I should just focus on this topic. I decided to say something about how many people die from the effects of air pollution. As morbid as that is, I felt that it was a fact that few people were aware of but was important. I believed that if I said that, everything else would flow.

When I finally sat down and the music died down, I was still jittery but felt more under control. I also remembered what the producer had told me: "Look toward Dave and keep your hands planted." Mr. Letterman made me feel at ease and threw me a softball statement to respond to: "... tonight, you have something positive that you can present to all Americans." My response was, "So Dave, we're developing science-based plans to eliminate global warming, air pollution, including the 2.5 to 4 million deaths that occur worldwide each year ..." His immediate response was, "Due to air pollution? That many people are dying due to air pollution?" Once he asked that, I knew I had made the right decision to focus on that topic, because it grabbed his attention and likely the attention of those watching. After that, I relaxed, and everything flowed for the rest of the interview (Figure 9.2).

The response to my being on the show was overwhelmingly positive. The New York governor's staff was watching, as were people interested in energy and the environment across America. Our Solutions Project student group had organized a viewing party, which humbled me further. The show brought 100 percent clean, renewable energy to the large-scale public sphere. One hundred percent WWS was no longer just a niche idea. It had begun to capture the public's imagination and was accelerating into a movement.

Soon after the show, a new star entered The Solutions Project scene. On November 26, 2013, Marco Krapels hosted a dinner in San Francisco, where he introduced me and others to Leilani Munter. Leilani was a racecar driver who was passionate about clean, renewable energy. She enthusiastically embraced the 100 percent goal and wanted to bring it to 80 million U.S. racecar fans who normally would not care about energy. Over the years, since that meeting, she has been a spokesperson for electric racing and electric-powered racing stadiums. She also graciously volunteered her time at events and to speak out whenever possible about 100 percent.

## 9.1.13 Impact of the California Roadmap on California Passing a 100 Percent Law

The California roadmap was ultimately published on July 22, 2014 (Jacobson et al., 2014b). Once it became public, the press began writing articles about it, and some articles reached the office of Governor Jerry Brown. On August 22, Mr. Brown's senior advisor on energy and environmental issues, Cliff Rechtschaffen, emailed me stating,

*We read with interest about your recent paper on a 2050 renewables strategy for California. I'm wondering if you would be interested in coming up to brief a group of advisors and policymakers who work on climate and energy issues in the administration.*

Needless to say, I was ecstatic over the thought, because the whole goal of our work was to perform science in order to inform policymakers better about solving air pollution and climate problems. Air pollution alone in California caused at the time about 13,000 premature mortalities per year. This was also a chance to inform the staff of the governor of the fifth largest economy in the world about a pressing global and local issue.

On October 27, 2014, Marco Krapels and I trekked to Sacramento to present our California 100 percent WWS roadmap. The group we presented to included Mr. Rechtschaffen, Mr. Brown's senior policy advisor Ken Alex, California Energy Commissioner David Hochschild, and others. I laid out the 100 percent roadmap for California in detail. The roadmap called for an 80 percent transition in all energy sectors by 2030 and 100 percent by 2050. Marco and I were peppered with lots of good questions. We left feeling we had given it our best shot but had no idea at the time what the impact might be. We did know that nothing would happen right away, if at all, because Mr. Brown was running for reelection, and the election was coming up in 11 days.

Mr. Brown was re-elected. During his inauguration speech, on January 5, 2015, he pleasantly surprised us by proposing several laws that followed logically from our discussion and California roadmap. He proposed that at least 50 percent of all electricity in the state of California should come from WWS sources by 2030. He also proposed reducing petroleum use in vehicles by 50 percent by 2030, which meant electrifying 50 percent of the vehicle fleet by 2030. We had proposed 80 percent by 2030. He finally proposed doubling the efficiency of existing buildings by 2030. We had proposed additional efficiency

measures and providing heating, cooling, and electricity in buildings with WWS. Whereas the oil industry gutted his proposal to reduce petroleum use, California eventually passed the 50 percent WWS portfolio standard and the energy efficiency standard through California SB 350.

In another surprise, on September 19, 2016, Governor Brown signed California SB 1383, which required a 50 percent reduction in black carbon (and other short-lived climate forcers) emissions below 2013 levels by 2030. After two decades of work on black carbon, the scientific results were finally paying off with specific, enforceable legislation.

Between 2016 and 2018, the 100 percent movement swelled nationally. The Solutions Project was central to this expansion. They, along with the Sierra Club and a few other key groups, mobilized almost 100 nonprofits (Table 9.2) in a group, now called the **100 Percent Network**, around the 100 WWS percent goal. The main purposes of The Solutions Project had shifted slightly, not only to raise awareness that 100 percent WWS is possible, but also to bring 100 percent WWS to 100 percent of the people.

To amplify the latter point, The Solutions Project created a subgroup called **100.org**. The purpose of 100.org is to provide small grants for small organizations that have creative ideas about how to engage people, especially people who might not otherwise prioritize a transition as a goal. The group 100.org also provides awards to honor transition leaders and people with innovative ideas. It also seeks an **equitable transition** to 100 percent WWS. This involves, for example, proposing that at least a certain fraction of new jobs during a transition are reserved for disadvantaged communities. Through The Solutions Project and 100.org, communities in many inner cities suddenly became part of the 100 percent movement.

Due to the groundswell of public support toward 100 percent WWS in California following the passage of SB 350, state senator Kevin DeLeon proposed a follow-up law, **SB 100**, for the state to move to 60 percent eligible renewables by 2030 and to obtain the remaining 40 percent by 2045 from either eligible renewables, large hydropower, or other zero-carbon technologies not yet invented.

In California, an eligible renewable includes primarily WWS technologies (e.g., wind, solar, geothermal). While small hydropower (e.g., run-of-the-river hydropower) is also considered an eligible renewable, large hydropower (with a dam) is not. However, because California has so much existing large hydropower and imports more from

**Table 9.2 Non-governmental organizations that were part of the U.S. 100 percent network as of January 26, 2019.**

| | | | |
|---|---|---|---|
| Solutions Project | Inst for Local Self Reliance | Demos | MA Climate Action |
| Sierra Club | Vote Solar | NAACP | Scope LA |
| Earth Justice | Sierra Club Student Coalition | Partnership for Working Families | Juntos NM |
| Mosaic | People's Action | OPAL | Blue Planet Foundation |
| Center for Community Change | USCAN | MIT | Green Muslims |
| Green Spaces Chattanooga | Green For All | Sustain U.S. | Interfaith Power and Light |
| Renewable Cities | Simon Frasier University | Grassroots Glo Justice Alliance | Climate Solutions |
| Renewables 100 | Community Power MN | People's Climate Movement | One Northside |
| Environment America | Environmental Action | The Point | We Own It |
| APEN | Ctr for Popular Democracy | NYC Env Justice Alliance | Sustainable CO |
| We Mean Business | Conservation Voters NM | Neighbor to Neighbor MA | Grid Alternatives |
| Avaaz | Toxics Action Coalition | WEACT | Puget Sound Sage |
| Greenpeace | Sunrise Movement | Iowa CCL | Ctr Comm Action & Env Justice |
| Purpose | Venner Consulting | Environmental Council of MI | Communities for a Better Env |
| 350.org | Energy and Policy Institute | Project South | CA Nurses |
| Renewable Energy Long Island | Front and Centered Coalition | Got Green | Partnership for Southern Equity |
| Race Forward | Local Clean Energy Alliance | UPROSE | NY Renews |
| Emerald Cities Collaborative | Environment MA | Better Future Project | Cleo Institute |
| Center for Working Families | Environment NM | PUSH Buffalo | |
| World Wildlife Fund | Climate Solutions | Power Shift | |
| Cooperative Energy Futures | Coalition for Comm of Color | Food and Water Watch | |

Washington State, and because state lawmakers wanted to maintain existing large hydropower as part of the 100 percent goal, it was necessary to create a separate category from eligible renewables in which to put large hydropower. So, the last 40 percent of SB 100 allows for eligible renewables and large hydropower. It also allows for other zero-carbon technologies that have not yet been invented. California Senate (2018) defines the reason for including other possible zero-carbon technologies: it "leaves the door open to new technologies we may not know about today." The term was not intended for nuclear power or technologies, such as CCS, that are not zero-carbon (Section 3.2.3). California Senate (2018) states that nuclear is being phased out of California and no new nuclear plants are being planned. The remaining 40 percent will likely be filled by hydropower plus other renewables.

## Transition highlight

SB 100 was signed into law on September 10, 2018. This was landmark legislation mandating the transition to effectively 100 percent WWS in the electric power section. The passage of this law was the kind of result hoped for when we embarked on the first 100 WWS percent roadmap in 2009. However, much more work is needed in other sectors and in other locations.

### 9.1.14 50-State and 139-Country Roadmaps, New York Climate March, and Paris Climate Conference

While working on individual Washington State and California roadmaps during October 2013, I came to the realization that, at the rate we were creating one state roadmap at a time, it could take decades to finish all 50 states. I then became determined to automatize the process. I took the Washington State roadmap spreadsheets and expanded them to include all 50 U.S. states. This began the process of developing 50 individual state roadmaps, which were ultimately published 20 months later, in June 2015 (Jacobson et al., 2015a).

Once we produced the first draft of the 50-state roadmaps in February 2014, Jon Wank created infographics for each state. These were ultimately posted on a clickable map on a website (Solutions Project, 2019). These maps were invaluable because they were simple and informative.

A ripe opportunity then came up to use the infographics. On August 27, 2014, Leilani Munter, Marco Krapels, Mark Ruffalo, Brandon Hurlbut (a Solutions Project board member), and Andres Lopez (a Solutions Project staff member), and I went to the White House to meet with Vice President Joe Biden. The purpose of our visit was to provide information to the vice president about our 100 percent roadmaps. He was taking a leading role in issues related to the U.S. renewable energy infrastructure. Ahead of the meeting, we provided his staff

with a document that included an infographic of our 100 percent roadmap for Delaware. We were told we had only 30 minutes to speak with him, so we each prepared short remarks.

Mr. Biden came in 15 minutes late and apologized for being late. The first thing he said, before we could put in a word, was that he fully supported our 100 percent roadmaps. He cautioned, though, that the political landscape was such that it would be difficult for the federal government at the time to take any action on them. We later chuckled, because Mr. Biden ended up staying an hour with us but talked almost the entire time. He was endearing and easy to listen to. Halfway through, he said he had to go and started to leave. I asked a parting question, and he came back in the room and spoke for another 30 minutes. For me, the best thing that came out of it was the feeling he was on board with solving the problem in a big way.

On September 21, 2014, New York held a climate march, attended by about 400,000 people. I was not able to attend, but The Solutions Project had a large and central presence there. Leonardo DiCaprio joined Mark Ruffalo, Marco Krapels, Jon Wank, Leilani Munter, and Brandon Hurlbut from The Solutions Project (Figure 9.3). The United Nations building itself was even lit up with the sign Solutions Exist.

Mr. DiCaprio, who was good friends with Mark Ruffalo, was passionate about what The Solutions Project stood for. He wanted to play a key role in helping to

**Figure 9.3** (a) Mark Ruffalo (left), a tribal leader (center), Leonardo DiCaprio (right), Marco Krapels (back left), Brandon Hurlbut (back center), and Leilani Munter (back right) at the New York City climate march on September 21, 2014. © Jon Wank. (b) Advertisement taken out in the *New York Times* on September 23, 2014, to raise mass awareness about 100 percent WWS. © Jon Wank.

disseminate research results on our behalf. To that end, on September 23, 2014, Leonardo DiCaprio spoke in front of the United Nations General Assembly. The week before, the U.N. secretary-general had designated Mr. DiCaprio as a United Nations Messenger of Peace.

Leonardo spoke about climate change and about the solutions we had developed. He stated in front of the world body, "New research shows that by 2050 clean, renewable energy could supply 100 percent of the world's energy needs using existing technologies, and it would create millions of jobs." A month later, on October 29, 2014, Leo DiCaprio came to Stanford to interview me and several students working with me on The Solutions Project roadmaps. He was producing a documentary, *Before the Flood*. He and director Fisher Stevens spent the day with us, filming. Marco Krapels was there as well. We all saw firsthand the passion that the students had for trying to solve the climate, pollution, and energy problems we face. The documentary ended up taking a different direction so did not focus on solutions, but some of the interviews were used as a postscript on the DVD version of the film (Stevens and DiCaprio, 2014). This event motivated our students for months afterward.

During 2015, I was not only completing the 50-state roadmaps but had embarked on an even more ambitious project: developing roadmaps for most countries of the world. Although we had developed world roadmaps in 2009 and 2011, individual countries could not practically implement a world plan, so country roadmaps were needed. The work on country roadmaps began during August 2014. Our group at Stanford first put together a sample roadmap for Ukraine. We expanded this in earnest to 139 countries by March 2015. This was the number of countries for which we could find raw energy data from the International Energy Agency. Work on the country roadmaps continued all year but accelerated feverishly to meet a deadline at the end of November 2015, just before the Paris climate conference.

In December, Paris was hosting the United Nations Climate Change Conference, referred to as the **Conferences of the Parties (COP) 21**. This was an important international stage and a ripe opportunity to disseminate our country roadmaps to world leaders. The event lasted two weeks, but I could stay for only a few days. I met Marco Krapels there and gave several talks but only one major one. Due to a quirk in scheduling, I was fortunate to be able to give a talk that was also recorded (Jacobson, 2015) at the Petit Palais in between a talk by U.N.

Secretary-General Ban Ki-moon and another one by U.S. Secretary of State John Kerry. I later quipped to myself that I was the only speaker who didn't have a security detail. I had only a few minutes to talk but made the most of it. I laid out to the world, for the first time, our 100 percent all-sector WWS energy roadmaps for 139 countries. These countries emit more than 99 percent of all greenhouse gases and air pollutants worldwide.

In Paris, many non-governmental organizations (NGOs), including those that helped build the 100 percent network (Table 9.2), were supportive of the 100 percent goal that The Solutions Project had been disseminating. Paris was filled with the spirit of 100 percent. The Eiffel Tower was lit up with the words "100% Renewable," and a group even formed a human sign (Figure 9.4). I felt that the 100 percent movement had hit the world stage.

### 9.1.15 Impacts of Roadmaps on U.S. Policies, Public Opinion, and International Business Commitments

Shortly after completing and publishing the 50-state roadmaps in 2015, we published a companion paper that examined whether it was possible to match power demand among all energy sectors with 100 percent WWS supply continuously over time among the 48 contiguous U.S. states for which roadmaps had been developed (Jacobson et al., 2015b). The new paper concluded that it was possible across all energy sectors, just as other papers have now found (Section 8.2.1). In fact, several countries have shown this in reality (Table 9.1). The paper received an award, the Cozzarelli Prize, from the *Proceedings of the National Academy of Sciences*.

The combination of the 50-state roadmaps and grid reliability study supporting those roadmaps gave confidence to lawmakers and politicians to legislate 100 percent clean, renewable energy at the national level in the United States.

On November 30, 2015, the first of several pieces of federal U.S. legislation calling for 100 percent WWS was introduced into the U.S. Congress. U.S. House Resolution (H.Res.) 540 proposed that the United States transition to 100 percent renewable energy for all energy sectors (House, 2015). The summary statement of the resolution was

*Expressing the sense of the House of Representatives that the policies of the United States should support a transition to near*

Table 9.3 **Proposed U.S. House of Representatives and Senate resolutions and bills between 2015 and 2019 for the United States to go to 100 percent WWS in one or more energy sectors. None has been voted on as of April 2020.**

| Name | Resolution or bill? | Date introduced | Scope of proposed resolution or law |
|------|---------------------|-----------------|-------------------------------------|
| [a]H.Res. 540 | Resolution | Nov. 30, 2015 | 100 percent clean, renewable energy for all purposes by 2050 |
| [b]S.Res. 632 | Resolution | Dec. 7, 2016 | 100 percent clean, renewable electricity for all purposes by 2050 |
| [c]S. 987 | Bill | Apr. 27, 2017 | 100 percent clean and renewable electricity and public transport by 2050 |
| [d]H.R. 3314 | Bill | Jul. 19, 2017 | 100 percent clean and renewable electricity by 2050 |
| [e]H.R. 3671 | Bill | Sep. 8, 2017 | 100 percent clean (only renewable) electricity by 2035 |
| [f]H.R. 330 | Bill | Jan. 8, 2019 | 100 percent renewable electricity by 2035 |
| [g]H.Res. 109 | Resolution | Feb. 7, 2019 | 100 percent clean, renewable, 0-emission energy for all purposes by 2030 |
| [h]S.Res. 59 | Resolution | Feb. 7, 2019 | 100 percent clean, renewable, 0-emission energy for all purposes by 2030 |

*Sources*: [a]House (2015); [b]Senate (2016); [c]Senate (2017); [d]House (2017a); [e]House (2017b); [f]House (2019a); [g]House (2019b); [h]Senate (2019).

(b)

(a)

Figure 9.4 **(a) The Eiffel tower is lit up with the words "100% Renewable" during the Paris climate conference of December 2015. ©
Naziha Mestaoui. (b) Dozens of people at the Paris climate conference form the words "100% Renewable." © Yann Arthus-Bertrand/
Spectral Q.**

*zero greenhouse gas emissions, 100 percent clean renewable energy, . . .*

The text itself acknowledged our work as the scientific basis for this resolution. It stated,

*Whereas a Stanford University study concludes that the United States energy supply could be based entirely on renewable energy by the year 2050 using current technologies . . .*

Policymakers were realizing on the U.S. national stage that a solution to the horrendous problems of air pollution, global warming, and energy security was possible. This realization continued. Between 2015 and 2019, seven more resolutions and bills were introduced into the U.S. House of Representatives and Senate proposing that the United States go to 100 percent clean, renewable WWS energy in one or more energy sectors (Table 9.3). None of these resolutions or bills has been voted on to date, but they have educated a larger segment of the public and policymakers, furthering the 100 percent movement. In the meantime, during 2016, the United States was going through a presidential election process. On the Democratic side, the three major candidates were Governor Martin O'Malley, Senator Bernie Sanders, and Senator Hillary Clinton. The 100 percent movement had taken off to such a degree that all three candidates supported it.

Governor O'Malley was the first. On July 2, 2015, he issued a press release, where the number one item on his platform was "a complete transition to renewable energy . . . by 2050" (O'Malley, 2015).

Next, Senator Sanders formulated a platform calling for a transition "toward a completely nuclear-free clean energy system for electricity, heating, and transportation." To illustrate, he put The Solutions Project 50-state infographic map on his campaign website (Sanders, 2016). Each state infographic shows what a 100 percent WWS system for all purposes looks like in the state in 2050.

Third, although Senator Clinton also supported the current use of natural gas, she stated publicly on October 16, 2015,

*We need to be moving as quickly as possible to 100 percent clean, renewable energy. We have a long way to go, but that should be our goal, and we should do nothing that interferes with or undermines our efforts to reach that goal as soon as possible.* (Clinton, 2015)

The support for 100 percent by the three Democratic presidential candidates culminated in the U.S. National Democratic Party platform echoing this sentiment by stating, "We believe America must be running entirely on clean energy by mid century" (DNC, 2016). Thus, the 100 percent movement had grown to a point where the largest U.S. party had adopted the 100 percent goal.

Just two weeks before the November 7, 2016, presidential election, I received a phone call from Senator Sanders. He wanted me to provide him with more details about our 100 percent roadmaps, which were prominent on his campaign website, because he was planning to submit a bill to the Senate calling for the United States to go to 100 percent renewables. He subsequently submitted a bill on April 27, 2017. The bill (S.987) was co-sponsored with Senators Merkley, Markey, Booker, and Schatz (Table 9.3).

The phone call with Senator Sanders lasted about 45 minutes. I asked him whether he wanted me to suggest possible policies that might help a 100 percent law be effective. His reply, in his characteristic gravelly Brooklyn accent, was, "No, no, no. That's our job." We further discussed other action that could be taken to educate the public about 100 percent renewables. One such suggestion he had was to write a joint op-ed together. Two days after he submitted S.987, we published a joint op-ed in the *Guardian* (Sanders and Jacobson, 2017).

Meanwhile, the 139-country roadmaps were published on September 6, 2017 (Jacobson et al., 2017), and reported widely in the press. The Solutions Project then developed infographics for each country summarizing the roadmaps (Solutions Project, 2019).

## Transition highlight

The international roadmaps helped to give countries the confidence that they need to reach 100 percent clean, renewable energy in all energy sectors. As of 2019, at least 61 countries have committed to 100 percent renewables in the electric power sector (Table 9.1). Denmark has also committed to 100 percent by 2050 in all energy sectors. Since many of these commitments are for 2050, which is far in the future, it is necessary for effective policies to be put in place in these countries to help them reach the goal. Some such policies are discussed in Section 9.4.

On November 13, 2017, results from an international public opinion poll about renewable energy were reported (Orsted, 2017). Approximately 26,000 people in 13 countries (Canada, China, Denmark, France, Germany, Japan, the Netherlands, Poland, South Korea, Sweden, Taiwan, the United Kingdom, and the United States) were queried. The poll found that 82 percent of all the people who were asked responded that it is "important to create a world fully powered by renewable energy." In the United States, 83 percent supported that statement.

Interestingly, only 66 percent of people in the same poll believed climate change was a significant international problem. The reasons more people believed in renewable energy solutions than in the impacts of climate change could be extracted as follows: (a) 75 percent of people polled believed their country's renewable technology leadership gave them a sense of pride in their county; (b) 73 percent believed that building and producing renewable energy increased economic growth and jobs; and (c) 69 percent believed that renewables made their country more energy independent. However, only 53 percent believed renewable energy reduced health problems. Ironically, that is their greatest benefit. The results of this poll suggest that **people don't need to believe in climate change (although they should) to believe in clean, renewable energy.**

In fact, because it is so strong, support for renewables crosses political party lines. In the United States, for example, 9 out of the 10 states with the most wind power installed have a population that generally votes for Republicans, whose base has traditionally been the most skeptical of climate change and renewables. Wind power is still installed in those states because of its low cost and abundance.

Following the publication of the 139-country-roadmap paper, we performed a follow-up study to determine whether matching continuous all-sector power demand with 100 percent WWS supply was possible in 20 world regions encompassing the 139 countries. This study was published in early 2018 (Jacobson et al., 2018a). It supported the previous grid integration study from 2015 for the United States, but it also showed that matching demand with supply and storage on the grid was possible under additional conditions beyond those shown in the earlier study, and for many more world regions (Table 8.5).

Both the 139-country-roadmap study and 20-world-region grid integration study were subsequently updated with roadmaps for 143 countries and grid integration studies for 24 world regions (Jacobson et al., 2019). These studies supported further the potential for a low-cost transition throughout the world. The latter study forms the basis of the data and case studies provided in Chapters 7 and 8.

During 2018, the policy momentum continued toward 100 percent WWS. During the November 2018 U.S. election, five new governors (in Colorado, Maine, Connecticut, Illinois, and Nevada) were elected whose platforms called for taking their respective states to 100 percent renewable electricity.

Subsequently, on December 18, 2018, the District of Columbia joined two states (Hawaii and California) that had enforceable 100 percent laws. Hawaii had passed a law on June 8, 2015, to go to 100 percent renewable electricity for the entire state by 2045. California had passed SB 100 to do the same on September 10, 2018. Washington, DC's law, though, was to go to 100 percent renewable electricity for the district sooner, by 2032.

Then, on March 22, 2019, New Mexico passed a law to go to 80 percent renewable electricity by 2040 and up to 100 percent by 2045 (Table 9.1).

A few weeks later, on April 11, 2019, the U.S. territory Puerto Rico passed a law to go to 100 percent renewable electricity by 2050.

On April 19, 2019, Nevada passed a law to go to 50 percent renewable electricity by 2030 and up to 100 percent renewable (termed "carbon-free") electricity by 2050.

On May 8, 2019, Washington State enacted a law to go to 100 percent renewables for all new electricity by 2045. The bill states (in Section 6.a.iii) that, to meet the law, electric utilities, when acquiring new resources, must "rely on renewable resources and energy storage . . ."

On May 23, 2019, New Jersey's governor signed an executive order requiring the development of a plan for the state to reach 50 percent renewable electricity by 2030 and "100 percent clean energy by 2050." Clean energy can be renewable electricity or electricity from another technology defined by the state to be "clean."

On June 26, 2019, Maine passed a law to go to 80 percent renewable electricity by 2030 and 100 percent by 2050.

Not to be outdone, New York passed a law on July 18, 2019, to go to 70 percent renewable electricity by 2030 and effectively 100 percent by 2045. The law also called for an 85 percent across-the-board greenhouse gas emission reduction in the state by 2050.

On August 16, 2019, the governor of Wisconsin signed an executive order expressing a goal for the state to go 100 percent carbon-free electricity, thus up to 100 percent renewable electricity, by 2050, and ordered state agencies to develop a plan to do this.

On September 3, 2019, the governor of Connecticut signed an executive order for the state to develop a plan to go to 100 percent zero-carbon electricity, thus up to 100 percent renewable electricity, by 2040.

The domino effect continued. On September 17, 2019, the governor of Virginia signed an executive order expressing a goal for the state to reach 30 percent renewable electricity by 2030 and 100 percent carbon-free electricity, thus up to 100 percent renewable electricity, by 2050. On January 17, 2020, the governor of Rhode Island signed an executive order for the state to go to 100 percent renewable electricity by 2030.

In 2019, new laws for a state to go up to 100 percent renewable electricity were also proposed in Florida, Illinois, Minnesota, Colorado, and Oregon. These laws and proposed laws are on top of legislation passed in dozens of U.S. towns and cities (Table 9.1).

## Transition highlight

The 100 percent fever caught on not only with state and local governments, but also with utilities. Seven U.S. states (Massachusetts, Ohio, California, Illinois, New Jersey, New York, and Rhode Island) have laws permitting community choice aggregation (CCA) utilities (Section 2.8.8). These utilities take over the electricity generation portion of a utility bill. Another utility handles the transmission and distribution portion.

Starting on March 7, 2010, the CCA Marin Clean Energy began offering the option for customers to purchase up to 100 percent of their electricity from clean, renewable WWS sources. The number of CCAs offering 100 percent WWS electricity accelerated rapidly across not only California but also across most other states with CCAs.

Not to be outdone, many traditional utilities began offering 100 percent WWS options in 2019. For example, in 2019, Xcel Energy, which services Minnesota, Michigan, Wisconsin, North Dakota, South Dakota, Colorado, Texas, and New Mexico, set up an option for customers, called *Windsource*, that allows them to obtain "all of their (electrical) energy from renewables." On August 20, 2019, Appalachian Power even opened a "Wind, Water, and Sunlight (WWS) service" to supply 100 percent renewable power to any customer in Virginia (Boyer, 2019). On August 21, 2019, Duke Energy obtained approval to allow military, University of North Carolina, and large non-residential customers to procure 100 percent WWS electricity directly from WWS suppliers (Duke Energy, 2019).

On another front, during the week of the September 21, 2014, climate march in New York City, and five years after the *Scientific American* 100 percent roadmap, a campaign called RE100 was launched by two nonprofits, the Climate Group and CDP. The original purpose of the campaign was to commit 100 companies to 100 percent renewable electricity in the annual average for their global operations by 2020. The first companies to commit were Ikea, Swiss Re, Mars, and H&M. By January 16, 2020, 221 companies had joined (Table 9.4), but with different compliance dates.

## Transition highlight

On a third front, during September through November 2019, the cities of Berkeley, Menlo Park, and Mountain View in California banned the use of natural gas in new residential and/or commercial buildings. Mountain View further required that 100 percent of parking spaces in multi-unit housing and commercial developments have the electrical infrastructure in place to set up electric vehicle charging. Banning gas and facilitating vehicle charging are both necessary steps toward the full electrification of society. Full electrification is needed before energy for all purposes can be provided by WWS.

The first company to reach the 100 percent renewable goal for electric power generation was Google, in 2017. This was fitting, because Mark Ruffalo, Marco Krapels, and I had presented our first public Solutions Project vision of connecting science, business, and culture to reach 100 percent WWS at Google on June 20, 2012 (Section 9.1.10). Apple met the 100 percent goal for its global operations during April 2018.

Between September 2017 and September 2019, the Climate Group went further, galvanizing 52 international companies to commit to transitioning their 2 million company vehicles to 100 percent electric by 2030 (Climate Group, 2019). Companies committed not only to changing their vehicle fleet but also to installing charging stations. On September 19, 2019, an additional company, Amazon, committed to replacing 100,000 delivery trucks with electric trucks. Transitioning company vehicles is important, given that up to half of all vehicles on the road in many countries are company owned.

At the beginning of 2019, a new law (H.R. 330) was proposed in the U.S. House of Representatives calling for the U.S. electric power sector to go to 100 percent WWS by 2030 (Table 9.3). Shortly thereafter, a broader **Green New Deal** resolution (rather than a law) was proposed in both the U.S. House (H.Res. 109) and Senate (S.Res. 59) (Table 9.3). The Green New Deal resolutions contain proposals related to energy, jobs, health care, education, and social justice. In terms of energy, they call for "100 percent clean, renewable, and zero-emission energy sources" for all energy sectors by 2030 throughout the United States. The 100 percent goal and the 2030 deadline both originate from our *Scientific American* article (Jacobson and Delucchi, 2009) and New York State energy roadmap (Jacobson et al., 2013) (Green Party US, 2018). Both articles conclude that 100 percent WWS by 2030 is technically and economically feasible, but the 2009 article cautioned that, for social and political reasons, a complete transition by 2030 was unlikely and may take up to a couple of decades longer.

Prior to the publication of the 2009 *Scientific American* article, the public discussion was whether it was possible to shift from fossil fuels to about 20 percent renewables. The subsequent country, state, city, and town roadmaps and grid studies, together with the 100 percent WWS movement spawned by The Solutions Project, its founders, members, and partner organizations, all reinforced by additional scientific papers from many independent

Table 9.4 International businesses (221, in alphabetical order) that have committed to transition their global operations to 100 percent renewable electricity in the annual average, as of January 16, 2020. The businesses may either produce the electricity themselves or source it from the market. Renewables include wind, solar, geothermal, hydroelectric, or biogas. Companies in blue are among the top 10 biggest companies in the world in terms of total value of a company's shares of stock.

| | | | | | | |
|---|---|---|---|---|---|---|
| 3M | Bozzuto | Derwent London | Heathrow Airport | Lululemon | Rackspace | Tetra Pak |
| ABInBev | British Land | Deutsche Tel. | Helvetia | Lyft | Radio Flyer Rakuten | T-Mobile |
| Adobe | Broad Group | Dexus | HP Enterprise | Mace | Ralph Lauren | Toda Corp. |
| Accenture | BT | Diageo | HP Inc. | Mahindra | RB | Tokyu Corp. |
| AEON | Burberry | DNB | HSBC | M&S | Roy. Bank Scot. | Tokyu Land |
| AkzoNobel | CaixaBank | Ebay | Hulic Co., Ltd. | Macquarie | RELX Group | Trane |
| Allianz | Califia Farms | Elion | IFF | Mars | Royal DSM | TRIDL |
| Alstria | Canary Wharf | Elopak | IHS Markit | Marui Group | Royal Philips | UBS |
| Amalgamated Bk | Capital One | Envipro | Ikea | McKinsey & Co. | Ricoh | Unilever |
| Amazon | Carlsberg Group | Envision | Infosys | Microsoft | Salesforce | Vail Resorts |
| Anthem | Citi | Equinix | ING | Mirvac | SAP | Vaisala |
| ANZ | City of London | Estee Lauder | Interface | Morgan Stanley | Schneider Elec. | Vestas |
| Apple | Clif Bar | Etsy | Iron Mountain | Nat Austral Bank | Schroders | VF Corp. |
| Asahi Kasei | Coca Cola | Facebook | JC Decaux | Nestle | Sekisui House | Virgin Media |
| Askul | Colruyt Group | Formula-E | JD Sports Fashion | Next | SGS | Visa |
| Asset Man. One | Commerzbank | Fifth Third Bank | Jinko Solar | New Balance | Signify | VM Ware |
| Astra Zeneca | Commonwealth Bk | Firmenich | Johnson & Johnson | Nike | Sky | Vodafone |
| Atlassian | Coop Sapporo | Fujifilm | Johnan Shinkin Bank | Nordea | Slaughter & May | Voya |
| Aurora | Corbion | Fujikura | JPMorgan Chase | Novo Nordisk | Sony | Walmart |
| Autodesk | Credit Agricole | Fujitsu | Jupiter | NREP | Starbucks | Watami |
| Aviva | Crown Estate | Fuyo Lease | Kellogg's | NRI | Steelcase | Westpac |
| Axa | Crown Holdings | Gatwick Airport | Keurig Dr. Pepper | O'right, Hair | Swisscom | WeWork |
| B of A | Dai-ichi Life Group | General Motors | Kingspan | Organic Valley | Swiss Post | Wells Fargo |
| Bank Australia | Daito Trust | Givaudan | Konica Minolta | Panasonic | Swiss Re | Wonderful Co. |

Table 9.4 (*cont.*)

| Bankia | Daiwa House | Goldman Sachs | KPN | Pearson Pernod Ricard | Symrise | Workday |
|--------|-------------|---------------|-----|------------------------|---------|---------|
| Barclays | Dalmia | Google | L'Occitane Group | P&G | Takashimaya | Yoox |
| BayWa | Danone | Grape King Bio | La Poste | PNC | Target | Zurich Insurance |
| BBVA | Danske Bank | Grupo Bimbo | Landsec | Proximus | Tata Motors | |
| Bestseller | DBS | Gurmengroup | Lego Group | PVH | TCl Co. | |
| Biogen | Decathalon | H&M | Lixil Group | PwC | TD Bank Group | |
| Bloomberg | Dell Techs. | HAP | Lloyds Banking | QBE | Telefonica | |
| BMW Group | Dentsu Aegis | Hazama Ando | Logitech | QTS | Tesco | |

Source: RE100 (2019).

groups (Section 8.2.1), changed the discussion to *when* 100 percent would be implemented.

I am under no illusion that we are finished solving the problem. Although progress has been made in the electric power sector, transitioning other sectors (transportation, building heat, and industrial heat) must be addressed in full force as well. However, the goal posts have shifted to where they should be, and this was due to cooperation among people with diverse expertise but with a like-minded passion and goal. The progress made to date is much more than I could have imagined when I first aspired to help solve air pollution problems in the late 1970s.

## 9.2 Timeline for a Transition

A critical step in implementing a transition to 100 percent WWS is to develop a timeline for a transition. The perfect timeline is one in which a 100 percent transition occurs in all energy sectors immediately. That won't happen. Jacobson and Delucchi (2009) and in subsequent papers postulated that transitioning all world energy by 2030 was technically and economically feasible but noted that, for social and political reasons, a date two decades later, such as 2050, would be a possible end point. Since then, all 100 percent WWS roadmaps developed, such as Jacobson et al. (2015a, 2017, 2019), have proposed an 80 percent transition by 2030 and 100 percent by no later than 2050. This does not mean that a faster transition for the last 20 percent can't or won't occur earlier. Ideally, it will

occur earlier. In some sectors, such as the electricity and building heat sectors, the transition may occur much faster in some countries, by 2035 or 2040. Because some technologies (e.g., long-distance hydrogen fuel cell-electric hybrid aircraft) may not be commercially available until after 2030, a transportation transition may take longer to fully implement.

Figure 9.5 shows one transition timeline for moving from BAU to 100 percent WWS in 143 countries. This timeline assumes an 80 percent conversion to WWS by 2030 and 100 percent by no later than 2050. Whereas new WWS infrastructure will be installed upon natural retirement of BAU infrastructure, policies are needed to force the remaining existing infrastructure to retire early to allow the conversion to 100 percent WWS at the pace necessary to eliminate air pollution mortality and avoid 1.5 °C net global warming as much as possible.

### 9.2.1 Timelines for Individual Technologies to Transition

Whereas Figure 9.5 shows an overall timeline for a transition, some technologies can transition faster than others. Below is a list of proposed transformation timelines for some individual technologies.

*Development of super grids and smart grids:* As soon as possible, countries should expand the transmission-and-distribution systems to supply geographically dispersed WWS energy to load centers. The grid should be

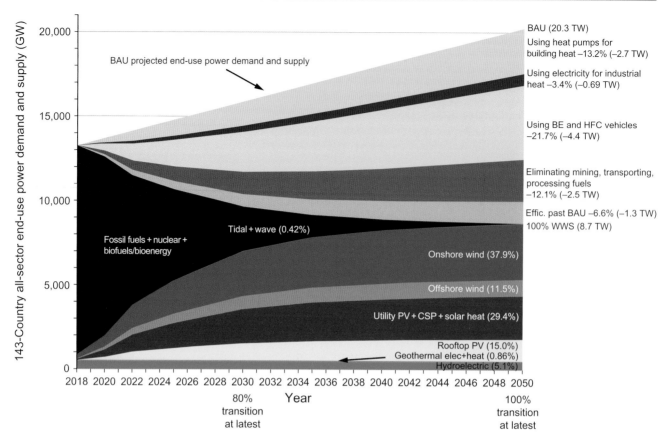

**Figure 9.5 Timeline for 143 countries, representing 99.7 percent of anthropogenic carbon dioxide world emissions, to transition from conventional fuels (BAU) to 100 percent wind-water-solar (WWS) in all energy sectors. Also shown are the annually averaged end-use power demand reductions that occur along the way. The energy sectors transitioned include the electricity, transportation, building heating/cooling, industrial, agriculture/forestry/fishing, and military sectors. The percentages next to each WWS energy source are the 2050 estimated percent supply of end-use power by the source. The 100 percent demarcation in 2050 indicates that 100 percent of end-use power in the annual average will be provided by WWS among all energy sectors by no later than 2050, but ideally sooner. An 80 percent transition is proposed to occur by no later than 2030. End-use power demand reductions occur for five reasons: (1) the efficiency of moving low-temperature building heat with heat pumps instead of creating heat with combustion; (2) the efficiency of electricity over combustion for high-temperature industrial heat; (3) the efficiency of electricity in battery-electric (BE) vehicles and in electrolytic hydrogen fuel cell (HFC) vehicles over combustion vehicles for transportation; (4) eliminating the energy to mine, transport, and process fossil fuels, biofuels, bioenergy, and uranium; and (5) improving end-use energy efficiency and reducing energy use beyond that in the BAU case. The total demand reduction due to these factors is 57.1 percent (Table 7.1). From Jacobson et al. (2019).**

managed with modern **smart grid** digital communication technologies that detect and react to local changes in supply and use by invoking storage, demand response, geographically dispersed generation, and backup WWS power generation.

*Power plants:* By 2020 to 2022, no more construction of new coal, nuclear, natural gas, oil, or biomass-fired power plants; all new power plants built should be WWS.

*Heating, drying, and cooking in the residential and commercial sectors:* By 2020 to 2022, all new devices, machines, and equipment for heating, drying, and cooking should be powered by electricity, direct heat, and/or district heating.

*Boats and ships:* By 2020 to 2030, all new boats (e.g., speed boats, yachts, barges, towing vessels, fishing boats, patrol vessels, ferries, military boats) and ships (e.g., container ships, cruise ships, military ships) should be electrified and/or use electrolytic hydrogen, and all new and existing port operations should be electrified. This should be feasible for ports and ships because ports are centralized, and few ships are built each year. Policies are needed to incentivize the early retirement of fossil-fuel-powered ships that do not naturally retire before 2050.

*Industrial heat:* By 2025, all new high-, mid-, and low-temperature industrial process technologies should be electric technologies, such as arc furnaces, induction furnaces, resistance furnaces, dielectric heaters, electron beam heaters, and steam from CSP plants and heat pumps.

*Rail and bus transport:* By 2025, all new trains and buses should be electric.

*Non-road vehicles:* By 2025, all new non-road vehicles (agricultural machines, construction vehicles, military vehicles) should be electric.

*Heavy-duty truck transport:* By 2025 to 2030, all new heavy-duty trucks and buses should be BE, HFC, or BE-HFC hybrids.

*Light-duty on-road transport:* By 2025, all new light-duty on-road vehicles should be BE vehicles.

*Short-haul aircraft:* By 2025, all new small, short-range aircraft should be BE, HFC, or BE-HFC hybrids.

*Long-haul aircraft:* By 2030 to 2040, all remaining new aircraft should be HFC or BE-HFC hybrids.

Whereas much new WWS infrastructure can be installed upon natural retirement of BAU infrastructure, new policies are needed to force remaining existing infrastructure to retire early in order to allow the complete conversion to WWS to occur on time.

Averaged worldwide, the social cost per kWh of the existing BAU system (predominantly fossil fuels) is about 20 percent that of a new WWS system (Table 8.9). However, because a WWS system uses only 43 percent of the energy of a BAU system, the aggregate social cost of a fossil-fuel system is about 9 percent that of a BAU system (Table 8.9). As such, replacing a BAU plant before the end of its life saves a large amount of money. It also increases the number of jobs (Table 8.13). As such, society benefits by stopping immediately the operation of existing BAU fossil-fuel plants and replacing them with new WWS plants.

During the transition, BAU fuels and existing WWS technologies are needed to produce the remaining WWS infrastructure. However, if no WWS infrastructure and machines were produced, much of the BAU energy would be used in any case to produce BAU infrastructure and machines. Further, as the fraction of WWS energy increases, BAU energy generation decreases, ultimately to zero, at which point, all new WWS devices will be produced with existing WWS energy. In sum, the time-dependent transition to WWS infrastructure may result in a temporary increase in emissions before emissions are ultimately reduced to zero.

## 9.2.2 How the Proposed Timeline May Impact Global $CO_2$ Levels into the Future

The WWS roadmaps proposed here will eliminate energy-related emissions of carbon dioxide ($CO_2$), black carbon (BC), and methane ($CH_4$), the first through third leading causes of global warming, respectively (Table 1.1). They will also reduce tropospheric ozone ($O_3$) precursors, carbon monoxide (CO), and nitrous oxide ($N_2O$), all of which are greenhouse gases.

An important question is, How long will $CO_2$ levels take to recover once $CO_2$ emissions have been halted? Figure 9.6 attempts to answer this question. It compares computer-model-simulated changes in near-surface $CO_2$ mixing ratios from 1751 to 2014 with data. It then predicts future changes in $CO_2$ under different emission scenarios, including the main WWS scenario proposed in Figure 9.5 (80 percent transition to WWS by 2030 and 100 percent, by 2050). The figure indicates that the model was able to match measured $CO_2$ over time, remarkably from 1751 to 2014. It also suggests that such a transition to 100 percent WWS may decrease $CO_2$ levels down to about 350 ppmv by 2100. Less aggressive transitions, such as under IPCC (2000) scenarios, appear likely to increase $CO_2$ levels substantially in comparison.

## 9.2.3 How the Proposed Timeline May Impact Global Temperatures into the Future

Figure 9.6 gives an indication of the possible change in the globally averaged $CO_2$ mixing ratio due to transitioning to 100 percent WWS. $CO_2$ changes, though, do not translate proportionally to near-surface air temperature changes over the same timescale.

Instead, the global near-surface air temperature change due to greenhouse gases may be more linearly proportional to cumulative carbon emissions, regardless of the timing of those emissions (e.g., Allen et al., 2009; Matthews et al., 2009; Meinshausen et al., 2009).

Currently, global greenhouse gas emissions are increasing rapidly. If this trajectory continues, global temperatures are expected to rise 2 °C above those in 1870 by somewhere between 2035 and 2060 and by 4.5 +/−1.2 °C above those in 1870 by 2100.

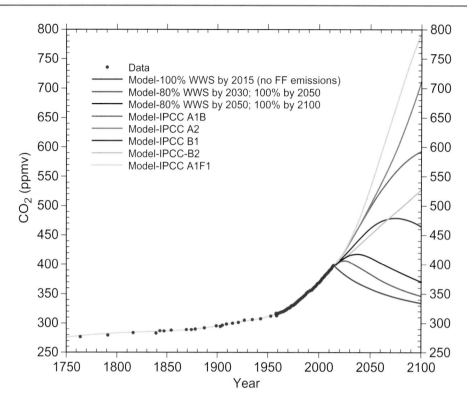

**Figure 9.6** Comparison of historic (1751–2014) observed CO₂ mixing ratios (ppmv) from the Siple ice core (Neftel et al., 1994) and the Mauna Loa Observatory (Tans and Keeling, 2015) with GATOR-GCMOM model results (Jacobson, 2014) for the same period, plus model projections from 2015 to 2100 for five Intergovernmental Panel on Climate Change (IPCC) scenarios (IPCC, 2000) and three WWS scenarios: an unobtainable case of 100 percent WWS by 2015; a case of 80 percent WWS by 2030 and 100 percent by 2050 (from Figure 9.5); and a less-aggressive case of 80 percent WWS by 2050 and 100 percent by 2100. The model was set up as in Jacobson (2005a) with two columns (one atmospheric box over 38 ocean layers plus one atmospheric box over land). It treated full ocean chemistry in all layers, vertical ocean diffusion with canonical diffusion coefficients, ocean removal of calcium carbonate for shell and rock formation, ocean photosynthesis by phytoplankton and the sinking of its detritus, gas-ocean transfer, and emissions from fossil fuels and land use change. The net carbon sink over land was calculated accounting for time-dependent green-plant photosynthesis, plant and soil respiration, and weathering. No data assimilation or nudging of model results to observations was performed. Preindustrial CO₂ was 276 parts-per-million-volume (ppmv). Fossil-fuel emissions from 1751 to 1958 were from Boden et al. (2011), from 1959 to 2014 were from Le Quere et al. (2015), and for 2015 onward were from three WWS scenarios scaled from 2014 emission and from five individual IPCC scenarios. Land use change emissions per year were held constant at 300 Tg-C/y for 1751 to 1849; from Houghton (2015) for 1850 to 1958; from Le Quere et al. (2015) for 1959 to 2014; from the IPCC (2000) A1B scenario for the WWS cases from 2015 to 2100; and from the individual IPCC scenarios for the remaining cases. Thus, some land use change emissions continued in all scenarios. The average e-folding lifetime of CO₂ in the air upon a decrease in CO₂ (estimated from the green 100 percent WWS curve in the figure) is about 90 years. This is longer than the data-constrained e-folding lifetime, which is based upon CO₂ increasing, of 30 to 60 years (Jacobson, 2012a). The reason for the difference is that, as CO₂ increases, the air is always supersaturated with respect to the ocean, so the ocean serves as a welcome sink for excess atmospheric CO₂. As CO₂ begins to decrease in the air, it reaches equilibrium with the ocean, making it difficult for more CO₂ in the air to dissolve in ocean surface water, increasing the atmospheric lifetime of CO₂. However, photosynthesis by phytoplankton and the sinking of phytoplankton detritus, formation of ocean shells and rocks from dissolved CO₂, and land removal of CO₂ still occur.

Friedlingstein et al. (2014) estimate that, for the globally averaged temperature change since 1870 to increase by less than 2 °C with a 67 percent probability, cumulative CO₂ emissions since 1870 must be kept below 3,200 (2,900 to 3,600) Gt-CO₂. This accounts for non-CO₂ warming agents that must also be reduced proportionately to CO₂. The cumulative CO₂ emission limit to keep warming below 2 °C with a 50 percent probability is 3,500 (3,100 to 3,900) Gt-CO₂. Matthews (2016) estimated (by scaling the 2 °C numbers by 1.5 °C / 2 °C) that the corresponding limits to keeping temperatures under 1.5 °C are 2,400 Gt-CO$_2$ (67 percent probability) and 2,625 Gt-CO$_2$ (50 percent probability), respectively.

By the end of 2018, about 2,155 Gt-CO$_2$ from fossil-fuel combustion, cement manufacturing, and land use change had already been emitted cumulatively since 1870, suggesting that no more than 244 to 469 Gt-CO$_2$ can be emitted for a 67 to 50 percent probability of keeping warming since 1870 under 1.5 °C, and that no more than 1,044 to 1,344 Gt-CO$_2$ can be emitted for a 67 to 50 percent probability of keeping warming under 2 °C.

In 2018, global CO$_2$ emissions from fossil-fuel burning and concrete production were about 37.1 Gt-CO$_2$/yr and those from land use change were about 4.0 Gt-CO$_2$/yr, for a total of 41.1 Gt-CO$_2$/yr. If a linear 80 percent decrease in emissions occurs between 2018 and 2030 and the remaining 20 percent decreases linearly by 2050, an additional cumulative 399 Gt-CO$_2$ will be emitted to the atmosphere. This is in the range of the maximum allowable after 2018 to keep warming under 1.5 °C.

Thus, aggressive policies that strive to reduce emissions of CO$_2$, other greenhouse gases, and dark particles at least 80 percent by 2030 and 100 percent by no later than 2050 may limit warming to 1.5 °C. However, this requires reducing not only energy emissions but also land use change emissions, agricultural emissions, and halocarbon emissions (Section 2.9).

## 9.3 Obstacles to Overcome for a Transition

While technically and economically feasible, a transition of the entire world's energy infrastructure to clean, renewable energy in all sectors is a daunting task that faces significant social and political hurdles. The following discusses several of these challenges.

### 9.3.1 Vested Interests in the Current Energy Infrastructure

Possibly the greatest challenge to overcome in transitioning the world to 100 percent clean, renewable energy is the challenge of overcoming vested interests. Fossil-fuel companies have accumulated enormous wealth and political influence since the start of the Industrial Revolution. As a result, they have been able to implement legislation that has resulted in their having an entrenched financial benefit in many countries through tax code breaks, direct

Table 9.5 **Cumulative tax code breaks, grants, and subsidies given by the U.S. federal government to the coal, oil, and gas industries between 2002 and 2008.**

| Tax code break, grant, or subsidy | Amount (USD billion) |
|---|---|
| Foreign tax credit for overseas production of oil | 15.3 |
| Credit for production of nonconventional fuels | 14.1 |
| Oil and gas exploration and developing expensing | 7.1 |
| Reduced government take on oil and gas leasing | 7.0 |
| Oil and gas excess percentage over cost depletion | 5.4 |
| Credit for enhanced oil recovery | 1.6 |
| Characterizing coal royalties as capital gains | 1.0 |
| Exclusion of benefits payments to disabled miners | 0.44 |
| Other tax code breaks | 0.22 |
| Direct grants and subsidies | 18.3 |
| **Total** | **72.5** |

Source: ELI, (2009).

grants, and subsidies. Table 9.5, for example, shows that between 2002 and 2008, the fossil-fuel industry in the United States received about $72 billion in such benefits. In addition, most of the existing energy infrastructure is a fossil-fuel infrastructure, and most people's day-to-day lives depend on that infrastructure (e.g., through transportation; building heating, cooling, and refrigeration; lighting). Further, over 25 million jobs worldwide depend on the fossil-fuel infrastructure (Table 7.7). Given the great financial resources available to the fossil-fuel industry and the dependency of many people's jobs and livelihoods on the current infrastructure, great care needs to be taken to ensure that a transition to renewables brings along the same or better standard of living, the same number of or more jobs, and the same or better comfort to as many people as possible. This means encouraging shareholders of fossil-fuel companies to move their investments to renewables, setting policies to retrain workers, and setting policies to ensure WWS electricity supply is at least as reliable as BAU supply on the grid.

Movement is already occurring in all three areas. Investors are moving capital resources into WWS. The WWS infrastructure is creating more jobs per unit energy than is the fossil infrastructure. The WWS system is also proving to be even more reliable and less expensive than was the preceding fossil-fuel system in the few places that WWS has completely replaced BAU.

For example, after one year of construction, a large (100-MW, 129-MWh) battery system that was installed in South Australia to fill in gaps in electric power supply for a 317-MW wind farm saved AUD 40 million in grid stabilization costs due to the rapid speed (100 milliseconds) at which batteries can react to shortages (Macdonald-Smith, 2018). In another example, during July 2019, NextEra energy determined that investing in a combined wind (250-MW), solar PV (250-MW), and battery storage (200-MW, 800-MWh) system is less expensive than investing in a natural gas peaker plant (Spector, 2019).

### 9.3.2 Zoning Issues (NIMBYism)

The second obstacle that needs to be overcome to implement a 100 percent WWS system is the **not-in-my-backyard** syndrome (**NIMBYism**). NIMBYism is an objection to the siting of something that a person thinks is unpleasant or dangerous in their neighborhood while not objecting to the development of the same thing somewhere else. Classic examples of NIMBYism are the siting of a landfill or a hazardous waste site near a neighborhood. However, NIMBYism extends to almost every type of infrastructure development. People generally do not want to see additional buildings or facilities, including energy facilities, near them.

Whereas the installation of WWS onshore and offshore wind and solar farms and transmission lines will face opposition in many locations, they will likely face less opposition than the building of more coal and gas plants, nuclear plants, oil and gas wells, refineries, and pipelines. The reason is that most people realize that WWS technologies do not bring air pollution or catastrophic risk along with them, whereas the fossil technologies bring both, and nuclear plants bring the risk of catastrophic failure and exposure to radiation above background levels. A second reason is that, whereas no one wants to add anything to the landscape, the addition of WWS often reduces land use relative to a fossil-fuel infrastructure (Section 8.3). Similarly, whereas land used for a coal or nuclear plant cannot be used for another purpose at the same time, land occupied by a wind farm can be used for multiple purposes simultaneously:

agriculture, animal grazing, open space, or even solar PV and batteries.

In addition, many people are becoming accustomed to rooftop solar panels, so they object to them less so than in the past. Some types of solar panels are also integrated into building design, so it is difficult to tell whether the panels are even there. Most wind farms are located away from buildings because winds are fastest where no obstacles exist on the ground. With the advent of floating offshore wind turbines, visual objections to siting offshore wind is virtually eliminated because such offshore turbines are usually sited beyond people's views.

Probably the most difficult WWS infrastructure to site is transmission. Siting transmission lines and pipelines today already results in NIMBYism, for good reason. Transmission lines and pipelines are not pretty. However, with 100 percent WWS, adding transmission lines will enable the elimination of all oil and gas pipelines. In addition, the land under transmission lines can still be used. More and more transmission lines are being buried, alleviating the problem near where people live. Finally, many new transmission lines can piggyback on the same transmission pathways that already exist, so new rights-of-way may not be needed in many cases.

Nevertheless, because of the delays caused by zoning requirements, the installation of many new transmission pathways for a 100 percent WWS system may be limited in some countries. Fortunately, though, an alternative to transmission is more energy storage. Because storage costs are declining rapidly, the future WWS system may be dominated by storage over transmission.

A key issue to address with respect to transmission and distribution is their triggering of wildfires. In 2018 and 2019 in particular, wildfires in California caused by transmission line malfunctions caused enormous damage, including loss of life. Whereas pipeline ruptures and explosions also cause damage and mortality, and transmission line fires were a problem long before WWS, so the issue needs to be addressed. One obvious solution is to bury transmission lines underground. Although underground lines are more costly than overground lines, the additional cost may pale in comparison with the cost savings of eliminating wildfires and their damage. Another partial solution is to reduce reliance on transmission lines by using more building rooftop solar PV and batteries in fire-prone regions and to use more storage in general instead of transmission. A third partial solution is improved safety measures (e.g., clearing more brush and learning from the causes of previous fires).

### 9.3.3 Countries Engaged in Conflict

A third potential obstacle is the building of a new energy infrastructure in countries suffering from conflict, such as civil war, terrorism, or war with another country. Table 9.6 summarizes the major conflicts worldwide in 2018. Dozens of additional countries suffered fewer than 1,000 casualties from conflict during the year as well.

Because millions of people live in countries with ongoing conflict, it is even more essential to bring distributed, clean renewable energy to these countries. Oil pipelines and transmission lines are often targeted for destruction or theft, so local microgrids may be the best way to provide energy safely to communities in countries engaged in conflict until the conflict is resolved. A **microgrid** (Section 2.2.2.2) is a grid isolated from outside sources of electricity. It may consist, in its simplest form, of solar panels plus batteries. The next step may be to add heat storage in a water tank. The electricity may be combined with a water filtration system to provide clean water from waste. It may also be combined with a greenhouse or food-growing container to provide food year-round. In locations with terrain, the microgrid may be combined with a small pumped hydropower storage facility and small wind turbines.

Major problems in a war-torn country are famine and poverty. Microgrids, if set up properly and maintained, can help produce food, water, and energy together, mitigating both problems. One way to help alleviate some of the difficulty of setting up a microgrid in a war-torn country is for safer countries to provide aid to them in the form of clean, renewable WWS microgrid technology.

Some conflicts between countries arise over energy. For example, one country may provide the natural gas or oil for another and is withholding it. In another example, one country may need more energy resources so decides to invade another to take control of oil and gas wells or coal mines. Transitioning a country entirely to clean, renewable energy, where most or all the raw WWS resources are obtained within the country, will make the country more energy independent, reducing the reason for and chance of conflict. However, many countries, even if they produce most of their own energy in the annual average, will benefit from trading energy with their neighbors to reduce the cost of matching power

Table 9.6 **Conflicts with more than 10,000 deaths and between 1,000 and 10,000 deaths in 2018. The number in parentheses is the year the conflict started.**

| Conflicts with >10,000 deaths in 2018 | Conflicts with 1,000 to 10,000 deaths in 2018 |
| --- | --- |
| Afghanistan (1978) | Somalia (1991) |
| Syria (2011) | Nigeria (1998) |
| Yemen (2011) | The Maghreb (Algeria, Morocco, Tunisia, Libya, Mali, Niger, Chad) (2002) |
| | Iraq (2003) |
| | Boko Haram insurgency (Nigeria, Cameroon, Niger, Chad) (2009) |
| | South Sudan (2011) |
| | Northern Mali (2012) |

demand with supply over time. The reason is, whereas the wind may not be blowing in one country, it is more likely blowing in another. The same applies to solar.

### 9.3.4 Countries with Substantial Poverty

Countries with a large segment of its population in poverty greatly need to transition to 100 percent clean, renewable energy. As discussed in Section 7.6.2, millions of people die from indoor plus outdoor air pollution each year. Of these, about 20 percent are children under the age of five. Almost all the indoor air pollution deaths per year (about 2.6 million in 2016) in developing countries are due to the burning of biomass and coal for home heating and cooking.

The main step in eliminating indoor air pollution deaths is to eliminate indoor burning of fuel for cooking and heating. This can be accomplished most readily with electric induction burners (Section 2.5.1) and electric heat pumps (Section 2.3). For remote communities, the electricity would be obtained from a microgrid that combines solar and wind with batteries or pumped hydropower (Section 9.3.3). Such an infrastructure costs money, and impoverished communities may not have access to such funds. However, costs of WWS generation and storage continue to decline. In addition, national governments may help financially. International aid will help with this solution too.

Impoverished countries will also benefit from a large-scale transition to WWS. Their end-use energy requirements will go down by an average of 57 percent; their direct aggregate costs of energy will consequently drop by the same percent. Their social costs of energy will decrease by over 90 percent (Table 8.9).

However, the main barrier these countries need to overcome is the capital cost of their investment. WWS generators have no fuel cost, but they have an upfront capital cost. Although that capital cost pays for itself relatively quickly, raising upfront capital is not a trivial matter. To that end, wealthier countries may need to work with more impoverished countries to help finance the upfront cost of a transition.

### 9.3.5 Transitioning Long-Distance Aircraft and Long-Distance Ships

Whereas the technology for 90 to 95 percent of a transition to 100 percent WWS is currently commercialized, some technologies are not yet available for use. The two most obvious are long-distance aircraft and long-distance ships. About 46 percent of commercial flight distance traveled worldwide is for long-haul flights (those longer than 3 hours) (Wilkerson et al., 2010). The best WWS solution for such flights may be hydrogen fuel cell or HFC-battery-electric hybrid aircraft. However, for such aircraft to work cost effectively, fuel cell sizes and efficiencies may need to improve more. Section 9.2.1 indicates that a transition to long-distance WWS aircraft may not occur until 2035 to 2040 because of the improvements and testing needed to commercialize long-distance HFC and HFC-BE aircraft.

Long-distance ships have a similar issue as aircraft and are also proposed to be HFC or HFC-BE hybrids. However, ships can stop during their journey to recharge or refuel hydrogen. In addition, ships have less mass constraints than aircraft, so they should be easier to design. As such, a transition of ships may be implemented between 2020 and 2030 (Section 9.2.1).

### 9.3.6 Competition among Solutions

One more obstacle is the competition among solutions to the problems of air pollution, global warming, and energy security. Given the severity of the problems facing the world and the short time needed to fix them, competition

among energy plans can result in poor solutions being implemented (Chapter 3), thereby slowing down the solution to the problems. The main competitors for solutions to date have been a WWS solution and an **all-of-the-above (AOTA)** solution. An AOTA solution includes WWS, as well as nuclear, fossil fuels with carbon capture, biofuels, and direct air capture, among other technologies discussed in Chapter 3. However, as shown in Chapter 3, those technologies are opportunity costs. They result in either more air pollution, more greenhouse gas emissions, more energy security risk, more land use risk, higher costs, or longer planning-to-operation times, or all of these, relative to WWS. As such, spending money on those solutions results in less benefit and a longer delay before we can eliminate air pollution, global warming, and energy insecurity. The best way to overcome this obstacle is by educating the public and policymakers about it and guiding them to solutions we know work and that can be implemented on a large scale quickly and at low cost.

## 9.4 Policy Mechanisms

The policy pathways necessary to transform each country to 100 percent WWS will differ by country, depending largely on the willingness of the government and people in each country to affect a rapid change. This section first defines several types of policy options that each country can consider implementing. Policy options for different energy sectors are then discussed in more detail.

### 9.4.1 Policy Options for a Transition

Here, some policy options that have been used in the past are provided. The list is by no means complete.

**Renewable portfolio standards (RPSs)**, also called renewable electricity standards, are policy mechanisms requiring a certain fraction of electric power generation to come from specified clean energy sources by a certain date. Thus, for example, a mandate of 80 percent WWS by 2030 and 100 percent no later than 2050 is an RPS. One hundred percent RPSs have been enacted in many countries, states, cities, and towns to date (Table 9.1).

**Financial incentives and laws for increasing energy efficiency and reducing energy use** are policy methods to reduce the demand for energy. For example, a law requiring the substitution of low-energy-consuming LED light bulbs for high-energy-consuming incandescent ones reduces energy demand. Demand reduction reduces the pressure on energy supply, which makes it easier for WWS supply to match demand.

**Laws requiring demand response** are helpful because they force utilities to incentivize customers to shift the time of their electricity use from a time of peak electricity demand to a time of lower demand during the day or night. Demand response is usually accomplished in one of two ways. The first way is for utilities to increase the price of electricity during times of peak electricity use. The higher cost of electricity incentivizes customers to use less electricity during those times of day. The second way is for utilities to pay customers to encourage them to use less electricity during times of high electricity use. In both cases, the customer can react manually to the change in rate or payment by reducing or stopping energy consumption during times of peak demand. Alternatively, the customer can agree to have internet- or radio-controlled switches installed on air conditioners or other devices to automatically reduce energy use during times of peak demand.

**Feed-in tariffs (FITs)** are subsidies to cover the difference between electricity generation cost (ideally including grid connection cost) and wholesale electricity prices. FITs have been an effective policy tool for stimulating the market for renewable energy. To encourage innovation and the large-scale implementation of WWS, which will itself lower costs, FITs should be reduced gradually. Otherwise, technology developers have little incentive to improve. One way to reduce a FIT gradually is with a **declining clock auction**, in which the right to sell power to the grid goes to bidders willing to do it at the lowest price, providing continuing incentive for developers and generators to reduce costs. A risk of any auction, however, is that the developer will underbid and be left unable to develop the proposed project at a profit. Regardless of whether the declining clock auction is used, reducing and eventually phasing out a FIT as WWS costs decline (referred to as **tariff regression**) is an important goal.

**Output subsidies** are payments by governments to energy producers per unit energy produced. For clean renewable energy producers, the justification of such subsidies is to correct the market because fossil-fuel and biofuel energy producers are not paying for the pollution they emit, which has health and climate costs to society. In other words, the subsidies attempt to address the **tragedy of the commons**, which arises because the air is not privately owned. As such, the air has been polluted without polluters paying the pollution's cost to health, climate, and the environment.

**Investment subsidies** are direct or indirect payments by government to energy producers for research and development. Historically, such subsides have been given mostly to conventional fuel producers through legislation and clauses in the tax code, such as deductions and credits for specified activities. Conventional generators have benefited historically from such subsidies by not paying externality costs of the pollution that their energy creates. Investment subsidies are now available to WWS energy sources in many countries.

One type of investment subsidy is a **loan guarantee**, whereby the government guarantees a loan to a company for constructing a facility. Without such guarantees, many large energy projects would not be approved for a construction loan. Such loan guarantees, historically provided to the conventional fuels industry, benefit the WWS industry as well.

On a smaller scale, **municipal financing** of residential energy efficiency retrofits and solar installations helps to overcome the financial barrier of the high upfront cost to individual homeowners. **Purchase incentives and rebates** can also help to stimulate the market for electric vehicles.

A potential policy tool that has not been used widely to date is a **revenue-neutral carbon tax or pollution tax**. This is a tax on polluting energy sources, with the revenue transferred directly to nonpolluting energy sources. In this way, no net tax is collected, so the cost to the public is zero in theory. If the tax is a revenue-neutral carbon tax, it may not address air pollution. For example, a biomass or coal-with-carbon-capture electricity-generating plant can claim they are low carbon (which itself is not accurate – Chapter 3), and thus avoid much of the tax, yet still emit substantial health-affecting air pollutants along with carbon. A second problem is that heavy polluters (e.g., coal and gas plants) can choose to pay the tax and withstand a lower profit margin or raise their prices, yet still emit.

A related tool is a straight **pollution tax** (e.g., a **carbon tax**). Such a tax is less popular; it is perceived as a cost to the public since the money collected from the tax doesn't necessarily go back to reducing energy prices. Instead, polluting companies merely increase their prices to pay the tax and pass the cost to consumers.

Another noneconomic policy is to **mandate emission limits** (usually tailpipe or stack exhaust emission limits) for technologies. This is a **command-and-control** policy option implemented widely under the U.S. Clean Air Act Amendments and in many countries to reduce vehicle emissions. By tightening emission standards, including for carbon dioxide, policymakers can force the adoption of cleaner vehicles. Such emission limits can and have also been set for other pollution sources. A disadvantage of emission limits (unless they are zero) is that they allow the fossil-fuel infrastructure to persist while they pursue incremental reductions in tailpipe or stack emissions and ignore upstream emissions and the other problems associated with fossil fuels. However, a mandate of absolutely zero tailpipe emissions (aside from water vapor for fuel cell cars) would mean that only electric and hydrogen fuel cell cars could meet these limits. That would effectively lead to the phaseout of combustion vehicles.

Related to mandatory emission limits is **cap and trade**. Under this policy mechanism, mandatory emission limits lower than current emissions are set for an entire industry, and pollution permits are issued corresponding to the total pollution emissions allowed. Polluters in the industry can then buy and sell pollution permits. The net result is lower overall emissions and a payment by the polluters for the remaining emissions. The problem with this mechanism is similar to that with emission limits. Unless the cap is zero, pollution, including upstream pollution, will persist long into the future.

## 9.4.2 Policy Options by Sector

Current energy markets, institutions, and policies have been developed to support the production and use of fossil fuels, biofuels, bioenergy, nuclear power, and clean renewable energy. New policies are needed to ensure that a 100 percent clean, renewable energy system develops quickly and broadly in each country and that dirtier energy systems are not promoted. In the following, several of the policy mechanisms just discussed are proposed for each energy sector to accomplish these goals. For each sector, the policy options are listed roughly in order of proposed priority.

### 9.4.2.1 Energy Efficiency and Building Energy Measures
- Expand energy efficiency standards.
- Incentivize conversion from natural gas water and air heaters to electric heat pumps (air-, ground-, water-, and waste-source).

- Promote, through municipal financing, incentives and rebates for implementing energy efficiency measures in buildings and other infrastructure. Section 2.6 gives examples of such efficiency measures.
- Revise building codes to incorporate "green building standards" based on best practices for building design, construction, and energy use.
- Incentivize landlord investment in energy efficiency. Allow owners of multi-family buildings to take a property tax exemption for energy efficiency improvements in their buildings that provide benefits to their tenants.
- Create energy performance rating systems with minimum performance requirements to assess energy efficiency levels and pinpoint areas of improvement.
- Create a green building tax credit program for the corporate sector.

### 9.4.2.2 Energy Supply Measures
- Increase renewable portfolio standards (RPSs).
- Extend or create WWS production tax credits (a tax credit given for every kWh of electricity produced).
- Invest in job retraining from BAU to WWS energy.
- Incentivize electricity storage expansion (Section 2.7).
- Streamline the permit approval process for large-scale WWS power generators and high-capacity transmission lines. Work with local and regional governments to manage zoning and permitting issues within existing planning efforts or pre-approve sites to reduce the cost and uncertainty of projects and expedite their physical build-out.
- Streamline the small-scale solar and wind installation permitting process. Create common codes, fee structures, and filing procedures across a country.
- Lock in fossil-fuel and nuclear power plants to retire under enforceable commitments. Implement taxes on air pollution and carbon emissions by current utilities to encourage their phaseout.
- Incentivize home or community energy storage (through garage electric battery systems, for example) that accompanies rooftop solar.

### 9.4.2.3 Utility Planning and Incentive Structures
- Incentivize district heating and community seasonal heat and cold storage (Section 2.8).
- Incentive the development of utility-scale grid electric power storage, such as in CSP, pumped hydropower, more efficient large hydropower, and large battery systems.

- Require utilities to use demand response grid management to reduce the need for short-term energy backup on the grid.
- Incentivize the use of excess WWS electricity to produce and store hydrogen, heat, and cold to help manage the grid.
- Develop programs to use electric-vehicle batteries, after the end of their useful life in vehicles, for stationary storage.
- Implement **virtual net metering** (**VNM**), whereby rooftop solar owners can sell electricity back to the grid to offset a part or all of the cost of the electricity that they buy from the grid.

### 9.4.2.4 Transportation Measures
- Mandate battery-electric vehicles for government transportation (e.g., mail delivery, service vehicles, military base vehicles) and use incentives and rebates to encourage the transition of commercial and personal vehicles to BE and HFC vehicles.
- Promote more public transit by increasing its availability and providing compensation to commuters for not purchasing parking passes.
- Increase safe biking and walking infrastructure, such as dedicated bike lanes, sidewalks, crosswalks, and timed walk signals.
- Set up time-of-use electricity rates to encourage vehicle charging during off-peak hours.
- Use incentives or mandates to stimulate the growth of private fleets of electric and/or hydrogen fuel cell buses.
- Incentivize electric and hydrogen fuel cell ferries, riverboats, and other local shipping.
- Adopt zero-emission standards for all new on-road and off-road vehicles for all purposes, with 100 percent of new production required to be zero-emission by 2030.
- Ease permitting for installing electric charging stations in public parking lots, hotels, suburban metro stations, on streets, and in residential and commercial garages.
- Incentivize the electrification of freight rail and shift freight from trucks to rail.

### 9.4.2.5 Industrial-Sector Measures
- Provide financial incentives for industry to convert to electricity for high-temperature and manufacturing processes (Section 2.4).

- Provide financial incentives to encourage industry to use WWS electric power generation for on-site electric power (private) generation.
- Encourage industry to take part in demand response measures to help match power demand with supply and storage on the grid.

The foregoing measures are a limited list. Yet, many are necessary to speed a transition, which would otherwise drag on long past 2050. Each town, city, state, province, or country must select its own policies based on what works best in that region. Yet, the reduction in cost that has already occurred combined with economies of scale due to further WWS expansion are cause for optimism. These factors suggest that, if effective policies are put in place, a transition to 100 percent WWS by no later than 2050 for all sectors and earlier for some sectors is possible.

## 9.5 Conclusion: Where Do We Go from Here?

A solution to the problems of air pollution, global warming, and energy security requires a large-scale conversion of the world's energy infrastructure. This book described and quantified many of the elements of such a conversion.

The main components of a transition are electrifying and providing some direct heat for all energy sectors, and sourcing that electricity and heat with clean, renewable wind, water, and solar (WWS) energy. The energy sectors that must be transitioned include electricity, transportation, building heating and cooling, industry, agriculture/forestry/fishing, and the military. This book also detailed methods of reducing or eliminating non-energy emissions.

WWS generators proposed for such a transition include onshore and offshore wind turbines, solar photovoltaics, solar heat collectors, concentrated solar power (CSP) plants, geothermal electricity plants, geothermal heat collectors, tidal turbines, wave devices, and hydroelectric facilities. Storage includes electricity, heat, cold, and hydrogen storage. Additional short- and long-distance transmission is also needed. By combining electricity, heat, cold, and hydrogen generation with storage and demand response, it appears possible to match power demand with supply on the electric power grid, thereby avoiding blackouts with a 100 percent WWS system.

The resulting system will use an average of 57 percent less end-use energy than the current system, and the direct energy cost that consumers will pay will be about 61 percent lower, on average. However, because a WWS system eliminates health and climate costs, which together are four times energy costs in the global average, a WWS system reduces the social (economic) cost that consumers pay by about 91 percent.

A worldwide transition to WWS will not only reduce 4 to 9 million air pollution deaths annually and slow, then reverse global warming, but it will also create far more jobs than those lost, use about half the land as the current energy infrastructure does, increase energy security and energy independence, and reduce terrorism risk to the power grid.

Given the limited time and funding to solve climate, pollution, and security problems, it is essential we focus on known, effective solutions. We should not waste money and allow more damage with inferior options. Such poorer options include nuclear power, fossil fuels or bioenergy with or without carbon capture, biofuels, direct air capture, and geoengineering.

In sum, a concerted international effort can lead to a conversion of the energy infrastructure so that by between 2020 and 2030, the world will no longer build new fossil-fuel electricity generation power plants or new transportation equipment with internal combustion engines. Rather, it will manufacture new wind turbines, solar power plants, and electric and fuel cell vehicles. One exception is long-distance aviation, which may take until 2035 or 2040 to transition. Once this WWS power plant and electric-vehicle manufacturing and distribution infrastructure is in place, the remaining existing fossil-fuel and nuclear power plants and internal combustion devices will gradually be retired and replaced with WWS-based systems, so that by 2050, the world will be powered by 100 percent WWS.

The main barriers to a conversion to WWS worldwide are not technical, resource based, or economic. Instead, they are political and social. The most difficult places to transition may be countries beset by conflict and countries that have high poverty rates. NIMBYism is also an issue but not a barrier that can't be overcome. Whereas recycling will be beneficial for keeping costs down, there is no materials limit to a 100 percent transition in all energy sectors.

Because most polluting energy technologies and resources currently receive government subsidies or tax breaks or are not required to eliminate their emissions, many can run inexpensively for a long time while competing economically with WWS technologies. Wisely implemented policy mechanisms can promote such a conversion; however, useful mechanisms can be implemented only if policymakers are willing to make changes.

Policymakers in democratic countries are elected, whereas those in autocratic countries make decisions dictated by one or a few leaders. Thus, in democratic countries, a large number of people need to be convinced that changes are beneficial; in autocratic countries, only a few need to be convinced. In both cases, those who need to be convinced require a social evolution in their thought. Such an evolution comes from a better understanding of global warming and air pollution science, consequences, and solutions. If the public and policymakers can become confident in understanding the problems and the large-scale solution needed to solve them, they will gravitate toward the solution. Thus, it is important for individuals, advocates, business leaders, scientists, and policymakers to come together to affect a change.

When the 100 percent WWS solution is finally implemented worldwide, the air pollution and climate problems outlined in this book will be relegated to the annals of history.

## Further Reading

RE100, 2019. The world's most influential companies committed to 100% renewable power, 2019, http://there100.org (accessed January 26, 2019).

REN21 (Renewable Energy Policy Network for the 21st Century), 2019. Renewables 2019 *Global Status Report*, www.ren21.net/gsr-2019/, www.ren21.net/gsr-2019/tables/table_06/table_06/ (accessed July 27, 2019).

Sierra Club, 2019. 100% commitments in cities, counties, and states, www.sierraclub.org/ready-for-100/commitments (accessed January 26, 2019).

The Solutions Project, 2019. Our 100% Clean Energy Vision, www.thesolutionsproject.org/why-clean-energy/ (accessed February 5, 2019).

## 9.6 Problems and Exercises

. . . . . . . . . . . . . . . . . . . . . . . . . . . . . . . . . . . . . . . . . . . . . . . . . . . . . . . . . . . . . . . . . . . . . . . . . . . . . . . . . . . . . . . . . . . . . .

9.1.  List four policy measures that could be implemented to encourage the expansion of WWS energy systems.

9.2.  List four barriers that could slow the large-scale implementation of clean, renewable energy.

9.3.  What fossil-fuel-based technologies will likely take the longest to transition to electricity or electrolytic-hydrogen alternatives?

9.4.  If the cumulative anthropogenic carbon emissions at the end of 2018 were 2,155 Gt-$CO_2$, and the limit to avoid 1.5 °C global warming is 2,400 Gt-$CO_2$, by what year will global warming reach 1.5 °C if the anthropogenic emission rate from fossil-fuel combustion, cement manufacturing, and land use change is 40 Gt-$CO_2$/y?

9.5.  List five types of tax code breaks, grants, or subsidies that fossil-fuel companies have received historically in the United States.

9.6.  If you were in charge of the world, what five policies would you put in place to push the world to transition to 100 percent WWS?

9.7.  What is the main issue with an all-of-the-above energy policy with respect to its potential impact on air pollution, global warming, and energy security?

9.8.  What are four challenges to implementing a 100 percent WWS energy system? For each, identify one way the challenge can be overcome.

9.9.  Group or Individual Project A

   Develop a "white paper" describing and analyzing the effects on a selected region (town, city, state, province, or country) of three methods of addressing global warming and air pollution simultaneously. In the paper, address the following components:

   1.  Discuss at least three methods of reducing emission of global warming gases and particles and air pollution gases and particles simultaneously using existing or emerging technologies. Such technologies can include emission control technologies, non- or low-emitting technologies that replace current technologies, energy efficiency measures, or policies encouraging lower emissions.

   2.  Discuss whether these methods have been implemented to date, and, if so, what the result of their implementation has been.

   3.  Quantify the estimated emission reductions (total mass per year) of the major pollutants during the next 20 years if the methods are implemented in the region.

   4.  Quantify the estimated reductions in the outdoor mixing ratio or concentration of major primary or secondary pollutants in the region if the methods are implemented. (This requires finding data on current concentrations or mixing ratios and using estimates of species lifetimes and of a mean mixing height.)

5. Quantify the health benefits (e.g., reduction in cardiovascular and respiratory disease, asthma, and mortality) of such a transition.

6. Discuss which industries will and will not benefit as a result of implementing the methods through regulation. Cost estimates may be given here.

7. Briefly summarize the selected methods, results, and overall recommendations in an abstract.

9.10. Group or Individual Project B

The goal of this project is to develop an energy roadmap to convert an individual town, city, company, state, province, or country entirely to clean, renewable energy supplied by wind, water, and sunlight (WWS) together with heat, cold, electricity, and/or hydrogen storage. One or more energy sectors (electricity, transportation, heating/cooling, industry, agriculture/forestry/fishing, and the military) should be transitioned. The plan should contain some of the following elements:

1. Current and estimated future (e.g., 2030, 2040, or 2050) energy mix by sector;

2. Energy required upon conversion of non-electric energy to electricity powered by WWS;

3. Number of power generators, heating/cooling devices, and vehicles needed;

4. Footprint and spacing requirements for the generator technologies;

5. Transmission requirements;

6. Determination of the reliability of the grid with the power generators chosen;

7. Battery-charging infrastructure requirements;

8. Cost of installed infrastructure, devices, vehicles, etc.;

9. Changes in number of temporary and permanent jobs;

10. Changes in temporary and permanent revenue streams;

11. Changes in air pollution mortality and morbidity and resulting cost savings;

12. Changes in global warming costs;

13. Changes in other environmental costs;

14. Change in tax consequence to state and local governments; and

15. Policies needed to make the transition.

It is not necessary to include all or even most of the elements listed above, and groups may elect to add elements not on the list. Each group should decide which set of elements to include. This may depend on the available information for a given region, the number of students in the group, etc. Each group has an option to cover at least one element in the above list per student in detail or to cover more than one element per student in less detail.

# GLOSSARY OF ACRONYMS

| | | | | |
|---|---|---|---|---|
| AC | = Alternating current | | FIT | = Feed-in tariff |
| ACfF | = Apparent centrifugal force | | FTE | = Full-time equivalent |
| ACoF | = Apparent Coriolis force | | GATOR- | = Gas, Aerosol, TranspOrt, Radiation- |
| AGL | = Above ground level | | GCMOM | General Circulation, Mesoscale, and |
| AGW | = Anthropogenic global warming | | | Ocean Model |
| AOTA | = All-of-the-above | | GCR | = Ground cover ratio |
| ATES | = Aquifer thermal energy storage | | GDP | = Gross domestic product |
| AU | = Astronomical unit | | GHG | = Greenhouse gas |
| BAU | = Business-as-usual | | GND | = Green New Deal |
| BC | = Black carbon | | GPE | = Gravitational potential energy |
| BE | = Battery electric | | GPS | = Global positioning satellite |
| BECCS | = Bioenergy with carbon capture and storage | | GW | = Gigawatt |
| BF | = Blast furnace | | GWP | = Global warming potential |
| BOF | = Basic oxygen furnace | | HAWT | = Horizontal axis wind turbine |
| BOS | = Basic oxygen steelmaking | | HDD | = Heating degree days |
| BrC | = Brown carbon | | HDR | = Hydrogen direct reduction |
| BTES | = Borehole thermal energy storage | | HFC | = Hydrogen fuel cell |
| BTU | = British thermal unit | | HHV | = Higher heating value |
| BWR | = Boiling water reactor | | HVAC | = High voltage alternating current |
| CAES | = Compressed air energy storage | | HVDC | = High voltage direct current |
| CCA | = Community choice aggregation | | HW | = Hot water |
| CCGT | = Combined cycle gas turbine | | ICE | = Internal combustion engine |
| CCS | = Carbon capture and storage | | ICE | = Ice storage |
| CCU | = Carbon capture and use | | IR | = Infrared |
| CDD | = Cooling degree days | | IRH | = Indirect resistance heating |
| CF | = Capacity factor | | ITCZ | = Intertropical convergence zone |
| CNG | = Compressed natural gas | | I-V | = Current-voltage |
| COP | = Coefficient of performance | | J | = Joule |
| COP | = Conference of the parties | | kW | = Kilowatt |
| COPD | = Chronic obstruction pulmonary disease | | LCOE | = Levelized cost of energy |
| CRF | = Capital recovery factor | | LED | = Light-emitting diode |
| CSP | = Concentrated solar power | | LHV | = Lower heating value |
| CURES | = Cancer unit risk estimates | | LNG | = Liquefied natural gas |
| CW | = Cold water | | LOLE | = Loss of load expectation |
| DC | = Direct current | | LWR | = Light water reactor |
| DR | = Demand response | | MOE | = Molten oxide electrolysis |
| DRH | = Direct resistance heating | | MPP | = Maximum power point |
| DRM | = Demand response management | | MW | = Microwave |
| FF | = Friction force | | MW | = Megawatt |

| | | | | |
|---|---|---|---|---|
| NDACCS | = Natural direct air carbon capture and storage | | SESI | = Stanford Energy Systems Innovation |
| NG | = Natural gas | | SCPC | = Supercritical pulverized coal |
| NIMBY | = Not-in-my-backyard | | SDR | = Social discount rate |
| O&M | = Operation and maintenance | | SMR | = Small modular reactor |
| OCGT | = Open cycle gas turbine | | STC | = Standard test conditions |
| PCM | = Phase change material | | STES | = Sensible-heat thermal energy storage |
| PDR | = Private discount rate | | SUV | = Sport utility vehicle |
| PEM | = Polymer electrolyte membrane | | SWPP | = Saturation wind power potential |
| PFC | = Perfluorocarbon | | T&D | = Transmission and distribution |
| PGF | = Pressure gradient force | | TIR | = Thermal-infrared |
| PGM | = Platinum group metals | | TSR | = Tip speed ratio |
| PHS | = Pumped hydropower storage | | TTES | = Tank thermal energy storage |
| POC | = Primary organic carbon | | TWR | = Thrust-to-weight ratio |
| PPP | = Purchasing power parity | | UHI | = Urban heat island |
| PTES | = Pit thermal energy storage | | UTES | = Underground thermal energy storage |
| PTO | = Planning-to-operation | | UV | = Ultraviolet |
| PV | = Photovoltaic | | V2G | = Vehicle-to-grid |
| P-V | = Power-voltage | | VAr | = Volt-Amperes reactive |
| PWR | = Power-to-weight ratio | | VAWT | = Vertical axis wind turbine |
| PWR | = Pressurized water reactor | | VNM | = Virtual net metering |
| RF | = Radio frequency | | VOSL | = Value of statistical life |
| RPS | = Renewable portfolio standard | | WWS | = Wind-water-solar |
| SDACCS | = Synthetic direct air carbon capture and storage | | | |

# APPENDIX
## Unit Conversions and Constants

### A.1 Distance Conversions

| 1 m | = 100 cm | = 1,000 mm | = $10^6$ μm |
|---|---|---|---|
| | = $10^9$ nm | = $10^{10}$ Å | = 0.001 km |
| | = 39.370 in | = 3.2808 ft | = 1.0936 yd |
| | = $6.2138 \times 10^{-4}$ mi | | |

### A.2 Volume Conversions

| 1 m³ | = 1,000 L | = $10^6$ cm³ | = $10^{18}$ μm³ |
|---|---|---|---|
| | = 264.172 U.S. gallon | = 35.313 ft³ | |

### A.3 Mass Conversions

| 1 kg | = 1,000 g | = $10^6$ mg | = $10^9$ μg |
|---|---|---|---|
| | = $10^{12}$ ng | = 0.001 tonne (metric) | = 0.0011023 short ton |
| | = 2.20462 lb | = 35.2739 oz | = $6.022 \times 10^{26}$ amu |
| 1 tonne | = 1,000 kg | = $10^6$ g | = 1.1023 short ton |
| | = 2204.623 lb | | |
| 1 Tg | = $10^{12}$ g | = $10^9$ kg | = $10^6$ Mg (megagram) |
| | = 1,000 Gg (gigagram) | = 0.001 Pg (petagram) | = $10^6$ tonne (metric) |
| | = 1 Mt (megatonne) | = 0.001 Gt (gigaton) | |

## A.4 Temperature Conversions

| °C | = K −273.15 | = (°F −32)/ 1.8 | |
|----|-------------|-----------------|--|

## A.5 Force Conversions

| 1 N | = 1 kg m s$^{-2}$ | = 10$^5$ g cm s$^{-2}$ | 10$^5$ dyn |
|-----|-------------------|------------------------|------------|
|     | = 0.2248 lbf (pound-force) | | |

## A.6 Pressure Conversions

| 1 bar | = 10$^3$ mb | = 0.986923 atm | = 10$^5$ N m$^{-2}$ |
|-------|-------------|----------------|---------------------|
|       | = 10$^5$ J m$^{-3}$ | = 10$^5$ Pa | = 10$^3$ hPa |
|       | = 10$^5$ kg m$^{-1}$ s$^{-2}$ | = 10$^6$ dyn cm$^{-2}$ | = 10$^6$ g cm$^{-1}$ s$^{-2}$ |
|       | = 750.06 torr | | = 750.06 mm Hg |
| 1 atm | = 1.01325 bar | = 760 torr | = 760 mm Hg |
|       | = 29.92 in Hg | = 1013.25 hPa | = 14.696 psi (lbf in$^{-2}$) |

## A.7 Energy Conversions

| 1 J | = 1 N m | = 1 W s | = 2.77778 × 10$^{-7}$ kWh |
|-----|---------|---------|---------------------------|
|     | = 0.001 kJ | = 10$^{-6}$ MJ | = 10$^{-9}$ GJ |
|     | = 1 kg m$^2$ s$^{-2}$ | = 10$^7$ g cm$^2$ s$^{-2}$ | = 1 C V |
|     | = 10$^7$ dyn cm | = 10$^4$ hPa cm$^3$ | = 10$^{-5}$ bar m$^3$ |
|     | = 10$^4$ hPa cm$^3$ | = 6.2415 × 10$^{18}$ eV | = 0.238902 cal |
|     | = 10$^7$ erg | = 0.009869 L atm | = 0.7373 lbf ft |
|     | = 9.4782 × 10$^{-4}$ Btu | = 9.4782 × 10$^{-10}$ MMBtu | |

## A.8  Power Conversions

| 1 W | $= 1 \text{ J s}^{-1}$ | $= 0.001 \text{ kW}$ | $= 10^{-6} \text{ MW}$ |
|---|---|---|---|
| | $= 10^{-9} \text{ GW}$ | $= 10^{-12} \text{ TW}$ | $= 1 \text{ V A}$ |
| | $= 1 \text{ A}^2 \, \Omega$ | $= 1 \text{ V}^2 / \Omega$ | $= 3.41252 \text{ Btu hr}^{-1}$ |
| | $= 0.013407$ horsepower | | |

## A.9  Speed Conversions

| 1 m s$^{-1}$ | $= 100 \text{ cm s}^{-1}$ | $= 3.6 \text{ km h}^{-1}$ | $= 1.94384$ knots |
|---|---|---|---|
| | $= 2.23694 \text{ mi hr}^{-1}$ | | |

## A.10  Electricity Conversions

| 1 V | $= 1 \text{ J C}^{-1}$ | $= 1 \text{ J A}^{-1} \text{ s}^{-1}$ | $= 1 \text{ A } \Omega$ |
|---|---|---|---|
| | $= 1 \text{ W A}^{-1}$ | $= 1 \text{ C F}^{-1}$ | $= 1 \text{ A G}^{-1}$ |

## A.11  Constants

| $c$ | = speed of light | $= 2.99792 \times 10^{8} \text{ m s}^{-1}$ |
|---|---|---|
| $c_W$ | = specific heat of liquid water | $= 4{,}185.5 \text{ J kg}^{-1} \text{ K}^{-1}$ |
| | $= 4.1855 \text{ J g}^{-1} \text{ K}^{-1}$ | $= 4{,}185.5 \text{ m}^2 \text{ s}^{-2} \text{ K}^{-1}$ |
| | $= 1.00 \text{ cal g}^{-1} \text{ K}^{-1}$ | |
| $\varepsilon_0$ | = permittivity of a vacuum | $= 8.854 \times 10^{-12} \text{ F m}^{-1}$ |
| $F_s$ | = solar constant | $= 1{,}365 \text{ W m}^{-2}$ |
| $g$ | = effective gravity at surface of Earth | $= 9.80665 \text{ m s}^{-2}$ |
| | $= 980.665 \text{ cm s}^{-2}$ | $= 32.17403 \text{ ft s}^{-2}$ |
| $h$ | = Planck's constant | $= 6.6260755 \times 10^{-34} \text{ J s cycle}^{-1}$ |
| | | |
| $m_d$ | = molecular weight of dry air | $= 28.966 \text{ g mol}^{-1}$ |
| $R^*$ | = universal gas constant | $= 8.31451 \text{ J mol}^{-1} \text{ K}^{-1}$ |
| | $= 8.31451 \text{ kg m}^2 \text{ s}^{-2} \text{ mol}^{-1} \text{ K}^{-1}$ | $= 0.0831451 \text{ m}^3 \text{ hPa mol}^{-1} \text{ K}^{-1}$ |
| | $= 8.31451 \times 10^{7} \text{ g cm}^2 \text{ s}^{-2} \text{ mol}^{-1} \text{ K}^{-1}$ | $= 0.08206 \text{ L atm mol}^{-1} \text{ K}^{-1}$ |

| | | |
|---|---|---|
| | $= 8.31451 \times 10^4$ cm$^3$ hPa mol$^{-1}$K$^{-1}$ | $= 8.31451 \times 10^7$ erg mol$^{-1}$ K$^{-1}$ |
| | $= 82.06$ cm$^3$ atm mol$^{-1}$ K$^{-1}$ | $= 1.98635$ cal mol$^{-1}$ K$^{-1}$ |
| $R'$ | $=$ gas constant for dry air $(R^\star/m_d)$ | $= 287.04$ J kg$^{-1}$ K$^{-1}$ |
| | $= 0.28704$ J g$^{-1}$K$^{-1}$ | $= 2.8704$ m$^3$ hPa kg$^{-1}$ K$^{-1}$ |
| | $= 2870.4$ cm$^3$ hPa g$^{-1}$ K$^{-1}$ | $= 287.04$ m$^2$ s$^{-2}$ K$^{-1}$ |
| | $= 2.8704 \times 10^6$ cm$^2$ s$^{-2}$ K$^{-1}$ | |
| $R_e$ | $=$ radius of the Earth | $= 6.371 \times 10^6$ m |
| $R_p$ | $=$ radius of the sun's photosphere | $= 6.936 \times 10^8$ m |
| $R_{es}$ | $=$ mean Earth-sun distance | $= 1.5 \times 10^{11}$ m |
| $\sigma_B$ | $=$ Stefan–Boltzmann constant | $= 5.67051 \times 10^{-8}$ W m$^{-2}$ K$^{-4}$ |
| $\Omega$ | $=$ rotation rate of the Earth | $= 7.292 \times 10^{-5}$ radians s$^{-1}$ |

# REFERENCES

ABB, 2004. HVDC-an ABB specialty, https://library.e.abb.com/public/d4863a9b0f77b74ec1257b0c00552758/HVDC%20Cable%20Transmission.pdf (accessed December 31, 2018).

ABB, 2005. HVDC technology for energy efficiency and grid reliability, www02.abb.com/global/abbzh/abbzh250.nsf/0/27c2fdbd96a879a4c12575ee00487a77/$file/HVDC+-+efficiency+and+reliability.pdf (accessed December 31, 2018).

ABC (American Bird Conservancy), 2019. https://abcbirds.org (accessed January 4, 2019).

AFDC, 2014. Public retail gas stations by year, U.S. Department of Energy, https://afdc.energy.gov/files/u/data/data_source/10333/10333_gasoline_stations_year.xlsx (accessed January 3, 2020).

Aghahosseini, A., D. Bogdanov, and C. Breyer, 2016. A techno-economic study of an entirely renewable energy-based power supply for North America for 2030 conditions, *Energies*, *10*, 1171, doi:10.3390/en10081171.

Aghahosseini, A., D. Bogdanov, L. S. N. S. Barbosa, and C. Breyer, 2019. Analyzing the feasibility of powering the Americas with renewable energy and inter-regional grid interconnections by 2030, *Renew. Sust. Energ. Rev.*, *105*, 187–205.

Allain, R., 2018. How much energy can you store in a stack of cement blocks?, *Wired*, www.wired.com/story/battery-built-from-concrete/ (accessed April 10, 2019).

Allanore, A., L. Yin, and D. Sadoway, 2013. A new anode material for oxygen evolution in molten oxide electrolysis, *Nature*, *497*, 353–356.

Allen, M. R., D. J. Frame, C. Huntingford, C. D. Jones, J. A. Lowe, M. Meinshausen, and N. Meinhausen, 2009. Warming caused by cumulative carbon emissions towards the trillionth tonne, *Nature*, *458*, 1163–1166.

Allred, B. W., W. K. Smith, D. Twidwell, J. H. Haggerty, S. W. Running, D. E. Naugle, and S. D. Fuhlendorf, 2015. Ecosystem services lost to oil and gas in North America, *Science*, *348*, 401–402.

Alvarez, R. A., S. W. Pacala, J. J. Winebrake, W. L. Chameides, and S. P. Hamburg, 2012. Greater focus needed on methane leakage from natural gas infrastructure, *Proc. Nat. Acad. Sci.*, *109*(17), 6435–6440, doi: 10.1073/pnas.1202407109.

Alvarez, R. A., D. Zavala-Araiza, D. R. Lyon, D. T. Allen, Z. R. Barkley, A. R. Brandt, et al., 2018. Assessment of methane emissions from the U.S. oil and gas supply chain, *Science*, *361*, 186–188.

Archer, C. L., and M. Z. Jacobson, 2003. Spatial and temporal distributions of U.S. winds and wind power at 80 m derived from measurements, *J. Geophys. Res.*, *108*(D9), 4289.

Archer, C. L., and M. Z. Jacobson, 2005. Evaluation of global wind power, *J. Geophys. Res.*, *110*, D12110, doi:10.1029/2004JD005462.

Archer, C. L., and M. Z. Jacobson, 2007. Supplying baseload power and reducing transmission requirements by interconnecting wind farms, *J. Applied Meteorol. Climatol.*, *46*, 1701–1717, doi:10.1175/2007JAMC1538.1.

Arcon/Sunmark, 2017. Large-scale showcase projects, http://arcon-sunmark.com/uploads/ARCON_References.pdf (accessed November 25, 2018).

Barasa, M., D. Bogdanov, A. S. Oyewo, and C. Breyer, 2018. A cost optimal resolution for sub-Saharan Africa powered by 100% renewables in 2030, *Renew. Sust. Energ. Rev.*, *92*, 440–457.

Barber, H., 1982. Electric heating fundamentals. In *The Efficient Use of Energy*, 2nd ed., ed. I. G. C. Dryden. Oxford: Butterworth-Heinemann, pp. 94–114, doi:10.1016/B978-0-408-01250-8.50016-7.

Barbosa, L. S. N. S., D. Bogdanov, P. Vainikka, and C. Breyer, 2017. Hydro, wind, and solar power as a base for a 100% renewable energy supply for South and Central America, *PLoS One*, doi:10.1371/journal.pone.0173820.

BBC News, September 5, 2018. Japan confirms first Fukushima worker death from radiation, www.bbc.com/news/world-asia-45423575 (accessed June 9, 2019).

Becker, S., B. A. Frew, G. B. Andresen, T. Zeyer, S. Schramm, M. Greiner, and M. Z. Jacobson, 2014. Features of a fully renewable U.S. electricity-system: optimized mixes of wind and solar PV and transmission grid extensions, *Energy*, *72*, 443–458.

Becker, S., B. A. Frew, G. B. Andresen, M. Z. Jacobson, S. Schramm, and M. Greiner, 2015. Renewable build-up pathways for the U.S.: generation costs are not system costs, *Energy*, *81*, 437–445.

Bellevrat, E., and K. West, 2018. Clean and efficient heat for industry (Commentary), International Energy Agency (IEA), www.iea.org/newsroom/news/2018/january/commentary-clean-and-efficient-heat-for-industry.html (accessed November 17, 2018).

Berthelemy, M., and L. E. Rangel, 2015. Nuclear reactors' construction costs: the role of lead-time, standardization, and technological progress, *Energy Policy*, *82*, 118–130.

Bistak, S., and S. Y. Kim, 2017. AC induction motors vs. permanent magnet synchronous motors, http://empoweringpumps.com/ac-induction-motors-versus-permanent-magnet-synchronous-motors-fuji/ (accessed January 5, 2018).

Bizee, 2019. Custom degree day data, www.degreedays.net (accessed January 21, 2019).

Blakers, A., B. Lu, and M. Socks, 2017. 100% renewable electricity in Australia, *Energy*, **133**, 471–482.

Blakers, A., B. Lu, M. Stocks, K. Anderson, and A. Nadolny, 2018. Pumped hydro storage to support 100% renewable power, *Energy News*, **36**, 11–14.

Blakers, A., M. Stocks, B. Lu, C. Cheng, and A. Nadolny, 2019. *Global Pumped Hydro Atlas*, RE100 Group, Australian National University, http://re100.eng.anu.edu.au/global/ (accessed March 31, 2019).

BloombergNEF, 2019. A behind-the-scenes take on lithium-ion battery prices, https://about.bnef.com/blog/behind-scenes-take-lithium-ion-battery-prices/ (accessed June 28, 2019).

Boden, T., B. Andres, and G. Marland, 2011. Global $CO_2$ emissions from fossil-fuel burning, cement manufacture, and gas flaring: 1751–2011, http://cdiac.ornl.gov/ftp/ndp030/global.1751_2011.ems (accessed January 24, 2019).

Boeing, 2012. 747-8 airplane characteristics for airport planning, www.boeing.com/assets/pdf/commercial/airports/acaps/747_8.pdf (accessed March 27, 2019).

Bogdanov, D., and C. Breyer, 2016. North-east Asian super grid for 100% renewable energy supply: optimal mix of energy technologies for electricity, gas, and heat supply options, *Energy Convers. Manag.*, **112**, 176–190.

Bogdanov, D., J. Farfan, K. Sadovskaia, A. Aghahosseini, M. Child, A. Gulagi, et al., 2019. Radical transformation pathway towards sustainable electricity via evolutionary steps, *Nat. Commun.*, **10**, 1077, doi: 10.1038/s41467-019-08855-1.

Boiocchi, R., K. V. Gemaey, and G. Sin, 2016. Control of wastewater $N_2O$ emission by balancing the microbial communities using a fuzzy-logic approach, *IFAC-PapersOnLine*, **49**, 1157–1162.

Bond, T. C., D. G. Streets, K. F. Yarber, S. M. Nelson, J.-H. Woo, and Z. Klimont, 2004. A technology-based global emission inventory of black and organic carbon emissions from combustion, *J. Geophys. Res.*, **109**, D14203, doi:10.1029/2003JD003697.

Bond, T. C., S. J. Doherty, D. W. Fahey, P. M. Forster, T. Berntsen, O. Boucher, et al., 2013. Bounding the role of black carbon in the climate system: a scientific assessment, *J. Geophys. Res.*, **118**, 5380–5552, doi: 10.1002/jgrd.50171.

Boukhalf, S., and N. Kaul, 2019. 10 disruptive battery technologies trying to compete with lithium-ion batteries, *Solar Power World*, www.solarpowerworldonline.com/2019/01/10-disruptive-battery-technologies-trying-to-compete-with-lithium-ion/ (accessed February 2, 2019).

Boyer, G., 2019. Appalachian power rolls out 100% renewable option for customers, WFXR News, www.wfxrtv.com/news/appalachian-power-rolls-out-100-renewable-option-for-va-customers/ (accessed August 21, 2019).

Bremner, S. P., M. Y. Levy, and C. B. Honsberg, 2008. Analysis of tandem solar cell efficiencies under {AM1.5G} spectrum using a rapid flux calculation method, *Prog. Photovoltaics*, **16**, 225–233.

Breyer, C., 2012. *Economics of Hybrid Photovoltaic Power Plants*, Pro Business, ISBN: 978-3863863906.

British Petroleum, 2018. BP statistical review of world energy, www.bp.com/content/dam/bp/business-sites/en/global/corporate/pdfs/energy-economics/statistical-review/bp-stats-review-2018-co2-emissions.pdf (accessed December 15, 2018).

Brown, T. W., T. Bischof-Niemz, K. Blok, C. Breyer, H. Lund, and B. V. Mathiesen, 2018. Response to "Burden of proof: a comprehensive review of the feasibility of 100% renewable electricity systems," *Renew. Sust. Energ. Rev.*, **92**, 834–847.

Bruckner T., I. A. Bashmakov, Y. Mulugetta, H. Chum, A. de la Vega Navarro, J. Edmonds, et al., 2014. Energy systems. In *Climate Change 2014: Mitigation of Climate Change. Contribution of Working Group III to the Fifth Assessment Report of the Intergovernmental Panel on Climate Change*, ed. O. Edenhofer, R. Pichs-Madruga, Y. Sokona, E. Farahani, S. Kadner, K. Seyboth, et al. Cambridge: Cambridge University Press.

BTS (Bureau of Transportation Statistics), 2018. U.S. oil and gas pipeline mileage, www.bts.gov/content/us-oil-and-gas-pipeline-mileage (accessed December 3, 2018).

Budischak, C., D. Sewell, H. Thompson, L. Mach, D. E. Veron, and W. Kempton, 2013. Cost-minimized combinations of wind power, solar power, and electrochemical storage, powering the grid up to 99.9% of the time, *J Power Sources*, **225**, 60–74.

Build Abroad, 2016. Ferrock: a stronger, more flexible and greener alternative to concrete, https://buildabroad.org/2016/09/27/ferrock/ (accessed November 20, 2018).

Burke, M., S. M. Hsiang, and E. Miguel, 2015. Global non-linear effect of temperature on economic production, *Nature*, **527**, 235–239.

Burnett, R., H. Chen, M. Szyszkowica, N. Fann, B. Hubbell, C. Arden Pope III, et al., 2018. Global estimates of mortality associated with long-term exposure to outdoor fine particulate matter, *Proc. Natl. Acad. Sci.*, **115**, 9592–9597.

Business Dictionary, 2019. Installed capacity: definition, www.businessdictionary.com/definition/installed-capacity.html (accessed March 15, 2019).

Caldera, U., and C. Breyer, 2018. Role that battery and water storage play in Saudi Arabia's transition to an integrated 100% renewable energy power system, *J. Energy Storage*, **17**, 299–310.

California Senate, 2018. SB 100 FAQs, https://focus.senate.ca.gov/sb100/faqs (accessed February 6, 2019).

CARB (California Air Resources Board), 2010. Estimate of premature deaths associated with fine particle pollution ($PM_{2.5}$) in California using a U.S. Environmental Protection Agency methodology, www.arb.ca.gov/research/health/pm-mort/pm-report_2010.pdf (accessed January 15, 2019).

Carbon Cure, 2018., CarbonCure concrete, www.carboncure.com (accessed November 20, 2018).

CDC (Centers for Disease Control and Prevention), 2000. Research on long-term exposure: uranium miners, www.cdc.gov/niosh/pgms/worknotify/uranium.html (accessed December 9, 2018).

Cebulla, F., and M. Z. Jacobson, 2018. Alternative renewable energy scenarios for New York, *J. Clean. Prod.*, **205**, 884–894.

Chakratek, 2019. Flywheel specifications and comparison, www.chakratec.com/technology/ (accessed March 18, 2019).

Chang, T. P., 2009. The sun's apparent position and the optimal tilt angle of a solar collector in the northern hemisphere, *Sol. Energy*, **83**, 1274–1284.

Child, M., and C. Breyer, 2016. Vision and initial feasibility analysis of a decarbonized Finnish energy system for 2050, *Renew. Sust. Energ. Rev.*, **66**, 517–536.

Child, M., A. Nordling, and C. Breyer, 2018. The impacts of high V2G participation in a 100% renewable Aland energy system, *Energies*, **11**, 2206, https://doi.org/10.3390/en11092206.

Climate Group, 2019. EV100 members, www.theclimategroup.org/ev100-members (accessed September 22, 2019).

Clinton, H., October 16, 2015, Hillary is ready for 100, C-SPAN, online video clip, www.c-span.org/video/?c4557641/hillary-readyfor100 (accessed February 7, 2019).

Cockburn, H., 2019. Scotland generating enough wind energy to power two Scotlands, www.independent.co.uk/environment/scotland-wind-power-on-shore-renewable-energy-climate-change-uk-a9013066.html (accessed July 26, 2019).

Colella, W. G., M. Z. Jacobson, and D. M. Golden, 2005. Switching to a U.S. hydrogen fuel cell vehicle fleet: the resultant change in emissions, energy use, and global warming gases, *J. Power Sources*, **150**, 150–181.

Connolly, D., and B. V. Mathiesen, 2014. Technical and economic analysis of one potential pathway to a 100% renewable energy system, *IJSEPM*, **1**, 7–28.

Connolly, D., H. Lund, B. V. Mathiesen, and M. Leahy, 2011. The first step to a 100% renewable energy-system for Ireland, *Appl. Energy*, **88**, 502–507.

Connolly, D., H. Lund, and B. V. Mathiesen, 2016. Smart energy Europe: the technical and economic impact of one potential 100% renewable energy scenario for the European Union, *Renew. Sust. Ener. Rev.*, **60**, 1634–1653.

Corcoran, B. A., N. Jenkins, and M. Z. Jacobson, 2012. Effects of aggregating electric load in the United States. *Energy Policy*, **46**, 399–416.

Cornell University, 2019. How lake source cooling works, Ithaca, NY: Cornell University, Facilities and Campus Services, https://fcs.cornell.edu/departments/energy-sustainability/utilities/cooling-home/cooling-distribution/lake-source-cooling-home/how-lake-source-cooling-works (accessed May 1, 2019).

Crossley, I, 2018. Simplifying and lightening offshore turbines with compressed air energy storage, *Wind Power Monthly*, www.windpoweroffshore.com/article/1463030/simplifying-lightening-offshore-turbines-compressed-air-energy-storage (accessed March 18, 2019).

Czisch, G., 2005. Szenarien zur zukünftigen Stromversorgung, kostenoptimierte Variationen zur Versorgung Europas und seiner Nachbarn mit Strom aus erneuerbaren Energien, PhD dissertation, University of Kassel, https://kobra.uni-kassel.de/handle/123456789/200604119596 (accessed February 24, 2019).

Czisch, G., and G. Giebel, 2007. Realisable scenarios for a future electricity supply based 100% on renewable energies, Riso-R-1608 (EN).

Damkjaer, L., 2016. Gram Fjernvarme 2016, online video clip, www.youtube.com/watch?v=PdF8e1t7St8 (accessed November 25, 2018).

Dandelion, 2018. Geothermal heating and air conditioning is so efficient, it pays for itself, https://dandelionenergy.com (accessed November 17, 2018).

de Coninck, H., A. Revi, M. Babiker, P. Bertoldi, M. Buckeridge, A. Cartwright, et al., 2018. Strengthening and implementing the global response. In *Global Warming of 1.5°C. An IPCC Special Report on the Impacts of Global Warming of 1.5°C above Pre-Industrial Levels and Related Global Greenhouse Gas Emission Pathways, in the Context of Strengthening the Global Response to the Threat of Climate Change, Sustainable Development, and Efforts to Eradicate Poverty*, ed. V. MassonDelmotte, P. Zhai, H.-O. Pörtner, D. Roberts, J. Skea, P. R. Shukla, et al. Geneva: IPCC.

de Gouw, J. A., D. D. Parrish, G. J. Frost, and M. Trainer, 2014. Reduced emissions of $CO_2$, $NO_x$, and $SO_2$ from U.S. power plants owing to switch from coal to natural gas with combined cycle technology, *Earth's Future*, **2**, 75–82.

de Gracia, A., and L. F. Cabeza, 2015. Phase change materials and thermal energy storage for buildings, *Energy Build.*, **103**, 414–419.

De Stercke, S., 2014. *Dynamics of Energy Systems: A Useful Perspective*. IIASA Interim Report No. IR-14-013, International Institute for Applied Systems Analysis, IIASA, Laxenburg, Austria.

Delucchi, M., 2011. A conceptual framework for estimating the climate impacts of land-use change due to energy crop programs, *Biomass Bioenergy*, **35**, 2337–2360.

Delucchi, M. Z., and M. Z. Jacobson, 2011. Providing all global energy with wind, water, and solar power, Part II: reliability, system and transmission costs, and policies, *Energy Policy*, **39**, 1170–1190, doi:10.1016/j.enpol.2010.11.045.

Denholm, P., Y.-H. Wan, M. Hummon, and M. Mehos, 2014. The value of CSP with thermal energy storage in the western United States, *Energy Procedia*, **49**, 1622–1631.

Denyer, S., February 20, 2019. Eight years after Fukushima's meltdown, the land is recovering, but public trust is not, *Washington Post*, www.washingtonpost.com/world/asia_pacific/eight-years-after-fukushimas-meltdown-the-land-is-recovering-but-public-trust-has-not/2019/02/19/0bb29756-255d-11e9-b5b4-1d18dfb7b084_story.html?utm_term=.8344c816d5bb (accessed December 21, 2019).

Diesendorf, M., and B. Elliston, 2018. The feasibility of 100% renewable electricity systems: a response to critics, *Renew. Sust. Energ. Rev.*, **93**, 318–330.

DNC (Democratic National Committee), 2016. Democratic Party Platform, July 9, 2016, https://democrats.org/wp-content/uploads/2018/10/2016_DNC_Platform.pdf (accessed February 7, 2019).

DOE (U.S. Department of Energy), 2008. *Concentrating Solar Power Commercial Application Study: Reducing Water Consumption of Concentrating Solar Power Electricity Generation*, Report to Congress, www1.eere.energy.gov/solar/pdfs/csp_water_study.pdf (accessed December 28, 2019).

DOE (U.S. Department of Energy), 2014. 2014: the year of concentrating solar power, www.energy.gov/sites/prod/files/2014/05/f15/2014_csp_report.pdf (accessed March 30, 2019).

DOE (U.S. Department of Energy), 2015a. Fuel cells, Washington, DC: U.S. Department of Energy, Fuel Cell Technologies Office, www.energy.gov/sites/prod/files/2015/11/f27/fcto_fuel_cells_fact_sheet.pdf (accessed January 11, 2019).

DOE (U.S. Department of Energy), 2015b. Chapter 6: Innovating clean energy technologies in advanced manufacturing. Technology assessments. In *Quadrennial Technology Review: An Assessment of Energy Technologies and Research Opportunities*. Washington, DC: U.S. Department of Energy, www.energy.gov/sites/prod/files/2016/06/f32/QTR2015-6I-Process-Heating.pdf (accessed November 17, 2018).

DOE (U.S. Department of Energy), 2019. How do wind turbines work?, www.energy.gov/eere/wind/wind-energy-technologies-office (accessed March 27, 2019).

Douglas, C. A., G. P. Harrison, and J. P. Chick, 2008. Life cycle assessment of the Seagen marine current turbine, *Proc. Inst. Mech. Eng. Part M*, *222*, 1–12.

Drake Landing, 2016. Drake Landing Solar Community: how it works, www.dlsc.ca/how.htm (accessed March 26, 2019).

Drupp, M., M. Freeman, B. Groom, and F. Nesje, 2015. *Discounting Disentangled: An Expert Survey on the Determinants of the Long-Term Social Discount Rate*, The Centre for Climate Change Economics and Policy Working Paper No. 195 and Grantham Research Institute on Climate Change and the Environment Working Paper No. 172 (CCCEP and Grantham Research Institute).

Duan, Y., and D. C. Sorescu, 2010. $CO_2$ capture properties of alkaline earth metal oxides and hydroxides: a combined density functional theory and lattice phonom dynamics study, *J. Chem. Phys.*, *133*, 074508.

Duke Energy, 2019. More renewable energy options available under Duke Energy's Green Source Advantage, https://news.duke-energy.com/releases/more-renewable-energy-options-available-under-duke-energys-green-source-advantage?_ga=2.88266651.1875174277.1566405614-658711925.1566405614 (accessed August 21, 2019).

Dvorak, M., C. L. Archer, and M. Z. Jacobson, 2010. California offshore wind energy potential, *Renew. Energy*, *35*, 1244–1254, doi:10.1016/j.renene.2009.11.022.

Dvorak, M. J., E. D. Stoutenburg, C. L. Archer, W. Kempton, and M. Z. Jacobson, 2012. Where is the ideal location for a U.S. East Coast offshore grid?, *Geophys. Res. Lett.*, *39*, L06804, doi:10.1029/2011GL050659.

Dvorak, M. J., B. A. Corcoran, J. E. Ten Hoeve, N. G. McIntyre, and M. Z. Jacobson, 2013. U.S. East Coast offshore wind energy resources and their relationship to peak-time electricity demand, *Wind Energy*, *16*, 977–997, doi:10.1002/we.1524.

Earthworks, 2019. *Responsible Minerals Sourcing for Renewable Energy*, Report prepared for Earthworks by the Institute for Sustainable Futures, University of Technology, Sydney, https://earthworks.org/publications/responsible-minerals-sourcing-for-renewable-energy/ (accessed May 1, 2019).

Eddington, S. A., 1916. On the radiative equilibrium of the stars, *Mon. Not. Roy. Astronom. Soc.*, *77*, 16–35.

EIA (U.S. Energy Information Administration), 2016a. *International Energy Outlook 2016*, DOE/EIA-0484, www.eia.gov/forecasts/ieo/pdf/0484(2016).pdf, www.eia.gov/forecasts/ieo/, www.eia.gov/forecasts/ieo/ieo_tables.cfm (accessed January 10, 2019).

EIA (U.S. Energy Information Administration), 2016b. Hydraulically fractured wells provide two-thirds of U.S. natural gas production, www.eia.gov/todayinenergy/detail.php?id=26112 (accessed December 2, 2018).

EIA (U.S. Energy Information Administration), 2017. Today in energy, www.eia.gov/todayinenergy/detail.php?id=33552 (accessed December 4, 2018).

EIA (U.S. Energy Information Administration), 2018a. Table 1. Coal production and number of mines by state and mine type, 2017 and 2016, www.eia.gov/coal/annual/pdf/table1.pdf (accessed December 3, 2018).

EIA (U.S. Energy Information Administration), 2018b. Frequently asked questions, www.eia.gov/tools/faqs/faq.php?id=29&t=6 (accessed December 3, 2018).

EIA (U.S. Energy Information Administration), 2018c. Table 4.1. Count of electric power industry power plants by sector, by predominant energy sources within plant, 2007 through 2017, www.eia.gov/electricity/annual/html/epa_04_01.html (accessed December 3, 2018).

EIA (U.S. Energy Information Administration), 2018d. Table 1.1. Total electric power industry summary statistics, 2017 and 2016, www.eia.gov/electricity/annual/html/epa_01_01.html (accessed December 5, 2018).

EIA (Energy Information Administration), 2018e. How much electricity is lost in transmission and distribution in the United States, www.eia.gov/tools/faqs/faq.php?id=105&t=3 (accessed December 31, 2018).

Electrical Systems, 2019. Principle of electricity generation, www.skm-eleksys.com/2010/02/practical-power-system.html (accessed April 5, 2019).

Electronics Hub, 2015. Characteristics and working of PN junction diode, www.electronicshub.org/characteristics-and-working-of-p-n-junction-diode/ (accessed March 31, 2019).

Electronics Tutorials, 2019a. Capacitor tutorial summary, www.electronics-tutorials.ws/capacitor/cap_9.html (accessed April 3, 2019).

Electronics Tutorials, 2019b. Reactive power, www.electronics-tutorials.ws/accircuits/reactive-power.html (accessed April 3, 2019).

ELI (Environmental Law Institute), 2009. Estimating U.S. government subsidies to energy sources: 2002–2008, www.eli.org/research-report/estimating-us-government-subsidies-energy-sources-2002-2008 (accessed January 25, 2019).

Elliston, B., M. Diesendorf, and I. MacGill, 2012. Simulations of scenarios with 100% renewable electricity in the Australian National Electricity Market, *Energy Policy*, *45*, 606–613.

Elliston, B., I. MacGill, and M. Diesendorf, 2013. Least cost 100% renewable electricity scenarios in the Australian National Electricity Market, *Energy Policy*, *59*, 270–282.

Elliston, B., I. MacGill, and M. Diesendorf, 2014. Comparing least cost scenarios for 100% renewable electricity with low emission fossil fuel scenarios in the Australian National Electricity Market, *Renew. Energy*, *66*, 196–204.

Enevoldsen, P., and M. Z. Jacobson, 2020. Data investigation of installed and output power densities of onshore and offshore wind turbines worldwide, *Wind Energy*.

ENTSO-E (European Network of Transmission System Operators for Electricity), 2016. European load data, www.entsoe.eu/dbquery/country-packages/production-consumption-exchange-package, 2016 (accessed January 29, 2019).

Erlich, I., and M. Wilch, 2010. Frequency control by wind turbines, paper presented at the IEEE PES General Meeting, July 25–29, doi: 10.1109/PES.2010.5589911, https://ieeexplore.ieee.org/document/5589911 (accessed March 2, 2019).

Esteban, M., J. Portugal-Pereira, B. C. Mclellan, J. Bricker, H. Farzaneh, N. Djalikova, et al., 2018. 100% renewable energy system in Japan: smoothening and ancillary services, *Appl. Energy*, **224**, 698–707.

Etminan, M., G. Myhre, E. J. Highwood, and K. P. Shine, 2016. Radiative forcing of carbon dioxide, methane, and nitrous oxide: a significant revision of the methane radiative forcing, *Geophys. Res. Lett.*, **43**, 12614–12623.

European Commission, 2019. EDGAR: Emissions Database for Global Atmospheric Research, https://edgar.jrc.ec.europa.eu/background.php (accessed May 26, 2019).

Evarts, E. C., 2019. The world's largest EV never has to be recharged, *Green Car Rep.*, www.greencarreports.com/news/1124478_world-s-largest-ev-never-has-to-be-recharged (accessed August 20, 2019).

EVWind, 2018. Current status of concentrated solar power globally, www.evwind.es/2018/07/25/current-status-of-concentrated-solar-power-csp-globally/64041 (accessed January 9, 2019).

Faulstich, S., B. Hahn, and P. J. Tavner, 2011. Wind turbine downtime and its importance for offshore deployment, *Wind Energy*, **14**, 327–337.

Feng, Z., 2018. Stationary high-pressure hydrogen storage, U.S. Department of Energy, Oak Ridge National Laboratory, www.energy.gov/sites/prod/files/2014/03/f10/csd_workshop_7_feng.pdf (accessed November 28, 2018).

FERC (Federal Energy Regulatory Commission), 2004. Current state of and issues concerning underground natural gas storage, www.ferc.gov/EventCalendar/Files/20041020081349-final-gs-report.pdf (accessed December 3, 2018).

Fischer, D., and H. Madani, 2017. On heat pumps in smart grids: a review, *Renew. Sustain. Energy Rev.*, **70**, 342–357.

Flury, K., and R. Frischknecht, 2012. Lifecycle inventories of hydroelectric power generation, ESU Services, http://esu-services.ch/fileadmin/download/publicLCI/flury-2012-hydroelectric-power-generation.pdf (accessed December 8, 2018).

Frangoul, A., 2019. Scandinavia's biggest offshore wind farm is officially open, CNBC, www.cnbc.com/2019/08/23/scandinavias-biggest-offshore-wind-farm-is-officially-open.html?__source=sharebar%7Ctwitter&par=sharebar (accessed August 23, 2019).

Free Dictionary, 2019. Installed capacity: definition, https://encyclopedia2.thefreedictionary.com/Installed+Capacity (accessed March 15, 2019).

Frew, B. A., and M. Z. Jacobson, 2016. Temporal and spatial tradeoffs in power system modeling with assumptions about storage: an application of the POWER model, *Energy*, **117**, 198–213.

Frew, B. A., S. Becker, M. J. Dvorak, G. B. Andresen, and M. Z. Jacobson, 2016. Flexibility mechanisms and pathways to a highly renewable U.S. electricity future, *Energy*, **101**, 65–78.

Friedlingstein, P., R. M. Andrew, J. Rogelj, G. P. Peters, J. G. Canadell, R. Knutti, et al., 2014. Persistent growth of $CO_2$ emissions and implications for reaching climate targets, *Nat. Geosci.*, **7**. 709–715.

Fthenakis, V., and M. Raugei, 2017. Environmental life-cycle assessment of photovoltaic systems. In *The Performance of Photovoltaic (PV) Systems: Modelling, Measurement, and Assessment*, ed. N. Pearsall. Duxford: Woodhead 209–232,

Fuhrmann, M., 2009. Spreading temptation: proliferation and peaceful nuclear cooperation agreements. *Int. Secur.*, **34**(1), 7–41, available at SSRN: https://ssrn.com/abstract=1356091 (accessed September 9, 2019).

Garthwaite, J., 2018. What should we do with nuclear waste?, *Stanford Earth*, https://earth.stanford.edu/news/qa-what-should-we-do-with-nuclear-waste#gs.1sfx0x (accessed March 20, 2019).

GBD (Global Burden of Disease 2013 Risk Factors Collaborators), 2015. Global, regional, and national comparative risk assessment of 79 behavioral, environmental and occupational, and metabolic risks or clusters of risks in 188 countries, 1990–2013: a systematic analysis for the Global Burden of Disease Study 2013, *Lancet*, **386**, 2287–2323.

GE (General Electric), 2018. Haliade-X offshore wind turbine platform, www.ge.com/renewableenergy/wind-energy/turbines/haliade-x-offshore-turbine (accessed November 16, 2018).

Gerber, H., Y. Takano, T. J. Garrett, and P. V. Hobbs, 2000. Nephelometer measurements of the asymmetry parameter, volume extinction coefficient, and backscatter ratio in Arctic clouds, *J. Atmos. Sci.*, **57**, 3021–3033,

Geuss, M., 2019. Florida utility to close natural gas plants, build massive solar-powered battery, Ars Technica, https://arstechnica.com/information-technology/2019/03/florida-utility-to-close-natural-gas-plants-build-massive-solar-powered-battery/ (accessed April 1, 2019).

Ginnebaugh, D. L., and M. Z. Jacobson, 2012. Examining the impacts of ethanol (E85) versus gasoline photochemical production of smog in a fog using near-explicit gas- and aqueous-chemistry mechanisms, *Environ. Res. Lett.*, **7**, 045901, doi:10.1088/1748-9326/7/4/045901.

Ginnebaugh, D. L., J. Liang, and M. Z. Jacobson, 2010. Examining the temperature dependence of ethanol (E85) versus gasoline emissions on air pollution with a largely-explicit chemical mechanism, *Atmos. Environ.*, **44**, 1192–1199, doi:10.1016/j.atmosenv.2009.12.024.

Green Party US, 2018. Green New Deal – Full Language, www.gp.org/gnd_full (accessed September 16, 2019).

Gross, B., @Bill_Gross, 2019. Efficiency of gravitational mass storage system, Twitter Page, https://twitter.com/Bill_Gross/status/1164617097927806976/photo/1 (accessed August 22, 2019).

Gulagi, A., D. Bogdanov, and C. Breyer, 2017a. A cost optimized fully sustainable power system for Southeast Asia and the Pacific Rim, *Energies*, **10**, 583, doi:10.3390/en10050583.

Gulagi, A., P. Choudhary, D. Bogdanov, and C. Breyer, 2017b. Electricity system based on 100% renewable for India and SAARC, *PLoS One*, doi:10.1371/journal.pone.0180611.

Hampson, S. E., J. A. Andres, M. E. Lee, L. S. Foster, R. E. Glasgow, and E. Lichtenstein, 1998. Lay understanding of synergistic risk:

the case of radon and cigarette smoking, *Risk Anal.*, **18**, 343–350.

Hanley, S., 2018. Energy Vault proposes and energy storage system using concrete blocks, *Cleantechnica*, https://cleantechnica.com/2018/08/21/energy-vault-proposes-an-energy-storage-system-using-concrete-blocks/ (accessed April 10, 2019).

Hansen, K., C. Breyer, and H. Lund, 2019a. Status and perspectives on 100% renewable energy systems, *Energy*, **175**, 471-480.

Hansen, K., B. Mathiessen, and I. R. Skov, 2019b. Full energy system transition towards 100% renewable energy in Germany in 2050, *Renew. Sust. Energ. Rev.*, **102**, 1–13.

Harrabin, R., 2019. How liquid air could help keep the lights on, *BBC News*, www.bbc.com/news/business-50140110 (accessed October 30, 2019).

Hart, E. K., and M. Z. Jacobson, 2011. A Monte Carlo approach to generator portfolio planning and carbon emissions assessments of systems with large penetrations of variable renewables, *Renew. Energy*, **36**, 2278–2286, doi:10.1016/j.renene.2011.01.015.

Hart, E. K., and M. Z. Jacobson, 2012. The carbon abatement potential of high penetration intermittent renewables, *Energ. Environ. Sci.*, **5**, 6592–6601, doi:10.1039/C2EE03490E.

Hart, E. K., E. D. Stoutenburg, and M. Z. Jacobson, 2012. The potential of intermittent renewables to meet electric power demand: a review of current analytical techniques, *Proc. IEEE*, **100**, 322–334, doi:10.1109/JPROC.2011.2144951.

Harvey, L. D. D., 2018. Resource implications of alternative strategies for achieving zero greenhouse gas emissions from light-duty vehicles by 2060, *Appl. Energy*, **212**, 663–679.

Hasager, C. B., L. Rasmussen, A. Pena, L. E. Jensen, and P.-E. Rethore, 2013. Wind farm wake: the Horns Rev photo case, *Energies*, **6**, 696–716.

Henshaw, D. L., J. P. Eatough, and R. B. Richardson, 1990. Radon as a causative factor in induction of myeloid leukaemia and other cancers, *Lancet*, **335**, 1008–1012.

Hope, P. April 6, 2017. Electric lawn mowers that rival gas models, *Consumer Reports*, www.consumerreports.org/push-mowers/electric-lawn-mowers-that-rival-gas-models/ (accessed November 21, 2018).

Hoste, G. R. G., M. J. Dvorak, and M. Z. Jacobson, 2009. *Matching Hourly and Peak Demand by Combining Different Renewable Energy Sources*, Stanford University Technical Report, https://web.stanford.edu/group/efmh/jacobson/Articles/I/CombiningRenew/HosteFinalDraft (accessed January 27, 2019).

Houghton, R. A., 2015. Annual net flux of carbon to the atmosphere from land-use change: 1850–2005, http://cdiac.ornl.gov/trends/landuse/houghton/1850-2005.txt (accessed January 24, 2019).

House (U.S. House of Representatives), 2015. H.Res.540, www.congress.gov/bill/114th-congress/house-resolution/540/text (accessed February 7, 2019).

House (U.S. House of Representatives), 2017a. H.R.3314 – 100 by'50 Act, www.congress.gov/bill/115th-congress/house-bill/3314 (accessed February 7, 2019).

House (U.S. House of Representatives), 2017b. H.R.3671 – Off Fossil Fuels For A Better Future Act, www.congress.gov/bill/115th-congress/house-bill/3671/text (accessed February 7, 2019).

House (U.S. House of Representatives), 2019a. H.R.330 – Climate Solutions Act of 2019, www.congress.gov/bill/116th-congress/house-bill/330/text (accessed February 7, 2019).

House (U.S. House of Representatives), 2019b. Resolution recognizing the duty of the federal government to create a green new deal, www.congress.gov/bill/116th-congress/house-resolution/109 https://apps.npr.org/documents/document.html?id=5729035-Green-New-Deal-FAQ (accessed February 10, 2019).

Howarth, R. W., 2014. A bridge to nowhere: methane emissions and the greenhouse gas footprint of natural gas, *Energy Sci. Eng.*, **2**, 47–60.

Howarth, R. W., 2019. Is shale gas a major driver of recent increase in global atmospheric methane, *Biogeosciences*, **16**, 3033–3046.

Howarth, R. W., R. Santoro, and A. Ingraffea, 2011. Methane and the greenhouse gas footprint of natural gas from shale formations, *Clim. Change*, **106**, 679–690.

Howarth, R. W., R. Santoro, and A. Ingraffea, 2012. Venting and leaking of methane from shale gas development: response to Cathles et al., *Clim. Change*, **113**(2).

HSMag, 2016. Simulation permanent magnet generators, www.hsmagnets.com/blog/simulation-permanent-magnet-generators/ (accessed July 23, 2019).

Hu, S-y, and J.-h Cheng, 2007. Performance evaluation of pairing between sites and wind turbines, *Renew. Energy*, **32**, 1934–1947.

Hulls, P. J., 2016. Development of the industrial use of dielectric heating in the United Kingdom, *J. Microwave Power*, **17**, 28–38.

Hunt, J. D., B. Zakeri, G. Falchetta, A. Nascimento, Y. Wada, and K. Riahi, 2020. Mountain gravity energy storage: a new solution for closing the gap between existing short- and long-term storage technologies, *Energy*, **190**, doi:10.1016/j.energy.2019.116419.

IEA (International Energy Agency), 2018a. *Integrated Cost-Effective Large-Scale Thermal Energy Storage for Smart District Heating and Cooling*, www.iea-dhc.org/fileadmin/documents/Annex_XII/IEA_DHC_AXII_Design_Aspects_for_Large_Scale_ATES_PTES_draft.pdf (accessed November 25, 2018).

IEA (International Energy Agency), 2018b. Wind energy, www.iea.org/topics/renewables/wind/ (accessed January 9, 2019).

IEA (International Energy Agency), 2018c. Geothermal energy, www.iea.org/topics/renewables/geothermal/ (accessed January 9, 2019).

IEA (International Energy Agency), 2018d. Solar PV, www.iea.org/tcep/power/renewables/solar/ (accessed January 9, 2019).

IEA (International Energy Agency), 2019. *Statistics*, www.iea.org/statistics/ (accessed January 5, 2019).

IEC (International Electrotechnical Commission), 2007. Efficient electrical energy transmission and distribution, https://basecamp.iec.ch/download/efficient-electrical-energy-transmission-and-distribution/ (accessed December 31, 2018).

IHA (International Hydropower Association), 2018. 2018 hydropower status report, www.hydropower.org/publications/2018-hydropower-status-report (accessed January 9, 2019).

IPCC (Intergovermental Panel on Climate Change), 2000. *Special Report on Emission Scenarios (SRES) final data*, http://sres.ciesin.org/final_data.html (accessed January 23, 2019).

IPCC (Intergovernmental Panel on Climate Change), 2005. *IPCC Special Report on Carbon Dioxide Capture and Storage.* Prepared by working group III, B. Metz, O. Davidson, H. C. de Coninck, M. Loos, and L. A. Meyer, eds. Cambridge: Cambridge University Press, 442 pp., http://arch.rivm.nl/env/int/ipcc/ (accessed June 26, 2019).

IPCC (Intergovernmental Panel on Climate Change), 2018. Special report: global warming of 1.5 $^{\circ}$C, www.ipcc.ch/sr15/ (accessed June 26, 2019).

IranWatch, 2015. Iran's nuclear potential before the implementation of the nuclear agreement, www.iranwatch.org/our-publications/articles-reports/irans-nuclear-timetable (accessed December 9, 2018).

IRENA (International Renewable Energy Agency), 2013. *Thermal Energy Storage*, IEA-ETSAP and IRENA Technology Brief E17, Abu Dhabi: IRENA.

Jacobson, M. Z., 1994. Developing, coupling, and applying a gas, aerosol, transport, and radiation model to study urban and regional air pollution, PhD dissertation, Department of Atmospheric Sciences, University of California, Los Angeles, 436 pp.

Jacobson, M. Z., 1997. Development and application of a new air pollution modeling system. Part III: aerosol-phase simulations, *Atmos. Environ.*, **31A**, 587–608.

Jacobson, M. Z., 1998. Studying the effects of aerosols on vertical photolysis rate coefficient and temperature profiles over an urban airshed, *J. Geophys. Res.*, **103**(10), 593–510, 604.

Jacobson, M. Z., 1999. Isolating nitrated and aromatic aerosols and nitrated aromatic gases as sources of ultraviolet light absorption, *J. Geophys. Res.*, **104**, 3527–3542.

Jacobson, M. Z., 2000. A physically-based treatment of elemental carbon optics: implications for global direct forcing of aerosols, *Geophys. Res. Lett.*, **27**, 217–220.

Jacobson, M. Z., 2001a. Strong radiative heating due to the mixing state of black carbon in atmospheric aerosols, *Nature*, **409**, 695–697.

Jacobson, M. Z., 2001b. GATOR-GCMM: a global through urban scale air pollution and weather forecast model. 1. Model design and treatment of subgrid soil, vegetation, roads, rooftops, water, sea ice, and snow, *J. Geophys. Res.*, **106**, 5385–5401.

Jacobson, M. Z., 2002. Control of fossil-fuel particulate black carbon plus organic matter, possibly the most effective method of slowing global warming, *J. Geophys. Res.*, **107**(D19), 4410, doi:10.1029/ 2001JD001376.

Jacobson, M. Z., 2004. The short-term cooling but long-term global warming due to biomass burning, *J. Clim.*, **17**(15), 2909–2926.

Jacobson, M. Z., 2005a. Studying ocean acidification with conservative, stable numerical schemes for nonequilibrium air-ocean exchange and ocean equilibrium chemistry, *J. Geophys. Res.*, **110**, D07302, doi:10.1029/2004JD005220.

Jacobson, M. Z., 2005b. *Fundamentals of Atmospheric Modeling*, 2nd ed., New York: Cambridge University Press, 813 pp.

Jacobson, M. Z., 2007. Effects of ethanol (E85) versus gasoline vehicles on cancer and mortality in the United States, *Environ. Sci. Technol.*, **41**(11), 4150–4157, doi:10.1021/es062085v.

Jacobson, M.Z, 2008a. On the causal link between carbon dioxide and air pollution mortality, *Geophys. Res. Lett.*, **35**, L03809, doi:10.1029/2007GL031101.

Jacobson, M. Z., 2008b. Effects of wind-powered hydrogen fuel cell vehicles on stratospheric ozone and global climate, *Geophys. Res. Lett.*, **35**, L19803, doi:10.1029/2008GL035102.

Jacobson, M. Z., 2009. Review of solutions to global warming, air pollution, and energy security, *Energy Environ. Sci.*, **2**, 148–173, doi:10.1039/b809990c.

Jacobson, M. Z., 2010a. Short-term effects of controlling fossil-fuel soot, biofuel soot and gases, and methane on climate, Arctic ice, and air pollution health, *J. Geophys. Res.*, **115**, D14209, doi:10.1029/2009JD013795.

Jacobson, M. Z., 2010b. The enhancement of local air pollution by urban $CO_2$ domes, *Environ. Sci. Technol.*, **44**, 2497–2502, doi:10.1021/es903018m.

Jacobson, M. Z., 2012a. *Air Pollution and Global Warming: History, Science, and Solutions*, 2nd ed., Cambridge: Cambridge University Press, 375 pp.

Jacobson, M. Z., 2012b. Investigating cloud absorption effects: global absorption properties of black carbon, tar balls, and soil dust in clouds and aerosols, *J. Geophys. Res.*, **117**, D06205, doi:10.1029/2011JD017218.

Jacobson, M. Z., 2014. Effects of biomass burning on climate, accounting for heat and moisture fluxes, black and brown carbon, and cloud absorption effects, *J. Geophys. Res.*, **119**, 8980–9002, doi:10.1002/2014JD021861.

Jacobson, M. Z., 2015. 100% WWS plans for countries and states, United Nations Foundation Earth to Paris Social Good Event, UNFCC, Petit Palais, Paris, France, December 7, http://livestream.com/unfoundation/earthtoparisENG/videos/106549410 (accessed February 6, 2019).

Jacobson, M. Z., 2019. The health and climate impacts of carbon capture and direct air capture, *Energy Environ. Sci.*, **12**, 3567–3574, doi:10.1039/C9EE02709B.

Jacobson, M. Z., and C. L. Archer, 2012. Saturation wind power potential and its implications for wind energy, *Proc. Nat. Acad. Sci.*, **109**(15), 679–684, doi:10.1073/pnas.1208993109.

Jacobson, M. Z., and M. A. Delucchi, November 2009. A path to sustainable energy by 2030, *Scientific American*.

Jacobson, M. Z., and M. A. Delucchi, 2011. Providing all global energy with wind, water, and solar power. Part I: technologies, energy resources, quantities and areas of infrastructure, and materials, *Energy Policy*, **39**, 1154–1169, doi:10.1016/j.enpol.2010.11.040.

Jacobson, M. Z., and V. Jadhav, 2018. World estimates of PV optimal tilt angles and ratios of sunlight incident upon tilted and tracked PV panels relative to horizontal panels, *Sol. Energy*, **169**, 55–66.

Jacobson, M. Z., and G. M. Masters, 2001. Exploiting wind versus coal, *Science*, **293**, 1438.

Jacobson, M. Z., and J. E. Ten Hoeve, 2012. Effects of urban surfaces and white roofs on global and regional climate, *J. Climate*, **25**, 1028–1044, doi:10.1175/JCLI-D-11-00032.1.

Jacobson, M. Z., R. Lu, R. P. Turco, and O. B. Toon, 1996. Development and application of a new air pollution modeling system. Part I: gas-phase simulations, *Atmos. Environ.*, **30B**, 1939–1963.

Jacobson, M. Z., J. H. Seinfeld, G. R. Carmichael, and D. G. Streets, 2004. The effect on photochemical smog of converting the U.S. fleet of gasoline vehicles to modern diesel vehicles, *Geophys. Res. Lett.*, **31**, L02116, doi:10.1029/2003GL018448.

Jacobson, M. Z., W. G. Colella, and D. M. Golden, 2005. Cleaning the air and improving health with hydrogen fuel cell vehicles, *Science*, **308**, 1901–1905.

Jacobson, M. Z., Y. J. Kaufmann, and Y. Rudich, 2007. Examining feedbacks of aerosols to urban climate with a model that treats 3-D clouds with aerosol inclusions, *J. Geophys. Res.*, **112**, D24205, doi:10.1029/2007JD008922.

Jacobson, M. Z., R. W. Howarth, M. A. Delucchi, S. R. Scobies, J. M. Barth, M. J. Dvorak, et al. 2013. Examining the feasibility of converting New York State's all-purpose energy infrastructure to one using wind, water, and sunlight, *Energy Policy*, **57**, 585–601.

Jacobson, M. Z., C. L. Archer, and W. Kempton, 2014a. Taming hurricanes with arrays of offshore wind turbines, *Nat. Clim. Change*, **4**, 195–200, doi: 10.1038/NCLIMATE2120.

Jacobson, M. Z., M. A. Delucchi, A. R. Ingraffea, R. W. Howarth, G. Bazouin, B. Bridgeland, et al., 2014b. A roadmap for repowering California for all purposes with wind, water, and sunlight, *Energy*, **73**, 875–889, doi:10.1016/j.energy.2014.06.099.

Jacobson, M. Z., M. A. Delucchi, G. Bazouin, Z. A. F. Bauer, C. C. Heavey, E. Fisher, et al., 2015a. 100 percent clean and renewable wind, water, sunlight (WWS) all-sector energy roadmaps for the 50 United States, *Energy Environ. Sci.*, **8**, 2093–2117, doi:10.1039/C5EE01283J.

Jacobson, M. Z., M. A. Delucchi, M. A. Cameron, and B. A. Frew, 2015b. A low-cost solution to the grid reliability problem with 100 percent penetration of intermittent wind, water, and solar for all purposes, *Proc. Nat. Acad. Sci.*, **112**(49), 15,060–15,065, doi: 10.1073/pnas.1510028112.

Jacobson, M. Z., M. A. Delucchi, G. Bazouin, M. J. Dvorak, R. Arghandeh, Z. A. F. Bauer, et al., 2016. A 100 percent wind, water, sunlight (WWS) all-sector energy plan for Washington State, *Renew. Energy*, **86**, 75–88.

Jacobson, M. Z., M. A. Delucchi, Z. A. F. Bauer, S. C. Goodman, W. E. Chapman, M. A. Cameron, et al., 2017. 100 percent clean and renewable wind, water, and sunlight (WWS) all-sector energy roadmaps for 139 countries of the world, *Joule*, **1**, 108–121, doi:10.1016/j.joule.2017.07.005.

Jacobson, M. Z., M. A. Delucchi, M. A. Cameron, and B. V. Mathiesen, 2018a. Matching demand with supply at low cost among 139 countries within 20 world regions with 100 percent intermittent wind, water, and sunlight (WWS) for all purposes, *Renew. Energy*, **123**, 236–248.

Jacobson, M. Z., M. A. Cameron, E. M. Hennessy, I. Petkov, C. B. Meyer, T. K. Gambhir, et al., 2018b. 100 percent clean, and renewable wind, water, and sunlight (WWS) all-sector energy roadmaps for 53 towns and cities in North America, *Sustain. Cities Soc.*, **42**, 22–37, doi:10.1016/j.scs.2018.06.031.

Jacobson, M. Z., M. A. Delucchi, M. A. Cameron, S. J. Coughlin, C. Hay, I. P. Manogaran, et al., 2019. Impacts of Green-New-Deal energy plans on grid stability, costs, jobs, health, and climate in 143 countries, *One Earth*, **1**, 449–463, doi:10.1016/j.oneear.2019.12.003.

Johnson, G., September 21, 2015. When radiation isn't the real risk, *New York Times*, www.nytimes.com/2015/09/22/science/when-radiation-isnt-the-real-risk.html (accessed December 8, 2018).

Kadiyala, A., R. Kommalapati, and Z. Huque, 2016. Evaluation of the lifecycle greenhouse gas emissions from different biomass feedstock electricity generation systems, *Sustainability*, **8**, 1181–1192.

Kahn, E., 1979. The reliability of distributed wind generators, *Electr. Pow. Syst.*, **2**, 1–14.

Kaldelis, J. K., and D. Apostolou, 2017. Life cycle energy and carbon footprint of offshore wind energy. Comparison with onshore counterpart, *Renew. Energy*, **108**, 72–84.

Kane, M., 2019. CATL breaks into 300+ Wh/kg energy density on battery cell level, INSIDEEVs, https://insideevs.com/news/343690/catl-breaks-into-300-wh-kg-energy-density-on-battery-cell-level/ (accessed July 17, 2019).

Karam, P. A., October 2006. How do fast breeder reactors differ from regular nuclear power plants?, *Scientific American*.

Katalenich, S. M., and M. Z. Jacobson, 2020. Toward battery electric and hydrogen fuel cell vehicles for land, air, and sea military missions, in review,

Keith, D. W., G. Holmes, D. St. Angelo, and K. Heidel, 2018. A process for capturing $CO_2$ from the atmosphere, *Joule*, **2**, 1573–1594.

Kempton, W., and J. Tomic, 2005a. Vehicle-to-grid power fundamentals: calculating capacity and net revenue, *J. Power Sources*, **144**, 268–279.

Kempton, W., and J. Tomic, 2005b. Vehicle-to-grid power implementation: from stabilizing the grid to supporting large-scale renewable energy, *J. Power Sources*, **144**, 280–294.

Kirby, B.J., 2004. *Frequency Regulation Basics and Trends*, ORNL/TM-2004/291, www.consultkirby.com/files/TM2004-291_Frequency_Regulation_Basics_and_Trends.pdf (accessed January 28, 2019).

Ko, N., M. Lorenz, R. Horn, H. Krieg, and M. Baumann, 2018. Sustainability assessment of concentrated solar power (CSP) tower plants – integrating LCA, LCC, and LCWE in one framework, *Procedia CIRP*, **69**, 395–400.

Koomey, J., and N. E. Hultman, 2007. A reactor-level analysis of busbar costs for U.S. nuclear plants, 1970–2005, *Energy Policy*, **35**, 5630–5642.

Kougias, I., K. Bodis, A. Jager-Waldau, M. Moner-Girona, F. Monforti-Ferrario, H. Ossenbrink, and S. Szabo, 2016. The potential of water infrastructure to accommodate solar PV systems in Mediterranean islands, *Sol. Energy*, **136**, 174–182, doi:10.1016/j.solener.2016.07.003.

Krewski, D., M. Jerrett, R. T. Burnett, R. Ma, E. Hughes, Y. Shi, et al., 2009. *Extended Follow-Up and Spatial Analysis of the American Cancer Society Study Linking Particulate Air Pollution and Mortality*, Health Effects Institute, Report No. 140.

Kuphaldt, T., 2019. What is alternating current (AC)?, www.allaboutcircuits.com/textbook/alternating-current/chpt-1/what-is-alternating-current-ac/ (accessed March 31, 2019).

Lackner, K.S., H.-J. Ziock, and P. Grimes, 1999. *Carbon Dioxide Extraction from Air: Is It an Option?* Report LA-UR-99-583, Los Alamos National Laboratory.

Lagarde, F., G. Pershagen, G. Akerblom, O. Axelson, U. Baverstam, L. Damber, et al., 1997. Residential radon and lung cancer in Sweden: risk analysis accounting for random error in the exposure assessment, *Health Physi.*, **72**, 269–276.

Lazard, 2018. Lazard's levelized cost of energy analysis – version 12.0, www.lazard.com/media/450784/lazards-levelized-cost-of-energy-version-120-vfinal.pdf (accessed January 16, 2019).

Le Quere, C., R. Moriarty, R. M. Andrew, G. P. Peters, P. Ciais, P. Friedlingstein, et al., 2015. Global carbon budget 2014, *Earth Syst. Sci. Data*, **7**, 47–85.

Lenzen, M., 2008. Life cycle energy and greenhouse gas emissions of nuclear energy: a review, *Energy Convers. Manag.*, **49**, 2178–2199.

Letterman, D., host, 2013, *Late Show with David Letterman*, New York City, October 9, online video clip, www.youtube.com/watch?v=AqIu2J3vRJc (accessed February 6, 2019).

Li, X., K. J. Chalvatzis, and D. Pappas, 2017. China's electricity emission intensity in 2020-an analysis at provincial level, *Energy Procedia*, **142**, 2779–2785.

Liou, K. N., 2002. *An Introduction to Atmospheric Radiation*. Amsterdam: Academic Press.

Liu, H., G. B. Andresen, and M. Greiner, 2018. Cost-optimal design of a simplified highly renewable Chinese network, *Energy*, **147**, 534–546.

Lu, B., A. Blakers, and M. Stocks, 2017. 90–100% renewable electricity for the South West Interconnected System of Western Australia, *Energy*, **122**, 663–674.

Lund, H., 2006. Large-scale integration of optimal combinations of PV, wind, and wave power into the electricity supply, *Renew. Energy*, **31**, 503–515.

Lund, H., and B. V. Mathiesen, 2009. Energy system analysis of 100% renewable energy systems – the case of Denmark in years 2030 and 2050, *Energy*, **34**, 524–531.

Lund, H., and B. V. Mathiesen, 2012. The role of carbon capture and storage in a future sustainable energy system, *Energy*, **44**, 469–476.

Macdonald-Smith, A., 2018. South Australia's big battery slashes $40m from grid control costs in first year, *Australian Financial Review*, www.afr.com/business/energy/solar-energy/south-australias-big-battery-slashes-40m-from-grid-control-costs-in-first-year-20181205-h18ql1 (accessed January 25, 2019).

Marine Energy, 2018. Global installed ocean energy power doubles in 2017, https://marineenergy.biz/2018/03/12/global-installed-ocean-energy-power-doubles-in-2017/ (accessed January 9, 2019).

Mason, I. G., S. C. Page, and A. G. Williamson, 2010. A 100% renewable energy generation system for New Zealand utilizing hydro, wind, geothermal, and biomass resources, *Energy Policy*, **38**, 3973–3984.

Masters, G., 2013. *Renewable and Efficient Electric Power Systems*, 2nd ed., Hoboken, NJ: Wiley, 712 pp.

Mathiesen, B. V., H. Lund, and K. Karlsson, 2011. 100% renewable energy systems, climate mitigation, and economic growth, *App. Energy*, **88**, 488–501.

Mathiesen, B. V., H. Lund, D. Connolly, H. Wenzel, P. Z. Ostergaard, B. Moller, et al, 2015. Smart energy systems for coherent 100% renewable energy and transport solutions, *Appl. Energy*, **145**, 139–154.

Matthews, H. D., 2016. *Montreal's emissions targets for 1.5 °C and 2 °C global warming*, http://ocpm.qc.ca/sites/ocpm.qc.ca/files/pdf/P80/7.2.19_damon_matthews.pdf (accessed January 13, 2018).

Matthews, H. D., N. P. Gillett, P. A. Stott, and K. Zickfeld, 2009. The proportionality of global warming to cumulative carbon emissions, *Nature*, **459**, 829–832.

McFadyen, S., 2012. Three phase power simplified, https://myelectrical.com/notes/entryid/172/three-phase-power-simplified (accessed April 3, 2019).

Meador, W. E., and W. R. Weaver, 1980. Two-stream approximations to radiative transfer in planetary atmospheres: a unified description of existing methods and a new improvement, *J. Atmos. Sci.*, **37**, 630–643.

Meinshausen, M., N. Meinshausen, W. Hare, S. C. Rapter, K. Frieler, R. Knutti, et al., 2009. Greenhouse-gas emission targets for limiting global warming to 2°C, *Nature*, **458**, 1158–1162.

Meyers, S., V. Franco, A. Lekov, L. Thompson, and A. Sturges, 2010. Do heat pump clothes dryers makes sense for the U.S. market?, *ACEEE Summer Study on Energy Efficiency in Buildings*, 9-240–9-251, https://aceee.org/files/proceedings/2010/data/papers/2224.pdf (accessed October 29, 2019).

Miceli, F., 2012. Offshore wind turbines foundation types, www.windfarmbop.com/offshore-wind-turbines-foundation-types/ (accessed April 4, 2019).

MIT (Massachusetts Institute of Technology), 2011. *The Future of Natural Gas*, 287 pp., https://energy.mit.edu/wp-content/uploads/2011/06/MITEI-The-Future-of-Natural-Gas.pdf (accessed December 2, 2018).

Monitoring Analytics, 2015. Quarterly state of the market report for PJM: January through June, www.monitoringanalytics.com/reports/PJM_State_of_the_Market/2015/2015q2-som-pjm-sec5.pdf (accessed January 19, 2019).

Moore, F. C., and D. B. Diaz, 2015. Temperature impacts on economic growth warrant stringent mitigation policy, *Nat. Clim. Change*, **5**, 127–131.

Moore, M. A., A. E. Boardman, A. R. Vining, D. L. Weimer, and D. H. Greenberg, 2004. Just give me a number! Practical values for the social discount rate, *J. Policy Anal. Manage.*, **23**, 789–812.

Morris, C., 2015. French nuclear power history – the unknown story, https://energytransition.org/2015/03/french-nuclear-power-history/ (accessed June 16, 2019).

Myhre, G., D. Shindell, F.-M. Breon, W. Collins, J. Fuglestvedt, J. Huang, et al., 2013. Anthropogenic and natural radiative forcing. In *Climate Change 2013: The Physical Science Basis. Contribution of Working Group I to the Fifth Assessment Report of the Intergovernmental Panel on Climate Change*, ed. T. F. Stocker, D. Qin, G.-K. Plattner, M. Tignor, S. K. Allen, J. Boschung, et al. Cambridge: Cambridge University Press.

NACAG (Nitric Acid Climate Action Group), 2014. Nitrous oxide emissions from nitric acid production, www.nitricacidaction .org/about/nitrous-oxide-emissions-from-nitric-acid-production/ (accessed December 1, 2018).

NAO (Nautical Almanac Office) and Her Majesty's Nautical Almanac Office, 1993. *Astronomical Almanac.* Washington, DC: U.S. Government Printing Office.

NASA (National Aeronautics and Space Administration), 2018. GISS surface temperature analysis (GISTEMP), https://data.giss .nasa.gov/gistemp/maps/ (accessed November 30, 2018).

NCEE (National Center for Environmental Economics), 2014. *Guidelines for Preparing Economic Analyses* (U.S. Environmental Protection Agency),

Neftel, A., H. Friedli, E. Moor, H. Lötscher, H. Oeschger, U. Siegenthaler, and B. Stauffer, 1994. Historical CO2 record from the Siple Station ice core. In *Trends: A Compendium of Data on Global Change.* Oak Ridge, TN: Carbon Dioxide Information Analysis Center, Oak Ridge National Laboratory, U.S. Department of Energy.

Neocarbon Energy, 2016. Future energy system, http:// neocarbonenergy.fi/internetofenergy/ (accessed December 6, 2016).

*New York Times*, June 11, 2001. Text of President Bush's remarks on global climate, www.nytimes.com/2001/06/11/world/text-of-president-bushs-remarks-on-global-climate.html (accessed February 2, 2019).

Ni, J., 2002. Carbon storage in grasslands of China, *J. Arid Environ.*, **50**, 205–218.

Nithyanandam, K., and R. Pitchumani, 2014. Cost and performance analysis of concentrating solar power systems with integrated latent thermal energy storage, *Energy*, **64**, 793–810.

Nonbol, E., 2013. Load-following capabilities of nuclear power plants, VBR Seminar, Technical University of Denmark, http:// orbit.dtu.dk/files/64426246/Load_following_capabilities.pdf (accessed November 22, 2018).

NRC (U.S. National Research Council), 2010. Real prospects for energy efficiency in the United States, National Academies Press, p. 251, www.nap.edu/read/12621/chapter/6#251 (accessed February 2, 2019).

NREL (National Renewable Energy Laboratory), 2017. *Jobs and Economic Development Impact Models (JEDI)*, www.nrel.gov/ analysis/jedi (accessed January 17, 2019).

NREL (National Renewable Energy Laboratory), 2018. PVWatts Calculator, http://pvwatts.nrel.gov (accessed December 25, 2018).

NWCC (National Wind Coordinating Collaborative), 2010. Wind turbine interactions with birds, bats, and their habitats, www1.eere.energy.gov/wind/pdfs/birds_and_bats_fact_sheet .pdf (accessed January 4, 2018).

O'Malley, M., 2015. A jobs agenda for our renewable energy future, www.p2016.org/omalley/omalley070215climate.html (accessed February 7, 2019).

Oil and Gas, 2018. Threat map, https://oilandgasthreatmap.com/ threat-map/ (accessed December 3, 2018).

OMB (U.S. Office of Management and Budget), September 17, 2003. Circular A-4, Regulatory Analysis, the White House, Washington, DC, www.whitehouse.gov/sites/whitehouse.gov/ files/omb/circulars/A4/a-4.pdf (accessed January 16, 2019).

Orsted, November 13, 2017. New survey shows strong global support for green energy, https://orsted.com/en/Barometer (accessed February 7, 2019).

Ostro, B. D., H. Tran, and J. I. Levy, 2006. The health benefits of reduced tropospheric ozone in California, *J. Air Waste Manage. Assoc.*, **56**, 1007–1021.

Pavel, C. C., R. Lacal-Arantegui, A. Marmier, D. Schuler, E. Tzimas, M. Buchert, et al., 2017. Substitution strategies for reducing the use of rare earths in wind turbines, *Resour. Policy*, **52**, 349–357.

Pires, O., X. Munduate, O. Ceyhan, M. Jacobs, and H. Snel, 2016. Analysis of high Reynolds numbers effects on a wind turbine airfoil using 2D wind tunnel test data, *J. Phys. Conf. Ser.*, **753**, 022047,

Polpong, P., and S. Bovornkitti, 1998. Indoor radon, *J. Med. Assoc. Thai.*, **81**, 47–57.

Puiu, T., 2019. Solar and wind supply more than 10% of electricity in 18 U.S. states, ZME Science, www.zmescience.com/science/ news-science/solar-wind-electricity-us-stated-04232/ (accessed March 25, 2019).

Rahi, O.P., and A. Kumar, 2016. Economic analysis for refurbishment and uprating of hydropower plants, *Renew. Energy*, **86**, 1197–1204.

Rahman, D., A. J. Morgan, Y. Xu, R. Gao, W. Yu, D. C. Hopkins, and I. Husain, 2016. Design methodology for a planarized high power density EV/HEV traction drive using SiC power modules, paper presented at the 2016 IEEE Energy Conversion Congress and Exhibition, September 18–22, doi: 10.1109/ ECCE.2016.7855018, https://ieeexplore.ieee.org/document/ 7855018/authors#authors (accessed March 2, 2019).

Ram, M., D. Bogdanov, A. Aghahosseini, A. Gulagi, A. S. Oyewo, M. Child, et al, 2019. *Global Energy System Based on 100% Renewable Energy – Power, Heat, Transport, and Desalination Sectors*, Lappeenranta University of Technology Research Reports 91, ISSN: 2243-3376, Lappeenranta, Berlin, http:// energywatchgroup.org/wp-content/uploads/EWG_LUT_ 100RE_All_Sectors_Global_Report_2019.pdf (accessed September 6, 2019).

Ramaiah, R., and K. S. S. Shekar, 2018. Solar thermal energy utilization for medium temperature industrial process heat applications, *IOP Conf. Ser.: Mater. Sci. Eng.*, **376**, 010235.

Ramboll, 2016. World's largest thermal heat storage pit in Vojens, State of Green, https://stateofgreen.com/en/partners/ramboll/ solutions/world-largest-thermal-pit-storage-in-vojens/ https:// ramboll.com/projects/re/south-jutland-stores-the-suns-heat-in-the-worlds-largest-pit-heat-storage (accessed November 25, 2018).

Rasmussen, M. G., G. B. Andresen, and M. Greiner, 2012. Storage and balancing synergies in a fully or highly renewable pan-European power system, *Energy Policy*, **51**, 642–651.

RE100, 2019. The world's most influential companies committed to 100% renewable power, 2019, http://there100.org (accessed January 26, 2019).

REN21 (Renewable Energy Policy Network for the 21st Century), 2019a. *Renewables 2019 Global Status Report*, www.ren21.net/gsr-2019/, www.ren21.net/gsr-2019/tables/table_06/table_06/ (accessed July 27, 2019).

REN21 (Renewable Energy Policy Network for the 21st Century), 2019b. *Renewables in Cities: 2019 Global Status Report*, www.ren21.net/wp-content/uploads/2019/05/REC-2019-GSR_Full_Report_web.pdf (accessed December 29, 2019).

Renewables Now, 2019. Chile's Coquimbo region nears 100% renewables share in H1 2019, https://renewablesnow.com/news/chiles-coquimbo-region-nears-100-renewables-share-in-h1-2019-663136/ (accessed July 26, 2019).

Roberts, D., April 28, 2016. The train goes up, the train goes down: a simple way to store energy, *Vox*, www.vox.com/2016/4/28/11524958/energy-storage-rail (accessed April 10, 2019).

Roselund, C., March 2, 2019. Inertia, frequency regulation and the grid, *PV Magazine*, https://pv-magazine-usa.com/2019/03/01/inertia-frequency-regulation-and-the-grid/ (accessed March 2, 2019).

Ruffalo, M. A., M. Krapels, and M. Z. Jacobson, 2012a. A plan to power the world with wind, water, and sunlight, Talks at Google, Google, Inc., Mountain View, California, June 20, online video clip, www.youtube.com/watch?v=N_sLt5gNAQs (accessed February 5, 2019).

Ruffalo, M. Z., M. Z. Jacobson, M. Krapels, and video from J. Fox, 2012b. Powering the world, U.S., and New York with wind, water, and sunlight, The Nantucket Project, Nantucket, Massachusetts, October 6, online video clip, http://vimeo.com/52038463 (accessed February 5, 2019).

Russell, L. M., C. D. Cappa, M. J. Kleeman, and M. Z. Jacobson, 2018. Characterizing the climate impacts of brown carbon, Final report to the California Air Resources Board Research Division, Project 13-330, November 30.

Sadiqa, A., A. Gulagi, and C. Breyer, 2018. Energy transition roadmap towards 100% renewable energy and role of storage technologies for Pakistan by 2050, *Energy*, **147**, 518–533.

Sadovskaia, K., D. Bogdanov, S. Honkapuro, and C. Breyer, 2019. Power transmission and distribution losses – a model based on available empirical data and future trends for all countries globally, *Int. J. Elec. Power*, **107**, 98–109.

Sanders, B., 2016. Combatting climate change to save the planet, https://berniesanders.com/people-before-polluters/ (accessed February 7, 2019).

Sanders, B., and M. Jacobson, April 29, 2017. The American people, not big oil, must decide our climate future, *Guardian*, www.theguardian.com/commentisfree/2017/apr/29/bernie-sanders-climate-change-big-oil (accessed February 7, 2019).

Santin, I., M. Barbu, C. Pedret, and R. Vilanova, 2017. Control strategies for nitrous oxide emissions reduction on wastewater treatment plants operation, *Water Res.*, **125**, 466–477.

Sanz-Perez, E. S., C. R. Murdock, S. A. Didas, and C. W. Jones, 2016. Direct capture of $CO_2$ from ambient air, *Chem. Rev.*, **116**(11), 840–876.

Schubel, P. J., and R. J. Crossley, 2012. Wind turbine blade design, *Energies*, **5**, 3425–3449.

Scottmadden, 2017. Billion dollar Petra Nova coal carbon capture project a financial success but unclear if it can be replicated, www.scottmadden.com/insight/billion-dollar-petra-nova-coal-carbon-capture-project-financial-success-unclear-can-replicated/ (accessed December 3, 2018).

Searchinger, T., R. Heimlich, R. A. Houghton, F. Dong, A. Elobeid, J. Fabiosa, et al., 2008. Use of U.S. cropland for biofuels increases greenhouse gases through emissions from land-use change, *Science*, **319**, 1238–1240.

Senate (U.S. Senate), 2016. S.Res.632, www.congress.gov/bill/114th-congress/senate-resolution/632 (accessed February 7, 2019).

Senate (U.S. Senate), 2017. S.987 – 100 by '50 Act, www.congress.gov/bill/115th-congress/senate-bill/987/text?r=1 (accessed February 7, 2019).

Senate (U.S. Senate), 2019. S.Res.59 – a resolution recognizing the duty of the Federal Government to create a Green New Deal, www.congress.gov/bill/116th-congress/senate-resolution/59?q=%7B%22search%22%3A%5B%22green+new+deal%22%5D%7D&s=1&r=2 (accessed March 26, 2019).

Sibbitt B., D. McClenahan, R. Djebbar, J. Thornton, B. Wong, J. Carriere, and J. Kokko, 2012. The performance of a high solar fraction seasonal storage district heating system – five years of operation, *Energy Procedia*, **30**, 856–865.

Sierra Club, 2019. 100% commitments in cities, counties, and states, www.sierraclub.org/ready-for-100/commitments (accessed January 26, 2019).

Sirnivas, S., W. Musial, B. Bailey, and M. Filippelli, 2014. *Assessment of Offshore Wind System Design, Safety, and Operation Standards*, NREL/TP-5000-60573.

Skone, T. J., 2015. Lifecycle greenhouse gas emissions: natural gas and power production, paper presented at the 2015 EIA Energy Conference, Washington, DC, June 15, www.eia.gov/conference/2015/pdf/presentations/skone.pdf (accessed December 2, 2018).

Smallwood, K. S., 2013. Comparing bird and bat fatality rate estimates among North American wind energy projects, *Wildl. Soc. Bull.*, **37**, 19–33.

Socaciu, L., 2011. Seasonal sensible thermal energy storage solutions, *Leonardo El. J. Pract. Technol.*, **10**, 49–68.

Solutions Project, 2019. Our 100% clean energy vision, www.thesolutionsproject.org/why-clean-energy/ (accessed February 5, 2019).

Sorensen, B., 1975. A plan is outlined to which solar and wind energy would supply Denmark's needs by the year 2050, *Science*, **189**, 255–260.

Sorensen, B., 1996. Scenarios of greenhouse warming mitigation, *Energy Convers. Manag.*, **37**, 693–698.

Sorensen, P. A., and T. Schmidt, 2018. Design and construction of large scale heat storages for district heating in Denmark, paper presented at the 14th International Conference on Energy Storage, April 25–28, Adana, Turkey, http://planenergi.dk/wp-content/uploads/2018/05/Soerensen-and-Schmidt_Design-and-

Construction-of-Large-Scale-Heat-Storages-12.03.2018-004.pdf (accessed November 25, 2018).

Sourcewatch, 2011. The footprint of coal, www.sourcewatch.org/index.php/The_footprint_of_coal (accessed December 3, 2018).

Sovacool, B. K., 2008. Valuing the greenhouse gas emissions from nuclear power: a critical survey, *Energy Policy*, **36**, 2940–2953.

Sovacool, B. K., 2009. Contextualizing avian mortality: a preliminary appraisal of bird and bat fatalities from wind, fossil-fuel, and nuclear electricity, *Energy Policy*, **37**, 2241–2248.

Spakovsky, Z. S., 2008. Section 11.5: trends in thermal and propulsive efficiency, lecture notes, Unified: Thermodynamics and Propulsion 16, MIT, http://web.mit.edu/16.unified/www/FALL/thermodynamics/notes/node84.html (accessed March 27, 2019).

Spector, J., 2019. "Cheaper than a peaker": NextEra inks massive wind+solar+storage deal in Oklahoma, *gtm*, www.greentechmedia.com/articles/read/nextera-inks-even-bigger-windsolarstorage-deal-with-oklahoma-cooperative#gs.s8lb02 (accessed July 26, 2019).

Stagner, J., 2016. Stanford University's "fourth-generation" district energy system, *District Energy*, Fourth Quarter, 19–42, https://sustainable.stanford.edu/sites/default/files/IDEA_Stagner_Stanford_fourth_Gen_DistrictEnergy.pdf (accessed November 27, 2018).

Stagner, J., 2017. Stanford Energy System Innovations, efficiency and environmental comparison, https://sustainable.stanford.edu/sites/default/files/SESI_Efficiency_Environmental_Comparisons.pdf (accessed December 31, 2019).

Statistica, 2017. Number of retail fuel stations in California from 2009 to 2016, by type, www.statista.com/statistics/818462/california-fueling-stations-by-type/ (accessed December 3, 2018).

Steinke, F., P. Wolfrum, and C. Hoffmann, 2013. Grid vs. storage in a 100% renewable Europe, *Renew. Energy*, **50**, 826–832.

Stevens, F., and L. DiCaprio, 2014. Interviews at Stanford University for *Before the Flood*, https://cee.stanford.edu/programs/atmosphere-energy-program (accessed February 7, 2019).

Stone, D., 2017. Ferrock basics, http://ironkast.com/wp-content/uploads/2017/11/Ferrock-basics.pdf (accessed November 20, 2018).

Stoutenburg, E. D., and M. Z. Jacobson, 2011. Reducing offshore transmission requirements by combining offshore wind and wave farms, *IEEE J. Ocean. Eng.*, **36**, 552–561, doi:10.1109/JOE.2011.2167198.

Stoutenburg, E. D., N. Jenkins, and M. Z. Jacobson, 2010. Power output variations of co-located offshore wind turbines and wave energy converters in California, *Renew. Energy*, **35**, 2781–2791, doi:10.1016/j.renene.2010.04.033.

Strata, 2017. The footprint of energy: land use of U.S. electricity production, www.strata.org/pdf/2017/footprints-full.pdf (accessed December 3, 2018).

Streets, D. G., K. Jiang, X. Hu, J. E. Sinton, X.-Q. Zhang, D. Xu, et al., 2001. Recent reductions in China's greenhouse gas emissions, *Science*, **294**, 1835–1836.

Talebizadeh, P., M. A. Mehrabian, and M. Abdolzadeh, 2011. Determination of optimum slope angles of solar collectors based on new correlations, *Energy Sources Part A.*, **33**, 1567–1580.

Tans, P., and R. F. Keeling, 2015. Trends in atmospheric carbon dioxide, www.esrl.noaa.gov/gmd/ccgg/trends/#mlo_full (accessed January 24, 2019).

Ten Hoeve, J. E., and M. Z. Jacobson, 2012. Worldwide health effects of the Fukushima Daiichi nuclear accident, *Energ. Environ. Sci.*, **5**, 8743–8757.

Teske, S., D. Giurco, T. Morris, K. Nagrath, F. Mey, C. Briggs, et al., 2019. Achieving the Paris Climate Agreement, https://oneearth.app.box.com/s/hctp4qlk34ygd0mw3yjdtctsymsdtaqs (accessed September 6, 2019).

Tomasini-Montenegro, C., E. Santoyo-Castelazo, H. Gujba, R. J. Romero, and E. Santoyo, 2017. Life cycle assessment of geothermal power generation technologies: an updated review, *Appl. Therm. Eng.*, **114**, 1119–1136.

Toon, O. B., and T. P. Ackerman, 1981. Algorithms for the calculation of scattering by stratified spheres, *Appl. Opt.*, **20**, 3657–3660.

Toon, O. B., C. P. McKay, and T. P. Ackerman, 1989. Rapid calculation of radiative heating rates and photodissociation rates in inhomogeneous multiple scattering atmospheres, *J. Geophys. Res.*, **94**(16), 287–301,

Uddin, K., T. Jackson, W. D. Widange, G. Chouchelamane, P. A. Jennings, and J. Marco, 2017. On the possibility of extending the lifetime of lithium-ion batteries through optimal V2G facilitated by an integrated vehicle and smart-grid system, *Energy*, **133**, 710–722.

Union Gas, 2018. Chemical composition of natural gas, www.uniongas.com/about-us/about-natural-gas/chemical-composition-of-natural-gas (accessed December 5, 2018).

U.S. DOI (U.S. Department of the Interior), 2005. Reclamation: managing water in the west; hydroelectric power, www.usbr.gov/power/edu/pamphlet.pdf (accessed November 22, 2018).

U.S. EPA (U.S. Environmental Protection Agency), 2011. 2008 U.S. National Emissions Inventory (NEI), www.epa.gov/air-emissions-inventories/2008-national-emissions-inventory-nei-data (accessed December 2, 2018).

U.S. EPA (U.S. Environmental Protection Agency), 2017. Revision under consideration for the 2018 GHGI: abandoned wells, www.epa.gov/sites/production/files/2017-06/documents/6.22.17_ghgi_stakeholder_workshop_2018_ghgi_revision_-_abandoned_wells.pdf (accessed December 3, 2018).

USGS (U.S. Geological Survey), 2018a. *Mineral Commodities Summaries 2011*, Washington, DC: U.S. Government Printing Office, https://minerals.usgs.gov/minerals/pubs/mcs/2018/mcs2018.pdf (accessed January 18, 2019).

USGS (U.S. Geological Survey), 2018b. Lithium statistics and information, Washington, DC: USGS, National Minerals Information Center, https://minerals.usgs.gov/minerals/pubs/commodity/lithium/mcs-2018-lithi.pdf (accessed November 23, 2018).

Ussiri, D., and R. Lal, 2012. Global sources of nitrous oxide. In *Soil Emission of Nitrous Oxide and Its Mitigation*. Dordrecht: Springer, pp. 131–175.

Van den Bergh, J. C. J. M., and W. J. W. Botzen, 2014. A lower bound the social cost of carbon emissions, *Nat. Clim. Change*, **4**, 253–258.

Vavrin, J. 2010. Power and energy considerations at forward operating bases (FOBs). United States Army Corps of Engineers, Engineer Research and Development Center, Construction Engineering Research Laboratory, www.dtic.mil/dtic/tr/fulltext/u2/a566876.pdf (accessed February 13, 2019).

Viking Heat Engines, 2019. Heat booster, Viking Development Group, www.vikingheatengines.com/news/vikings-industrial-high-temperature-heat-pump-is-available-to-order (accessed January 13, 2019).

Vogl, V., M. Ahman, and L. J. Nilsson, 2018. Assessment of hydrogen direct reduction for fossil-free steelmaking, *J. Cleaner Production*, **203**, 736–745.

Wank, J., M. A. Ruffalo, Z. Saldana, and M. Z. Jacobson, 2012. *Tommy and the Professor*, online video clip, www.youtube.com/watch?v=AqTID6Wv_xk&feature=youtu.be (accessed July 25, 2019).

WEC (World Energy Council), 2016. World energy resources: marine energy, www.worldenergy.org/wp-content/uploads/2017/03/WEResources_Marine_2016.pdf (accessed January 9, 2019).

Werner, S., 2017. International review of district heating, *Energy*, **15**, 617–631.

Wiencke, J., H. Lavelaine, P.-J. Panteix, C. Petijean, and C. Rapin, 2018. Electrolysis of iron in a molten oxide electrolyte, *J. Appl. Electrochem.*, **48**, 115–126.

Wigley, T. M. L., 2011. Coal to gas: the influence of methane leakage, *Clim. Change*, **108**, 601–608.

Winther, M., D. Balslev-Harder, S. Christensen, A. Prieme, B. Elberling, E. Crosson, and T. Blunier, 2018. Continuous measurements of nitrous oxide isotopomers during incubation experiments, *Biogeosciences*, **15**, 767–780.

WHO (World Health Organization), 2017a. Health statistics and information systems, www.who.int/healthinfo/global_burden_disease/estimates/en/ (accessed July 26, 2019).

WHO (World Health Organization), 2017b. Global health observatory data, www.who.int/gho/phe/outdoor_air_pollution/en/ (accessed July 26, 2019).

WHO (World Health Organization), 2017c. Mortality from environmental pollution, http://apps.who.int/gho/data/node.sdg.3-9-data?lang=en (accessed July 26, 2019).

Wilkerson, J. T., M. Z. Jacobson, A. Malwitz, S. Balasubramanian, R. Wayson, G. Fleming, et al., 2010. Analysis of emission data from global commercial aviation: 2004 and 2006, *Atmos. Chem. Phys.*, **10**, 6391–6408.

Winnefeld, C., T. Kadyk, B. Bensmann, U. Krewer, and R. Hanke-Rauschenback, 2018. Modelling and designing cryogenic hydrogen tanks for future aircraft applications, *Energies*, **11**, 105, doi:10.3390/en11010105.

Wiser, R., M. Bolinger, G. Barbose, N. Darghouth, B. Hoen, A. Mills, et al., 2019. 2018 wind technologies market report, U.S. Department of Energy, https://eta-publications.lbl.gov/publications/2018-wind-technologies-market-report (accessed August 16, 2019).

Woodford, C., 2019. Lithium ion batteries, EXPLAINTHATSTUFF!, www.explainthatstuff.com/how-lithium-ion-batteries-work.html (accessed March 26, 2019).

World Bank, 2017. Agricultural and rural development, https://data.worldbank.org/indicator/ (accessed September 16, 2019).

World Bank, 2018. Electric power transmission and distribution losses (% of output), https://data.worldbank.org/indicator/EG.ELC.LOSS.ZS?end=2014&start=2009 (accessed January 1, 2019).

World Nuclear Association, 2019. World uranium mining production, www.world-nuclear.org/information-library/nuclear-fuel-cycle/mining-of-uranium/world-uranium-mining-production.aspx (accessed March 30, 2019).

World Nuclear News, 2018. Green light for next Darlington refurbishment, http://world-nuclear-news.org/Articles/Green-light-for-next-Darlington-refurbishment (accessed December 7, 2018).

Worldwatch Institute, 2019. Energy poverty remains a global challenge for the future, www.worldwatch.org/energy-poverty-remains-global-challenge-future (accessed February 13, 2019).

Zapata, S., M. Casteneda, M. Jiminez, A. J. Aristizabel, C. J. Franco, and I. Dyner, 2018. Long-term effects of 100% renewable generation on the Colombian power market, *Sustain. Energy Techn. Assess.*, **30**, 183–191.

Zhang, J., S. Chowdhury, and J. Zhang, 2012. Optimal preventative maintenance time windows for offshore wind farms subject to wake losses, *AIAA*, 2012–5435.

Zhou, L., Y. Tian, S. B. Roy, C. Thorncroft, L. F. Bosart, and Y. Hu, 2012. Impacts of wind farms on land surface temperature, *Nat. Clim. Change*, **2**, 539–543.

# INDEX